T0270909

LONDON MATHEMATICAL SOCIETY STUDENT TEXTS

Managing Editor: Ian J. Leary,
Mathematical Sciences, University of Southampton, UK

50 A brief guide to algebraic number theory, H. P. F. SWINNERTON-DYER
51 Steps in commutative algebra: Second edition, R. Y. SHARP
52 Finite Markov chains and algorithmic applications, OLLE HÄGGSTRÖM
53 The prime number theorem, G. J. O. JAMESON
54 Topics in graph automorphisms and reconstruction, JOSEF LAURI & RAFFAELE SCAPELLATO
55 Elementary number theory, group theory and Ramanujan graphs, GIULIANA DAVIDOFF, PETER SARNAK & ALAIN VALETTE
56 Logic, induction and sets, THOMAS FORSTER
57 Introduction to Banach algebras, operators and harmonic analysis, GARTH DALES et al
58 Computational algebraic geometry, HAL SCHENCK
59 Frobenius algebras and 2-D topological quantum field theories, JOACHIM KOCK
60 Linear operators and linear systems, JONATHAN R. PARTINGTON
61 An introduction to noncommutative Noetherian rings: Second edition, K. R. GOODEARL & R. B. WARFIELD, JR
62 Topics from one-dimensional dynamics, KAREN M. BRUCKS & HENK BRUIN
63 Singular points of plane curves, C. T. C. WALL
64 A short course on Banach space theory, N. L. CAROTHERS
65 Elements of the representation theory of associative algebras I, IBRAHIM ASSEM, DANIEL SIMSON & ANDRZEJ SKOWROŃSKI
66 An introduction to sieve methods and their applications, ALINA CARMEN COJOCARU & M. RAM MURTY
67 Elliptic functions, J. V. ARMITAGE & W. F. EBERLEIN
68 Hyperbolic geometry from a local viewpoint, LINDA KEEN & NIKOLA LAKIC
69 Lectures on Kähler geometry, ANDREI MOROIANU
70 Dependence logic, JOUKU VÄÄNÄNEN
71 Elements of the representation theory of associative algebras II, DANIEL SIMSON & ANDRZEJ SKOWROŃSKI
72 Elements of the representation theory of associative algebras III, DANIEL SIMSON & ANDRZEJ SKOWROŃSKI
73 Groups, graphs and trees, JOHN MEIER
74 Representation theorems in Hardy spaces, JAVAD MASHREGHI
75 An introduction to the theory of graph spectra, DRAGOŠ CVETKOVIĆ, PETER ROWLINSON & SLOBODAN SIMIĆ
76 Number theory in the spirit of Liouville, KENNETH S. WILLIAMS
77 Lectures on profinite topics in group theory, BENJAMIN KLOPSCH, NIKOLAY NIKOLOV & CHRISTOPHER VOLL
78 Clifford algebras: An introduction, D. J. H. GARLING
79 Introduction to compact Riemann surfaces and dessins d'enfants, ERNESTO GIRONDO & GABINO GONZÁLEZ-DIEZ
80 The Riemann hypothesis for function fields, MACHIEL VAN FRANKENHUIJSEN
81 Number theory, Fourier analysis and geometric discrepancy, GIANCARLO TRAVAGLINI
82 Finite geometry and combinatorial applications, SIMEON BALL
83 The geometry of celestial mechanics, HANSJÖRG GEIGES
84 Random graphs, geometry and asymptotic structure, MICHAEL KRIVELEVICH et al
85 Fourier analysis: Part I – Theory, ADRIAN CONSTANTIN
86 Dispersive partial differential equations, M. BURAK ERDOĞAN & NIKOLAOS TZIRAKIS
87 Riemann surfaces and algebraic curves, R. CAVALIERI & E. MILES
88 Groups, languages and automata, DEREK F. HOLT, SARAH REES & CLAAS E. RÖVER
89 Analysis on Polish spaces and an introduction to optimal transportation, D. J. H. GARLING

London Mathematical Society Student Texts 89

Analysis on Polish Spaces

and an Introduction to Optimal Transportation

D. J. H. GARLING

Emeritus Reader in Mathematical Analysis, University of Cambridge, and Fellow of St John's College, Cambridge

CAMBRIDGE
UNIVERSITY PRESS

University Printing House, Cambridge CB2 8BS, United Kingdom

One Liberty Plaza, 20th Floor, New York, NY 10006, USA

477 Williamstown Road, Port Melbourne, VIC 3207, Australia

314-321, 3rd Floor, Plot 3, Splendor Forum, Jasola District Centre, New Delhi - 110025, India

79 Anson Road, #06-04/06, Singapore 079906

Cambridge University Press is part of the University of Cambridge.

It furthers the University's mission by disseminating knowledge in the pursuit of education, learning and research at the highest international levels of excellence.

www.cambridge.org
Information on this title: www.cambridge.org/9781108421577
DOI: 10.1017/9781108377362

© D. J. H. Garling 2018

This publication is in copyright. Subject to statutory exception and to the provisions of relevant collective licensing agreements, no reproduction of any part may take place without the written permission of Cambridge University Press.

First published 2018

A catalogue record for this publication is available from the British Library

Library of Congress Cataloging in Publication data
Names: Garling, D. J. H., author.
Title: Analysis on Polish spaces and an introduction to optimal transportation / D.J.H. Garling (University of Cambridge).
Other titles: London Mathematical Society student texts ; 89.
Description: Cambridge, United Kingdom ; New York, NY : Cambridge University Press, 2018. | Series: London Mathematical Society student texts ; 89 | Includes bibliographical references and index.
Identifiers: LCCN 2017028186 | ISBN 9781108421577 (hardback ; alk. paper) | ISBN 1108421571 (hardback ; alk. paper) | ISBN 9781108431767 (pbk. ; alk. paper) | ISBN 1108431763 (pbk. ; alk. paper)
Subjects: LCSH: Polish spaces (Mathematics) | Mathematical analysis. | Transportation problems (Programming) | Topology.
Classification: LCC QA611.28 .G36 2018 | DDC 514/.32–dc23
LC record available at https://lccn.loc.gov/2017028186

ISBN 978-1-108-42157-7 Hardback
ISBN 978-1-108-43176-7 Paperback

Cambridge University Press has no responsibility for the persistence or accuracy of URLs for external or third-party internet websites referred to in this publication, and does not guarantee that any content on such websites is, or will remain, accurate or appropriate.

Contents

	Introduction	1
	PART ONE TOPOLOGICAL PROPERTIES	7
1	**General Topology**	9
1.1	Topological Spaces	9
1.2	Compactness	15
2	**Metric Spaces**	18
2.1	Metric Spaces	18
2.2	The Topology of Metric Spaces	21
2.3	Completeness: Tietze's Extension Theorem	24
2.4	More on Completeness	27
2.5	The Completion of a Metric Space	29
2.6	Topologically Complete Spaces	31
2.7	Baire's Category Theorem	33
2.8	Lipschitz Functions	35
3	**Polish Spaces and Compactness**	38
3.1	Polish Spaces	38
3.2	Totally Bounded Metric Spaces	39
3.3	Compact Metrizable Spaces	41
3.4	Locally Compact Polish Spaces	47
4	**Semi-continuous Functions**	50
4.1	The Effective Domain and Proper Functions	50
4.2	Semi-continuity	50
4.3	The Brézis–Browder Lemma	53
4.4	Ekeland's Variational Principle	54

5 Uniform Spaces and Topological Groups 56
5.1 Uniform Spaces 56
5.2 The Uniformity of a Compact Hausdorff Space 59
5.3 Topological Groups 61
5.4 The Uniformities of a Topological Group 64
5.5 Group Actions 66
5.6 Metrizable Topological Groups 67

6 Càdlàg Functions 71
6.1 Càdlàg Functions 71
6.2 The Space $(D[0,1], d_\infty)$ 72
6.3 The Skorohod Topology 73
6.4 The Metric d_B 75

7 Banach Spaces 79
7.1 Normed Spaces and Banach Spaces 79
7.2 The Space *BL(X)* of Bounded Lipschitz Functions 82
7.3 Introduction to Convexity 83
7.4 Convex Sets in a Normed Space 86
7.5 Linear Operators 88
7.6 Five Fundamental Theorems 91
7.7 The Petal Theorem and Daneš's Drop Theorem 95

8 Hilbert Spaces 97
8.1 Inner-product Spaces 97
8.2 Hilbert Space; Nearest Points 101
8.3 Orthonormal Sequences; Gram–Schmidt Orthonormalization 104
8.4 Orthonormal Bases 107
8.5 The Fréchet–Riesz Representation Theorem; Adjoints 108

9 The Hahn–Banach Theorem 112
9.1 The Hahn–Banach Extension Theorem 112
9.2 The Separation Theorem 116
9.3 Weak Topologies 118
9.4 Polarity 119
9.5 Weak and Weak* Topologies for Normed Spaces 120
9.6 Banach's Theorem and the Banach–Alaoglu Theorem 124
9.7 The Complex Hahn–Banach Theorem 125

10 Convex Functions 128
10.1 Convex Envelopes 128
10.2 Continuous Convex Functions 130

11 Subdifferentials and the Legendre Transform 133
11.1 Differentials and Subdifferentials 133
11.2 The Legendre Transform 134
11.3 Some Examples of Legendre Transforms 137
11.4 The Episum 139
11.5 The Subdifferential of a Very Regular Convex Function 140
11.6 Smoothness 143
11.7 The Fenchel–Rockafeller Duality Theorem 148
11.8 The Bishop–Phelps Theorem 149
11.9 Monotone and Cyclically Monotone Sets 151

12 Compact Convex Polish Spaces 155
12.1 Compact Polish Subsets of a Dual Pair 155
12.2 Extreme Points 157
12.3 Dentability 160

13 Some Fixed Point Theorems 162
13.1 The Contraction Mapping Theorem 162
13.2 Fixed Point Theorems of Caristi and Clarke 165
13.3 Simplices 167
13.4 Sperner's Lemma 168
13.5 Brouwer's Fixed Point Theorem 170
13.6 Schauder's Fixed Point Theorem 171
13.7 Fixed Point Theorems of Markov and Kakutani 173
13.8 The Ryll–Nardzewski Fixed Point Theorem 175

PART TWO MEASURES ON POLISH SPACES 177

14 Abstract Measure Theory 179
14.1 Measurable Sets and Functions 179
14.2 Measure Spaces 182
14.3 Convergence of Measurable Functions 184
14.4 Integration 187
14.5 Integrable Functions 188

15 Further Measure Theory 191
15.1 Riesz Spaces 191
15.2 Signed Measures 194
15.3 $M(X)$, L^1 and L^∞ 196
15.4 The Radon–Nikodym Theorem 199
15.5 Orlicz Spaces and L^p Spaces 203

16 Borel Measures 210
16.1 Borel Measures, Regularity and Tightness 210
16.2 Radon Measures 214
16.3 Borel Measures on Polish Spaces 215
16.4 Lusin's Theorem 216
16.5 Measures on the Bernoulli Sequence Space $\Omega(\mathbf{N})$ 218
16.6 The Riesz Representation Theorem 222
16.7 The Locally Compact Riesz Representation Theorem 225
16.8 The Stone–Weierstrass Theorem 226
16.9 Product Measures 228
16.10 Disintegration of Measures 231
16.11 The Gluing Lemma 234
16.12 Haar Measure on Compact Metrizable Groups 236
16.13 Haar Measure on Locally Compact Polish Topological Groups 238

17 Measures on Euclidean Space 243
17.1 Borel Measures on $\mathbf{R}$ and $\mathbf{R}^d$ 243
17.2 Functions of Bounded Variation 245
17.3 Spherical Derivatives 247
17.4 The Lebesgue Differentiation Theorem 249
17.5 Differentiating Singular Measures 250
17.6 Differentiating Functions in bv_0 251
17.7 Rademacher's Theorem 254

18 Convergence of Measures 257
18.1 The Norm $\|.\|_{TV}$ 257
18.2 The Weak Topology w 258
18.3 The Portmanteau Theorem 260
18.4 Uniform Tightness 264
18.5 The β Metric 266
18.6 The Prokhorov Metric 269
18.7 The Fourier Transform and the Central Limit Theorem 271
18.8 Uniform Integrability 276
18.9 Uniform Integrability in Orlicz Spaces 278

19 Introduction to Choquet Theory 280
19.1 Barycentres 280
19.2 The Lower Convex Envelope Revisited 282
19.3 Choquet's Theorem 284
19.4 Boundaries 285
19.5 Peak Points 289
19.6 The Choquet Ordering 291
19.7 Dilations 293

PART THREE INTRODUCTION TO OPTIMAL TRANSPORTATION 297

20 Optimal Transportation 299
20.1 The Monge Problem 299
20.2 The Kantorovich Problem 300
20.3 The Kantorovich–Rubinstein Theorem 303
20.4 c-concavity 305
20.5 c-cyclical Monotonicity 308
20.6 Optimal Transport Plans Revisited 310
20.7 Approximation 313

21 Wasserstein Metrics 315
21.1 The Wasserstein Metrics W_p 315
21.2 The Wasserstein Metric W_1 317
21.3 W_1 Compactness 318
21.4 W_p Compactness 320
21.5 W_p-Completeness 322
21.6 The Mallows Distances 323

22 Some Examples 325
22.1 Strictly Subadditive Metric Cost Functions 325
22.2 The Real Line 326
22.3 The Quadratic Cost Function 327
22.4 The Monge Problem on $\mathbf{R}^d$ 329
22.5 Strictly Convex Translation Invariant Costs on $\mathbf{R}^d$ 331
22.6 Some Strictly Concave Translation–Invariant Costs on $\mathbf{R}^d$ 336

Further Reading 339
Index 342

Introduction

Analysis is concerned with continuity and convergence. Investigation of these ideas led to the notions of topology and topological spaces. Once these had been introduced, they became subjects in their own right, which were investigated in fine detail to see how far the theory might lead (an excellent illustration of this is given by the fascinating book by Steen and Seebach [SS]).

In practice, however, a great deal of analysis is concerned with what happens on a very restricted class of topological spaces, namely, the Polish spaces. A Polish space is a separable topological space whose topology is defined by a complete metric. Important examples include Euclidean space, pathwise-connected Riemannian manifolds, compact metric spaces and separable Banach spaces.

The purpose of this book is to develop the study of analysis on Polish spaces. It consists of three parts. The first considers topological properties of Polish spaces, and the second deals with the theory of measures on Polish spaces. In the third part, we give an introduction to the theory of optimal transportation. This makes essential use of the results of the first two parts, or modifications of them. It was, in fact, study of optimal transportation that led to the realization of how much its study required properties of Polish spaces, and measures on them.

There are three important advantages of restricting attention to Polish spaces. First, many of the curious complications of the general topological theory disappear. For example, a subspace of a separable topological space need not be separable, whereas a subspace of a separable metric space is always separable. Secondly, the proofs of standard results are frequently much easier in this restricted setting. For example, Urysohn's lemma for normal topological spaces is quite delicate, whereas it is very easy for metric spaces. Thirdly, Polish spaces enjoy some very important properties. Thus it follows from Alexandroff's theorem that a topological space is a Polish space if and only if it is homeomorphic to a G_δ subset of the Hilbert cube $\mathbf{H} = [0, 1]^{\mathbf{N}}$, which is a compact metrizable space. From this, or directly, it follows that a Borel measure on a Polish space is tight (Ulam's theorem: the measure of a Borel set can be approximated from below by the measures of compact sets contained in it). It also means that we can push forward a Borel measure on a Polish space X to a Borel measure on a compact metric space containing X. This greatly simplifies both the measure theory and also the construction of measures. In fact, I believe that almost all the probability measures that arise in practice are Borel measures on Polish spaces; one important exception, which we do not consider or need, is the theory of uniform central limit theorems.

One major advantage of restricting attention to Polish spaces is that it is not necessary to appeal to the axiom of choice. Instead, we proceed by induction, using the axiom of dependent choice; we make an infinite sequence of decisions, each possibly dependent on what has gone before.

In analysis, there are a few fundamental results which require the axiom of choice. The first is Tychonoff's theorem, which states that an arbitrary product of compact topological spaces, with the product topology, is compact. We do not prove this, or use it. On the other hand, we do prove, and use, the fact that a countable product of compact metrizable spaces is compact and metrizable.

Secondly, there are two fundamental results of linear analysis which need the axiom of choice, using Zorn's lemma. The first of these is the Hahn–Banach theorem (together with the separation theorem). Using induction, we prove weak forms of these, for separable normed spaces; this is sufficient for our purposes.

But for completeness' sake we also give the classical results, using Zorn's lemma; Here we first prove the separation theorem, showing that it essentially depends upon the connectedness of the unit circle $\mathbf{T}$, and then derive the Hahn–Banach theorem from it.

The other fundamental result which requires the axiom of choice is the Krein–Mil'man theorem, which states that every weakly compact convex subset K has an extreme point. Again, we only need, and use, the result in the case where K is metrizable, and we prove this without the axiom of choice.

The fact that we avoid using the axiom of choice suggests that the proofs should, in some sense, be less abstract and more constructive. Unfortunately, this is not the case; the arguments that are used are frequently indirect (consider the collection of all sets with a particular property), so that for example a typical Borel subset of a Polish space does not have a simple description.

Let us now describe the contents of the three parts of this book in more detail.

Part I: Topological Properties

Although it is assumed that the reader has some knowledge of general topology and metric spaces, the first two chapters give an account of these topics, including Tietze's extension theorem, Baire's category theorem and Lipschitz functions.

This leads to the notion of a Polish space, a separable topological space whose topology is given by a complete metric. A fundamental example is given by a compact metrizable space, and Alexandroff's theorem is used to show that a topological space is a Polish space if and only if it is homeomorphic to a G_δ subspace of a compact metric space, and in particular homeomorphic to a G_δ subspace of the Hilbert cube.

We shall need to consider suprema of sets of real-valued continuous functions. Such functions are lower semi-continuous, and we consider such functions in Chapter 4. A lower semi-continuous function on a compact space attains its infimum, but this is not necessarily true for lower semi-continuous functions on a complete metric space. We establish its replacement, Ekeland's variational principle, together with two of its corollaries, the petal theorem and Daneš's drop theorem, and various other applications.

Metric spaces have more structure than a topological one, and Chapter 5 contains an account of uniform spaces; uniformity is particularly important when we consider locally compact topological groups, in Part II.

Chapter 6 is devoted to showing that the space of càdlàg functions is a Polish space under the Skorohod topology; many stochastic processes, and their underlying measures, lie on such spaces, and this helps justify the claim that almost all probability measures of interest lie on Polish spaces. Further examples are given by separable Banach spaces and Hilbert spaces; these are principally used to introduce the notion of convexity.

The rest of Part I is concerned with convexity. The Hahn–Banach theorem is one of the key results here, and we give proofs of appropriate results, both without and with the axiom of choice. For us, the Hahn–Banach theorem is essentially a geometric theorem showing that two suitable convex sets can be separated by a hyperplane. It also leads onto the notion of weak topology.

The Legendre transform provides an important duality theory for convex functions, and this leads naturally to the concept of subdifferentials and subdifferentiability. We prove the Bishop–Phelps theorem, and also introduce the notion of cyclic monotonicity.

The rest of Part I is concerned with convex sets which are compact and metrizable in some suitable topology. We prove versions of the Krein–Mil'man theorem, Krein's theorem and a swathe of fixed point theorems, many of which are used later.

Part II: Measures on Polish Spaces

We expect that the reader has some knowledge of abstract measure theory, but Chapter 14 contains a survey of the basic results. Chapter 15 contains some further results: we introduce the Banach space $M(X)$ of finite measures on a Polish space X, its subspaces $L^1(\mu)$ and Orlicz spaces (with the use of Legendre duality). We give von Neumann's Hilbert space proof of the Radon–Nikodym property and a proof of the strong law of large numbers (to be used later).

In Chapter 16, we investigate Borel measures on Polish spaces. We prove regularity and tightness properties; we may not know what a typical Borel set looks like, but we can approximate the Borel measure of a Borel set from the outside by open sets, and on the inside by compact sets. This leads to Lusin's theorem, which says that if μ is a Borel measure on a Polish space X then a Borel measurable function on X is continuous on a large compact subset.

So far, all is theory, and no measures, other than trivial ones, have been shown to exist. We remedy this by showing how to construct Borel measures on the Bernoulli space $\Omega(\mathbf{N})$, and then, pushing forward, constructing measures on compact metric spaces and Polish spaces. We prove the Riesz representation theorem, and use this to give a measure-theoretic proof of the Stone–Weierstrass theorem.

We then show how Borel measures can be disintegrated, and establish the existence of Haar measure on compact and locally compact Polish spaces; we follow an account by Pedersen to show that this last result is relatively straightforward.

In Chapter 17, we come down to earth and consider Borel measures on Euclidean space, where the point at issue is the differentiation of measures and of Borel measurable functions. We establish Lebesgue's differentiation theorem and Rademacher's theorem on the differentiability almost everywhere of Lipschitz functions.

We now proceed to study one of the key points of this chapter, namely, the weak convergence of measures. We show that there are various metrics which define the weak topology w, and show that although the unit ball $M_1(X)$ is generally not metrizable, the space of probability measures $P(X)$ is a Polish space. Examples of weak convergence include the central limit theorem and the empirical law of large numbers. Finally, uniform integrability is investigated.

Part II ends with an introduction to Choquet theory on a metrizable compact convex set. The theory is notoriously difficult for general weakly compact convex sets, but the difficulties disappear in the metrizable case.

Parts I and II contain more than two hundred exercises. These are usually very straightforward, but most are an essential part of the text; do them.

Part III: Introduction to Optimal Transportation

The setting is this; μ and ν are Borel probability measures on Polish spaces X and Y, and c is a lower semi-continuous cost function on $X \times Y$. We consider two problems. Kantorovich's problem is to find a measure π on $X \times Y$ with marginals μ and ν with minimal cost $\int_{X\times Y} c\, d\pi$. Monge's problem is a special case of this; find a measurable mapping $T : X \to Y$ which pushes forward μ to ν with minimal cost $\int_X c(x, T(x))\, d\mu(x)$. The results of Parts I and II are used, or modified, to tackle these problems. For example we can push forward μ and ν to measures on metrizable compactifications. We also consider the concepts of c-cyclic monotonicity and c-concavity. It is quite easy to show that Kantorovich's problem has a solution, but with more care we introduce a 'maximal Kantorovich potential', which with its c-transform can give a great deal of information.

When $X = Y$ and $c = d^p$, where d is a metric on X, we introduce and investigate the Wasserstein metric W_p, which is the minimal cost of transforming μ into ν. Similarly, we introduce the Mallows distance, which does the same for distributions of random variables. As an example, we prove a metric version of the central limit theorem.

In the last chapter, we consider special cases. For example, we consider the case when $X = Y = \mathbf{R}$, and the case where the cost is a quadratic function on a separable Hilbert space. Finally, following Gangbo and McCann [GMcC], we consider the cases when the cost on $\mathbf{R}^d$ is given by a strictly convex or strictly concave function.

This only scratches the surface: for more, see the two large volumes by Villani, [V I] and [V II].

Although I have checked the proofs carefully, no doubt errors remain. Please consult my home page at **www.dpmms.cam.ac.uk** where a list of comments and corrections will be found, together with my email address, to which corrections should be sent.

PART ONE

Topological Properties

1
General Topology

This chapter contains a brief account of topological spaces and their properties. It contains definitions and statements of fundamental results, and describes the notation that is used. Proofs are generally not given; they can be found in [G II] (and elsewhere).

1.1 Topological Spaces

A *topological space* (X, τ) is a set X together with a set τ of subsets of X, the *topology*, which satisfies

(i) $\emptyset \in \tau$ and $X \in \tau$;
(ii) if F is a finite subset of τ, then $\cap_{U \in F} U \in \tau$; and
(iii) if G is any subset of τ, then $\cup_{U \in G} U \in \tau$.

The elements of τ are called *open sets*.

Here are some examples. The set $P(X)$ of *all* subsets of X is a topology on X, the *discrete* topology, and the set $\{\emptyset, X\}$ is also a topology on X, the *trivial topology*.

The *usual topology* on the real line $\mathbf{R}$ is defined by saying that U is open if whenever $x \in U$ there exists $\delta > 0$ such that

$$(x - \delta, x + \delta) = \{y : x - \delta < y < x + \delta\} \subseteq U.$$

Similarly, the *usual topology* on $\mathbf{C}$ is defined by saying that U is open if whenever $z \in U$ there exists $\delta > 0$ such that $\{w : |w - z| < \delta\} \subseteq U$.

Let $\overline{\mathbf{R}} = [-\infty, \infty]$. The *usual topology* on $\overline{\mathbf{R}}$ is defined by saying that U is open if

(i) $U \cap \mathbf{R}$ is open in $\mathbf{R}$ in the usual topology,
(ii) if $\infty \in U$ there exists $R \in \mathbf{R}$ such that $(R, \infty] \subseteq U$, and
(iii) if $-\infty \in U$ there exists $R \in \mathbf{R}$ such that $[-\infty, R) \subseteq U$.

The *right half-open* topology τ_r on $\mathbf{R}$ is defined by saying that $U \in \tau_r$ is open if whenever $x \in U$ there exists $\delta > 0$ such that

$$[x, x+\delta) = \{y : x \leq y < x+\delta\} \subseteq U.$$

Suppose that (X, τ) is a topological space. A subset $\mathcal{B}$ of τ is a *base* for the topology if every $U \in \tau$ is the union of sets in $\mathcal{B}$. Thus the open intervals (r, s) with $r, s \in \mathbf{Q}$ form a countable base for the usual topology on $\mathbf{R}$. A subset C of X is *closed* if $X \setminus C$ is open. If $A \subseteq X$, the *interior* A^{int} of A is the union of the open sets contained in A, the *closure* $\overline{A}$ of A is the intersection of the closed sets containing A, and the *boundary*, or *frontier*, ∂A of A is the set $\overline{A} \setminus A^{int}$. A^{int} is the largest open set contained in A, and $\overline{A}$ is the smallest closed set containing A. Elements of A^{int} are called *interior points* of A, and elements of $\overline{A}$ are called *closure points* of A. A subset B of A is *dense* in A if $A \subseteq \overline{B}$.

A subset N of X is a *neighbourhood* of an element x of X if $x \in N^{int}$. The set of neighbourhoods of x is denoted by $\mathcal{N}_x$. A subset $\mathcal{B}_x$ of $\mathcal{N}_x$ is a *base of neighbourhoods* of x if every $N \in \mathcal{N}_x$ contains an element of $\mathcal{B}_x$.

A *punctured neighbourhood* of x is a set of the form $N \setminus \{x\}$, where $N \in \mathcal{N}_x$. An element x of X is an *accumulation point*, or *limit point*, of A if $N^* \cap A \neq \emptyset$, for each punctured neighbourhood N^* of x. A point of A is an *isolated point* of A if it is not an accumulation point of A.

Here are two easy ways of constructing topological spaces. Suppose that Y is a subset of a topological space (X, τ). The *subspace topology* on Y is defined by taking the set $\{U \cap Y : U \in \tau\}$ of subsets of Y as the topology on Y. Suppose that f is a surjective mapping of X onto a set Z. The *quotient topology* on Z is defined by taking the sets $\{V : f^{-1}(V) \in \tau\}$ as the topology on Z.

A most important construction is the construction of topological product spaces. Suppose that $(X_\alpha, \tau_\alpha)_{\alpha \in A}$ is a family of topological spaces. Let $X = \prod_{\alpha \in A} X_\alpha$, and for each $\alpha \in A$, let $\pi_\alpha : X \to X_\alpha$ be the *co-ordinate projection*; $\pi_\alpha(x) = x_\alpha$. The *product topology* on X is defined by taking the collection

$$\{\cap_{\alpha \in F} \pi_\alpha^{-1}(U_\alpha) : F \text{ a finite subset of } A, U_\alpha \in \tau_\alpha \text{ for } \alpha \in F\}$$

as base of open sets. One special case occurs when $(X_\alpha, \tau_\alpha) = (X, \tau)$ for each $\alpha \in A$. In this case, $\prod_{\alpha \in A} X_\alpha = X^A$, the space of mappings from A to X. In particular, $X^{\mathbf{N}}$ (where $\mathbf{N}$ is the set $\{1, 2, 3, \ldots\}$ of natural numbers) is the space of sequences in X. A product of the form $[0, 1]^A$, where $[0, 1]$ is given its usual subspace topology, as a subspace of $\mathbf{R}$, is called a *hypercube*. The space $[0, 1]^{\mathbf{N}}$ is called the *Hilbert cube*.

Suppose that f is a mapping from a topological space (X, τ) into a topological space (Y, σ), and that $x \in X$. Then f is *continuous at* x if $f^{-1}(N) \in \mathcal{N}_x$, for each $N \in \mathcal{N}_{f(x)}$. Note that this agrees with the usual definition of continuity, when $X = Y = \mathbf{R}$, with the usual topology. f is *continuous on* X if it is continuous at each point of X. Then f is continuous on X if and only if $f^{-1}(V) \in \tau$, for each $V \in \sigma$, and if and only if $f^{-1}(C)$ is closed in X for each closed subset C of Y. The composition of two continuous functions is continuous.

If τ_1 and τ_2 are two topologies on X, then we say that τ_1 is *finer*, or *stronger*, than τ_2, and that τ_2 is *coarser*, or *weaker*, than τ_1, if the identity mapping $i : (X, \tau_1) \to (X, \tau_2)$ is continuous; that is, if $\tau_2 \subseteq \tau_1$.

A bijective mapping $f : (X, \tau) \to (Y, \sigma)$ is a *homeomorphism* if f and f^{-1} are both continuous; that is, if $f(\tau) = \sigma$.

Suppose that $(X, \tau) = \prod_{\alpha \in A}(X_\alpha, \tau_\alpha)$, and that $\alpha \in A$. The *co-ordinate projection* π_α is a continuous mapping from (X, τ) onto (X_α, τ_α). A mapping f from a topological space (Y, σ) into (X, τ) is continuous if and only if $\pi_\alpha \circ f : (Y, \sigma) \to (X_\alpha, \tau_\alpha)$ is continuous, for each $\alpha \in A$.

Suppose that $(X, \tau) = \prod_{\alpha \in A}(X_\alpha, \tau_\alpha)$, that $\alpha \in A$ and that $x \in X$. We define a mapping $k_{x,\alpha}$, the *cross-section mapping*, from X_α into X. If $y \in X_\alpha$, let $(k_{x,\alpha}(y))_\alpha = y$, and let $(k_{x,\alpha}(y))_\beta = x_\beta$ if $\beta \neq \alpha$. $k_{x,\alpha}$ is a homeomorphism of (X_α, τ_α) onto $(k_{x,\alpha}(X_\alpha), \tau)$.

A mapping f from a set X to a set Y is defined as a relation on $X \times Y$ which satisfies certain conditions. It is therefore natural to consider the corresponding *graph mapping* $G(f)$ from X to $X \times Y$, by setting $G(f)(x) = (x, f(x))$, for $x \in X$. The set $\Gamma_f = G(f)(X)$ is the *graph* of X.

Exercise 1.1.1 Suppose that X and Y are topological spaces, and that f is a mapping from X into Y. Then f is continuous if and only if the graph mapping $G(f)$ is a homeomorphism of X onto Γ_f.

Suppose that $(x_n)_{n=1}^\infty$ is a sequence in a topological space (X, τ) and that $x \in X$. Then x_n *converges* to x if for each $N \in \mathcal{N}_x$ there exists n_0 such that $x_n \in N$ for each $n \geq n_0$; if so, we write that $x_n \to x$ as $n \to \infty$. x is an *accumulation point* or *limit point* of the sequence if for each $N \in \mathcal{N}_x$ and each $n \in \mathbf{N}$ there exists $m \geq n$ such that $x_m \in N$. If $x_n \to x$ as $n \to \infty$, then x is a limit point of the sequence.

Suppose that f is a continuous mapping from a topological space (X, τ) into a topological space (Y, σ), and that $x \in X$. f is *sequentially continuous at* x if $f(x_n) \to f(x)$ as $n \to \infty$ whenever $x_n \to x$ as $n \to \infty$, and is *sequentially continuous on* X if it is sequentially continuous at each point of X. A continuous mapping is sequentially continuous; as we shall see, the converse is generally not true.

There are conditions that control the size of a topological space, and of topologies. A topological space is *separable* if there is a countable dense subset. It is *first countable* if every point has a countable base of neighbourhoods, and it is *second countable* if there is a countable base for the topology. These notions are related in the following way.

Exercise 1.1.2 (i) A second countable space is first countable and separable.

(ii) A subspace of a first countable space is first countable, and the product of countably many first countable spaces is first countable.

(iii) A subspace of a second countable space is second countable, and the product of countably many second countable spaces is second countable.

(iv) The product of countably many separable topological spaces is separable. (But see Proposition 1.1.3.)

(v) A mapping from a first countable topological space into a topological space is continuous if and only if it is sequentially continuous.

(vi) The topological space $(\mathbf{R}, \tau_r)$ is first countable and separable, but is not second countable.

(Proofs can be found in [G II], Propositions 13.5.1 and 13.5.3.)

Products of separable spaces behave remarkably well; this illustrates the fact that a product topology is a weak topology.

Proposition 1.1.3 *Suppose that (X, τ) is a separable topological space. Then $X^{(0,1]}$, with the product topology, is separable.*

Proof Let C be a countable dense subset of X, and let Y be the space of elements of $X^{(0,1]}$ which take constant values in C on each of the intervals $(i/k, (i+1)/k]$ for $1 \leq i \leq k$, for some k. Then Y is countable, and dense in $X^{(0,1]}$. □

There are also conditions which ensure that points and closed sets can be distinguished topologically. A topological space (X, τ) is

- a T_1 *space* if singleton sets are closed, so that finite sets are closed;
- a T_2 *space*, or *Hausdorff space*, if whenever x and y are distinct points of X there exist disjoint open sets U and V with $x \in U$ and $y \in V$;
- a T_3 *space* if whenever A is a closed subset of X and $x \notin A$ there exist disjoint open sets U and V with $x \in U$ and $A \subseteq V$;
- a T_4 *space* if whenever A and B are disjoint closed sets there exist disjoint open sets U and V with $A \subseteq U$ and $B \subseteq V$.

A topological space is a T_3 space if and only if every point has a base of neighbourhoods consisting of closed sets. A Hausdorff T_3 space is called a *regular space* and a Hausdorff T_4 space is called a *normal space*. A normal space is regular.

Exercise 1.1.4 Show that a topological space (X, τ) is Hausdorff if and only if the diagonal $\Delta = \{(x, x) : x \in X\}$ is closed in $X \times X$ (with the product topology).

If f is a continuous mapping from a topological space (X, τ) into a Hausdorff topological space (Y, σ) then the graph Γ_f is closed in $X \times Y$.

Theorem 1.1.5 *A second countable regular space is normal.*

Proof Let $\mathcal{B}$ be a countable base for the topology. Suppose that C and D are disjoint closed sets. Let $(V_i)_{i=1}^\infty$ be an enumeration of $\{U \in \mathcal{B} : \overline{U} \cap C = \emptyset\}$ and let $(W_j)_{j=1}^\infty$ be an enumeration of $\{U \in \mathcal{B} : \overline{U} \cap D = \emptyset\}$. Let $P_j = W_j \setminus (\cup_{i=1}^j \overline{V}_i)$ and let $Q_k = V_k \setminus (\cup_{i=1}^k \overline{W}_i)$. Then P_j and Q_k are disjoint open sets, for all j, k. Let $P = \cup_{j=1}^\infty P_j$ and $Q = \cup_{k=1}^\infty Q_k$, so that P and Q are disjoint open sets. If $x \in C$ then there exists W_j such that $x \in W_j$. But $x \notin \overline{V}_i$ for $1 \leq i \leq j$. Thus $x \in P_j \subseteq P$, and so $C \subseteq P$. Similarly, $D \subseteq Q$, and so X is normal. □

The words 'normal' and 'regular' are sadly overused in a mathematical context. Later, we shall use the term 'regular', with a quite different meaning, in a measure-theoretic setting.

In spite of their name, normal topological spaces can behave very badly.

Exercise 1.1.6 (i) Show that the subspace $L = \{(x, y) : x + y = 0\}$ of $(X, \tau) = (\mathbf{R}, \tau_r) \times (\mathbf{R}, \tau_r)$ has the discrete topology, and so is not separable.

(ii) Show that every subset of L is closed in (X, τ), and that (X, τ) is not normal.

In analysis, another property is important. A topological space (X, τ) is *completely regular* if it is Hausdorff, and whenever A is a closed subset of X and $x \notin A$ there exists a continuous function f on X taking values in $[0, 1]$, with $f(x) = 0$ and $f(a) = 1$ for each $a \in A$. A completely regular space is regular.

Exercise 1.1.7 Show that a subspace of a T_1 space (respectively Hausdorff space, regular space, completely regular space) is a T_1 space (respectively Hausdorff space, regular space, completely regular space), and the product of T_1 spaces (respectively Hausdorff spaces, regular spaces, completely regular spaces) is a T_1 space (respectively Hausdorff space, regular space, completely regular space).

Here is a more difficult result; we shall see that it is almost trivial when τ is given by a metric.

Theorem 1.1.8 (Urysohn's lemma) *If A and B are disjoint closed subsets of a T_4 space (X, τ), there exists a continuous function f on X taking values in $[0, 1]$, with $f(a) = 0$ for each $a \in A$ and $f(b) = 1$ for each $b \in B$.*

Proof Let $A_0 = A$. There exists a closed set A_1 such that

$$A_0 \subseteq A_1^{int} \subseteq A_1 \subseteq X \setminus B.$$

Let D be the set of dyadic fractions in $[0, 1]$. Arguing inductively, if $r = p/2^n$, with p odd, define a closed set A_r such that

$$A_{(p-1)/2^n} \subseteq A_r^{int} \subseteq A_r \subseteq A_{(p+1)/2^n}^{int} \subseteq A_{(p+1)/2^n};$$

($p - 1$ and $p + 1$ are even, so that $A_{(p-1)/2^n}$ and $A_{(p+1)/2^n}$ have already been defined). Now if $x \in X$ let $f(x) = \inf\{r \in D : x \in A_r\}$ (where $\inf(\emptyset) = 1$). Then $f(x) = 0$ for $x \in A$, $f(x) = 1$ for $x \in B$ and $0 \leq f(x) \leq 1$. It remains to show that f is continuous. But if $0 < s \leq 1$ then $f(x) < s$ if and only if $x \in C_s = \cup_{s<r\leq 1}(X \setminus A_r)$, and if $0 \leq t < 1$ then $f(x) > t$ if and only if $x \in D_t = \cup_{0\leq r\leq t} A_r^{int}$. Then C_s and D_t are open sets, from which it follows that f is continuous. □

Consequently, a normal space is completely regular.

Theorem 1.1.9 *A topological space (X, τ) is completely regular if and only if it is homeomorphic to a subspace of a hypercube.*

Proof The condition is certainly sufficient. Suppose that (X, τ) is completely regular. Let

$$A = \{(x, F) : x \in X, F \text{ closed in } X, x \notin F\}.$$

If $\alpha = (x, F) \in A$ there exists a continuous function $f_\alpha : X \to [0, 1]$ such that $f_\alpha(x) = 0$ and $f(y) = 1$ for $y \in F$. Let $f(x) = \{f_\alpha(x) : \alpha \in A\}$. If $x \neq y$ then $\beta = (x, \{y\}) \in A$, and $f_\beta(x) = 0 \neq 1 = f_\beta(y)$, and so f is injective. Since each f_α is continuous, $f : X \to [0, 1]^A$ is continuous. Conversely if $x \in X$ and U is an open neighbourhood of x then $\gamma = (x, X \setminus U) \in A$, $V = \{g \in f(X) : g_\gamma < 1\}$ is a neighbourhood of $f(x)$ in $f(X)$, and $V \subseteq f(U)$, so that $f^{-1} : f(X) \to X$ is also continuous. □

Corollary 1.1.10 *A second countable normal space is homeomorphic to a subspace of the Hilbert cube $\mathcal{H}$.*

Proof Replace A by the countable set of pairs $\{(x_n, C_m)\}$, where $(C_m)_{m=1}^\infty$ is a countable base of closed sets and $(x_n)_{n=1}^\infty$ is a dense sequence, and $x_n \notin C_m$. □

Note that the theorem uses the axiom of choice, but the corollary does not.

Let us end this section by observing that the definition of a topological space that we have given is short and easy, but is a little misleading. The important fact is that topology is a *local* phenomenon; it is therefore often appropriate to define a topology in terms of neighbourhoods. The neighbourhoods N_x of a point x form a *filter*; that is,

(i) $\emptyset \notin N_x$;
(ii) if $N \in N_x$ and $N \subseteq M$ then $M \in N_x$;
(iii) if $N_1, N_2 \in N_x$ then $N_1 \cap N_2 \in N_x$.

Exercise 1.1.11 Suppose that (X, τ) is a topological space and that $\{N_x\}_{x \in X}$ is the set of neighbourhood filters. Show that if $N \in N_x$ then $x \in N$ and there exists $O \in N_x$ such that $O \subseteq N$ and $O \in N_y$ for each $y \in O$.

Conversely, if $\{N_x\}_{x \in X}$ is a set of filters on a set X which has these properties, show that this defines a topology on X for which $\{N_x\}_{x \in X}$ is the set of neighbourhood filters.

1.2 Compactness

Suppose that A is a subset of a topological space (X, τ). A collection $\mathcal{O}$ of open sets is an *open cover* of A if $A \subseteq \cup_{O \in \mathcal{O}} O$. An open cover is *finite* if it has finitely many members. The set A is *compact* if every open cover of A has a finite subcover. Compact sets are a topological approximation to finite sets. By considering complements, it follows that (X, τ) is compact if and only if it has the *finite intersection property*; if $\mathcal{A}$ is a collection of closed sets with the property that $\cap_{A \in \mathcal{A}_f} A$ is not empty for each finite collection $\mathcal{A}_f$ of sets in $\mathcal{A}$, then the total intersection $\cap_{A \in \mathcal{A}} A$ is not empty.

Here are some basic properties of compact sets.

Exercise 1.2.1 (i) The union of a finite set of compact sets is compact.
(ii) A compact Hausdorff space is normal.
(iii) A subset of a compact Hausdorff space is compact if and only if it is closed.
(iv) The continuous image of a compact set is compact.
(v) A continuous bijection of a compact space onto a Hausdorff space is a homeomorphism.

For details, see [G II], Propositions 15.1.3, 15.1.4, Corollary 15.1.7.

Here is one of the fundamental results of general topology. It requires the axiom of choice, and indeed is equivalent to it; Tychonoff's theorem is true if and only if the axiom of choice holds.

Theorem 1.2.2 (Tychonoff's theorem) *The product of a set of compact topological spaces is compact.*

Proof See [G II], Appendix D. We shall only use this in Proposition 1.2.3. We shall however prove (and use) the fact that a countable product of compact metric spaces is compact. □

Corollary 1.2.3 *A topological space is a compact Hausdorff space if and only if it is homeomorphic to a closed subset of a hypercube.*

Proof This follows from Theorem 1.1.9 and Exercise 1.2.1 (iii). □

Let us give an example of this. The *Helly space* H is the subset of functions h in $[0,1]^{[0,1]}$ which are non-decreasing; if $0 \le s < t \le 1$ then $h(s) \le h(t)$. H is closed in $[0,1]^{[0,1]}$, and is therefore compact. It is first countable and separable, but is not second countable. For more details, see [SS: 107].

There are two related definitions. A is *sequentially compact* if every sequence in A has a convergent subsequence, convergent to an element of A, and A is *countably compact* if every sequence in A has a limit point in A. A compact set is countably compact, and a sequentially compact set is countably compact. In general, there are no further implications.

Proposition 1.2.4 *Let* $\Omega(\mathbf{N}) = \{0,1\}^{\mathbf{N}}$ *and let* $X = \{0,1\}^{\Omega(\mathbf{N})}$, *with the product topology* τ, *when* $\{0,1\}$ *is given the discrete topology.*

(i) (X,τ) *is compact and separable, but not sequentially compact.*
(ii) Let $X_0 = \{x \in X : \{\omega : x(\omega) = 1 \text{ is countable}\}\}$. *Then* X_0 *is sequentially compact, but is not compact or separable.*

Proof (i) (X,τ) is compact, by Tychonoff's theorem, and is separable, by Proposition 1.1.3, since there is a bijection from $\Omega(\mathbf{N})$ onto $(0,1]$. If $\omega \in X$, let $e_n(\omega) = \omega_n$ for $n \in \mathbf{N}$. The sequence $(e_n)_{n=1}^{\infty}$ in X has no convergent subsequence; (X,τ) is not sequentially compact.

(ii) A diagonal argument shows that X_0 is sequentially compact. It is a dense proper subspace of the Hausdorff space (X,τ), and so it is not compact. If C is a countable subset of X_0, the set $\{\omega : c(\omega) = 1 \text{ for some } c \in C\}$ is countable, and so there exists ω' such that $c(\omega') = 0$ for all $c \in C$. Then $I_{\{\omega'\}}$ is not in the closure of C, and so X_0 is not separable. □

A topological space (X,τ) is *locally compact* if every point has a base of neighbourhoods consisting of compact sets.

Exercise 1.2.5 A Hausdorff topological space is locally compact if and only if each point has a compact neighbourhood.

A *compactification* of a topological space (X, τ) is a compact space (X^*, τ^*), together with a homeomorphism j of X onto a dense subspace $j(X)$ of X^*. There are usually many compactifications of a topological space (X, τ); the one-point compactification is perhaps the simplest. If (X, τ) is not compact, the *one-point compactification* of (X, τ) is defined by taking $X^* = X \cup \{\infty\}$, where ∞ is a point not in X, and defining a subset U to be open if $U \cap X \in \tau$, and, if $\infty \in U$, then $X^* \setminus U$ is a compact subset of X.

The one-point compactification of (X, τ) is Hausdorff if and only if (X, τ) is Hausdorff and locally compact. Thus a topological space (X, τ) is Hausdorff and locally compact if and only if it is homeomorphic to an open subset of a compact Hausdorff space.

A topological space (X, τ) is σ*-compact* if it is the union of a sequence of compact subsets. If so, then by considering finite unions, X is the union of an increasing sequence of compact subsets.

Proposition 1.2.6 *A locally compact space (X, τ) is a σ-compact if and only if there is an increasing sequence $(K_n)_{n=1}^\infty$ of compact subsets such that $\cup_{n \in \mathbf{N}} K_n = X$ and $K_n \subseteq K_{n+1}^{int}$, for each $n \in \mathbf{N}$.*

Proof The condition is certainly sufficient. Suppose that (X, τ) is σ-compact and locally compact. Let $(L_n)_{n=1}^\infty$ be an increasing sequence of compact subsets of X whose union is X. Let $K_1 = L_1$. Suppose that we have found $K_1, \ldots, K_n$ such that $L_j \subseteq K_j$ for $1 \leq j \leq n$ and that $K_j \subseteq K_{j+1}^{int}$ for $1 \leq j < n$. For each $x \in K_n$ there exists an open neighbourhood N_x of x such that $\overline{N_x}$ is compact. The sets $\{N_x : x \in K_n\}$ are an open cover of the compact set K_n, and so there is a finite subcover $\{N_x : x \in F\}$. Let $K_{n+1} = (\cup_{x \in F} \overline{N_x}) \cup L_{n+1}$. Then $K_n \subseteq K_{n+1}^{int}$ and $L_{n+1} \subseteq K_{n+1}$, so that $X = \cup_{n \in \mathbf{N}} K_n$. □

A separable locally compact topological space need not be σ-compact. Let $\{0, 1\}$ be given the discrete topology, and let $X = \{0, 1\}^{[0,1]}$ be given the product topology τ. Then (X, τ) is compact, by Tychonoff's theorem, and is separable, but is not first countable. Let $x \in X$ and let $Y = X \setminus \{x\}$. Then Y is locally compact and separable, but is not σ-compact, since x does not have a countable base of neighbourhoods in (X, τ).

2
Metric Spaces

In this chapter, we introduce the idea of a metric space. Metric spaces have a natural topology, which we describe in the first two sections of the chapter, but they have more structure than that. They also have a uniform structure, which we shall consider in detail in Chapter 5, and which is the setting for the notion of completeness, which we discuss in Sections 2.3 to 2.7, a Lipschitz structure, which we consider in Section 2.8, and a geometric structure.

2.1 Metric Spaces

A *metric space* is a set X, together with a function $d : X \times X \to \mathbf{R}^+$, the *metric*, which satisfies

(i) $d(x, y) = d(y, x)$ for all $x, y \in X$ *(symmetry)*;
(ii) $d(x, z) \leq d(x, y) + d(y, z)$ for all $x, y, z \in X$ *(the triangle inequality)*; and
(iii) $d(x, y) = 0$ if and only if $x = y$.

The following inequality is a useful consequence of axioms (i) and (ii):

(iv) $|d(x_1, y_1) - d(x_2, y_2)| \leq d(x_1, x_2) + d(y_1, y_2)$ for all x_1, x_2, y_1, y_2 *(the quadrilateral inequality)*.

A function $p : X \times X \to \mathbf{R}^+$ which satisfies (i) and (ii) is called a *pseudometric*. If p is a pseudometric, set $x \sim y$ if $p(x, y) = 0$. Then $\sim$ is an equivalence relation on X. Let $X/\sim$ be the quotient space, and let $q : X \to X/\sim$ be the quotient mapping. If $x \sim x'$ and $y \sim y'$ then it follows from the quadrilateral inequality that $p(x, y) = p(x', y')$. Thus the function $d(q(x), q(y)) = p(x, y)$ is well-defined, and is easily seen to be a metric on $E/\sim$, the *associated metric*.

A metric defines a topology in a natural way. Suppose that (X, d) is a metric space. If $x \in X$ and $\epsilon > 0$, the *open ϵ-neighbourhood* $N_\epsilon(x)$ is defined as $\{y \in X : d(x, y) < \epsilon\}$, and the *closed ϵ-neighbourhood* $M_\epsilon(x)$ is defined as $\{y \in X : d(x, y) \leq \epsilon\}$. A subset U of X is *open* if whenever $x \in U$ there exists $\epsilon > 0$ such that $N_\epsilon(x) \subseteq U$. The collection of open sets is then a topology τ on X, the *metric topology*. Open ϵ-neighbourhoods are open in the topology, and closed ϵ-neighbourhoods are closed. The sets $\{N_\epsilon(x) : \epsilon > 0\}$ form a base of neighbourhoods of x, as does the collection $\{M_\epsilon(x) : \epsilon < 0\}$. Since the sets $\{N_{1/n}(x) : n \in \mathbf{N}\}$ form a base of neighbourhoods, the metric topology is first countable. Thus if f is a mapping of (X, d) into a topological space (Y, σ), f is continuous at x if and only if it is sequentially continuous; that is, $f(x_n) \to f(x)$ whenever $d(x_n, x) \to 0$ as $n \to \infty$.

If (X, d) and (Y, ρ) are metric spaces, then a mapping $f:(X, d) \to (Y, \rho)$ is an *isometry* if $\rho(f(x), f(y)) = d(x, y)$ for all $x, y \in X$. An isometry is clearly a homeomorphism of (X, d) onto $(f(X), \rho)$.

Let us give some examples.

(i) The function $d(x, y) = |x-y|$ is the *usual metric* on $\mathbf{R}$ and the *usual metric* on $\mathbf{C}$. Similarly the function $d(x, y) = (\sum_{j=1}^n |x_j - y_j|^2)^{\frac{1}{2}}$ is the *usual metric*, or *Euclidean metric*, on $\mathbf{R}^n$ and the *usual metric*, or *Hermitian metric*, on $\mathbf{C}^n$.

(ii) If X is any set, the function defined as $d(x, y) = 1$ if $x \neq y$ and $d(x, y) = 0$ if $x = y$ is the *discrete metric* on X.

(iii) If Y is a subset of a metric space (X, d) then the restriction of d to $Y \times Y$ is a metric on Y, the *subspace metric*.

(iv) If A is a non-empty subset of a metric space, the *diameter* $\operatorname{diam}(A)$ is defined to be $\sup\{d(a, b) : a, b \in A\}$. Note that it follows from the quadrilateral inequality that $\operatorname{diam}(A) = \operatorname{diam}(\overline{A})$. A set A is *bounded* if A is empty or $\operatorname{diam}(A) < \infty$.

Suppose that S is a non-empty set, that (Y, ρ) is a metric space and that f is a mapping from S to Y. We define the *oscillation* $\omega_S(f)$ of f on S to be $\operatorname{diam}(f(S)) = \sup\{\rho(f(s), f(t)) : s, t \in S\}$. f is *bounded* on S if $\omega_S(f) < \infty$, and $B(S, Y)$ denotes the set of bounded mappings from S to Y. The function $d_\infty(f, g) = \sup_{s \in S} \rho(f(s), g(s))$ is then a metric on $B(S, Y)$, the *uniform metric* on $B(S, Y)$. A sequence $(f_n)_{n=1}^\infty$ converges to f in this metric if and only if $f_n(s) \to f(s)$ uniformly in s; $\sup_{s \in S} |f_n(s) - f(s)| \to 0$ as $n \to \infty$; thus convergence in d_∞ is called *uniform convergence*. We denote $B(S, \mathbf{R})$ by $B(S)$, and we denote $(B(\mathbf{N}, Y)$ by $l_\infty(Y)$; thus $l_\infty(Y)$ is the space of bounded sequences in Y.

Exercise 2.1.1 (Dini's theorem) Suppose $(f_n)_{n=1}^\infty$ is a sequence of continuous real-valued functions on a compact Hausdorff space (X, τ) which decreases pointwise to 0. Show that f_n converges uniformly to 0 as $n \to \infty$.

We also introduce the notion of local oscillation. We define the *local oscillation* $\Omega_f(x)$ of f at x to be

$$\Omega_f(x) = \inf\{\omega_U(f) : U \text{ a neighbourhood of } x\}.$$

A function from a topological space (X, τ) into $[-\infty, \infty]$ is *upper semi-continuous* at x if given $\epsilon > 0$ there exists a neighbourhood U of x such that if $y \in U$ then $f(y) < f(x) + \epsilon$, and it is upper semi-continuous on X if it is upper semi-continuous at every point of X.

Proposition 2.1.2 *Suppose that (X, τ) is a topological space, that (Y, ρ) is a metric space, that f is a mapping from X to Y and that $x \in X$. Then Ω_f is a non-negative upper semi-continuous function on X. f is continuous at x if and only if $\Omega_f(x) = 0$.*

Proof Given $\epsilon > 0$, there exists an open neighbourhood U of x such that if y and z belong to U then $\rho(f(y), f(z)) < \Omega_f(x) + \epsilon$; but U is a neighbourhood of y, and so $\Omega_f(y) < \Omega_f(x) + \epsilon$ for $y \in U$.

If $\Omega_f(x) = 0$ then if $\epsilon > 0$ there exists an open neighbourhood U' of x such that $\rho(f(y), f(x)) < \epsilon$ for $y \in U'$; that is, f is continuous at x. The converse is just as easy. □

We consider semi-continuity further in Chapter 4.

Suppose that (X, d) and (Y, ρ) are metric spaces. Then $C_b(X, Y)$ is the set of *bounded* continuous mappings from X to Y.

Proposition 2.1.3 *Suppose that (X, d) and (Y, ρ) are metric spaces. Then $C_b(X, Y)$ is a closed subset of $(B(X, Y), d_\infty)$.*

Proof Suppose that $f \in \overline{C_b(X, Y)}$, that $x \in X$ and that $\epsilon > 0$. Then there exists $g \in C_b(X, Y)$ such that $d_\infty(f, g) < \epsilon/3$. Since g is continuous, there exists $\delta > 0$ such that if $d(x, y) < \delta$ then $\rho(g(x), g(y)) < \epsilon/3$. If $d(x, y) < \delta$ then

$$\rho(f(x), f(y)) \leq \rho(f(x), g(x)) + \rho(g(x), g(y)) + \rho(g(y), f(y)) < \epsilon,$$

so that f is continuous at x. □

Exercise 2.1.4 Let $L(\mathbf{R}^n)$ be the set of linear mappings from $\mathbf{R}^n$ to itself. If $T \in L(\mathbf{R}^n)$ show that T is continuous. Let $\|T\| = \sup\{d(T(x), 0) : d(x, 0) \leq 1\}$.

If $S, T \in L(\mathbf{R}^n)$, let $d_{Op}(S,T) = \|S - T\|$. Show that d_{Op} is a metric on $L(\mathbf{R}^n)$, the *operator metric*.

A topological space (X, τ) is *metrizable* if there is a metric d on X such that τ is the corresponding metric topology. In general, if (X, τ) is metrizable, there are many metrics with metric topology τ. Two metrics are said to be *equivalent* if they have the same metric topology.

Exercise 2.1.5 Suppose that (X, d) is a metric space, that $c > 0$ and that $(d \wedge c)(x, y) = \min(d(x, y), c)$. Show that $d \wedge c$ is a bounded metric which is equivalent to d.

Exercise 2.1.6 A metric ρ on a set X is an *ultrametric* if $\rho(x, z) \leq \max(\rho(x, y), \rho(y, z))$ for $x, y, z \in X$. Show that if (X, ρ) is an ultrametric space, then any open ϵ-neighbourhood is closed.

Exercise 2.1.7 Suppose that p is a prime. If $x, y \in \mathbf{Q}$, let $d_p(x, y) = 0$ if $x = y$ and let $d_p(x, y) = p^{-r}$ if $|x - y| = ap^r/b \neq 0$, where p does not divide the integers a and b, and $r \in \mathbf{Z}$. Show that d_p is a metric (the *p-adic metric*) on $\mathbf{Q}$. Show that d_p is an ultrametric.

2.2 The Topology of Metric Spaces

Let us now consider some topological properties of metric spaces.

As we have seen, a metric space is first countable.

Theorem 2.2.1 *A metric space (X, d) is separable if and only if it is second countable.*

Proof Suppose first that (X, d) is separable. Let $(x_n)_{n=1}^\infty$ be a dense sequence in (X, d). Then the countable collection $\{N_{1/j}(x_n) : j, n \in \mathbf{N}\}$ of open sets is a basis for the topology. For if U is an open subset of X and $x \in U$ there exists $\epsilon > 0$ such that $N_\epsilon(x) \subseteq U$. There exists j such that $1/j < \epsilon$, and there exists n such that $d(x, x_n) < 1/4j$. Then $x \in N_{1/2j}(x_n) \subseteq U$, so that $U = \cup\{N_{1/j}(x_n) : N_{1/j}(x_n) \subseteq U\}$.

Conversely, if (X, d) is second countable, and $\mathcal{B}$ is a countable base for the topology, choose a point x_B from each non-empty B in $\mathcal{B}$. Then the countable set $\{x_B : B \in \mathcal{B}, B \neq \emptyset\}$ is dense in X. □

Corollary 2.2.2 *A subspace of a separable metric space is separable.*

Proof For the subspace is second countable. □

We have seen that the Helly space is separable and first countable, but not second countable, and so it is not metrizable – the topology is not given by a metric. We have also seen that a subspace of a separable topological space need not be separable.

Theorem 2.2.3 *Suppose that S is a dense subset of a metric space (X, d). Then there is an isometry of (X, d) onto a subset of $(B(S), d_\infty)$.*

Proof Pick a point $x_0 \in X$. If $x \in X$ and $s \in S$ let $f_x(s) = d(x, s) - d(x_0, s)$. By the quadrilateral inequality, $|d(x, s) - d(x_0, s)| \leq d(x, x_0)$, so that $f_x \in B(S)$. If $x, x' \in X$,

$$|f_x(y) - f_{x'}(y)| = |d(x, y) - d(x', y)| \leq d(x, x'),$$

so that $d_\infty(f_x, f_{x'}) \leq d(x, x')$. On the other hand if $\epsilon > 0$ there exists $s \in S$ with $d(x', s) < \epsilon/2$. Then $|d(x, s) - d(x, x')| < \epsilon/2$, so that

$$|(f_x(s) - f_{x'}(s)) - d(x, x')| = |d(x, s) - d(x', s) - d(x, x')| < \epsilon,$$

from which it follows that $d_\infty(f_x, f_{x'}) \geq d(x, x')$. □

Corollary 2.2.4 *If (X, d) is a separable metric space, then there exists an isometry of (X, d) onto a subset of (l_∞, d_∞).*

On the other hand, (l_∞, d_∞) is not separable. If $A \subseteq \mathbf{N}$, let I_A be the indicator function of A; then $I_A \in l_\infty$. If $A \neq B$, then $d_\infty(I_A, I_B) = 1$. Thus if C is a countable subset of l_∞ and $f \in C$, then there is at most one $A \in P(\mathbf{N})$ for which $d(f, I_A) < 1/3$. Since $P(\mathbf{N})$ is uncountable, there exists $A \in P(\mathbf{N})$ such that $d(f, I_A) \geq 1/3$ for all $f \in C$, and so C is not dense in l_∞.

If A is a non-empty closed subset of (X, d), let $d(x, A) = \inf_{a \in A} d(x, a)$ be the *distance* of x from A. Then $d(x, A) = 0$ if and only if $x \in \overline{A}$. If $y \in X$ and $a \in A$, then $d(y, A) \leq d(y, a) \leq d(x, y) + d(x, a)$, so that $d(y, A) \leq d(x, y) + d(x, A)$. Similarly, $d(x, A) \leq d(x, y) + d(y, A)$, so that $|d(x, A) - d(y, A)| \leq d(x, y)$. Thus the function $x \to d(x, A)$ is continuous on X. Suppose now that A and B are disjoint non-empty closed subsets of X. Let

$$f_{A,B}(x) = \frac{d(x, A)}{d(x, A) + d(x, B)}.$$

Then $f_{A,B}$ is a continuous function on X taking values in $[0, 1]$, $f_{A,B} = 0$ if and only if $x \in A$ and $f_{A,B} = 1$ if and only if $x \in B$. This is so much easier than Urysohn's lemma! Thus a metric space is normal, and completely regular.

If A is a non-empty subset of (X, d), and $r > 0$ let $N_r(A) = \{y \in X : d(y, A) < r\}$. Then $N_r(A)$ is an open subset of X, the *open r-neighbourhood* of A. If A is closed, $A = \cap_{n \in \mathbf{N}} N_{1/n}(A)$, A is a G_δ set: the intersection of a decreasing sequence of open sets. As we shall see, this has important implications for measure theory.

Theorem 2.2.5 *The countable topological product (X, τ) of a sequence (X_n, d_n) of metric spaces is metrizable.*

Proof If $x = (x_n)_{n=1}^\infty$ and $y = (y_n)_{n=1}^\infty$ are elements of X, let

$$d(x, y) = \sum_{n=1}^{\infty} (d_n \wedge 2^{-n})(x_n, y_n).$$

Then the series converges, and it is easily verified that d is a metric on X. Let τ' be the corresponding metric topology. Let π_n be the co-ordinate map from X to X_n. Since the identity mapping $(X_n, d_n \wedge 2^{-n}) \to (X_n, d_n)$ is continuous, the identity mapping $(X, \tau') \to (X, \tau)$ is continuous. On the other hand, suppose that $N_\epsilon(x)$ is an open ϵ-neighbourhood of x for the metric d. There exists n_0 such that $2^{-n_0} < \epsilon/2$. Then

$$N = \{y \in X : d_n(x_n, y_n) < \epsilon/2n \text{ for } 1 \leq n \leq n_0\}$$

is a τ-neighbourhood, and if $y \in N$ then $d(x, y) < \epsilon$, and so the identity mapping $(X, \tau) \to (X, \tau')$ is also continuous. Thus $\tau = \tau'$. □

There are many other metrics that can be used, such as

$$\sup_{n \in \mathbf{N}} (d_n(x_n, y_n) \wedge 2^{-n}) \text{ or } \left(\sum_{n=1}^{\infty} (d_n(x_n, y_n) \wedge' 2^{-n})^2 \right)^{1/2}.$$

For finite products, there is no need to impose boundedness conditions on the metrics d_n.

There are two product spaces which will play an important role in what follows. First, there is the *Hilbert cube*, the product space $[0, 1]^{\mathbf{N}}$, which is denoted by $\mathcal{H}$. Secondly, the product space $\{0, 1\}^{\mathbf{N}}$, where $\{0, 1\}$ is given the discrete topology, is called the *Bernoulli sequence space*, and is denoted by $\Omega(\mathbf{N})$. Thus the points of $\Omega(\mathbf{N})$ are infinite sequences of 0*s* and 1*s*. Let us introduce some notions concerning $\Omega(\mathbf{N})$. If $n \in \mathbf{N}$ and either $\eta \in \Omega(\mathbf{N})$ or $\eta \in \{0, 1\}^m$ for some $m \geq n$, we set

$$C_{\eta,n} = \{\omega \in \Omega(\mathbf{N}) : \omega_j = \eta_j \text{ for } 1 \leq j \leq n\};$$

$C_{\eta,n}$ is a *cylinder set*, of *rank n*.

Exercise 2.2.6 Show that there are 2^n cylinder sets of rank n, which form a partition of $\Omega(\mathbf{N})$. Show that cylinder sets are open and closed, and that the cylinder sets form a base for the topology of $\Omega(\mathbf{N})$. Show that the countable collection of sets $\{C_{\eta,n} : n \in \mathbf{N}\}$ is a base of neighbourhoods of η.

The Hilbert cube $\mathcal{H}$ and the Bernoulli sequence space $\Omega(\mathbf{N})$ are metrizable. There are other metrics on them which define the product topology, and may be more natural to consider. For example, suppose that $(a_n)_{n=1}^\infty$ is a sequence of positive numbers for which $\sum_{j=1}^{\infty} a_j^2 < \infty$. If $x, y \in \mathcal{H}$ let $d(x, y) = (\sum_{n=1}^{\infty} a_n^2(x_n - y_n)^2)^{1/2}$. Then d is a metric on $\mathcal{H}$ which defines the product topology.

If $\omega, \omega' \in \Omega(\mathbf{N})$, let $d(\omega, \omega') = 2\sum_{n=1}^{\infty} |\omega_n - \omega'_n|/3^n$. Then d is a metric on $\Omega(\mathbf{N})$ which defines the product topology. If $\omega \in \Omega(\mathbf{N})$, let $f(\omega) = 2\sum_{n=1}^{\infty} \omega_n/3^n$. Then f is an isometry of $(\Omega(\mathbf{N}), d)$ onto the Cantor set C, and so we call d the *Cantor metric* on $\Omega(\mathbf{N})$. Note that the open and closed neighbourhood $M_{1/3^n}(\omega)$ of ω is the cylinder set $C_{\omega,n}$.

When (X, τ) is a separable topological space, there are necessary and sufficient conditions for (X, τ) to be metrizable.

Theorem 2.2.7 (Urysohn's metrization theorem) *Suppose that (X, τ) is a separable topological space, the following are equivalent:*

(i) (X, τ) is regular and second countable.
(ii) (X, τ) is homeomorphic to a subspace of the Hilbert cube $\mathcal{H}$.
(iii) (X, τ) is metrizable.

Proof Suppose that (i) holds. Then (X, τ) is normal (Theorem 1.1.5), and so (X, τ) is homeomorphic to a subspace of the Hilbert cube, by Corollary 1.1.10, and so (ii) is true. Condition (ii) certainly implies (iii). Finally suppose that (X, τ) is the topology defined by a metric d. Then τ is normal, and (X, τ) is second countable, by Theorem 2.2.1, so that (i) holds. □

2.3 Completeness: Tietze's Extension Theorem

For topological spaces, continuity is generally a local phenomenon, and it is not possible to compare what happens at one point with another. Metric spaces have other properties than topological ones, and here we consider uniform properties; we shall discuss uniform spaces in Chapter 5.

A sequence $(x_n)_{n=1}^{\infty}$ in a metric space (X, d) is a *Cauchy sequence* if whenever $\epsilon > 0$ there exists $N \in \mathbf{N}$ such that $d(x_m, x_n) < \epsilon$, for $m, n \geq N$. A convergent sequence is a Cauchy sequence; conversely, if every Cauchy sequence converges, then (X, d) is said to be *complete*. The real line $\mathbf{R}$, with its usual metric, is complete (the *general principle of convergence*): this property lies behind the construction of the real number system.

Proposition 2.3.1 *If S is a non-empty set and (Y, ρ) is a metric space then the metric space $(B(S, Y), d_\infty)$ is complete if and only if (Y, ρ) is complete.*

Proof Suppose that (Y, ρ) is complete and that $(f_n)_{n=1}^{\infty}$ is a Cauchy sequence in $B(S)$. If $s \in S$ then $\rho(f_m(s), f_n(s)| \leq d_\infty(f_m, f_n)$, and so $(f_n(s))_{n=1}^{\infty}$ is a Cauchy sequence in Y, which converges, to $f(s)$, say, as $n \to \infty$. There exists N such that $d_\infty(f_m, f_n) < 1$ for $m, n \geq N$, and so $\rho(f(s), f_N(s)) \leq 1$; thus $f \in B(S)$. Finally, $d_\infty(f, f_n) \leq \sup_{m \geq n} d_\infty(f_m, f_n) \to 0$ as $n \to \infty$.

Conversely, if (Y, ρ) is not complete then there is a Cauchy sequence $(y_n)_{n=1}^\infty$ in Y which does not converge. Let $f_n(s) = y_n$, for $s \in S$ and $n \in \mathbf{N}$. Then $(f_n)_{n=1}^\infty$ is a Cauchy sequence in $B(S, Y)$ which does not converge. □

Corollary 2.3.2 *If S is a non-empty set then $(B(S), d_\infty)$ is complete.*

Proof For $\mathbf{R}$, with its usual metric, is complete. □

Proposition 2.3.3 *A subset Y of a complete metric space (X, d) is complete in the subspace metric if and only if it is closed.*

Proof If Y is closed, and if $(y_n)_{n=1}^\infty$ is a Cauchy sequence in Y, then it is a Cauchy sequence in X, and so converges to an element x of X. Since Y is closed, $x \in Y$, and so Y is complete.

Conversely, if Y is complete and $x \in \overline{Y}$ then there exists a sequence $(y_n)_{n=1}^\infty$ in Y which converges to x. But then $(y_n)_{n=1}^\infty$ is a Cauchy sequence in Y, and so converges to an element y of Y, since Y is complete. But limits are unique, and so $x = y \in Y$. Thus $\overline{Y} = Y$, and so Y is closed. □

Corollary 2.3.4 *If (X, τ) is a topological space and (Y, ρ) is a metric space then the metric space $(C_b(X, Y), d_\infty)$ is complete if and only if (Y, ρ) is complete.*

Proof For $(C_b(X, Y), d_\infty)$ is closed in $(B(X, Y), d_\infty)$. □

We use this, and Urysohn's lemma, to prove an important extension theorem.

Theorem 2.3.5 (Tietze's extension theorem) *Suppose that (X, τ) is a normal topological space, that Y is a non-empty closed subset of X and that $f \in C_b(Y)$. Let $\Lambda = \sup_{y \in Y} f(y)$ and $\lambda = \inf_{y \in Y} f(y)$. There exists $g \in C_b(X)$ such that $f(y) = g(y)$ for $y \in Y$, $\sup_{x \in X} g(x) = \Lambda$ and $\inf_{x \in X} g(x) = \lambda$. If $f(y) < \Lambda$ for all $y \in Y$ then the extension g can be chosen so that $g(x) < \Lambda$ for all $x \in X$. Similarly, for λ.*

Proof The result is trivially true if f is constant. Otherwise, by adding a constant and scaling we can suppose that $\Lambda = 1$ and that $\lambda = -1$. We shall show by induction that there exists a sequence $(g_n)_{n=0}^\infty$ in $C_b(X)$ such that

$$\text{(i)} \quad \sup_{y \in Y} |f(y) - g_n(y)| \leq (2/3)^n \text{ for } n \in \mathbf{Z}^+$$

$$\text{and (ii)} \quad \sup_{x \in X} |g_{n-1}(x) - g_n(x)| \leq \tfrac{1}{2}(2/3)^n \text{ for } n \in \mathbf{N}.$$

We set $g_0 = 0$. Suppose that we have found $g_0, \dots, g_n$ satisfying the conditions. Let

$$A_n = \{y \in Y : f(y) - g_n(y) \geq 2^n/3^{n+1},$$
$$B_n = \{y \in Y : f(y) - g_n(y) \leq -2^n/3^{n+1}.$$

Then A_n and B_n are disjoint closed subsets of X, and so by Urysohn's lemma there exists $k_n \in C_b(X)$ such that

$$k_n(a) = 2^n/3^{n+1} \text{ for } a \in A_n,$$
$$k_n(b) = -2^n/3^{n+1} \text{ for } b \in B_n,$$
$$\text{and } |k_n(x)| \leq 2^n/3^{n+1} \text{ for } x \in X.$$

Let $g_{n+1} = g_n - k_n$. Then $|f(y) - g_{n+1}(y)| \leq (2/3)^{n+1}$ for $y \in Y$, and $|g_n(x) - g_{n+1}(x)| \leq \frac{1}{2}(2/3)^{n+1}$ for $x \in X$; the induction is established.

If $m < n$ and $x \in X$ then

$$|g_m(x) - g_n(x)| \leq \sum_{j=m+1}^{n} |g_{j-1}(x) - g_j(x)| \leq \sum_{j=m+1}^{n} \tfrac{1}{2}(2/3)^j < (2/3)^m,$$

so that $d_\infty(g_m, g_n) < (2/3)^m$. Thus $(g_n)_{n=0}^\infty$ is a Cauchy sequence in $(C_b(X), d_\infty)$. Since $(C_b(X), d_\infty)$ is complete, there exists $g \in C_b(X)$ such that $g_n \to g$ as $n \to \infty$. By (i), $g_n(y) \to f(y)$ as $n \to \infty$, for $y \in Y$, and so $f(y) = g(y)$ for $y \in Y$. Since $|g_n(x)| = |g_n(x) - g_0(x)| < 1$ for $x \in X$ and $n \in \mathbf{N}$, it follows that $|g(x)| \leq 1$ for $x \in X$, and so $\sup_{x \in X} g(x) = -\inf_{x \in X} g(x) = 1$.

Suppose now that $f(y) < \Lambda$ for all $y \in Y$. If g is an extension, as shown earlier, and $Z = \{x \in X : g(x) = \Lambda\}$ is not empty, then Y and Z are disjoint closed subsets of X, and so, by Urysohn's lemma there exists $h \in C_b(X)$ with $h(y) = 1$ for $y \in Y$, $h(z) = 0$ for $z \in Z$ and $0 \leq h \leq 1$. Then gh is an extension with the required property. Similarly for λ. □

If (X, τ) is a topological space, we denote the space of continuous real-valued functions on X by $C(X)$.

Corollary 2.3.6 *Suppose that (X, τ) is a normal topological space, that Y is a non-empty closed subset of X and that $f \in C(Y)$. Then there exists $g \in C(X)$ such that $g(y) = f(y)$ for $y \in Y$.*

Proof Let $F(y) = \tan^{-1} f(y)$, for $y \in Y$. Then $F \in C_b(Y)$ and $F(Y) \subseteq (-\pi/2, \pi/2)$. There exists an extension G with $G(X) \subseteq (-\pi/2, \pi/2)$. Let $g(x) = \tan G(x)$, for $x \in X$. Then g is a continuous extension of f. □

Exercise 2.3.7 Suppose that p is an odd prime, that a is not a square and that $x_0^2 = a (\bmod p)$. Show by induction that there is a sequence $(x_n)_{n=0}^\infty$ such that

$$x_n = x_{n-1} (\bmod p^n) \quad \text{and} \quad x_n^2 = a \left(\bmod p^{n+1}\right).$$

Show that $(x_n)_{n=0}^\infty$ is a Cauchy sequence in the p-adic metric which does not converge in $(\mathbf{Q}, d_p)$; $(\mathbf{Q}, d_p)$ is not complete.

2.4 More on Completeness

A mapping f from a metric space (X, d) to a metric space (Y, ρ) is *uniformly continuous* if whenever $\epsilon > 0$ there exists $\delta > 0$ such that if $d(x, y) < \delta$ then $\rho(f(x), f(y)) < \epsilon$. A bijection $j : (X, d) \to (Y, \rho)$ is a *uniform homeomorhism* if f and f^{-1} are uniformly continuous, and two metrics d and d' are *uniformly equivalent* if the identity mapping $i : (X, d) \to (X, d')$ is a uniform homeomorphism. As the names suggest, these ideas relate to the uniform structures of X and Y.

It is often convenient to replace a metric with a uniformly equivalent metric. The following exercise suggests how this can be done.

Exercise 2.4.1 Suppose that ϕ is an increasing real-valued function on $[0, \infty)$, that $\phi(t) = 0$ if and only if $t = 0$ and that ϕ is continuous at 0. Consider the following statements.

(i) ϕ is concave: if $a, b \geq 0$ and $0 \leq \lambda \leq 1$ then $\phi((1-\lambda)a + \lambda b) \geq (1-\lambda)\phi(a) + \lambda\phi(b)$.
(ii) Let $\psi(t) = \phi(t)/t$, for $t > 0$. Then ψ is a decreasing function on $(0, \infty)$.
(iii) If $a, b \geq 0$ then $\phi(a+b) \leq \phi(a) + \phi(b)$.
(iv) If (X, d) is a metric space, and $\rho(x, y) = \phi(d(x, y))$ then ρ is a metric on X which is uniformly equivalent to d.

Show that (i) implies (ii), (ii) implies (iii) and (iii) implies (iv). Show that there are no converse implications.

Thus if (X, d) is a metric space, and if $\phi(t) = t \wedge c$, where $c > 0$, then $d \wedge c$ is a bounded metric uniformly equivalent to d; another bounded uniformly equivalent metric is obtained by taking $\phi(t) = t/(1+t)$. If $0 < p < 1$ then the function $\phi(t) = t^p$ is concave on $[0, \infty)$, and so if $d^p(x, y) = d(x, y)^p$ then d^p is a metric on X uniformly equivalent to d. Note that d^p is *strictly subadditive*: if x, y, z are three distinct elements of X then $d^p(x, z) < d^p(x, y) + d^p(y, z)$.

If $f : (X, d) \to (Y, \sigma)$ is a uniformly continuous mapping and $(x_n)_{n=1}^\infty$ is a Cauchy sequence in X, then $(f(x_n))_{n=1}^\infty$ is a Cauchy sequence in Y. In particular, if d and d' are uniformly equivalent metrics on X, then a sequence $(x_n)_{n=1}^\infty$ in X is a d-Cauchy sequence if and only if it is a d'-Cauchy sequence; consequently (X, d) is complete if and only if (X, d') is complete. These properties are not topological ones. Let d be the usual metric on $[1, \infty)$ and let $d'(x, y) = |1/x - 1/y|$. Then d and d' are equivalent metrics. The sequence $(n)_{n=1}^\infty$ is a d'-Cauchy sequence,but not a d-Cauchy sequence, and $[1, \infty)$ is d-complete, but is not d'-complete.

In the previous section, we defined a product metric on the product of a sequence of metric spaces. If d is a product metric on $X = \prod_{i=1}^{\infty}(X_i, d_i)$, d is a *uniform product metric* if each cross-section mapping $k_{x,i}$ is a uniform homeomorphism and each co-ordinate projection π_i is uniformly continuous. The product metrics introduced in the previous section are all uniform product metrics, and we shall only consider uniform product metrics.

Theorem 2.4.2 *Suppose that $((X_n, d_n))_{n=1}^{\infty}$ is a sequence of complete metric spaces and that d is a uniform product metric on $X = \prod_{n=1}^{\infty} X_n$. Then (X, d) is complete.*

Proof Suppose that $(x^{(j)})_{j=1}^{\infty}$ is a d-Cauchy sequence in X. If $n \in \mathbf{N}$ then $(x_n^{(j)})_{j=1}^{\infty}$ is a d_n-Cauchy sequence in (X_n, d_n); since (X_n, d_n) is complete, it converges to an element x_n of X_n. But then $x^{(j)} \to x = (x_n)_{n=1}^{\infty}$ as $j \to \infty$ in (X, d). □

Completeness can be characterized in terms of decreasing sequences of closed sets.

Proposition 2.4.3 *Suppose that (X, d) is a metric space. the following are equivalent.*

(i) (X, d) is complete.
(ii) If $(A_n)_{n=1}^{\infty}$ is a decreasing sequence of non-empty closed subsets of X, and $\operatorname{diam}(A_n) \to 0$ as $n \to \infty$, then $\cap_{n\in\mathbf{N}} A_n$ is non-empty.

If so, then $\cap_{n\in\mathbf{N}} A_n$ is a singleton $\{a\}$, and if $a_n \in A_n$ for each n then $a_n \to a$ as $n \to \infty$.

Proof Suppose first that (ii) holds. Suppose that $(x_n)_{n=1}^{\infty}$ is a Cauchy sequence in X. Let $A_n = \overline{\{x_j : j \geq n\}}$. Then $(A_n)_{n=1}^{\infty}$ is a decreasing sequence of non-empty closed subsets of X, and $\operatorname{diam}(A_n) \to 0$ as $n \to \infty$. Thus $\cap_{n\in\mathbf{N}} A_n$ is non-empty. If $a \in \cap_{n\in\mathbf{N}} A_n$, then $x_n \to a$ as $n \to \infty$, and so (X, d) is complete.

Suppose conversely that (X, d) is complete and that $(A_n)_{n=1}^{\infty}$ is a decreasing sequence of non-empty closed subsets of X. Pick $a_n \in A_n$ for each $n \in \mathbf{N}$. If $n \geq m \geq N$ then $d(a_n, a_m) \leq \operatorname{diam}(A_N)$ so that (a_n) is a Cauchy sequence. Since (X, d) is complete, there exists $a \in X$ such that $a_n \to a$ as $n \to \infty$. Since each A_n is closed, $a \in \cap_{n\in\mathbf{N}} A_n$, so that $\cap_{n\in\mathbf{N}} A_n \neq \emptyset$. Finally, since $\operatorname{diam}(\cap_{n\in\mathbf{N}} A_n) = 0$, $\cap_{n\in\mathbf{N}} A_n = \{a\}$. □

We now prove a fundamental extension result.

Theorem 2.4.4 *Suppose that A is a dense subset of a metric space (X,d) and that f is a uniformly continuous mapping from A into a complete metric space (Y,ρ). Then there exists a uniformly continuous mapping $\tilde{f} : X \to Y$ which extends f: $\tilde{f}(a) = f(a)$ for $a \in A$.*

The extension is unique; indeed, if g is a continuous extension of f, then $g = \tilde{f}$.

Proof Suppose that $x \in X$. If $n \in \mathbf{N}$, let $A_n(x) = \overline{f(N_{1/n}(x) \cap A)}$. If $\epsilon > 0$, there exists $N \in \mathbf{N}$ such that if $a, b \in A$ and $d(a,b) < 2/N$ then $\rho(f(a), f(b)) < \epsilon/3$. Thus if $n \geq N$ then $\operatorname{diam}(A_n(x)) \leq \epsilon/3$. By Proposition 2.4.3, the set $\cap_{n\in\mathbf{N}} A_n(x)$ is a singleton $\{a(x)\}$: we set $\tilde{f}(x) = a(x)$. If $x \in A$, then $f(x) \in \cap_{n\in\mathbf{N}} A_n$, and so $\tilde{f}(a) = f(a)$: $\tilde{f}$ extends f.

Next we show that $\tilde{f}$ is uniformly continuous. Suppose that ϵ and N are as above, and that $d(x,y) < 1/N$. Choose $a_N \in N_{1/2N}(x) \cap A$ and $b_N \in N_{1/2N}(y) \cap A$. Then $f(a_N) \in A_N(x)$, so that $\rho(\tilde{f}(x), f(a_N)) \leq \epsilon/3$. Similarly, $\rho(\tilde{f}(y), f(b_N)) < \epsilon/3$. Further, $d(a_N, b_N) < 2/N$, so that $\rho(f(a_N), f(b_N)) < \epsilon/3$. Consequently $\rho(\tilde{f}(x), \tilde{f}(y)) < \epsilon$: $\tilde{f}$ is uniformly continuous.

Finally, if g is a continuous extension, then $\{x \in X : \tilde{f}(x) = g(x)\}$ is closed, and contains A, and so is the whole of X. □

Here is a useful test for completeness.

Theorem 2.4.5 *Suppose that (X,d) is a complete metric space, that $Y \subseteq X$ and that ρ is a metric on Y with the following properties.*

(i) If $(y_n)_{n=1}^\infty$ is a ρ-Cauchy sequence then it is a d-Cauchy sequence.
(ii) There exists $r > 0$ such that if $y_0 \in Y$ and $0 < \eta < r$ then the set $M_\eta(y_0) = \{y \in Y : \rho(y, y_0) \leq \eta\}$ is d-closed.

Then (Y,ρ) is complete.

Proof Suppose that $(y_n)_{n=1}^\infty$ is a ρ-Cauchy sequence in Y. Then it is a d-Cauchy sequence in X, and so converges in (X,d) to an element $x \in X$. If $0 < \epsilon < r$, there exists N such that $\rho(y_m, y_n) < \epsilon/2$ for $m, n \geq N$. Thus $y_n \in M_{\epsilon/2}(y_N)$ for $n \geq N$. Since $M_{\epsilon/2}(y_N)$ is d-closed, $x \in M_{\epsilon/2}(y_N) \subseteq Y$, and $\rho(x, y_n) \leq \rho(x, y_N) + \rho(y_N, y_n) < \epsilon$, for $n \geq N$. Thus $y_n \to x$ in (Y,ρ) as $n \to \infty$. □

2.5 The Completion of a Metric Space

If (X,d) is a metric space, a *completion* of (X,d) is a pair $((\hat{X},\hat{d}), j)$, where $(\hat{X},\hat{d})$ is a complete metric space, and j is an isometry of (X,d) onto a dense subspace of $(\hat{X},\hat{d})$.

Theorem 2.5.1 *Every metric space (X, d) has a completion, which is essentially unique: if $((\hat{X}_1, \hat{d}_1), j_1)$ and $((\hat{X}_2, \hat{d}_2), j_2)$ are two completions, there exists a unique isometry k of $(\hat{X}_1, \hat{d}_1)$ onto $(\hat{X}_2, \hat{d}_2)$ such that $j_2 = k \circ j_1$.*

Proof By Theorem 2.2.3, there exists an isometry j of (X, d) into $(B(X), d_\infty)$, and so $((\overline{j(X)}, d_\infty), j)$ is a completion of (X, d). Suppose that $((\hat{X}_1, \hat{d}_1), j_1)$ and $((\hat{X}_2, \hat{d}_2), j_2)$ are two completions of (X, d). It follows from Theorem 2.4.4 that there exists a unique uniform homeomorphism $k : (\hat{X}_1, \hat{d}_1) \to (\hat{X}_2, \hat{d}_2)$ such that $j_2 = k \circ j_1$. The set

$$\{(\hat{x}, \hat{y}) \in \hat{X}_1 \times \hat{X}_1 : \hat{d}_2(k(\hat{x}), k(\hat{y})) = \hat{d}_1(\hat{x}, \hat{y})\}$$

is closed in $\hat{X}_1 \times \hat{X}_1$, and contains the dense subset $j_1(X) \times j_1(X)$ of $\hat{X}_1 \times \hat{X}_1$, and so k is an isometry. □

As a result, we talk about *the* completion $(\hat{X}, \hat{d})$ of (X, d), and consider (X, d) as a subspace of $(\hat{X}, \hat{d})$.

This construction of a completion is short, but artificial. It is useful to give a more natural construction of the completion of a metric space. We need a preliminary result.

Proposition 2.5.2 *Suppose that Y is a dense subspace of a metric space (X, d), and that every Cauchy sequence in Y converges to an element of X. Then (X, d) is complete.*

Proof Suppose that $(x_n)_{n=1}^\infty$ is a Cauchy sequence in X. For each $n \in \mathbf{N}$ there exists $y_n \in Y$ with $d(y_n, x_n) < 1/n$. Then $d(y_m, y_n) \leq d(x_m, x_n) + 1/m + 1/n$, and so $(y_n)_{n=1}^\infty$ is a Cauchy sequence in Y. By hypothesis, it converges to an element x of X. But $d(x, x_n) \leq d(x, y_n) + d(y_n, x_n) < d(x, y_n) + 1/n$, and so $x_n \to x$ as $n \to \infty$. □

Suppose now that (X, d) is a metric space. Let $Ca(X)$ be the set of Cauchy sequences in X. If $x, y \in Ca(X)$, then

$$|d(x_m, y_m) - d(x_n, y_n)| \leq d(x_m, x_n) + d(y_m, y_n),$$

by the quadrilateral inequality, and so $(d(x_n, y_n))_{n=1}^\infty$ is a real Cauchy sequence, which converges to $\phi(x, y)$, say. Then it is easy to see that ϕ is a pseudometric on $Ca(X)$. Let $\hat{X}$ be the corresponding quotient space, q the quotient mapping and $\hat{d}$ the associated metric on $\hat{X}$. If $x \in X$, let $c(x)$ be the constant sequence with $c(x)_n = x$, for all n. Then $c(x)$ is a Cauchy sequence in X; let $j(x)$ be the corresponding element of $\hat{X}$.

Proposition 2.5.3 *$((\hat{X},\hat{d}),j)$ is a completion of (X,d).*

Proof First, it is clear that $\phi(c(x),c(y)) = d(x,y)$, so that $\hat{d}(j(x),j(y)) = d(x,y)$: j is an isometry of (X,d) into $(\hat{X},\hat{d})$. Secondly, $j(X)$ is dense in $\hat{X}$. For if $x = (x_n)_{n=1}^\infty \in Ca(X)$, then $\phi(x,c(x_n)) = \lim_{m\to\infty} d(x_m,x_n) \to 0$ as $n \to \infty$. Finally, if $(j(x_n))_{n=1}^\infty$ is a Cauchy sequence in $j(X)$, $(x_n)_{n=1}^\infty$ is a Cauchy sequence in X, and

$$\hat{d}(j(x_n),q(x)) = \lim_{m\to\infty} d(x_m,x_n) \to 0 \text{ as } n \to \infty,$$

so that $j(x_n) \to q(x)$ as $n \to \infty$. Thus $(\hat{X},\hat{d})$ is complete, by the preceding proposition. □

2.6 Topologically Complete Spaces

A topological space is *topologically complete* if it is homeomorphic to a complete metric space. The countable product of topologically complete spaces is topologically complete. A closed subspace of a topologically complete space is topologically complete.

A subset A of a topological space is a G_δ-set (F_σ-set) if it is the countable intersection of open sets (countable union of closed sets).

Theorem 2.6.1 *A G_δ-subset of a topologically complete space (X,τ) is topologically complete.*

Proof Let d be a complete metric on X which defines the topology on X.

First suppose that A is open. If $A = X$, then A is topologically complete. Otherwise, let $C(A) = X \setminus A$ and give $X \times \mathbf{R}$ a uniform product metric σ. Then $(X \times \mathbf{R}, \sigma)$ is complete. Consider the injective mapping f from A to $X \times \mathbf{R}$ defined by $f(a) = (a, 1/d(a,C(A)))$. Since $d(a,C(A)) > 0$ for $a \in A$, this is well-defined. Since the mapping $a \to 1/d(a,C(A))$ is continuous on A, f is continuous. We show that $f(A)$ is closed in $X \times \mathbf{R}$. Suppose that

$$f(a_n) \to y = (x,\lambda) \in X \times \mathbf{R} \text{ as } n \to \infty,$$

so that $a_n \to x$ and $1/d(a_n,C(A)) \to \lambda$ as $n \to \infty$. Since

$$|d(a_m,C(A)) - d(a_n,C(A))| \leq d(a_m,a_n),$$

the sequence $(d(a_n,C(A)))_{n=1}^\infty$ is bounded, so that $1/d(a_n,C(A))$ does not tend to 0 as $n \to \infty$. Thus $\lambda \neq 0$, and $d(a_n,C(A)) \to 1/\lambda$. Since $d(a_n,C(A)) \to d(x,C(A))$, it follows that $d(x,C(A)) = 1/\lambda$, and so $x \in A$. Consequently $f(a_n) \to f(x)$ as $n \to \infty$, so that $f(A)$ is closed in $X \times \mathbf{R}$, and $(f(A),\sigma)$ is complete. The mapping $f : (A,d) \to (X \times \mathbf{R},\sigma)$ is continuous, and the inverse mapping $f^{-1} : f(A) \to A$ is continuous, since $d(a,a') \leq \sigma(f(a),f(a'))$, so that f is a homeomorphism of (A,d) onto $(f(A),\sigma)$. Thus if we define $\rho(a,a') = \sigma(f(a),f(a'))$ then ρ is a complete metric on A equivalent to d.

Next, suppose that $A = \cap_{j=1}^{\infty} U_j$ is a G_δ set. For each j there is a metric σ_j on U_j, equivalent to the restriction of d to U_j, under which U_j is complete. Let $U = \prod_{j=1}^{\infty}(U_j,\sigma_j)$, and let σ be a uniform product metric on U. Then (U,σ) is complete. If $a \in A$, let $i_j : A \to U_j$ be the inclusion map and let $i : A \to U$ be defined as $i(a) = (i_j(a))_{j=1}^{\infty}$. Then i is a continuous injective map of (A,d) into (U,σ). We show that $i(A)$ is closed in U. Suppose that $i(a_n) \to u$ as $n \to \infty$. Then, for each $j \in \mathbf{N}$, $i_j(a_n) \to u_j$ as $n \to \infty$. Since the metric σ_j on U_j is equivalent to the metric d, $d(a_n,u_j) \to 0$ as $n \to \infty$. Since this is true for each j, there exists l in X such that $u_j = l$ for each $j \in \mathbf{N}$. Thus $l \in \cap_{n=1}^{\infty} U_j = A$, and $u = i(l)$. Thus $i(A)$ is closed in (U,σ), and so $(i(A),\sigma)$ is complete. If $i(a_n) \to i(a)$ in $(i(A),\sigma)$, then $i_1(a_n) \to i_1(a)$ in (U_1,d_1) as $n \to \infty$. But d and d_1 are equivalent on U_1, and so $a_n \to a$ in (A,d). Thus $i^{-1} : (i(A),\sigma) \to (A,d)$ is continuous, and i is a homeomorphism of (A,d) onto $(i(A),\sigma)$. Thus if we define $\rho(a,a') = \sigma(i(a),i(a'))$ then ρ is a complete metric on A equivalent to d. □

There is also a converse result.

Theorem 2.6.2 (Alexandroff) *Suppose that Y is a topologically complete subspace of a metric space (X,d). Then Y is a G_δ subset of X.*

Proof Let ρ be a complete metric on Y which defines the topology of Y. If $y \in Y$ and $\alpha > 0$, let $N_\alpha(y) = \{x \in X, d(x,y) < \alpha\}$ be the open α-neighbourhood of y in X. Since ρ and d agree on Y, for each y in Y there exists $0 < \delta_n(y) < 1/n$ such that if $z \in Y$ and $d(y,z) < \delta_n(y)$ then $\rho(y,z) < 1/n$. Let $U_n(y) = \{x \in X : d(x,y) < \delta_n(y)/2\}$, and let $O_n = \cup_{y \in Y} U_n(y)$. Then O_n is an open subset of X containing Y. Let $D = \cap_{n \in \mathbf{N}} O_n$. Then D is a G_δ-subset of X which contains Y. We show that $D = Y$.

Suppose that $z \in D$. For each $n \in \mathbf{N}$, $z \in O_n$, and so there exists $y_n \in Y$ such that $d(z,y_n) < \delta_n(y_n)/2 \leq 1/2n$. There exists $k_n \in \mathbf{N}$ such that $1/k_n < \delta_n(y_n)$. If $k \geq k_n$ then

$$d(y_n,y_k) \leq d(y_n,z) + d(z,y_k) < \delta_n(y_n)/2 + \delta_k(y_k)/2$$
$$\leq \delta_n(y_n)/2 + 1/2k \leq \delta_n(y_n),$$

and so $\rho(y_n,y_k) \leq 1/n$. Thus if $k,l \geq k_n$ then

$$\rho(y_k,y_l) \leq \rho(y_k,y_n) + \rho(y_n,y_l) \leq 2/n.$$

Thus $(y_n)_{n=1}^{\infty}$ is a ρ-Cauchy sequence in Y. Since (Y,ρ) is complete, there exists $y_\infty \in Y$ such that $\rho(y_n,y_\infty) \to 0$ as $n \to \infty$. Since the metrics d and ρ

are equivalent metrics on Y, $d(y_\infty, y_n) \to 0$ as $n \to \infty$. But $d(z, y_n) \to 0$ as $n \to \infty$, so that $z = y_\infty \in Y$. □

2.7 Baire's Category Theorem

We now come to one of the most useful theorems of analysis.

A topological space (X, τ) is a *Baire space* if whenever $(U_n)_{n=1}^\infty$ is a sequence of dense open subsets of X then $\bigcap_{n\in\mathbf{N}} U_n$ is dense in X.

Theorem 2.7.1 (Baire's category theorem) *(i) A topologically complete space (X, τ) is a Baire space.*
(ii) A locally compact Hausdorff space (X, τ) is a Baire space.

Proof Suppose that $(U_n)_{n=1}^\infty$ is a sequence of dense open subsets of X, and that V is a non-empty open subset of X. We must show that $V \cap (\bigcap_{n=1}^\infty U_n)$ is not empty.

(i) Let d be a complete metric on X which defines the topology τ. Since U_1 is dense in X, there exists $c_1 \in V \cap U_1$. Since $V \cap U_1$ is open, there exists $0 < \epsilon_1 \leq 1/2$ such that

$$N_{\epsilon_1}(c_1) \subseteq M_{\epsilon_1}(c_1) \subseteq V \cap U_1.$$

We now iterate the argument; for each $n \in \mathbf{N}$ there exist

$$c_n \in N_{\epsilon_{n-1}}(c_{n-1}) \cap U_n \text{ and } 0 < \epsilon_n < 1/2^n$$

such that

$$N_{\epsilon_n}(c_n) \subseteq M_{\epsilon_n}(c_n) \subseteq N_{\epsilon_{n-1}}(c_{n-1}) \cap U_n.$$

The sequence $(N_{\epsilon_n}(c_n))_{n=1}^\infty$ is decreasing, so that if $m, p \geq n$ then $c_m \in M_{\epsilon_n}(c_n)$ and $c_p \in M_{\epsilon_n}(c_n)$, so that

$$d(c_m, c_p) \leq d(c_m, c_n) + d(c_n, c_p) < 2/2^n;$$

thus $(c_n)_{n=1}^\infty$ is a Cauchy sequence in (X, d). Since (X, d) is complete, it converges to an element c of X. Suppose that $n \in \mathbf{N}$. Since $c_m \in M_{\epsilon_n}(c_n)$ for $m \geq n$ and since $M_{\epsilon_n}(c_n)$ is closed, $c \in M_{\epsilon_n}(c_n) \subseteq U_n$. Thus $c \in \bigcap_{n=1}^\infty U_n$. Further, $c \in M_{\epsilon_1}(c_1) \subseteq V$, and so $c \in V$.

(ii) The proof is similar, but easier. $V \cap U_1$ is non-empty; let $x_1 \in V \cap U_1$. There exists a compact neighbourhood K_1 of x contained in $V \cap U_1$. Let $W_1 = K_1^{int}$. Now iterate the argument. For each $n \in \mathbf{N}$ there exist $x_n \in W_{n-1} \cap U_n$, and a compact neighbourhood K_n of x_n contained in $W_{n-1} \cap U_n$. Let $W_n = K_n^{int}$. Then $(K_n)_{n=1}^\infty$ is a decreasing sequence of compact subsets of V, which has a non-empty intersection. If $x \in \bigcap_{n\in\mathbf{N}} K_n$ then $x \in V \cap (\bigcap_{n\in\mathbf{N}} U_n)$. □

The following corollary is particularly useful.

Corollary 2.7.2 *Suppose that $(C_n)_{n=1}^{\infty}$ is a sequence of closed subsets of a Baire space (X,τ) whose union is X. Then there exists n such that C_n has a non-empty interior.*

Proof Let $U_n = X \setminus C_n$. Then $(U_n)_{n=1}^{\infty}$ is a sequence of open sets and $\cap_{n=1}^{\infty} U_n$ is empty, and so is certainly not dense in X. Thus there exists U_n which is not dense in X; that is C_n has a non-empty interior. □

It is sometimes useful to have a local version of this corollary.

Corollary 2.7.3 *Suppose that (X,τ) is topologically complete or locally compact and Hausdorff. Suppose that $(C_n)_{n=1}^{\infty}$ is a sequence of closed subsets of X whose union contains a non-empty open set W. Then there exists n such that $C_n \cap W$ has a non-empty interior.*

Proof The space (W,τ) is also topologically complete, or locally compact and Hausdorff. The sets $C_n \cap W$ are closed subsets of W whose union is W, and so there exists n and a non-empty open subset V of W such that $V \subseteq C_n \cap W$. Since W is open in X, it follows that V is open in X. □

Why is Baire's category theorem a 'category' theorem? There is a collection of terminologies related to it. A subset A of a topological space (X,τ) is said to be *nowhere dense* if $\overline{A}$ has an empty interior, and is *meagre*, or *of the first category* if it is the union of a sequence of nowhere dense sets. Otherwise, A is of the *second category*. Thus the Baire category theorem says that a complete metric space is of the second category in itself.

Here is an application of Baire's category theorem.

Proposition 2.7.4 *Suppose that (X,τ) is a topologically complete σ-compact topological space. Then there exists a dense open subset Y such that Y, with the subspace topology, is locally compact.*

Proof Let $Y = \{y \in X : y$ has a compact neighbourhood in $X\}$. Then Y is open and locally compact. We must show that Y is dense in X. Let $x \in X$ and let V be an open neighbourhood of x. Then V is topologically complete. Let $(K_n)_{n=1}^{\infty}$ be an increasing sequence of compact subsets of X whose union is X, and let $L_n = V \cap K_n$, for each n. Each L_n is closed in V, and so, by Baire's category theorem, there exists n such that L_n has a non-empty interior in V. Let $y \in V_n^{int}$. Then $y \in Y$, and so Y is dense in X. □

A topological space (X,τ) is *locally homogeneous* if whenever $x, y \in X$ there exist open neighbourhoods U_x and U_y of x and y, and a homeomorphism $\phi : U_y \to U_x$ with $\phi(y) = x$.

Corollary 2.7.5 *A locally homogeneous σ-compact topologically complete space is locally compact.*

Proof For there exists a dense open locally compact subset Y. Let $y \in Y$ and $x \in X$, and let ϕ be a homeomorphism from an open neighbourhood U_y of y onto an open neighbourhood U_x of x. But U_y contains a compact neighbourhood of y, and so U_x contains a compact neighbourhood of x. □

On the other hand, let $X = \{z \in \mathbf{C} : |z| < 1\} \cup \{1\}$. Then X, with its usual topology, is σ-compact. It is topologically complete, by Theorem 2.6.1, and it is not locally compact.

Exercise 2.7.6 Suppose that (X, τ) is a countable locally homogeneous Baire space. Show that τ is the discrete topology. Give another proof that $(\mathbf{Q}, d_p)$ is not complete.

Exercise 2.7.7 Suppose that f is a continuous function on $[0, \infty)$ for which $f(nx) \to 0$ as $n \to \infty$ for each $x > 0$. Show that $f(x)) \to 0$ as $x \to \infty$.

2.8 Lipschitz Functions

Suppose that $f : X \to Y$, where (X, d) and (Y, ρ) are metric spaces, and that $L \geq 0$. f is an *L-Lipschitz function* if $\rho(f(x) - f(y)) \leq Ld(x, y)$ for all $x, y \in X$. A *Lipschitz function* is a function which is an L-Lipschitz function, for some $L > 0$. A Lipschitz function is uniformly continuous. The space $L(X)$ of real-valued Lipschitz functions on X is a linear subspace of $C(X)$. If f is a Lipschitz function, there is a least $L \geq 0$ such that f is an L-Lipschitz function: this is denoted by $p_L(f)$; $p_L(f) = 0$ if and only if f is constant.

Proposition 2.8.1 *Suppose that A is a non-empty subset of a metric space (X, d). Let $d(x, A) = \inf\{d(x, a) : a \in A\}$ be the distance from x to A. Then $d(\cdot, A)$ is a 1-Lipschitz function on X, and $d(x, A) = d(x, \overline{A})$.*

Proof If $y \in X$ and $a \in A$ then $d(x, A) \leq d(x, a) \leq d(x, y) + d(y, a)$. Taking the infimum over A, we see that $d(\cdot, A)$ is a 1-Lipschitz function. Clearly, $d(x, \overline{A}) \leq d(x, A)$. Given $\epsilon > 0$ there exist $a \in A$ and $a' \in \overline{A}$ such that $d(x, \overline{A}) \leq d(x, a') + \epsilon/2$ and $d(a, a') < \epsilon/2$. Then $d(x, A) \leq d(x, a) \leq d(x, a') + \epsilon/2 \leq d(x, A) + \epsilon$, so that $d(x, A) \leq d(x, \overline{A})$. □

Suppose now that $\epsilon > 0$. Recall that $N_\epsilon(A) = \{x \in X : d(x, A) < \epsilon\}$: $N_\epsilon(A)$ is the *ϵ-neighbourhood of A*. As usual, we write $N_\epsilon(x)$ for $N_\epsilon(\{x\})$.

We set $H_{(A,\epsilon)}(x) = (1 - d(x, A)/\epsilon)^+$.

Then $H_{(A,\epsilon)}$ is a $1/\epsilon$-Lipschitz function on X, which takes the value 1 on $\overline{A}$, the value 0 on $X \setminus N_\epsilon(A)$ and values in $(0,1)$ otherwise. $H_{(A,\epsilon)}$ converges pointwise to $I_{\overline{A}}$, the indicator function of $\overline{A}$, as $\epsilon \to 0$. We write $H_{(x,\epsilon)}$ for $H_{(\{x\},\epsilon)}$.

The set $L(X)$ of real-valued Lipschitz functions on a metric space (X,d) is a linear subspace of the vector space $C(X)$ of continuous functions on X.

Proposition 2.8.2 *Suppose that F is a set of real-valued L-Lipschitz functions on a metric space (X,d), and that* $\inf\{f(x_0) : f \in F\} > -\infty$ *for some $x_0 \in X$. Let $g(x) = \inf_F f(x)$. Then $g(x) > -\infty$ for all $x \in X$, and g is an L-Lipschitz function. Similarly for* $\sup_F f$.

Proof If $x \in X$ and $f \in F$ then

$$f(x) \geq f(x_0) - Ld(x,x_0) \geq g(x_0) - Ld(x,x_0),$$

so that $g(x) \geq g(x_0) - Ld(x,x_0) > -\infty$.

If $x' \in X$ then $g(x) \leq f(x) \leq Ld(x,x') + f(x')$, so that $g(x) \leq Ld(x,x') + g(x')$, and similarly $g(x') \leq Ld(x,x') + g(x)$, so that g is an L-Lipschitz function. □

Corollary 2.8.3 *If $(f_n)_{n=1}^\infty$ is a sequence of real-valued L-Lipschitz functions on (X,d) which converges pointwise to f, then f is an L-Lipschitz function.*

Proof For $f = \lim_{n\to\infty}(\inf_{m\geq n} f_m)$. □

Suppose that f is a real-valued function on a metric space (X,d) and that $L \geq 0$. Let $U_L(f) = \{h : h\ L\text{-Lipschitz}, h \geq f\}$. If $U_L(f)$ is not empty, we set $f^{d,L} = \inf\{h : h \in U_L\}$. It follows from Proposition 2.8.2 that $f^{d,L}$ is a real-valued L-Lipschitz function, and is the smallest L-Lipschitz function greater than or equal to f. $f^{d,L}$ is the *upper L-Lipschitz envelope* of f. The *lower L-Lipschitz envelope* $f_{d,L}$ is defined in a similar way.

We can characterize the upper L-Lipschitz envelope in the following way.

Theorem 2.8.4 *Suppose that f is a real-valued function on a metric space (X,d), that $L \geq 0$ and that $U_L(f) \neq \emptyset$. If $x \in X$ let $h(x) = \sup_{y\in X}(f(y) - Ld(x,y))$. Then $h = f^{d,L}$.*

Proof Setting $y = x$, it follows that $h \geq f$. If $x, y \in X$ then $f^{d,L}(x) \geq f(y) - Ld(x,y)$, so that $f^{d,L}(x) \geq h(x)$. But h is an L-Lipschitz function, and so $h = f^{d,L}$. □

Corollary 2.8.5 *If there exists a real-valued Lipschitz function less than or equal to f, then $f_{d,L}(x) = \inf_{y\in X}(f(y) + Ld(x,y))$.*

In the same vein, we have the following.

Theorem 2.8.6 (The McShane–Whitney extension theorem) *Suppose that f is a real-valued L-Lipschitz function on a non-empty subset A of X. Then there exists an L-Lipschitz function g on X which extends f.*

Proof Let $g(x) = \inf\{f(y) + Ld(x, y) : y \in A\}$. If $x, x' \in X$ and $a \in A$ then

$$g(x) \leq f(a) + Ld(x, a) \leq f(a) + Ld(x', a) + Ld(x, x') \leq g(x') + Ld(x, x').$$

Thus g is an L-Lipschitz function. If $a \in A$ then

$$g(a) = \inf_{b \in A}(f(b) + Ld(b, a)) \geq f(a) = f(a) + Ld(a, a) \geq g(a),$$

so that $g(a) = f(a)$. □

We denote by $BL(X)$ the vector space of bounded real-valued Lipschitz functions on X.

Corollary 2.8.7 *If A is a non-empty subset of X and $f \in BL(A)$, then there exists $g \in BL(X)$ with $p_L(g) = p_L(f)$, $\sup_{x \in X} g(x) = \sup_{a \in A} f(a)$ and $\inf_{x \in X} g(x) = \inf_{a \in A} f(a)$ such that $f = g_{|A}$.*

Proof By the McShane–Whitney theorem, there exists $h \in L(X)$ with $p_L(h) = p_L(f)$, such that $h_{|A} = f$. Let $g = (h \wedge \sup f) \vee (\inf f)$. □

Note that if $X = \mathbf{N}$, with the discrete metric, then $BL(\mathbf{N}) = C_b(\mathbf{N}) = l_\infty$.

3
Polish Spaces and Compactness

3.1 Polish Spaces

A topological space (X, τ) is a *Polish space* if it is separable and topologically complete; that is, it is separable, and there is a complete metric d on X which defines the topology τ. In particular, a complete separable metric space is called a *Polish metric space*. Let us bring earlier results together.

Theorem 3.1.1 *Suppose that (X, τ) is a topological space. The following are equivalent.*

(i) (X, τ) is a Polish space.
(ii) (X, τ) is homeomorphic to a G_δ subspace of the Hilbert cube $\mathcal{H}$.
(iii) (X, τ) is homeomorphic to a G_δ subspace of a Polish space.
(iv) (X, τ) is homeomorphic to a closed subspace of a Polish space.
(v) (X, τ) is homeomorphic to a closed separable subspace of (l_∞, d_∞).

Further, a Polish space is second countable, and the product of a sequence of Polish spaces, with the product topology, is a Polish space.

Here are some important examples.

(i) Euclidean space, and Hermitian space.
(ii) The Hilbert cube $\mathcal{H}$.
(iii) The Bernoulli sequence space $\Omega(\mathbf{N})$.
(iv) The set $\mathbf{I}$ of irrational numbers, with the subspace topology.

For $\mathbf{I} = \mathbf{R} \setminus \mathbf{Q}$ is a G_δ subset of the Polish space $\mathbf{R}$.

(v) The product $\mathbf{N}^{\mathbf{N}}$, with the product topology.

Exercise 3.1.2 Use continued fraction expansions to show that $\mathbf{I}$ and $\mathbf{N}^{\mathbf{N}}$ are homeomorphic.

Exercise 3.1.3 Show that it follows from Baire's category theorem that the set $\mathbf{Q}$ of rational numbers, with the subspace topology, is not a Polish space.

3.2 Totally Bounded Metric Spaces

We need some more definitions. Suppose that $\epsilon > 0$. If B is a subset of a metric space (X, d), then a subset A of B is an *ϵ-net* in B if $B \subseteq \cup_{a \in A} N_\epsilon(a)$. Thus A is an ϵ-net if and only if each element of B is within ϵ of a member of A. A subset B of X is *totally bounded* if there is a finite ϵ-net in B for each $\epsilon > 0$. Total boundedness is not a topological property: $(0, 1]$ is totally bounded under the usual metric, and is homeomorphic to $[1, \infty)$, which is not totally bounded under its usual metric. On the other hand, if f is a uniformly continuous mapping of a metric space (X, d) into a metric space (Y, ρ) and B is a totally bounded subset of X, then $f(B)$ is a totally bounded subset of Y. In particular, if f is a uniform homeomorphism then B is totally bounded if and only if $f(B)$ is.

Proposition 3.2.1 *A totally bounded subset B of a metric space is separable, and is therefore second countable.*

Proof For each $n \in \mathbf{N}$ there is a finite $1/n$-net A_n in B. Then $\cup_{n \in \mathbf{N}} A_n$ is a countable dense subset of B. □

Here is a characterization of sets that are not totally bounded.

Proposition 3.2.2 *A subset C of a metric space (X, d) is not totally bounded if and only if there exist $\epsilon > 0$ and a sequence $(c_n)_{n=1}^\infty$ in C such that $d(c_m, c_n) \geq \epsilon$ for $m \neq n$.*

Proof Suppose that C is not totally bounded. Thus there exists $\epsilon > 0$ such that C has no finite ϵ-net. We use an inductive argument to find a sequence $(c_n)_{n=1}^\infty \in C$ such that $d(c_m, n_n) \geq \epsilon$ for $m \neq n$. If we have found $c_1, \ldots, c_n$ satisfying this condition, then $\{c_j : 1 \leq j \leq n\}$ is not an ϵ-net in C, and so there exists $c_{n+1} \in C$ such that $d(c_j, c_{n+1}) \geq \epsilon$ for $1 \leq j \leq n$.

Conversely, if the condition is satisfied, and if $c \in C$ then $N_{\epsilon/2}(c)$ can contain at most one term of the sequence, and so C contains no finite $\epsilon/2$-net. □

Totally bounded sets can be characterized in terms of Cauchy sequences.

Theorem 3.2.3 *A subset B of a metric space (X, d) is totally bounded if and only if every sequence in B has a Cauchy subsequence.*

Proof Suppose first that B is totally bounded, and that $(x_n)_{n=1}^\infty$ is a sequence in B. We use a diagonal argument. For each $n \in \mathbf{N}$, let A_n be a finite $1/n$-net in B. First, there exists $a_1 \in A_1$ such that $x_n \in N_1(a_1)$ for infinitely many n; that is, there exists a subsequence $(x_{1n})_{n=1}^\infty$ such that $x_{1n} \in N_1(a_1)$ for $n \in \mathbf{N}$. Arguing inductively, for each $j \in \mathbf{N}$ there exist $a_j \in A_j$ and a subsequence $(x_{jn})_{n=1}^\infty$ of $(x_{j-1,n})_{n=1}^\infty$ such that $x_{jn} \in N_{1/j}(a_j)$ for $n \in \mathbf{N}$. Let $x_{k_j} = x_{jj}$. Then $(x_{k_n})_{n=1}^\infty$ is a subsequence of $(x_n)_{n=1}^\infty$, and $d(x_{k_m}, x_{k_p}) < 2/n$ for $m, p \geq n$, and so it is a Cauchy sequence.

Conversely, if B is not totally bounded, there exist $\epsilon > 0$ and a sequence $(b_n)_{n=1}^\infty$ in B such that $d(b_m, b_n) \geq \epsilon$ for $m \neq n$. This sequence clearly has no convergent subsequence. □

Let us establish a result about ϵ-nets that we shall need later, when we prove the existence of Haar measure on a compact metric group. If (X, d) is a totally bounded metric space, and $\epsilon > 0$, then there is a finite ϵ-net in X. A finite ϵ-net with as few elements as possible is called a *minimal* ϵ-net. (Beware! An ϵ-net which contains no proper ϵ-subnet need not be minimal.) The members of two minimal ϵ-nets can be paired so that members of a pair are close together. For this we need Hall's marriage theorem.

Theorem 3.2.4 (Hall's marriage theorem) *Suppose that A and B are finite sets and that $H \subset A \times B$. If $C \subset A$, let $h(C) = \{b \in B : (a, b) \in H$ for some $a \in C\}$. Then there exists an injective mapping $\psi : A \to B$ such that $(a, \psi(a)) \in H$ for all $a \in A$ if and only if $|h(C)| \geq |C|$ for all $C \subseteq A$.*

Proof The condition is certainly necessary. We prove sufficiency by induction on $|A|$. The result is true for $|A| = 1$. Suppose that $|A| = n$ and that the result is true for $1 \leq k < n$. We consider two cases.

First, suppose that $|h(C)| > |C|$ for each proper subset of A. Pick $a_1 \in A$, and choose $\psi(a_1) \in h(\{a_1\})$. Let $A_1 = A \setminus \{a_1\}$, and let $H_1 = \{(a, b) : a \in A_1, (a, b) \in H\}$. Then A_1, $B_1 = B \setminus \{\psi(a_1)\}$ and H_1 satisfy the conditions of the theorem, and $|A_1| < n$, so that we can define an injective mapping $\psi : A_1 \to B_1$ so that $(a, \psi(a)) \in H_1$ for each $a \in A_1$. Then $\psi : A \to B$ satisfies the theorem.

Secondly, suppose that there exists a proper subset C of A such that $|h(C)| = |C|$. Then by the inductive hypothesis we can define an injective mapping $\psi : C \to h(C)$ so that $(c, \psi(c)) \in H$ for $c \in C$. Let $D = A \setminus C$, $E = B \setminus h(C)$ and $H_D = \{(d, e) : (d, e) \in D \times E, (d, e) \in H\}$. Then $|D| < n$, and D, E, H_D satisfy the conditions, so that we can define an injective mapping $\psi : D \to E$ so that $(d, \psi(d)) \in H_D$ for each $d \in D$. Then $\psi : A \to B$ satisfies the theorem. □

Theorem 3.2.5 *Suppose that (X, d) is a totally bounded metric space, that $\epsilon > 0$ and that m_ϵ and n_ϵ are minimal ϵ-nets in X. Then there exists a bijective mapping $\chi : m_\epsilon \to n_\epsilon$ such that $d(m, \chi(m)) < 2\epsilon$ for all $m \in m_\epsilon$.*

Proof Define a relation H on $m_\epsilon \times n_\epsilon$ by setting $H(m, n)$ if and only if $d(m, n) < 2\epsilon$. If $m \in m_\epsilon$, let $h(m) = \{n \in n_\epsilon : H(m, n)\}$, and if $A \subseteq m_\epsilon$ let $h(A) = \cup_{m \in A} h(m)$. We shall show that $|h(A)| \geq |A|$, for each $A \subseteq m_\epsilon$. If not, let $L = (m_\epsilon \setminus A) \cup h(A)$. Then $|L| < |N|$. We shall show that L is an ϵ-net, giving a contradiction. If $x \in X$, then $d(x, m) < \epsilon$ for some $m \in m_\epsilon$. Suppose that $m \in A$. Then there exists $n \in n_\epsilon$ such that $d(x, n) < \epsilon$. But then $d(m, n) \leq d(m, x) + d(x, m) < 2\epsilon$ and so $n \in h(A)$. If $m \notin A$ then $x \in \cup_{m \in m_x \setminus A} N_\epsilon(m)$. Thus

$$X = (\cup_{m \in m_x \setminus A} N_\epsilon(m)) \cup (\cup_{n \in h(A)} N_\epsilon(n)) = \cup_{m \in L} N_\epsilon(m),$$

so that L is an ϵ-net.

We now apply Hall's marriage theorem, which ensures that a suitable mapping χ exists. □

3.3 Compact Metrizable Spaces

Compact metrizable spaces form one of the most fundamental and important classes of Polish spaces. When is a metric space compact?

Theorem 3.3.1 *Suppose that (X, d) is a metric space. The following are equivalent.*

(i) (X, d) is compact.
(ii) (X, d) is countably compact.
(iii) (X, d) is sequentially compact.
(iv) (X, d) is complete and totally bounded.

Proof A compact topological space is countably compact, so that (i) implies (ii). Since (X, d) is first countable, (ii) and (iii) are equivalent. Suppose that (X, d) is sequentially compact. Then every sequence in X has a convergent sequence, and this is a Cauchy sequence, so that (X, d) is totally bounded. Further, a Cauchy sequence in X has a convergent subsequence, and is therefore convergent, so that (X, d) is complete. Thus (iii) implies (iv). Conversely, if (X, d) is totally bounded and complete, then every sequence in X has a Cauchy subsequence (by total boundedness), and this converges (by completeness), and so (iv) implies (iii). Thus (ii), (iii) and (iv) are equivalent.

Finally we show that (iii) and (iv) together imply (i). Suppose that (iii) and (iv) hold. Let $\mathcal{O}$ be an open cover of X. First we show that there exists $\delta > 0$ such that if $x \in X$ then there exists $O \in \mathcal{O}$ such that $N_\delta(x) \subseteq O$. Suppose not. Then for each $n \in \mathbf{N}$ there exists $x_n \in X$ such that $N_{1/n}(x_n) \not\subseteq O$, for all $O \in \mathcal{O}$. Since (X,d) is sequentially compact, there is a convergent subsequence $(x_{n_k})_{k=1}^\infty$, convergent to x, say. Then there exists $O \in \mathcal{O}$ and $N \in \mathbf{N}$ such that $N_{1/N}(x) \subseteq O$. There exists $K \in \mathbf{N}$, with $n_K > 2N$, such that $d(x_{n_k}, x) < 1/2N$ for $k \geq K$. But then $N_{1/2n_K}(x_{n_K}) \subseteq O$, giving a contradiction. Since (X,d) is totally bounded, there exists a finite δ-net A_δ in X. For each $a \in A_\delta$ there exists $O_a \in \mathcal{O}$ such that $N_\delta(a) \subseteq O_a$. Then $X = \cup_{a \in A} N_\delta(a) \subseteq \cup_{a \in A} O_a$, so that $\{O_a : a \in A\}$ is a finite subcover of X; (X,d) is compact. □

Corollary 3.3.2 *A compact metrizable space is a Polish space.*

It is a remarkable fact that neither completeness nor total boundedness is a topological property, but that together they are equivalent to compactness, which is a topological property.

Corollary 3.3.3 *The completion $(\hat{X}, \hat{d})$ of a totally bounded metric space is compact.*

Proof For if $0 < \delta < \epsilon$ then a δ-net in X is an ϵ-net in $\hat{X}$. □

For this reason, a totally bounded set is sometimes called a *precompact* set.

Exercise 3.3.4 Suppose that $(U_n)_{n=1}^N$ is a finite open cover of a compact metric space. Show that there is a corresponding *partition of unity* $(f_n)_{n=1}^N$; each f_n is a continuous real-valued function on X, with $0 \leq f_n \leq 1$ and is zero outside U_n, and $\sum_{n=1}^N f_n = 1$.

Theorem 3.3.5 *A continuous mapping f from a compact metric space (X,d) into a metric space (Y,ρ) is uniformly continuous.*

Proof Suppose not. Then there exists $\epsilon > 0$ such that for each $n \in \mathbf{N}$ there exist $x_n, y_n \in X$ with $d(x_n, y_n) < 1/n$ and $\rho(f(x_n), f(y_n)) \geq \epsilon$. Since (X,d) is sequentially compact, there exists a subsequence $(x_{n_k})_{k=1}^\infty$ which converges to a point x of X. But then $y_{n_k} \to x$ as $k \to \infty$. Since f is continuous at x, $\rho(f(x_{n_k}), f(x)) \to 0$ and $\rho(f(y_{n_k}), f(x)) \to 0$ as $k \to \infty$, and so $\rho(f(x_{n_k}), f(y_{n_k})) \to 0$ as $k \to \infty$, giving a contradiction. □

Theorem 3.3.6 *The topological product (X,d) of a sequence (X_n, d_n) of compact metrizable spaces is compact and metrizable.*

Proof (X,d) is metrizable (Theorem 2.2.5), and so it is sufficient to show that it is countably compact. Suppose that $(x^{(j)})_{j=1}^\infty = ((x_n^{(j)})_{n=1}^\infty)_{j=1}^\infty$ is a sequence in X. For each n there is a limit point x_n of the sequence $(x_n^{(j)})_{j=1}^\infty$. Then $x = (x_n)_{n=1}^\infty$ is a limit point of the sequence $(x^{(j)})_{j=1}^\infty$. □

Of course, this is a special case of Tychonoff's theorem. But the proof does not use the axiom of choice.

Corollary 3.3.7 *The Hilbert cube $\mathcal{H}$ and the Bernoulli sequence space $\Omega(\mathbf{N})$ are compact metrizable spaces. A topological space (X, τ) is a compact metrizable space if and only if it is homeomorphic to a closed subset of the Hilbert cube.*

Proof A compact metrizable space is second countable, and so it is homeomorphic to a subspace Y of the Hilbert cube. Since a compact subspace of a Hausdorff space is closed, Y must be closed. Conversely, since the Hilbert cube is compact and metrizable, a topological space which is homeomorphic to a closed subspace of $\mathcal{H}$ must be compact and metrizable. □

Corollary 3.3.8 *A topological space (X, τ) is a Polish space if and only if it is a dense G_δ subset of a compact metric space $(\tilde{X}, \tilde{d})$. If so, there is a metric d on (X, τ) which defines the topology τ such that (X, d) is totally bounded.*

Proof For we can take $\tilde{X}$ as the closure of X, when (X, τ) is considered as a subspace of the Hilbert cube $\mathcal{H}$, take $\tilde{d}$ to be a metric defining the subspace topology on $\tilde{X}$, and take d to be the restriction of $\tilde{d}$ to X. □

Proposition 3.3.9 *The subspace*

$$S(\Omega) = \left\{ \sum_{j=1}^{n} a_j I_{C_j} : n \in \mathbf{N}, a_j \in \mathbf{R}, C_j \text{ a cylinder set} \right\}$$

of $C(\Omega)$ is dense in $C(\Omega)$.

Proof Let d be the Cantor metric on Ω. If $f \in C(\Omega)$ then f is uniformly continuous, and so if $\epsilon > 0$ there exists $n \in \mathbf{N}$ such that if $d(\omega, \omega') < 1/3^n$ then $|f(\omega) - f(\omega')| < \epsilon$. For each cylinder set C of rank n pick $\omega_C \in C$, and let

$$g = \sum \{f(\omega_C) I_C : C \text{ a cylinder set of rank } n\}.$$

Then $g \in S(\Omega)$ and $\|f - g\|_\infty < \epsilon$. □

Functions in $S(\Omega)$ are called *step functions*.

Suppose that A is a subset of a topological space (X, τ). A continuous mapping $r : X \to A$ is a *retraction* if $r(a) = a$ for $a \in A$. If there is a retraction of X onto A then A is a *retract* of X.

Theorem 3.3.10 *Suppose that K is a non-empty compact subset of a metric space (X,d) and that there is a mapping $n : X \to K$ such that $d(x,n(x)) < d(x,k)$ for $k \neq n(x)$. Then n is a retraction of X onto K.*

Proof We must show that n is continuous. Suppose that it is not continuous at x. Then there exists $\epsilon > 0$ and a sequence $(x_j)_{j=1}^{\infty}$ in X such that $x_j \to x$ as $j \to \infty$, while $d(n(x),n(x_j)) \geq \epsilon$ for all j. Since K is compact, the sequence $(n(x_j))_{j=1}^{\infty}$ has a subsequence $(n(x_{j_i}))_{i=1}^{\infty}$ which converges to a point y of K as $i \to \infty$. Then $d(n(x),y) \geq \epsilon$. Now

$$\begin{aligned} d(x,y) &= \lim_{i\to\infty} d(x_{j_i}, n(x_{j_i})) \leq \lim_{n\to\infty} d(x_{j_i}, n(x)) \\ &\leq \lim_{i\to\infty} d(x_{j_i}, x) + d(x,n(x)) = d(x,n(x)). \end{aligned}$$

But $n(x)$ is the unique nearest point to x in K and $y \neq n(x)$, giving a contradiction. □

We shall also need the following theorem.

Theorem 3.3.11 *A non-empty metric space X is compact if and only if there is a continuous surjective mapping of the Bernoulli sequence space $\Omega(\mathbf{N})$ onto X.*

Proof The condition is sufficient, since $\Omega(\mathbf{N})$ is compact, and the continuous image of a compact space is compact.

Suppose that (X,d) is compact. Then it is totally bounded, and so it is the union of finitely many sets of diameter at most ϵ; since the diameter of a set is the same as the diameter of its closure, we can suppose that the sets are closed.

First we find a finite set $\mathcal{A}$ of closed non-empty subsets of X, each of diameter at most 1 such that $X = \cup_{A\in\mathcal{A}} A$. We choose $k_1 = l_1$ such that $2^{k_1} \geq |\mathcal{A}|$, and a surjective mapping ϕ_0 of $\Omega(k_1)$ onto $\mathcal{A}$. We set $A_\omega = \phi_0(\omega)$.

We now repeat the procedure for each A_ω. For each $\omega \in \Omega(l_1)$, there exists a finite set $\mathcal{A}_\omega$ of non-empty closed subsets of A_ω of diameter at most $1/2$ such that $A_\omega = \cup_{A\in\mathcal{A}_\omega} A$. We choose k_2 such that $2^{k_2} \geq \max_{\omega\in\Omega(l_1)} |\mathcal{A}_\omega|$ and for each $\omega' \in \Omega(l_1)$ define a surjective mapping $\phi_{1,\omega'}$ of $\Omega(k_1)$ onto $\mathcal{A}_\omega$. We set $l_2 = l_1 + k_2$. If $\omega \in \Omega(l_2)$, we can write ω as (ω',ω''), with $\omega' \in \Omega(l_1)$ and $\omega'' \in \Omega(k_2)$. We set $A_\omega = \phi_{1,\omega'}(\omega'')$. Thus $A_{\omega'} = \cup_{\omega''\in\Omega(k_2)} A_{(\omega',\omega'')}$.

We now iterate. There exists a strictly increasing sequence $(l_n)_{n=1}^{\infty}$, with $l_{n+1} = l_n + k_n$, such that for each $n \in \mathbf{N}$ there exists a family $\{A_\omega : \omega \in \Omega(l_n)\}$ of closed non-empty subsets of X, each of diameter at most $1/n$, such that if $\omega' \in \Omega(l_n)$ then $A_{\omega'} = \cup_{\omega''\in\Omega(k_{n+1})}(A_{(\omega',\omega'')})$.

Suppose now that $\omega \in \Omega(\mathbf{N})$. For $n \in \mathbf{N}$, let $A_n(\omega) = A_{(\omega_1,\ldots\omega_{l_n})}$. Then $(A_n(\omega))_{n=1}^{\infty}$ is a decreasing sequence of closed subsets of X, and $\operatorname{diam}(A_n(\omega)) \to 0$ as $n \to \infty$. Since (X,d) is complete, it follows from

Proposition 2.4.3 that $\cap_{n\in\mathbf{N}}A_n$ is a singleton, $\{\psi(\omega)\}$, say. Thus ψ is a mapping of $\Omega(\mathbf{N})$ into X.

Next we show that ψ is continuous. If $\omega \in \Omega(\mathbf{N})$ and $n \in \mathbf{N}$ then the cylinder set C_{ω,l_n} is contained in $A_n(\omega)$, so that if $\omega' \in C_{\omega,l_n}$ then $d(\psi(\omega), \psi(\omega')) \leq 1/n$. Finally, we show that ψ is surjective. Since $\psi(\Omega(\mathbf{N}))$ is compact, and is therefore closed in X, it is enough to show that $\psi(\Omega(\mathbf{N}))$ is dense in X. If $x \in X$ and $n \in \mathbf{N}$ then $x \in A_n(\omega')$ for some $\omega' \in \Omega(l_n)$. There exists $\omega \in \Omega(\mathbf{N})$ such that $\omega' = (\omega_1, \ldots, \omega_{l_n})$. Then $\psi(\omega) \in A_n(\omega')$, so that $d(x, \psi(\omega)) \leq 1/n$. □

We use this theorem to give an easy proof of the following result.

Theorem 3.3.12 *Suppose that (X, d) is a compact metric space and that (Y, ρ) is a Polish space. Then $(C(X, Y), d_\infty)$ is a Polish space.*

Proof Certainly $(C(X, Y), d_\infty)$ is complete. We must show that it is separable. Let $\psi : \Omega(\mathbf{N}) \to X$ be a continuous surjection. If $f \in C(X, Y)$, let $T_\psi(f) = f \circ \psi$. Then T_ψ is an isometry of $(C(X, Y), d_\infty)$ into $(C(\Omega(\mathbf{N}), Y), d_\infty)$, and so it is sufficient to show that $(C(\Omega(\mathbf{N}), Y), d_\infty)$ is separable.

Let D be a countable dense subset of Y. If $\omega \in \Omega(\mathbf{N})$ and $n \in \mathbf{N}$, let $P_n(\omega) = (\omega_1, \ldots, \omega_n)$, so that P_n maps $\Omega(\mathbf{N})$ onto Ω_n. Let F_n be the (countable) set of mappings from Ω_n into D, let $G_n = \{g \circ P_n : g \in F_n\}$, and let $G = \cup_{n=1}^\infty G_n$. Then G is a countable subset of $C(\Omega(\mathbf{N}), Y)$: we show that G is dense in $(C(\Omega(\mathbf{N}), Y), d_\infty)$. Suppose that $f \in C(\Omega(\mathbf{N}), Y)$ and that $\epsilon > 0$. Since f is uniformly continuous, there exists $n \in \mathbf{N}$ such that if $P_n(\omega) = P_n(\omega')$, then $\rho(f(\omega), f(\omega')) < \epsilon/2$. Thus if $\tilde{\omega} \in \Omega_n$ there exists $\tilde{d}(\tilde{\omega}) \in D$ such that $\rho(\tilde{d}(\tilde{\omega}), f(\omega)) < \epsilon$ whenever $P_n(\omega) = \tilde{\omega}$. Let $g = \tilde{d} \circ P_n$. Then $g \in G$ and $d_\infty(f, g) \leq \epsilon$. □

Corollary 3.3.13 *If (X, d) is a compact metric space, then $(C(X), d_\infty)$ $(=(C(X, \mathbf{R}), d_\infty))$ is a Polish space.*

Corollary 3.3.13 characterizes those compact spaces which are metrizable.

Proposition 3.3.14 *Suppose that (X, τ) is a compact topological space for which $(C(X), d_\infty)$ is separable. Then X is metrizable.*

Proof Suppose that $(f_n)_{n=1}^\infty$ is a dense sequence in $(C(X), d_\infty)$. If $x \in X$, let $F(x) = (f_n(x))_{n=1}^\infty$. Then F is a continuous mapping of X into $(\mathbf{R}^\mathbf{N}, d)$, where d is a metric on $\mathbf{R}^\mathbf{N}$ which defines the product topology, so that $(\mathbf{R}^\mathbf{N}, d)$ is a Polish space. Suppose that $x \neq y$. Let $\mathrm{g}(\cdot) = \mathrm{d}(\cdot, \mathrm{y})$. There exists $n \in \mathbf{N}$ such that $d_\infty(f_n, g) < d(x, y)/2$. Then $f_n(x) \neq f_n(y)$, and so the mapping F is injective. By Exercise 1.2.1(iii), F is then a homeomorphism of X onto $F(X)$, and so (X, τ) is metrizable. □

Recall that the Helly space is compact, separable and first countable, but is not metrizable.

Let us consider compactness in $(C(X, Y)$, when X and Y are compact. First we consider total boundedness. We need some definitions. Suppose that (X, d) and (Y, ρ) are metric spaces, that $x \in X$ and that F is a set of mappings from X to Y. Then F is *equicontinuous* at x if whenever $\epsilon > 0$ there exists $\delta > 0$ such that if $d(x, y) < \delta$ then $\rho(f(x), f(y)) < \epsilon$ for all $f \in F$. F is *equicontinuous* on X if it is equicontinuous at each point of X. F is *uniformly equicontinuous* if whenever $\epsilon > 0$ there exists $\delta > 0$ such that if $d(y, z) < \delta$ then $\rho(f(y), f(z)) < \epsilon$ for all $f \in F$.

Proposition 3.3.15 *Suppose that (X, d) and (Y, ρ) are metric spaces and that F is a set of mappings from X to Y which is equicontinuous on X. If (X, d) is compact, then F is uniformly equicontinuous.*

Proof Make obvious changes to the proof of Theorem 3.3.5. □

Proposition 3.3.16 *Suppose that (X, d) and (Y, ρ) are totally bounded metric spaces and that $F \subseteq C(X, Y)$. If F is a uniformly equicontinuous set of mappings from X to Y, then F is totally bounded in $C(X, Y)$.*

Proof Suppose that $\epsilon > 0$. Then there exists $\delta > 0$ such that if $d(x, y) < \delta$ then $\rho(f(x), f(y)) < \epsilon$ for all $f \in F$. There is a finite partition $\mathcal{S}$ of X into non-empty sets of diameter at most δ, and a finite partition $\mathcal{T}$ of Y into non-empty sets of diameter at most ϵ. For each $S \in \mathcal{S}$ pick $x_S \in S$. For each ϕ in the finite set $\mathcal{T}^{\mathcal{S}}$, let $F_\phi = \{f \in F : f(x_S) \in \phi(S) \text{ for all } S \in \mathcal{S}\}$. Let $\Phi = \{\phi \in \mathcal{T}^{\mathcal{S}} : F_\phi \neq \emptyset\}$. Then $F = \cup_{\phi \in \Phi} F_\phi$ is a finite partition of F. If $f, g \in F_\phi$ and $x \in X$, then $x \in S$ for some $S \in \mathcal{S}$, and so $\rho(f(x), g(x)) < \epsilon$. Hence $\text{diam}(F_\phi) < \epsilon$. Thus F is totally bounded in $C_b(X, Y)$. □

Theorem 3.3.17 (The Arzelà–Ascoli theorem) *Suppose that (X, d) and (Y, ρ) are compact metric spaces and that $F \subseteq C(X, Y)$. Then $\overline{F}$ is compact if and only if F is equicontinuous on X.*

Proof If F is equicontinuous on X then it is uniformly equicontinuous, and so F is totally bounded. Then $\overline{F}$ is totally bounded. But it is also complete, since $(C(X, Y), d_\infty)$ is complete, and so $\overline{F}$ is compact.

Conversely, suppose that F is not equicontinuous. Then there exists $x \in X$ for which F is not equicontinuous at x, and so there exists $\epsilon > 0$ for which there is no suitable $\delta > 0$. Take $\delta_1 = 1$. Then there exists $x_1 \in X$ such that $d(x_1, x) < \delta_1$ and $f_1 \in F$ such that $\rho(f_1(x_1), f_1(x)) \geq \epsilon$. We show by induction that there exist a decreasing sequence $(\delta_n)_{n=1}^\infty$ of positive numbers, a sequence $(x_n)_{n=1}^\infty$ in X and a sequence $(f_n)_{n=1}^\infty$ in F such that $\rho(f_n(x_n), f_n(x)) \geq \epsilon$ and

$\rho(f_n(y), f_n(x)) < \epsilon/3$ if $d(y, x) < \delta_{n+1}$. Suppose that we have found δ_j, x_j and f_j for $1 \leq j \leq n$ which satisfy the conditions. Since f_n is continuous at x there exists δ_{n+1} such that $\rho(f_n(y), f_n(x)) < \epsilon/3$ for $d(y, x) < \delta_{n+1}$. Since F is not equicontinuous at x, there exist x_{n+1} with $d(x_{n+1}, x) < \delta_{n+1}$ and $f_{n+1} \in F$ such that $\rho(f_{n+1}(x_{n+1}, f_{n+1}(x))) \geq \epsilon$. This establishes the induction. If $m > n$ then either $\rho(f_m(x), f_n(x)) \geq \epsilon/3$ or

$$\rho(f_m(x_m), f_n(x_m)) \geq \\ \rho(f_m(x_m), f_m(x)) - \rho(f_m(x), f_n(x)) - \rho(f_n(x_m), f_n(x)) \geq \epsilon/3,$$

so that $d_\infty(f_m, f_n) \geq \epsilon/3$. Thus F is not totally bounded, and $\overline{F}$ is not compact. □

Corollary 3.3.18 *Suppose that (X, d) is a compact metric space and that $F \subseteq C(X)$, the space of continuous real-valued functions on X. The following are equivalent.*

(i) *F is totally bounded.*
(ii) *F is equicontinuous, and there exists $M > 0$ such that $|f(x)| \leq M$ for each $x \in X$ and $f \in F$.*
(iii) *F is equicontinuous, and $\{f(x) : f \in F\}$ is bounded, for each $x \in X$.*

Proof Suppose that F is totally bounded, so that F is equicontinuous. Suppose that N is a 1-net in F. Then $|f(x)| \leq \max_{g \in N} |g(x)| + 1 \leq \max_{g \in N} \|g\|_\infty + 1$, and so (i) implies (ii). Conversely, suppose that (ii) holds. Then we can consider F as a subset of $C(X, [-M, M]$, and so (i) holds.

(ii) certainly implies (iii). Suppose that (iii) holds. For each $x \in X$, there exists an open neighbourhood O_x of x in X such that if $y \in O_x$ then $|f(y) - f(x)| \leq 1$ for all $f \in F$. Since X is compact, there exists a finite set X_0 of X such that $X = \cup_{x \in X_0} O_x$. Then

$$\sup_{x \in X} \sup_{f \in F} |f(x)| \leq \max_{x \in X_0} \sup_{f \in F} |f(x)| + 1,$$

so that (iii) implies (ii). □

3.4 Locally Compact Polish Spaces

When is a metrizable locally compact space a Polish space?

Theorem 3.4.1 *Suppose that (X, τ) is a metrizable locally compact space. The following are equivalent.*

(i) (X,τ) *is a Polish space.*
(ii) (X,τ) *is separable.*
(iii) (X,τ) *is second countable.*
(iv) (X,τ) *is* σ*-compact.*
(v) *There is an increasing sequence* $(K_n)_{n=1}^{\infty}$ *of compact subsets of* X *such that* $\cup_{n\in\mathbf{N}}K_n = X$ *and* $K_n \subseteq K_{n+1}^{int}$*, for each* $n \in \mathbf{N}$.
(vi) *The one-point compactification* $(\tilde{X},\tilde{\tau})$ *of* (X,τ) *is metrizable.*
(vii) (X,τ) *is homeomorphic to an open subset of a compact metric space.*

Proof (i) trivially implies (ii), and (ii) implies (iii), by Theorem 2.2.1. Suppose that (X,τ) is second countable, and that $\mathcal{B}$ is a countable base for the topology. Let $\mathcal{B}_c = \{B \in \mathcal{B} : \overline{B} \text{ is compact}\}$. If $x \in X$, let $N(x)$ be a compact neighbourhood of X. There exists $B \in \mathcal{B}$ such that $x \in B \subseteq N(x)$. But then $B \in \mathcal{B}_c$, and so $X = \cup_{B\in\mathcal{B}_c}\overline{B}$, and X is σ-compact. If (iv) holds, then (v) holds by Proposition 1.2.6. If (v) holds then (X,d) is separable and therefore second countable. If $\mathcal{B}$ is a countable base for τ, then $\mathcal{B}\cup\{\tilde{X}\setminus K_n : n \in \mathbf{N}\}$ is a countable base for $\tilde{\tau}$. Thus $(\tilde{X},\tilde{\tau})$ is metrizable, by Urysohn's metrization theorem. Since X is an open subset of $(\tilde{X},\tilde{\tau})$, (vi) implies (vii), and (vii) implies that (X,τ) is a Polish space, by Theorem 2.6.1, since a compact metric space is a Polish space. □

Let $X = \{z \in \mathbf{C} : |z| < 1\} \cup \{1\}$. With its usual topology, X is an example of a σ-compact Polish space which is not locally compact.

Let (X,τ) be a locally compact Polish space, and suppose that $(K_n)_{n=1}^{\infty}$ is a sequence of compact subsets of X such that $\cup_{n\in\mathbf{N}}K_n = X$ and $K_n \subseteq K_{n+1}^{int}$, for each $n \in \mathbf{N}$. If $f \in C(X)$, the space of continuous real-valued functions on X, let $f_n = f_{|K_n}$ and let $j(f) = (f_n)_{n=1}^{\infty}$. Then it follows from Tietze's extension theorem that j maps $C(X)$ onto a closed subset of $\prod_{n=1}^{\infty} C(X_n)$. Thus if we give $C(X)$ a metric such as

$$d(f,g) = \sum_{n=1}^{\infty} \sup_{x\in K_n} (|f(x) - g(x)| \wedge 2^{-n}),$$

$(C(X),d)$ becomes a complete Polish metric space.

We also consider the space $C_0(X)$. Suppose that f is a continuous real-valued function on a locally compact space (X,τ), considered as a subspace of its one-point compactification $(X^* = X \cup \{\infty\}, \tau^*)$. Then $f \in C_0(X)$ if and only if $f(x) \to 0$ as $x \to \infty$; that is, given $\epsilon > 0$ there exists a compact subset K of X such that $|f(x)| < \epsilon$ for $x \in X \setminus K$. We give $C_0(X)$ the metric d_∞. Then $(C_0(X), d_\infty)$ is homeomorphic to the closed hyperplane $\{f \in C(X^*) : f(\infty) = 0\}$, and so is complete.

Proposition 3.4.2 *If (X, τ) is a locally compact space, then $(C_0(X), d_\infty)$ is separable if and only if X is σ-finite and metrizable.*

Proof This follows immediately from Corollary 3.3.13 and Proposition 3.3.14. □

In particular, $C_0(\mathbf{R}^d)$ is separable, for $d \in \mathbf{N}$.

4
Semi-continuous Functions

4.1 The Effective Domain and Proper Functions

We shall need to consider the suprema and infima of sets of real-valued functions defined on a set X. A supremum can however take the value $+\infty$ and an infimum take the value $-\infty$. It is therefore frequently convenient to consider functions taking values in the extended real line $\overline{\mathbf{R}} = [-\infty, \infty]$. If f is such a function then we define the *effective domain* Φ_f, or $\text{dom}(f)$ to be the set $\{x \in X : f(x) \in \mathbf{R}\}$, and say that f is a *proper* function if Φ_f is not empty.

4.2 Semi-continuity

Suppose that (X, τ) is a topological space, that f is a proper function on X taking values in the extended real line $\overline{\mathbf{R}} = [-\infty, \infty]$ and that $x \in X$. Then f is *lower semi-continuous* at x if whenever $\lambda < f(x)$ there exists a neighbourhood N of x such that $f(y) > \lambda$ for each $y \in N$. (Note that the condition is trivially satisfied if $f(x) = -\infty$.) f is lower semi-continuous on X if it is lower semi-continuous at each point of X. *Upper semi-continuity*, which we have met earlier, is defined in a similar way. There are corresponding results for upper semi-continuous functions, but we shall concentrate for the most part on lower semi-continuous functions.

Exercise 4.2.1 Let τ_- be the one-sided topology $\{\emptyset\} \cup \{(a, +\infty] : a < +\infty\} \cup \{\overline{\mathbf{R}}\}$ on $\overline{\mathbf{R}}$. Show that an $\overline{\mathbf{R}}$-valued function on X is lower semi-continuous on X if and only if it is continuous from X to $(\overline{\mathbf{R}}, \tau_-)$.

Suppose that f is a mapping from a set X into $\overline{\mathbf{R}}$. The *epigraph* A_f is the set $\{(x, \mu) \in X \times \mathbf{R} : f(x) \leq \mu\}$, and the *strict epigraph* S_f is the set $\{(x, \mu) \in X \times \mathbf{R} : f(x) < \mu\}$. (Note that we exclude the values $\pm\infty$ in the definitions of the epigraph and strict epigraph.)

Proposition 4.2.2 *Suppose that f is a proper function on a topological space (X, τ) taking values in $[-\infty, \infty]$. f is lower semi-continuous if and only if the epigraph A_f is closed in $X \times [-\infty, \infty)$.*

Proof Suppose that f is lower semi-continuous. If $x \in X$ and $\lambda < f(x)$, let $\mu = \frac{1}{2}(\lambda + f(x))$. Then there exists a neighbourhood N of x such that $f(y) > \mu$ for $y \in N$. Then $N \times (-\infty, \mu)$ is an open subset of $X \times \overline{\mathbf{R}}$ disjoint from A_f. Thus A_f is closed. Conversely, if A_f is closed, and $f(x) > \lambda$ then there is an open neighbourhood U of x and $0 < \epsilon < f(x) - \lambda$ such that $U \times (-\infty, \lambda + \epsilon)$ is disjoint from A_f, so that $f(y) > \lambda$ for $y \in U$; thus f is lower semi-continuous. □

This proposition implies many basic results.

Exercise 4.2.3 Suppose that F is a set of lower semi-continuous functions on X. Show the following.

(i) $g = \sup_{f \in F} f$ is lower semi-continuous.
(ii) If F is finite, then $h = \inf_{f \in F} f$ is lower semi-continuous.
(iii) If F is infinite, then $\inf_{f \in F} f$ need not be lower semi-continuous.
(iv) If $f, g \in F$, then $f + g$ is lower semi-continuous.
(v) If $f, g \in F$ and f and g are non-negative, then $f.g$ is lower semi-continuous (when $0.\infty$ is defined as 0).
(vi) If $f \in F$ and $f \geq 0$ then $1/f$ is upper semi-continuous.
(vii) f is upper semi-continuous if and only if S_f is open, and f is continuous if and only if $G(f)$ is closed.

Proposition 4.2.4 *Suppose that f is a proper lower semi-continuous function on a compact space (X, τ), taking values in $(-\infty, \infty]$. Then f attains its infimum.*

Proof Let $c = \inf\{f(x) : x \in X\}$, let $L = \{\lambda : \lambda > c\}$ and for each $\lambda \in L$ let $C_\lambda = \{x : f(x) \leq \lambda\}$. Then each C_λ is a closed non-empty set, by Proposition 4.2.2, and so since X is compact, $C = \cap_{\lambda \in L} C_\lambda$ is non-empty. If $x \in C$, then $f(x) = c$. Note that this implies that $c > -\infty$. □

Suppose that f is a proper function on a topological space (X, τ) taking values in $(R, \infty]$. We define

$$f_\dagger(x) = \sup\{\inf_{y \in U} f(y) : U \text{ a neighbourhood of x}\}.$$

Then $f_\dagger \in \overline{\mathbf{R}}$.

Proposition 4.2.5 *If f is a proper function on a topological space (X, τ) taking values in $(R, \infty]$, then $f_\dagger \leq f$, $f_\dagger$ is lower semi-continuous on Φ_f, and $f_\dagger$ is the largest lower semi-continuous function which is less than or equal to f.*

Proof Since $x \in U$ for each neighbourhood U of x, $f_\dagger \leq f$. If $x \in \Phi_f$ and $\alpha < f_\dagger(x)$, choose $\alpha < \alpha' < f(x)$. There exists an open neighbourhood U of x such that $f(y) > \alpha'$ for $y \in U$. If $z \in U$ then U is a neighbourhood of z, and so $f_\dagger(z) \geq \alpha' > \alpha$. Thus $f_\dagger$ is lower semi-continuous at x.

If f is lower semi-continuous at x and $\alpha < f(x)$, there exists an open neighbourhood U of x such that $f(y) > \alpha$ for $y \in U$. Thus $f_\dagger(x) \geq \alpha$. Since α is arbitrary, $f_\dagger(x) \geq f(x)$. But $f_\dagger \leq f$, and so $f_\dagger(x) = f(x)$.

If g is lower semi-continuous and $g \leq f$, then $g = g_\dagger \leq f_\dagger$, so that $f_\dagger$ is the largest lower semi-continuous function less than or equal to f. □

$f_\dagger$ is called the *lower semi-continuous envelope* of f. The *upper semi-continuous envelope* $f^\dagger$ is defined similarly.

Corollary 4.2.6 *(i) If f is lower semi-continuous at x and $x_n \to x$ as $n \to \infty$, then* $\liminf_{n\to\infty} f(x_n) \geq f(x)$.

(ii) If (X, τ) is first countable, then conversely f is lower semi-continuous at x if $\liminf_{n\to\infty} f(x_n) \geq f(x)$ *whenever $x_n \to x$ as $n \to \infty$.*

Proof (i) This follows, since $\liminf_{n\to\infty} f(x_n) \geq \liminf_{y\to x} f(y) \geq f_\dagger(x) = f(x)$.

(ii) Conversely, suppose that (X, τ) is first countable, and that f is not lower semi-continuous at x. Let $(U_n)_{n=1}^\infty$ be a base of neighbourhoods of x. There exists $\epsilon > 0$ such that $\inf\{f(y) : y \in U_n\} \leq f(x) - \epsilon$ for each $n \in \mathbf{N}$. Thus there exists $x_n \in U_n$ such that $f(x_n) < f(x) - \epsilon/2$ for $n \in \mathbf{N}$. Then $x_n \to x$ as $n \to \infty$ and $\liminf_{n\to\infty} f(x_n) \leq f(x) - \epsilon/2$. □

Suppose that f is a real-valued function on a topological space (X, τ), and that $x \in X$. Then the local oscillation Ω_f of X is $f^\dagger - f_\dagger$, so that Ω_f is a non-negative upper semi-continuous function. For clearly $f^\dagger \geq f_\dagger$, and $f^\dagger$ and $-f_\dagger$ are upper semi-continuous.

We can also use the epigraph of a function to construct its lower semi-continuous envelope. In fact, we do a bit more.

Proposition 4.2.7 *Suppose that Y is a subset of a topological space (X, τ) and that f is a function from Y to $\mathbf{R}$ which is bounded below. Let $A_f = \{(x, \mu) \in X \times \overline{\mathbf{R}} : f(x) \leq \mu\}$, and let $B_f = \overline{A_f} \cup (X \times \{\infty\})$.*

(i) There is a function $g : x \to (-\infty, \infty]$ such that $B_f = A_g$.

(ii) $g(y) = f_\dagger(y)$ for $y \in Y$.

Proof (i) Let $Z = \{z \in X :$ there exists $(z, t) \in B_f\}$, so that $Y \subseteq Z \subseteq X$. Suppose that $(z, t) \in B_f$ and that $s > t$. First we show that $(z, s) \in B_f$. If $U \times I$ is an open neighbourhood of (x, s), then $U \times (I - (s - t))$ is an open neighbourhood of (x,t), and there exists $(y,r) \in A_f \cap (U \times (I - (s - t)))$. Then $(y, r + s - t) \in A_f \cap (U \times I)$, so that $(x,s) \in B_f$. Thus if we set $g(z) = \inf(t : (z,t) \in B_f)$ for $z \in Z$ and $g(z) = \infty$ otherwise, then $B_f = A_g$.

(ii) Since A_g is closed, g is lower semi-continuous, by Proposition 4.2.2. Since $A_f \subseteq A_g$, $g \leq f$. If h is lower semi-continuous and $h \leq f$, then $A_f \subseteq A_h$, and so $A_g = \overline{A_f} \subseteq A_h$: $h \leq g$. □

Theorem 4.2.8 (The extension theorem for lower semi-continuous functions) *Suppose that Y is a subset of a topological space (X, τ) and that f is a lower semi-continuous function on Y which is bounded below. There exists a lower semi-continuous function g on X such that $g_{|Y} = f$.*

Proof Let g be the function of Proposition 4.2.7. Since A_f is closed in $Y \times \mathbf{R}$, $B_f \cap (Y \times \mathbf{R}) = A_f$, and so $g_{|Y} = f$. □

The extension of the function $f(x) = 1/x$ on $(0,1]$ to $[0,1]$ shows that the extension of a real-valued lower semi-continuous function need not be real-valued.

Here is a useful approximation result. If (X,d) is a metric space, let $BL(X)$ be the space of bounded Lipschitz functions on X.

Theorem 4.2.9 *Suppose that f is a non-negative lower semi-continuous proper function on a non-empty subset A of a metric space (X,d). Then there exists an increasing sequence $(f_n)_{n=1}^\infty$ of non-negative functions in $BL(X)$ such that $f_n(a) \to f(a)$ for each $a \in A$.*

Proof Let $f_n(x) = (\inf_{a \in A}\{f(a) + nd(x,a)\}) \wedge n$. Then $f_n \in BL(X)$, (f_n) is a pointwise increasing sequence of functions, and if $a \in A$ then $f_n(a) \leq f(a)$.

Suppose that $a \in A$. Then $f_n(a) \to f_\infty(a)$ as $n \to \infty$, for some $f_\infty(a) \leq f(a)$. There exists $a_n \in A$ such that

$$f(a_n) + nd(a, a_n) \leq f_n(a) + 1/n.$$

Thus $d(a, a_n) \leq (f_n(a) + 1/n)/n \leq (f_\infty(a) + 1/n)/n$, so that $a_n \to a$ as $n \to \infty$. Consequently $f(a) \leq \liminf_{n\to\infty} f(a_n) \leq f_\infty(a)$, and so $f(a) = f_\infty(a)$. □

4.3 The Brézis–Browder Lemma

We now consider abstract partially ordered sets. A partially ordered set $(X, \preceq)$ is said to be *countably inductive* if whenever $(x_n)_{n=1}^\infty$ is an increasing sequence in X then there exists an upper bound $y \in X$; $x_n \preceq y$ for each $n \in \mathbf{N}$.

Theorem 4.3.1 (The Brézis–Browder lemma) *Suppose that ϕ is a decreasing real-valued function on a countably inductive partially ordered space X, and that ϕ is bounded below. Then there exists $y \in X$ such that if $y \preceq z$ then $\phi(y) = \phi(z)$.*

Proof If $x \in X$, let $S(x) = \inf\{\phi(y) : x \prec y\}$. Pick x_1 arbitrarily in X. An inductive argument then shows that there is an increasing sequence $(x_n)_{n=1}^\infty$ such that $\phi(x_{n+1}) - S(x_n) \leq \frac{1}{2}(\phi(x_n) - S(x_n))$ for each $n \in \mathbf{N}$. The sequence of real numbers $(S(x_n))_{n=1}^\infty$ is increasing and bounded above by $\phi(x_1)$; let $\beta = \lim_{n\to\infty} S(x_n)$. There exists an upper bound y to the sequence $(x_n)_{n=1}^\infty$. Suppose that $y \preceq z$. Since $\phi(z) \geq S(x_n)$, for each n,

$$\phi(x_{n+1}) - \phi(z) \leq \tfrac{1}{2}(\phi(x_n) - \phi(z)) \text{ for each } n \in \mathbf{N}.$$

Taking the limit as $n \to \infty$, it follows that $\beta \leq \phi(z)$. But $\beta \geq \phi(y) \geq \phi(z)$, so that $\phi(y) = \phi(z) = \beta$. □

Note that this theorem does not use the axiom of choice, and can be thought of as a countable version of Zorn's lemma.

Exercise 4.3.2 Suppose that ϕ is a decreasing real-valued function on a countably inductive partially ordered space X for which, if $(x_n)_{n=1}^\infty$ is an increasing sequence then there exists an upper bound y for which $\phi(x_n) \to \phi(y)$. Suppose also that if $t < \phi(x)$ there exists $z > x$ for which $t < \phi(z) < \phi(x)$. By considering $X_0 = \{z \geq x : \phi(z) > t\}$, show that $\{\phi(z) : z \geq x\} = (-\infty, \phi(x))$.

Exercise 4.3.3 Suppose that $(X, \tau, \leq)$ is a partially ordered topological space for which each increasing sequence converges, and for which each set $\{z : z \geq x\}$ is closed. Show that if there exists a strictly decreasing real-valued function ϕ on X (if $x < y$ then $\phi(y) < \phi(x)$) then if $x \in X$ there exists $y \geq x$ which is maximal.

4.4 Ekeland's Variational Principle

If f is a real-valued lower semi-continuous function on a compact Hausdorff space (X, τ) then f is bounded below, and there exists $\tilde{x} \in X$ such that $f(\tilde{x}) = \inf_{x\in X} f(x)$. This result does not extend to more general spaces. *Ekeland's variational principle* provides a powerful substitute for proper lower semi-continuous functions on a complete metric space.

Theorem 4.4.1 (Ekeland's variational principle) *Suppose that f is a proper non-negative lower semi-continuous function on a complete metric space*

(X,d) and that $\inf_{x\in X} f(x) = 0$. Suppose that $\alpha > 0$, that $\epsilon > 0$ and that $f(x_0) \leq \epsilon$. Then there exists $\tilde{x} \in X$ such that

(i) $d(\tilde{x}, x_0) \leq \epsilon/\alpha$,
(ii) $0 \leq f(\tilde{x}) \leq f(x_0) - \alpha d(\tilde{x}, x_0) \leq \epsilon$, and
(iii) $f(\tilde{x}) < f(z) + \alpha d(z, \tilde{x})$ for $z \in X \setminus \{\tilde{x}\}$.

Proof Set $x \prec y$ if $f(y) < f(x) - \alpha d(x,y)$. It follows from the triangle inequality that this is a partial order: let us show that it is countably inductive. If $(x_n)_{n=1}^{\infty}$ is an increasing sequence, $f(x_n)$ is a decreasing sequence of non-negative real numbers, which is therefore convergent, to l say. If $m > n$ then $\alpha d(x_m, x_n) \leq f(x_n) - f(x_m)$, so that $(x_n)_{n=1}^{\infty}$ is a Cauchy sequence, which therefore converges to a point y of X. By lower semi-continuity, $l \leq f(y)$. Thus

$$l \leq f(y) \leq f(x_m) \leq f(x_n) - \alpha d(x_m, x_n).$$

Letting $m \to \infty$, $f(y) \leq f(x_n) - \alpha d(y, x_n)$, so that $y \preceq x_n$, and the partial order is countably inductive.

We now apply the Brézis–Browder lemma to $\{x : x \preceq x_0\}$; there exists $\tilde{x} \preceq x_0$ such that if $y \preceq \tilde{x}$ then $f(y) = f(\tilde{x})$. Since $\tilde{x} \preceq x_0$, (ii) is satisfied, and (i) follows from it. Suppose that $y \preceq \tilde{x}$. Then

$$f(\tilde{x}) = f(y) \leq f(\tilde{x}) - \alpha d(y, \tilde{x}),$$

so that $y = \tilde{x}$. Thus if $z \neq \tilde{x}$ then $f(y) < f(\tilde{x}) - \alpha d(\tilde{x}, y)$, giving (iii). □

This theorem has many applications. We shall use it later to prove the petal theorem (Theorem 7.7.2), Daneš's drop theorem (Theorem 7.7.3), the Bishop–Phelps theorem (Theorem 11.8.4), and the fixed point theorems of Caristi (Theorem 13.2.1) and Clarke (Theorem 13.2.3).

5
Uniform Spaces and Topological Groups

5.1 Uniform Spaces

Metric spaces not only have a topological structure, but also have a (more restrictive) uniform structure. Here we introduce the notion of a uniform space, show how a metric space has a natural uniform structure and prove a fundamental metrization theorem.

Suppose that X is a set. A *relation* A is a subset of the product $X \times X$. It is *reflexive* if $(x,x) \in A$ for all $x \in X$; that is A contains the *diagonal* $\Delta = \{(x,x) : x \in X\}$. The transposed relation A^T is defined as $\{(x,y) : (y,x) \in A\}$. The relation A is *symmetric* if $A = A^T$. If A and B are relations, then the product $A.B$ is defined as the relation

$$\{(x,y) \in X \times X : \text{there exists } z \in X \text{ with } (x,z) \in A, (z,y) \in B\}.$$

A *uniformity* $\mathcal{U}$ on X is a set of relations on X satisfying the following conditions.

(i) $\mathcal{U}$ is a *filter*; that is, $\emptyset \notin \mathcal{U}$, the intersection of two elements of $\mathcal{U}$ is in $\mathcal{U}$, and if $U \in \mathcal{U}$ and $U \subseteq V$, then $V \in \mathcal{U}$.
(ii) If $U \in \mathcal{U}$, then U is reflexive and $U^T \in \mathcal{U}$.
(iii) If $U \in \mathcal{U}$, then there exists $V \in U$ with $V.V \subseteq U$.

The relations in $\mathcal{U}$ are called *entourages*, or *vicinities*.

A *uniform space* is a pair $(X,\mathcal{U})$, where X is a set and $\mathcal{U}$ is a uniformity on X. A *base* $\mathcal{B}$ for the uniformity $\mathcal{U}$ is a subset of $\mathcal{U}$ such that if $U \in \mathcal{U}$ then there exists $B \in \mathcal{B}$ with $B \subseteq U$. For example, the symmetric elements of $\mathcal{U}$ form a base for $\mathcal{U}$. If $\mathcal{U}$ has a countable base, then there is a base $\{B_n : n \in \mathbf{N}\}$ consisting of symmetric sets, for which $U_{n+1}.U_{n+1}.U_{n+1} \subseteq U_n$, for all $n \in \mathbf{N}$.

For example, if (X,d) is a metric space and $n \in \mathbf{N}$, let

$$B_n = \{(x,y) \in X \times X : d(x,y) < 1/3^n\},$$

and let

$$\mathcal{U}_d = \{V \in X \times X : V \supset B_n \text{ for some } n \in \mathbf{N}\}.$$

Then $\mathcal{U}_d$ is a uniformity on X, the *metric uniformity*, and $\{B_n : n \in \mathbf{N}\}$ is a base satisfying the conditions of the previous paragraph.

As a more special example, let d be the usual metric on $\mathbf{R}$. The corresponding uniformity has a base $(C_n)_{n=1}^{\infty}$ of entourages, where

$$C_n = \{(x, y) : |x - y| < 3^{-n}\}.$$

If we give $\mathbf{R}$ the equivalent metric $d'(x, y) = |\tan^{-1}(x) - \tan^{-1}(y)|$, the corresponding uniformity has base $(C'_n)_{n=1}^{\infty}$, where

$$C'_n = \{(x, y) : |\tan^{-1}(x) - \tan^{-1}(y)| < 3^{-n}\},$$

which is not the same as the uniformity defined by d: equivalent metrics can define different uniformities.

A uniform space has a natural topology. If A is a relation on X and $x \in X$, we set $A(x) = \{y \in X : (x, y) \in A\}$. If $\mathcal{U}$ is a uniformity on X, we define a topology $\tau_{\mathcal{U}}$ on X by saying that a subset O of X is open if whenever $x \in O$ then there exists $U \in \mathcal{U}$ with $U(x) \subseteq O$. It follows immediately from the filter properties of $\mathcal{U}$ that $\tau_{\mathcal{U}}$ is a topology on X, and that the sets $\{U(x); U \in \mathcal{U}\}$ form a base of neighbourhoods of x, for $x \in X$. If d is a metric in X, then the topology $\tau_{\mathcal{U}_d}$ is simply the topology defined by the metric d.

Suppose that $(X, \mathcal{U})$ and $(Y, \mathcal{V})$ are uniform spaces, and that f is a mapping from X into Y. We define the mapping $\tilde{f} : X \times X \to Y \times Y$ by setting $\tilde{f}(x_1, x_2) = (f(x_1), f(x_2))$. Then f is said to be *uniformly continuous* if whenever $V \in \mathcal{V}$ then $\tilde{f}^{-1}(V) \in \mathcal{U}$. In the case where $\mathcal{U}$ is defined by a metric d and $\mathcal{V}$ is defined by a metric ρ, then it is immediate that f is uniformly continuous in the uniform space sense if and only if it is uniformly continuous in the metric space sense.

Theorem 5.1.1 *Suppose that $(X, \mathcal{U})$ and $(Y, \mathcal{V})$ are uniform spaces, and that f is a uniformly continuous mapping from X into Y. Then f is a continuous mapping from the topological space $(X, \tau_{\mathcal{U}})$ into the topological space $(Y, \tau_{\mathcal{V}})$.*

Proof Suppose that $x \in X$ and that O is a $\tau_{\mathcal{V}}$-open neighbourhood of $f(x)$. Then there exists $V \in \mathcal{V}$ such that $V(f(x)) \subseteq O$. Then $U = \tilde{f}^{-1}(V) \in \mathcal{U}$, and $U(x)$ is a $\tau_{\mathcal{U}}$-neigbourhood of x. If $x' \in U(x)$ then $(f(x'), f(x)) \in V$, so that $f(x') \in V(f(x)) \subseteq O$: f is continuous at x. □

A bijective mapping from a uniform space $(X, \mathcal{U})$ onto a uniform space $(Y, \mathcal{V})$ is a *uniform homeomorphism* if both f and f^{-1} are uniformly continuous.

Proposition 5.1.2 *If $\mathcal{U}$ is a uniformity on X then the $\tau_{\mathcal{U}} \times \tau_{\mathcal{U}}$-open entourages form a base for $\mathcal{U}$.*

Proof Suppose that $U \in \mathcal{U}$. Then there exists a symmetric entourage V with $V.V.V \subseteq U$. If $(x, y) \in V$ then $V(x) \times V(y) \subseteq U$, so that $V \subseteq U^{int}$, and so the sets U^{int} form a base for the uniformity. □

Suppose that $\mathcal{D}$ is a family of pseudometrics on X. We can then define a uniformity $\mathcal{U}(\mathcal{D})$ on X by setting

$$\mathcal{B}(\mathcal{D}) = \{\{(x, y) : d(x, y) < \epsilon\} : d \in D\} : \epsilon > 0, D \text{ a finite subset of } \mathcal{D}\},$$

and $\mathcal{U}(\mathcal{D}) = \{U : V \subseteq U \text{ for some } V \in \mathcal{B}(\mathcal{D})\}$. The sets in $\mathcal{B}(\mathcal{D})$ are then $\tau_{\mathcal{U}(\mathcal{D})}$ open, and form a base for $\mathcal{U}(\mathcal{B})$.

In fact, the converse is true; a uniformity can always be defined by a family of pseudometrics.

Theorem 5.1.3 *If U is a uniformity on X then there is a family $\mathcal{D}$ of continuous pseudometrics on X such that $\mathcal{U} = \mathcal{U}(\mathcal{D})$.*

Proof Suppose that B is a base for $\mathcal{U}$ consisting of symmetric entourages. If $U \in B$, there exists a decreasing sequence $(U_n)_{n=0}^{\infty}$ of symmetric entourages such that $U_0 = X \times X$, $U_1 = U$ and $U_{n+1}.U_{n+1}.U_{n+1} \subseteq U_n$, for $n \in \mathbf{N}$. If $(x, y) \in X \times X$, let $k_U(x, y) = \inf\{n : (x, y) \in U_n\}$, let $\rho_U(x, y) = 1/2^{k_U(x,y)}$ and let

$$d_U(x, y) = \inf\left\{\sum_{i=1}^{j} \rho_U(x_{i-1}, x_i) : j \in \mathbf{N}, x_i \in X, x_0 = x, x_j = y\right\}.$$

We show that d is a continuous pseudometric on X, and that as U and ϵ vary, the sets $\{(x, y) : d(x, y) < \epsilon\}$ form a base for the uniformity.

Since each U_n is symmetric, $d(x, y) = d(y, x)$, and it follows from the definition that $d(x, z) \leq d(x, y) + d(y, z)$, so that d is a pseudometric on X. If $(x, y) \in U_n$, then, taking $j = 1$, it follows that $d(x, y) \leq 1/2^n$, so that d is uniformly continuous on $X \times X$ and the sets $\{(x, y) : d(x, y) < \epsilon\}$ are entourages in $\mathcal{U}$.

Suppose that $x = x_0, x_1, \ldots, x_j = y$ and that $\sum_{i=1}^{j} d(x_{i-1}, x_i) < 1/2^n$. We prove by induction on j that $(x, y) \in U_n$. If $j = 1$ then certainly $(x, y) \in U_{n+1} \subseteq U_n$. Suppose that the result is true for $j - 1$. We consider two cases. First, suppose that $\rho(x_0, x_1) = 1/2^{n+1}$. Then $(x_0, x_1) \in U_{n+1}$ and $\sum_{i=2}^{j} \rho(x_{i-1}, x_i) < 1/2^{n+1}$. By the inductive hypothesis, $(x_1, y) \in U_{n+1}$, and so $(x, y) \in U_{n+1}.U_{n+1} \subseteq U_n$. Secondly, suppose that $\rho(x_0, x_1) < 1/2^{n+1}$. Let

$$k = \sup\left\{l \in \{1, \ldots, j-1\} : \sum_{i=1}^{l} d(x_{i-1}, x_i) < 1/2^{k+1}\right\}.$$

Then $(x, x_k) \in U_{n+1}$, by the inductive hypothesis. Further, $\rho(x_k, x_{k+1}) \leq 1/2^{n+1}$, so that $(x_k, x_{k+1}) \in U_{k+1}$. Thus if $k = j-1$ then $(x, y) \in U_{n+1}.U_{n+1} \subseteq U_n$. Otherwise, $\sum_{i=k+1}^{j} d(x_{i-1}, x_i) < 1/2^{k+1}$, so that, by the inductive hypothesis, $(x_k, y) \in U_{n+1}$, and $(x, y) \in U_{n+1}.U_{n+1}.U_{n+1} \subseteq U_n$.

Consequently, if $d(x, y) < 1/2^n$ then $(x, y) \in U_n$. Thus, as U varies, the sets $\{(x, y) : d(x, y) < \epsilon\}$ form a base for the uniformity $\mathcal{U}$. □

What about the topological properties of a uniform space?

A uniformity $\mathcal{U}$ is *Hausdorff* if $\cap_{U \in \mathcal{U}} U = \Delta$.

Proposition 5.1.4 *If $\mathcal{U}$ is Hausdorff then $(X, \tau_{\mathcal{U}})$ is completely regular.*

Proof Suppose that $x \in X$, that C is a closed subset of X and that $x \notin C$. There exists a symmetric entourage U such that $(x, c) \notin U$, for $c \in C$. Let d be a continuous pseudometric as defined earlier and let $f(y) = d(x, y)$. Then f is a continuous function on X taking values in $[0, 1]$, $f(x) = 0$ and $f(c) = 1$ for $c \in C$. □

Exercise 5.1.5 Show that a Hausdorff uniformity $\mathcal{U}$ has a base of $\tau_{\mathcal{U}} \times \tau_{\mathcal{U}}$ closed entourages.

We have the following fundamental metrization theorem.

Theorem 5.1.6 *If $\mathcal{U}$ is a Hausdorff uniformity on X with a countable base, then there is a metric D on X such that $\mathcal{U}$ is the corresponding metric uniformity.*

Proof We can find a sequence $(U_n)_{n=0}^{\infty}$ of symmetric entourages which satisfy the conditions of Theorem 5.1.3 and which form a base for the uniformity. Let d be the corresponding pseudometric. If $x \neq y$, there exists n such that $(x, y) \notin U_n$, since the uniformity is Hausdorff, and so $d(x, y) \geq 1/2^n$. Thus d is a metric, which defines the uniformity $\mathcal{U}$. □

5.2 The Uniformity of a Compact Hausdorff Space

Theorem 5.2.1 *Suppose that (X, τ) is a compact Hausdorff space. The collection $\mathcal{O}$ of symmetric open subsets of $X \times X$ which contain Δ is a base for a uniformity $\mathcal{U}$ on X with $\tau_{\mathcal{U}} = \tau$.*

Proof $\mathcal{O}$ is closed under finite intersections, and $O^T \in \mathcal{O}$ if and only if $O \in \mathcal{O}$. Suppose that $O \in \mathcal{O}$. We show that there exists $P \in \mathcal{O}$ with $P.P \subseteq O$. Suppose not. Let $C = (X \times X) \setminus O$. Then for each $P \in \mathcal{O}$, the set $S(P) = \overline{P.P} \cap C$ is non-empty. The sets $\{S(P) : P \in \mathcal{O}\}$ have the finite intersection property, and so $S = \cap_{P \in \mathcal{O}} S(P)$ is not empty. Suppose that $(x, y) \in S$. Then $x \neq y$. Since (X, τ) is normal, there exist open sets U_x, V_x, U_y and V_y in X such that $x \in U_x \subseteq \overline{U}_x \subseteq V_x$, $y \in U_y \subseteq \overline{U}_y \subseteq V_y$ and $V_x \cap V_y = \emptyset$. Let $W = X \setminus (\overline{U}_x \cup \overline{U}_y)$: W is open in X. Now let $P = (V_x \times V_x) \cup (V_y \times V_y) \cup (W \times W)$. Then $P \in \mathcal{O}$. Suppose if possible that $(x', y') \in (U_x \times U_y) \cap P.P$, so that there exists $z \in X$ such that $(x', z) \in P$ and $(y', z) \in P$. Since $(x', z) \in P$, $z \in V_x$, and similarly $z \in V_y$. But $V_x \cap V_y = \emptyset$, giving a contradiction. Thus $(x, y) \notin \overline{P.P}$, giving the required contradiction. Consequently, $\mathcal{O}$ is a base for a uniformity $\mathcal{U}$ on X, and clearly $\tau_{\mathcal{U}} = \tau$. □

Theorem 5.2.2 *If (X, τ) is a compact Hausdorff space, then the uniformity $\mathcal{U}$ of the preceding theorem is the unique uniformity on X which defines the topology τ.*

Proof Suppose that $\mathcal{V}$ is a uniformity on X for which $\tau_{\mathcal{V}} = \tau$. Since $\mathcal{V}$ has a base of open sets, $\mathcal{V} \subseteq \mathcal{U}$. Suppose if possible that $O \in \mathcal{O} \setminus \mathcal{V}$. Let $\mathcal{F}$ be the set of closed symmetric elements in $\mathcal{V}$. Then the sets $\{F \setminus O : F \in \mathcal{F}\}$ have the finite intersection property, and so there exists $(x, y) \in (\cap_{F \in \mathcal{F}} F) \setminus O$. But then $x \neq y$, contradicting the fact that $\mathcal{V}$ is Hausdorff. □

When is a compact Hausdorff space metrizable?

Theorem 5.2.3 *A compact Hausdorff space (X, τ) is metrizable if and only if the diagonal Δ is a G_δ subset of $(X, \tau) \times (X, \tau)$.*

Proof If (X, τ) is metrizable then so is $(X, \tau) \times (X, \tau)$, and so the closed set Δ is a G_δ subset of $X \times X$.

Conversely, suppose that Δ is a G_δ subset of $X \times X$. There therefore exists a decreasing sequence $(U_n)_{n=1}^\infty$ of open subsets of $X \times X$, with $\cap_{n \in \mathbf{N}} U_n = \Delta$. By considering $U_n \cap U_n^T$, we can suppose that each U_n is symmetric. Let $O_1 = U_1$. Since the symmetric reflexive open subsets of $X \times X$ form a base for the uniformity of X, there exists a symmetric reflexive open set V_1 such that $V_1.V_1 \subseteq O_1$. Let $O_2 = U_2 \cap V_1$, so that $O_2.O_2 \subseteq O_1$. Iterating this procedure, there exists a decreasing sequence $(O_n)_{n=1}^\infty$ of open symmetric reflexive sets, with $O_{n+1}.O_{n+1} \subseteq O_n \subseteq U_n$, for $n \in \mathbf{N}$. Consequently, $\cap_{n \in \mathbf{N}} O_n = \Delta$. Thus if we use the sequence $(O_n)_{n=1}^\infty$ to define a pseudometric d on X, then d is in fact a metric on X. The metric d is τ-continuous on X, and the identity mapping $(X, \tau) \to (X, d)$ is continuous. But (X, τ) is compact and (X, d) is Hausdorff, and so the identity mapping $(X, \tau) \to (X, d)$ is a homeomorphism. □

5.3 Topological Groups

A *topological group* (G, τ) is a group G, with identity element e, together with a topology τ on G which satisfies:

(a) the mapping $(g, h) \to gh : (G, \tau) \times (G, \tau) \to (G, \tau)$ is continuous;
(b) the mapping $g \to g^{-1} : (G, \tau) \to (G, \tau)$ is continuous.

Exercise 5.3.1 Suppose that (G, τ) is a topological group, and that $g \in G$. Establish the following, which follow immediately from the definition.

(i) Let $l_g(h) = gh$ and $r_g = hg$. Then l_g and r_g are homeomorphisms of (G, τ).
(ii) Let $i(h) = h^{-1}$. Then i is a homeomorphism of (G, τ).
(iii) If U is a neighbourhood of e then there exists a neighbourhood V of e such that $V^2 = \{gh : g, h \in V\} \subseteq U$.

A subset A of G is *symmetric* if $g^{-1} \in A$ whenever $g \in A$.

Exercise 5.3.2 Suppose that (G, τ) is a topological group. Establish the following.

(i) $\mathcal{N}(g) = \{gN : N \in \mathcal{N}(e)\} = \{Ng : N \in \mathcal{N}(e)\}$.
(ii) There is a base of symmetric neighbourhoods of e.
(iii) If σ is a topology on a group G which satisfies (a), and
(iv) the mapping $g \to g^{-1} : (G, \tau) \to (G, \tau)$ is continuous at e,

then (G, σ) is a topological group.

Let us give some examples.

First, the general linear group GL_n of invertible linear mappings of $\mathbf{R}^n$ into itself. To anticipate Chapter 7, if $S \in GL_n$, let

$$\|S\| = \sup\{d_\infty(S(x), 0) : d\infty(x, 0) \leq 1\}.$$

Suppose that $S_0, T_0 \in GL_n$, that $\epsilon > 0$ and that $\|S - S_0\| < \epsilon$ and $\|T - T_0\| < \epsilon$ then

$$\begin{aligned} \|ST - S_0T_0\| &= \|(S - S_0)T_0 + S_0(T - T_0) + (S - S_0)(T - T_0)\| \\ &\leq \|(S - S_0)T_0\| + \|g_0(T - T_0)\| + \|(S - S_0)(T - T_0)\| \\ &\leq \epsilon(\|S_0\| + \|T_0\| + \epsilon), \end{aligned}$$

from which it follows that multiplication is continuous at (S_0, T_0). If $S \in L(E)$ and $\|S - I\| < 1$ then $S \in GL(E)$ and $S^{-1} = I + \sum_{j=1}^{\infty}(I - S)^j$ (the *Neumann series*). Then

$$\|I - S^{-1}\| \leq \sum_{j=1}^{\infty} \|I - S\|^j = \frac{\|I - S\|}{(1 - \|I - S\|)},$$

from which it follows that the mapping $S \to S^{-1}$ is continuous at I. It therefore follows that $GL(E)$ is a topological group.

Since a subgroup of a topological group, with the subspace topology, is also a topological group, the *special linear group* $SL_n = \{T \in GL_n : \det T = 1\}$ is also a topological group, as is the *orthogonal group* $O_n = \{T \in GL_n : T'T = I\}$. Similarly the *unitary group* $U_n = \{T \in GL(\mathbf{C}^n) : T^*T = I\}$ is a topological group.

A group is *abelian* if $gh = hg$ for all $g, h \in G$. The group operation of an abelian group is often written as $+$, as in the group $(\mathbf{R}, +)$, which is a locally compact topological group under its usual topology. On the other hand, $\mathbf{T} = \{z \in \mathbf{C} : |z| = 1\}$ is an abelian group under multiplication, and is a compact topological group when it is given the subspace topology. The Bernoulli sequence space $\Omega(\mathbf{N})$ is also a compact abelian group, when addition is defined co-ordinatewise (mod 2); $\Omega(\mathbf{N}) = \prod_{n=1}^{\infty} (\mathbf{Z}_2)_n$.

Suppose that $(X, \mathcal{U})$ is a uniform space. The set Homeo(X) of uniform homeomorphisms of X onto itself is a group under composition. If $U \in \mathcal{U}$, let

$$H(U) = \{(f, g) \in \text{Homeo}(X)^2 : (f(x), g(x)) \in U \text{ for all } x \in X\}.$$

Then it is easy to check that $\{H(U) : U \in \mathcal{U}\}$ is the base for a uniformity $H(\mathcal{U})$ on Homeo(X).

Theorem 5.3.3 *The group* Homeo(X) *of homeomorphisms of a uniform space* $(X, \mathcal{U})$ *with the topology* $\tau_{H(\mathcal{U})}$ *is a topological group.*

Proof Suppose that $g_0, h_0 \in$ Homeo(X) and that U is a symmetric entourage in $\mathcal{U}$. In order to show that multiplication is continuous, we must show that there exist V and W in $\mathcal{U}$ such that if $(g, g_0) \in H(V)$ and $(h, h_0) \in H(W)$ then $(gh, g_0h_0) \in H(U)$. There exists $V \in \mathcal{U}$ such that $V.V \subseteq U$. Since g_0 is uniformly continuous, there exists $W \in \mathcal{U}$ such that if $(x, x') \in W$ then $(g(x), g(x')) \in V$. Now suppose that $(g, g_0) \in H(V)$ and that $(h, h_0) \in H(W)$. If $x \in X$ then $(gh(x), gh_0(x)) \in V$ and $(g_0h(x), g_0h_0(x)) \in V$, so that $(gh(x), g_0h_0(x)) \in U$. Thus $(gh, g_0h_0) \in H(U)$.

If $(g, e) \in H(U)$ and $x \in X$ then $(g(x), x) \in U$. Applying this to $g^{-1}(x)$, $(x, g^{-1}(x)) \in U$, and so $(e, g^{-1}) \in H(U)$. Thus inversion is continuous at e, and so Homeo(X), $\tau_{H(\mathcal{U})}$ is a topological group, by Exercise 5.3.2(iii). □

Recall that if (X, d) is a metric space then we give the space $C(X, X)$ the metric d_∞: $d_\infty(f, g) = \sup_{x \in X} d(f(x), g(x))$. Homeo$(X)$ is a subset of $C(X, X)$.

Corollary 5.3.4 *Suppose that* (X, d) *is a compact metric space. Then* $(\mathrm{Homeo}(X), d_\infty)$ *is a topological group. If* $k \in \mathrm{Homeo}(X)$*, the mapping* $g \to gk$ *is an isometry of* $(\mathrm{Homeo}(X), d_\infty)$.

Proof Let $\mathcal{U}$ be the metric uniformity defined by d. Then the collection of sets

$$\{(g,h) \in \mathrm{Homeo}(X) \times \mathrm{Homeo}(X) : d_\infty(g,h) < \alpha\}_{\alpha>0}$$

forms a basis for $H(\mathcal{U})$, and so $(\mathrm{Homeo}(X), d_\infty)$ is a topological group.

Further, if $g, h, k \in \mathrm{Homeo}(X)$ then $d(gk(x), hk(x)) \leq d_\infty(g,h)$, so that $d_\infty(gk, hk) \leq d_\infty(g,h)$. Similarly,

$$d(g(x), h(x)) = d(gkk^{-1}(x), hkk^{-1}(x)) \leq d_\infty(gk, hk),$$

so that $d_\infty(g,h) \leq d_\infty(gk, hk)$. Thus $d_\infty(g,h) = d_\infty(gk, hk)$. □

Proposition 5.3.5 *Suppose that* $T : X \to X$ *is an isometry of a compact metric space* (X, d) *into itself. Then* T *is surjective.*

Proof Suppose not, so that there exists $x \in X \setminus T(X)$. Since $T(X)$ is compact, it is closed, and so $\alpha = d(x, T(X)) > 0$. If $m < n$ then $d(T^m(x), T^n(x)) = d(x, T^{n-m}(x)) > \alpha$, and so the sequence $(T^n(x))_{n=1}^\infty$ has no convergent subsequence, contradictiong the sequential compactness of (X, d). □

Thus the set $\mathrm{Iso}(X)$ of isometries of (X, d) is a group, under composition.

Theorem 5.3.6 *The group* $\mathrm{Iso}(X)$ *of isometries of a compact metric space* (X, d) *is a compact subgroup of* $(\mathrm{Homeo}(X), d_\infty)$*. If* $R \in \mathrm{Iso}(X)$*, the mappings* $T \to RT$ *and* $T \to TR$ *and the mapping* $T \to T^{-1}$ *are isometries of* $(\mathrm{Iso}(X), d_\infty)$.

Proof $\mathrm{Iso}(X)$ is a closed subset of the complete metric space $(C(X, X), d_\infty)$, and is an equicontinuous set of functions, and so it is compact, by the Arzelà–Ascoli theorem.

If $x \in X$ then

$$\begin{aligned} d(S^{-1}(x), T^{-1}(x) &= d(x, ST^{-1}(x) = d(TT^{-1}(x), ST^{-1}(x) \\ &\leq d_\infty(T, S), \end{aligned}$$

so that $d_\infty(S^{-1}, T^{-1}) \leq d_\infty(S, T)$. Replacing S by S^{-1} and T by T^{-1}, it follows that $d_\infty(S, T) \leq d_\infty(S^{-1}, T^{-1})$. Thus $d_\infty(S^{-1}, T^{-1}) = d_\infty(S, T)$, and the mapping $T \to T^{-1}$ is an isometry. Finally,

$$d_\infty(RS, RT) = d_\infty(S^{-1}R^{-1}, T^{-1}R^{-1}) = d_\infty(S^{-1}, T^{-1}) = d(S, T).$$

□

5.4 The Uniformities of a Topological Group

Suppose that (G, τ) is a topological group. If U is a symmetric neighbourhood of the identity e, let

$$l(U) = \{(g, h) : g^{-1}h \in U\} \text{ and } r(U) = \{(g, h) : hg^{-1} \in U\}.$$

There exists a symmetric neighbourhood V of e such that $V^2 \subseteq U$. Then $l(V).l(V) \subseteq l(U)$ and $r(V).r(V) \subseteq r(U)$, and so the sets

$$\{l(U) : U \text{ a symmetric neighbourhood of } e\}$$

form a base for a uniformity $\mathcal{L}(G)$ on G, the *left uniformity* of G. Similarly, the sets

$$\{r(U) : U \text{ a symmetric neighbourhood of } e\}$$

form a base for a uniformity $\mathcal{R}(G)$ on G, the *right uniformity* of G. Further, $\tau = \tau_{\mathcal{L}(G)} = \tau_{\mathcal{R}(G)}$. If $k \in G$ and $(g, h) \in l(U)$, then $(kg)^{-1}(kh) = g^{-1}h \in U$, so that $(kg, kh) \in l(U)$: the mapping $g \to kg$ is a uniform homeomorphism of $(G, \mathcal{L}(G))$ onto itself. The mapping $g \to g^{-1}$ is a uniform homeomorphism of $(G, \mathcal{L}(G))$ onto $(G, \mathcal{R}(G))$.

Proposition 5.4.1 *The left uniformity $\mathcal{L}(G)$ and the right uniformity $\mathcal{R}(G)$ of a topological group are Hausdorff uniformities if and only if $\{e\}$ is closed in G.*

Proof If $\mathcal{L}(G)$ or $\mathcal{R}(G)$ is a Hausdorff uniformity, then G is completely regular, and so it is a Hausdorff topological space. Conversely, suppose that $\{e\}$ is closed. Since left multiplication is a homeomorphism, each point of G is closed. Thus if $g \neq h$ there exists a symmetric neighbourhood U of e such that $g^{-1}h \notin U$. Hence $(g, h) \notin l(U)$. Consequently the left uniformity is Hausdorff. Similarly for the right uniformity. □

Corollary 5.4.2 *A Hausdorff topological group is completely regular.*

Suppose that G is a topological group. Let $N = \overline{\{e\}}$. Then it is easy to see that N is a normal subgroup of G, and that the quotient group G/N, with the quotient topology, is a Hausdorff topological group.

If G is a compact topological group, then, since there is only one uniformity on G which defines the topology, the left and right uniformities are the same. This is not the case for locally compact groups, as the next example shows. This is one of the reasons why the representation theory of locally compact groups is in general much more complicated than the representation theory of compact groups.

Let

$$G = \left\{ \begin{bmatrix} a & b \\ 0 & 1 \end{bmatrix} \in GL_2 : a > 0, b \in \mathbf{R} \right\}.$$

G is a closed subgroup of GL_2, and so it is a locally compact topological group.

Exercise 5.4.3 Show that G acts on the line $\{(x, 1) : x \in \mathbf{R}\}$:

$$\begin{bmatrix} a & 0 \\ 0 & 1 \end{bmatrix} \text{ acts by dilation, and } \begin{bmatrix} 1 & b \\ 0 & 1 \end{bmatrix} \text{ acts by translation.}$$

Let

$$V = \left\{ \begin{bmatrix} a & b \\ 0 & 1 \end{bmatrix} \in G : 2/3 < a < 3/2 \text{ and } |b| < 2 \right\}.$$

V is a symmetric neighbourhood of e. We shall show that if U is any symmetric neighbourhood of e then $l(U) \not\subseteq r(V)$. There exists $\epsilon > 0$ such that if $|a - 1| < \epsilon$ then

$$\begin{bmatrix} a & 0 \\ 0 & 1 \end{bmatrix} \in U.$$

Choose a such that $|1 - a| < \epsilon$, and choose $b = 2/(1 - a)$. Now

$$\begin{bmatrix} a^{-1} & b \\ 0 & 1 \end{bmatrix}^{-1} \begin{bmatrix} 1 & b \\ 0 & 1 \end{bmatrix} = \begin{bmatrix} a & -ab \\ 0 & 1 \end{bmatrix} . \begin{bmatrix} 1 & b \\ 0 & 1 \end{bmatrix} = \begin{bmatrix} a & 0 \\ 0 & 1 \end{bmatrix},$$

so that

$$\left(\begin{bmatrix} a^{-1} & b \\ 0 & 1 \end{bmatrix}, \begin{bmatrix} 1 & b \\ 0 & 1 \end{bmatrix} \right) \in l(U).$$

On the other hand,

$$\begin{aligned} \begin{bmatrix} 1 & b \\ 0 & 1 \end{bmatrix} . \begin{bmatrix} a^{-1} & b \\ 0 & 1 \end{bmatrix}^{-1} &= \begin{bmatrix} 1 & b \\ 0 & 1 \end{bmatrix} . \begin{bmatrix} a & -ab \\ 0 & 1 \end{bmatrix} \\ &= \begin{bmatrix} a & (1-a)b \\ 0 & 1 \end{bmatrix} = \begin{bmatrix} a & 2 \\ 0 & 1 \end{bmatrix}, \end{aligned}$$

so that

$$\left(\begin{bmatrix} a^{-1} & b \\ 0 & 1 \end{bmatrix}, \begin{bmatrix} 1 & b \\ 0 & 1 \end{bmatrix} \right) \notin r(V).$$

5.5 Group Actions

Suppose that G is a group and that X is a set. Let bij(X) denote the set of bijective mappings of X onto itself; bij(X) is a group, under composition of mappings. An *action* of G on X is a homomorphism of G into bij(X). Thus $a(gh)(x) = a(g)(a(h)(x))$, for $g, h \in G$ and $x \in X$. The action is *transitive* if $\{a(g)(x) : g \in G\} = X$, for all $x \in X$. For example, GL_n acts transitively on $\mathbf{R}^n \setminus \{0\}$, and O_n and SO_n act transitively on the unit sphere $S^{n-1} = \{x : \|x\| = 1\}$.

A group G acts transitively on itself, on the left and on the right. If $g, h \in G$, let $\lambda_g(h) = gh$ and let $\rho_g(h) = hg^{-1}$; λ is the *left action* of G on itself, and ρ is the *right action*.

An action a of a topological group G on a topological space X is *continuous* if the mapping $(g, x) \to a(g)(x)$ from $G \times X$ to X is jointly continuous. Each of the examples of group actions that we have given is a continuous action.

The next result is technically very convenient.

Proposition 5.5.1 *If a is an action of a topological group G on a topological space X then the action is continuous if and only if the mapping $x \to a(g)(x) : X \to X$ is continuous for each $g \in G$, and the action is continuous at (e, x) for each $x \in X$.*

Proof The condition is certainly necessary. Suppose that it is continuous and that W is a neighbourhood of $a(g)(x)$. Then there is a neighbourhood $U \times V$ of $(e, a(g)(x))$ such that if $(h, y) \in U \times V$ then $a(h)(y) \in W$. But then $Ug \times a(g^{-1})V$ is a neighbourhood of (g, x), and if $(h, y) \in Ug \times a(g^{-1})V$ then $a(h)(y) \in W$. □

Suppose that a is a continuous action of a compact Hausdorff group G on a compact Hausdorff space X. Then G acts linearly on $C(X)$; if $g \in G$ and $f \in C(X)$ then we set $\pi_g(f)(x) = f(a(g)x)$. Then $\pi_g(f) \in C(X)$ and π_g is an isometry of $(C(X))$. For example, if we consider the left action of G on itself, then we get the *left regular representation* l of G, given by $l_g(f)(h) = f(gh)$; similarly for the *right regular representation* r, given by $r_g(f)(x) = f(xg^{-1})$. Is the action continuous? We need a lemma.

Lemma 5.5.2 *Suppose that a is a continuous action of a compact Hausdorff group G on a compact Hausdorff space X. If C is a closed subset of X, U an open subset of X and $C \subseteq U$ then there exists a neighbourhood V of the identity e of G such that $a(g)(C) \subseteq U$ for $g \in V$.*

Proof For each $x \in C$ there exist open sets V_x in G and W_x in X with $e \in V_x$, $x \in W_x \subseteq U$ such that $a(g)(y) \in U$ for $g \in V_x$, $y \in W_x$. Since C is compact, there is a finite subset F of C such that $C \subseteq \bigcup_{x \in F} W_x$. Take $V = \bigcap_{x \in F} V_x$. □

Proposition 5.5.3 *Suppose that a is a continuous action of a compact Hausdorff group G on a compact Hausdorff space X. Then the action π of G on $C(X)$ is continuous.*

Proof By Proposition 5.5.1, it is enough to show that the action is jointly continuous at (e,f), where $f \in C(X)$. Suppose that $\epsilon > 0$. For each x there is an open neighbourhood U_x of x such that $|f(x) - f(y)| < \epsilon/3$ for $y \in U_x$. Since X is normal, there exists a closed neighbourhood C_x of x contained in U_x, and by Lemma 5.5.2 there is a neighbourhood V_x of e such that $a(g)(C_x) \subseteq U_x$ for $g \in V_x$. Since X is compact, there is a finite set Z in X such that $X = \bigcup_{z \in Z} C_z$. Let $V = \bigcap_{z \in Z} V_z$.

Suppose that $g \in V$ and $d_\infty(f,h) < \epsilon/3$. If $x \in X$ then $x \in C_z$ for some $z \in F$, and so $a(g)(x) \in U_z$. Thus

$$\begin{aligned} |f(x) - \pi_g(h)(x)| = |f(x) - h(a(g)x)| &\leq |f(x) - f(a(gx)| + \epsilon/3 \\ &\leq |f(x) - f(z)| + |f(z) - f(a(g)x)| + \epsilon/3 \leq \epsilon. \end{aligned}$$

Thus $d_\infty(f, \pi_g(h)) \leq \epsilon$. □

5.6 Metrizable Topological Groups

Theorem 5.6.1 *Suppose that (G, τ) is a first countable Hausdorff topological group. Then there exists a left-invariant metric d_l on G (that is, $d_l(gh, gk) = d_l(h, k)$ for all $g, h, k \in G$), which defines the left uniformity of G. Similarly, there exists a right-invariant metric d_r on G which defines the right uniformity of G.*

Proof There exists a basic sequence (U_n) of left-invariant entourages in $\mathcal{L}(G)$. The construction of Theorem 5.1.6 now provides a left-invariant metric which defines $\mathcal{L}(G)$. Similarly for the right-invariant case. □

If (G, τ) is a compact Hausdorff topological group, then there is only one unifomity on G which defines the topology, and so the left and right uniformities are the same. There are other topological groups with this property.

Theorem 5.6.2 *Suppose that (G, τ) is a first countable Hausdorff topological group for which the left and right uniformities are the same. Then there exists a metric d on G which is both left- and right-invariant, and which satisfies $d(g, h) = d(g^{-1}, h^{-1})$ for all $g, h \in G$.*

Proof Let $\mathcal{U}$ be the left uniformity. The mapping $g \to g^{-1}$ is a uniform homeomorphism of $(G,\mathcal{U})$ onto itself. There therefore exists a basic sequence

(U_n) of left-invariant relations in $\mathcal{L}(G)$ for which $(g^{-1}, h^{-1}) \in U_n$ whenever $(g, h) \in U_n$. If d is the corresponding left-invariant metric, then $d(g, h) = d(g^{-1}, h^{-1})$ for all $g, h \in G$. In particular, if $g, h, k \in G$ then

$$\begin{aligned} d(hg, kg) &= d((hg)^{-1}, (kg)^{-1}) = d(g^{-1}h^{-1}, g^{-1}k^{-1}) \\ &= d(h^{-1}, k^{-1}) = d(h, k), \end{aligned}$$

so that d is also right-invariant. □

The topology of a topological group is determined by the neighbourhoods of the identity; we consider this in the metrizable case. A function $\nu : G \to \mathbf{R}^+$ on a group G is called a *group-norm* if

(a) $\nu(g) \geq 0$ if and only if $g = e$;
(b) $\nu(g) = \nu(g^{-1})$ for all $g \in G$;
(c) $\nu(gh) \leq \nu(g) + \nu(h)$ for all $g, h \in G$.

This is clearly similar to, but should not be confused with, a norm on a vector space.

Theorem 5.6.3 *(i) Suppose that ν is a group-norm on a group G. Let $d_L(g, h) = \nu(g^{-1}h)$ and let $d_R(g, h) = \nu(gh^{-1})$, for $g, h \in G$. Then d_L is a left-invariant metric on G, d_R is a right-invariant metric on G, and d_L and d_R define the same topology τ on G. (G, τ) is a topological group, and the collection of sets $\{\{g : \nu(g) < \alpha\} : \alpha > 0\}$ forms a base of τ-neighbourhoods of e.*
(ii) Suppose that d is a left-invariant or right-invariant metric on a group G. Let $\nu(g) = d(e, g)$. Then ν is a group-norm on G.

Proof (i) Suppose that $g, h, k \in G$. Then $d_L(g) = 0$ if and only if $g = h$,

$$\begin{aligned} d_L(g, k) &= \nu(g^{-1}k) = \nu((g^{-1}h)(h^{-1}k)) \\ &\leq \nu(g^{-1}h) + \nu(h^{-1}k) = d_L(g, h) + d_L(h, k), \end{aligned}$$

and $d_L(h, g) = \nu(h^{-1}g) = \nu(g^{-1}h) = d_L(g, h)$, so that d_L is a metric on G. Further,

$$d_L(kh, kg) = \nu(g^{-1}k^{-1}kh) = \nu(g^{-1}h) = d_L(g, h),$$

so that d_L is left-invariant. Similarly for d_R. If $\alpha > 0$ then

$$\{g : d_L(g, e) < \alpha\} = \{g : \nu(g) < \alpha\} = \{g : d_R(g, e) < \alpha\},$$

so that d_L and d_R define the same topology τ on G and (G, τ) is a topological group.

(ii) Suppose that d is left-invariant. (a) is trivially satisfied. Since d is left-invariant,

$$\nu(g^{-1}) = d(e, g^{-1}) = d(g, gg^{-1}) = d(g, e) = d(e, g) = \nu(g),$$

so that (b) holds. Further,

$$\nu(gh) = d(e, gh) \leq d(e, g) + d(g, gh) = \nu(g) + d(e, h) = \nu(g) + \nu(h),$$

so that (c) holds. Similarly if d is right-invariant. □

Exercise 5.6.4 Show that the function $t \to |\log t|$ is a group-norm on $((0, \infty), \times)$. Is the corresponding left- and right-uniformity on $((0, \infty), \times)$ the same as the uniformity defined by the usual metric on $(0, \infty)$?

We can also consider the problem of finding a group-norm on GL_n, when $n > 1$. GL_n contains subgroups of translations and dilations, and the previous exercise shows that corresponding group-norms are very different. As a consequence, any explicit group-norm on GL_n will be rather artificial. Here is one.

Proposition 5.6.5 *If $S \in GL_n$, let*

$$\nu(S) = \min(\max(\|S - I\|^{\frac{1}{2}}, \|S^{-1} - I\|^{\frac{1}{2}}), 1).$$

Then ν is a group-norm on GL_n.

Proof Certainly $\nu(S) = 0$ if and only if $S = I$, and $\nu(S^{-1}) = \nu(S)$. Suppose that $S, T \in GL(E)$. We must show that $\nu(ST) \leq \nu(S) + \nu(T)$. This is certainly the case if $\nu(S) = 1$ or $\nu(T) = 1$. Otherwise, $\|S\| < 2$, $\|T\| < 2$, $\left\|S^{-1}\right\| < 2$ and $\left\|T^{-1}\right\| < 2$. Without loss of generality, we can suppose that $\|S - I\| \leq \|T - I\|$. Then

$$\|ST - I\| \leq \|ST - S\| + \|S - I\| \leq 2\|T_I\| + \|S_I\|.$$

If $0 < a \leq b$ then $\sqrt{2a + b} \leq \sqrt{a} + \sqrt{b}$, and so

$$\|ST - I\|^{\frac{1}{2}} \leq \|S - I\|^{\frac{1}{2}} + \|T_I\|^{\frac{1}{2}}.$$

The same argument applies when we consider pairs (S, T^{-1}), (S^{-1}, T) and (S^{-1}, T^{-1}). □

We can also consider group-norms on the groups $\text{Homeo}(X, d)$ and, more particularly, on $\text{Homeo}^+([0, 1], d)$.

Proposition 5.6.6 *Suppose that (X, d) is a compact metric space. Then the function*

$$\nu(g) = \|g - e\|_\infty = \sup_{x \in X} d(g(x), x)$$

is a group-norm on $\text{Homeo}(X, d)$.

Proof Suppose that $g, h \in \text{Homeo}(X, d)$. Then clearly $\nu(g) = 0$ if and only if $g = e$. Next,

$$\nu(g) = \sup_{x \in X} d(g(x), x) = \sup_{x \in X} d(g(g^{-1}(x)), g^{-1}(x)) = d(e, g^{-1}(x)) = \nu(g^{-1}),$$

and finally

$$\nu(gh) = \sup_{x \in X} d(gh(x), x) \leq \sup_{x \in X} d(gh(x), h(x)) + \sup_{x \in X} d(h(x), x) = \nu(g) + \nu(h).$$

□

Thus the right-invariant metric on $\text{Homeo}(X)$ is

$$\begin{aligned} d_R(g, h) = \nu(gh^{-1}) &= \sup_{x \in X} d(gh^{-1}(x), x) \\ &= \sup_{x \in X} d(g(x), h(x)) = \|g - h\|_\infty \end{aligned}$$

and the left-invariant metric is $d_L(g, h) = \left\|g^{-1} - h^{-1}\right\|_\infty$.

Let us consider the important case where $X = [0, 1]$ with the usual metric. In this case, $\text{Homeo}[0, 1]$ has two pathwise-connected components:

$$\begin{aligned} \text{Homeo}^+[0, 1] &= \{g : g \text{ strictly increasing}, g(0) = 0, g(1) = 1\} \\ \text{and } \text{Homeo}^-[0, 1] &= \{g : g \text{ strictly decreasing}, g(0) = 1, g(1) = 0\}. \end{aligned}$$

$\text{Homeo}^+[0, 1]$ is a closed pathwise-connected subgroup of $\text{Homeo}[0, 1]$. If $n \in \mathbf{N}$, let

$$h_n(t) = \begin{cases} 2t/n & \text{for } 0 \leq t \leq \frac{1}{2}, \\ (2 - 2/n)t - (1 - 2/n) & \text{for } \frac{1}{2} \leq t \leq 1. \end{cases}$$

Then $(h_n)_{n=1}^\infty$ is a d_R-Cauchy sequence in $\text{Homeo}^+[0, 1]$ which converges in $(C[0, 1], d_\infty)$ to a function not in $\text{Homeo}^+[0, 1]$; thus $(\text{Homeo}^+[0, 1], d_R)$ is not complete. Further $d_L(h_n, h_{2n}) = \frac{1}{4}$, and so the left and right uniformities are different.

On the other hand, let $J = \{(r, s) : r, s \text{ rational}, 0 \leq r < s \leq 1\}$. Then

$$\begin{aligned} \text{Homeo}^+[0, 1] = \{f \in C[0, 1] : |f(0)| < \frac{1}{n}, f(r) < f(s), |f(1) - 1| < \frac{1}{n} \\ \text{for } n \in \mathbf{N}, (r, s) \in J\}, \end{aligned}$$

so that $\text{Homeo}^+[0, 1]$ is a G_δ-subset of $C[0, 1]$. Thus the topological group $\text{Homeo}^+[0, 1]$ is a Polish space. Dieudonné has however shown that there is no left-invariant metric on $\text{Homeo}^+[0, 1]$ under which it is complete.

6
Càdlàg Functions

6.1 Càdlàg Functions

As we shall see in Part II, measure theory and probability theory are greatly simplified when the measures under consideration are Borel measures, or their completions, defined on a Polish space. In fact, most, but not all, of probability theory takes place on a Polish space. For example, one of the most useful settings for stochastic processes is the space $D[0, 1]$ or $D[0, \infty)$ of càdlàg functions. In this chapter, we shall show that $D[0, 1]$ is naturally a Polish space (from which it follows easily that $D[0, \infty)$ is also a Polish space). This is not obvious, and shows that it is not always easy to show that a space is a Polish space; the details, though interesting, are quite technical, and this chapter can therefore be omitted on a first reading.

In the theory of stochastic processes, random functions are considered which need not be continuous, but which may have random jumps. We consider real-valued functions on $[0, 1]$. Such a function f is a *càdlàg function* (continue à droite, limite finie à gauche), or *Skorohod function* if it is continuous on the right at each $x \in [0, 1)$ and if $f(y)$ tends to a finite limit $f_-(x)$ as y increases to x, for each $x \in (0, 1]$. The space $D[0, 1]$ of càdlàg functions is a real vector space under pointwise addition and scalar multiplication, and contains $C[0, 1]$ as a linear subspace.

In order to study càdlàg functions, we need to consider dissections $D = (0 = t_0 < \ldots < t_k = 1)$ of $[0, 1]$. We denote the set of such dissections by $\mathcal{D}$. If f is a real-valued function on $[0, 1]$ and $D \in \mathcal{D}$, we set

$$\omega_{(D)}(f) = \max_{1 \le j \le k} \left(\sup\{|f(s) - f(t)| : t_{j-1} \le s < t < t_j\} \right).$$

Theorem 6.1.1 *A real-valued function f on $[0,1]$ is a càdlàg function if and only if whenever $\epsilon > 0$ there exists a dissection D in $\mathcal{D}$ such that $\omega_{(D)}(f) < \epsilon$.*

Note that the intervals that we consider are closed on the left and open on the right.

Proof Suppose that f is a càdlàg function and that $\epsilon > 0$. Let

$$\mathcal{T} = \{t \in [0,1] : \text{ there exists a suitable dissection of } [0,t]\},$$

and let $T = \sup \mathcal{T}$. Then $T > 0$, since f is continuous on the right at 0. Since f has finite limits on the left, $T \in \mathcal{T}$. Suppose that $T < 1$. Since f is continuous on the right at T, there exists $T < T_0 \leq 1$ such that $|f(x) - f(y)| < \epsilon$ for $x, y \in [T, T_0)$, and so $T_0 \in \mathcal{T}$, giving a contradiction. Thus $T = 1$.

Conversely, if f is not continuous on the right at $t \in [0,1)$, there exists $\epsilon > 0$ such that if $t < s \leq 1$, then there exist $x, y \in [t,s)$ such that $|f(x) - f(y)| \geq \epsilon$, so that $f \notin D[0,1]$. A similar argument applies if f does not have a finite limit on the left at some $x \in (0,1]$. Thus the condition is sufficient. □

Corollary 6.1.2 *If $f \in D[0,1]$ then f is bounded.*

If $f \in D[0,1]$ and $x \in (0,1]$, let $j_f(x) = f(x) - f_-(x)$. $j_f(x)$ is the *jump* of f at x. Thus $\Omega_f(x) = |j_f(x)|$, and f is continuous at x if and only if $j_f(x) = 0$.

Exercise 6.1.3 If $f \in D[0,1]$ and $\epsilon > 0$ show that $\{x \in (0,1] : |j_f(x)| > \epsilon\}$ is finite, deduce that f has only countably many points of discontinuity.

6.2 The Space $(D[0,1], d_\infty)$

The space $D[0,1]$ is a linear subspace of the space $(B[0,1], d_\infty)$ of bounded functions on $[0,1]$, and contains $C[0,1]$ as a linear subspace.

Theorem 6.2.1 *$(D[0,1], d_\infty)$ is complete.*

Proof It is enough to show that $D[0,1]$ is closed in $(B[0,1], d_\infty)$. Suppose that $f \in \overline{D[0,1]}$, that $t \in [0,1]$ and that $\epsilon > 0$. There exists $g \in D[0,1]$ with $d_\infty(f,g) < \epsilon/3$, and there exists $0 < \delta < 1 - t$ such that $|g(s) - g(t)| < \epsilon/3$ for $t \leq s < t + \delta$. If $t \leq s < t + \delta$, then

$$|f(s) - f(t)| \leq |f(s) - g(s)| + |g(s) - g(t)| + |g(t) - f(t)| < \epsilon,$$

so that f is continuous on the right at t. The existence of finite limits on the left is as easy to prove. □

A *step function* g on $[0, 1]$ is a function of the form

$$g = \sum_{j=1}^{k} g_j I_{[t_{j-1},t_j)} + g_{\{1\}} I_{\{1\}},$$

where $D = (0 = t_0 < \ldots < t_k = 1) \in \mathcal{D}$.

Exercise 6.2.2 The step functions are dense in $(D[0, 1], d_\infty)$.

Proof This follows easily from Theorem 6.1.1. □

Thus $D[0, 1]$ can be thought of as the completion of the space of step functions, with the uniform norm.

The uniform norm is however too strong to be useful. If $0 < s < t \leq 1$ then $d_\infty(I_{[0,s)}, I_{[0,t)}) = 1$. Thus $(D[0, 1], d_\infty)$ is not separable, and does not reflect the geometric properties of $[0, 1]$.

6.3 The Skorohod Topology

In this section, we introduce a topology τ_S, the *Skorohod topology*, on $D[0, 1]$. This is weaker than the topology of uniform convergence, but agrees with the topology of uniform convergence on $C[0, 1]$. The topology is defined by constructing a metric d_S, the *Skorohod metric* on $D[0, 1]$; τ_S is the corresponding metric topology. In terms of stochastic processes, the idea behind the construction is to allow small perturbations of the time variable; this is appropriate, since time cannot be measured with complete accuracy.

We consider the group $\mathrm{Homeo}^+[0, 1]$ of increasing homeomorphisms of $[0, 1]$, and its group-norm $\nu_S(h) = \sup\{|h(t) - t| : t \in [0, 1]\}$. If $a > 0$, let

$$V_a = \{h \in \mathrm{Homeo}^+[0, 1] : \nu_S(h) \leq a\}.$$

Let

$$U_a = \{(f, g) \in D[0, 1] \times D[0, 1] : \inf\{d_\infty(f, g \circ h) : h \in V_a\} \leq a\},$$

and let $d_S(f, g) = \inf\{a : (f, g) \in U_a\}$.

Theorem 6.3.1 *d_S is a metric on $D[0, 1]$.*

Proof Suppose that $f, g, k \in D[0, 1]$. Certainly $d_S(f, g) = 0$ if and only if $f = g$. If $a > d_S(f, g)$ there exists $h \in V_a$ such that $d_\infty(f, g \circ h) < a$. Then $h^{-1} \in V_a$, and $d_\infty(g, f \circ h^{-1}) = d_\infty(f, g \circ h) < a$, so that $d_S(g, f) \leq d_S(f, g)$. Exchanging f and g, it follows that $d_S(f, g) = d_S(g, f)$.

Suppose that $a' > d_S(g,k)$, so that there exists $h' \in V_{a'}$ such that $d_\infty(g, k \circ h') < a'$. Then $\nu_S(h'h) \leq \nu_S(h') + \nu_S(h) < a + a'$, and

$$\begin{aligned} d_\infty(f, k \circ (h'h)) &\leq d_\infty(f, g \circ h) + d_\infty(g \circ h, k \circ (h'h)) \\ &= d_\infty(f, g \circ h) + d_\infty(g, k \circ h') < a + a', \end{aligned}$$

so that $d_S(f,k) \leq d_S(f,g) + d_S(g,k)$. □

Clearly $d_S(f,g) \leq d_\infty(f,g)$, so that the identity mapping from $(D[0,1], d_\infty)$ to $(D[0,1], d_S)$ is a 1-Lipschitz mapping. In particular, the step functions are dense in $(D[0,1], \tau_S)$.

Theorem 6.3.2 *Suppose that $(f_n)_{n=1}^\infty$ is a sequence in $D[0,1]$ which converges in the Skorohod topology to f, and that $f \in C[0,1]$. Then f_n converges uniformly to f as $n \to \infty$.*

Proof Suppose that $\epsilon > 0$. Since f is uniformly continuous on $[0,1]$, there exists $0 < \delta < \epsilon/2$ such that if $h \in V_\delta$ then $d_\infty(f, f \circ h) < \epsilon/2$. There exists $N \in \mathbf{N}$ such that if $n \geq N$ then $d_S(f_n, f) < \delta$. If $n \geq N$, there exists $h_n \in V_\delta$ such that $\|f_n - f \circ h_n\|_\infty < \delta$. Thus

$$d_\infty(f_n, f) \leq d_\infty(f_n, f \circ h_n) + d_\infty(f \circ f_n, f) \leq \delta + \epsilon/2 < \epsilon.$$ □

Corollary 6.3.3 *The restriction of τ_S to $C[0,1]$ is the topology of uniform convergence.*

Proposition 6.3.4 *$(D[0,1], \tau_S)$ is separable.*

Proof Suppose that $g = \sum_{j=1}^k g_j I_{[t_{j-1}, t_j)} + g_{\{1\}} I_{\{1\}}$ is a step function, and that $\epsilon > 0$. There exist rational $0 = s_0 < \ldots < s_k = 1$ such that $t_j < s_j < t_{j+1}$ and $s_j - t_j < \epsilon/2$ for $1 \leq j \leq k-1$. There is a piecewise linear $h \in \mathrm{Homeo}^+[0,1]$ such that $h(t_j) = s_j$ for $0 \leq j \leq k$. Then $\nu_S(h) < \epsilon/2$, so that $d_S(g, g \circ h) < \epsilon/2$. There exist rational $f_1, \ldots f_k$ and f_r such that $|f_j - g_j| < \epsilon/2$ for $1 \leq j \leq k$ and $|f_{\{1\}} - g_{\{1\}}| < \epsilon/2$.

Let $f = \sum_{j=1}^k f_j I_{[s_{j-1}, s_j)} + f_{\{1\}} I_{\{1\}}$. Then $d_S(f, g \circ h) \leq d_\infty(f, g \circ h) < \epsilon/2$, and so $d_S(f,g) < \epsilon$. Since the step functions are dense in $(D[0,1], d_S)$, and since there are countably many step functions taking rational values, and with rational points of dissection, $(D[0,1], \tau_S)$ is separable. □

Is $(D[0,1], \tau_S)$ a Polish space? Unfortunately, $(D[0,1], d_S)$ is not complete.

Exercise 6.3.5 Let $f_n = I_{[\frac{1}{2}, \frac{1}{2} + \frac{1}{2n})}$. Show that $d_S(f_m, f_n) = |1/2m - 1/2n|$, so that $(f_n)_{n=1}^\infty$ is a d_S-Cauchy sequence. Show that $(f_n)_{n=1}^\infty$ is not d_S-convergent to any elememt of $D[0,1]$.

We need to introduce another metric.

6.4 The Metric d_B

In this section, we introduce another metric, d_B, on $D[0, 1]$. This is a complete metric equivalent to d_S, so that the Skorohod topology τ_S is the corresponding metric topology: thus $(D[0, 1], \tau_S)$ is a Polish space.

The metric d_S considers elements of $\text{Homeo}^+[0, 1]$ which are uniformly close to the identity. But such homeomorphisms may behave badly locally. We need to consider homeomorphisms with good local behaviour. For example, we could consider differentiable homeomorphisms h for which $\|h' - 1\|_\infty$ is small. But these homeomorphisms do not have good limiting properties, and so we consider a rather larger subgroup of $\text{Homeo}^+[0, 1]$; homeomorphisms for which the slope $(h(t) - h(s))/(t - s)$ of chords is uniformly bounded away from 0. For technical reasons, it is convenient to consider

$$l(h)(s, t) = \log\left(\frac{h(t) - h(s)}{t - s}\right) = \log(h(t) - h(s)) - \log(t - s).$$

We shall need the following elementary inequalities, which we shall use without comment.

Exercise 6.4.1 Show that if $0 < \alpha < \frac{1}{2}$ then

$$\alpha/2 < \log(1 + \alpha) < \alpha$$
$$-2\alpha < \log(1 - \alpha) < -\alpha.$$

If $h \in \text{Homeo}^+[0, 1]$, let

$$\nu_B(h) = \sup\{|l(h)(s, t)| : 0 \leq s < t \leq 1\},$$

and let $G_B = \{h \in \text{Homeo}^+[0, 1] : \nu_B(h) < \infty\}$.

Theorem 6.4.2 *G_B is a subgroup of* $\text{Homeo}^+[0, 1]$, *and ν_B is a group-norm on it.*

Proof $\nu_B(h) = 0$ if and only if $h = e$. If $h \in \text{Homeo}^+[0, 1]$ and $0 \leq s < t \leq 1$ then

$$\begin{aligned}|l(h^{-1})(s, t)| &= |\log(h^{-1}(t) - h^{-1}(s)) - \log(t - s)| \\ &= |\log(t - s) - \log(h(t) - h(s))| = |l(h)(s, t)|,\end{aligned}$$

so that $\nu_B(h^{-1}) = \nu_B(h)$, and $h^{-1} \in G_B$ if and only if $h \in G_B$. If $h, k \in \text{Homeo}^+[0, 1]$ and $0 \leq s < t \leq 1$ then

$$\begin{aligned}l(hk)(s, t) &\leq |\log(hk(t) - hk(s)) - \log(k(t) - k(s))| + \\ &\qquad\qquad |\log(k(t) - k(s)) - \log(t - s)| \\ &= l(h)(k(s), k(t)) + l(k)(s, t)\end{aligned}$$

so that $\nu_B(hk) \le \nu_B(h) + \nu_B(k)$. Thus G_B is a group, and ν_B is a group-norm on it. □

Let ρ_B be the right-invariant metric on G_B defined by ν_B. Thus

$$\begin{aligned}\rho_B(h,k) &= \nu_B(hk^{-1}) \\ &= \sup\{|\log(h(t)-h(s)) - \log(k(t)-k(s))| : 0 \le s < t \le 1\}.\end{aligned}$$

If $\alpha > 0$, let $M_\alpha = \{h \in G_B : \nu_B(h) \le \alpha\}$.

Proposition 6.4.3 *The inclusion* $(G_B, \rho_B) \to (\mathrm{Homeo}^+[0,1], \rho_S)$ *is uniformly continuous; if* $0 < \alpha < \frac{1}{2}$ *and* $h \in M_\alpha$ *then* $h \in V_{2\alpha}$.

Proof Set $s = 0$. If $0 < t \le 1$ then $|\log h(t) - \log t| \le \alpha$, so that $e^{-\alpha}t \le h(t) \le e^{\alpha}t$ and

$$|h(t) - t| \le (e^\alpha - 1)t \le e^\alpha - 1.$$

Thus $\nu_S(h) \le e^\alpha - 1 < 2\alpha$. □

Theorem 6.4.4 *The topological group* (G_B, ρ_B) *is complete.*

Proof It follows from Theorem 2.4.5 that it is sufficient to show that M_α is uniformly closed in $C[0,1]$, for $\alpha \le \frac{1}{2}$. Suppose that $h \in \overline{M_\alpha}$, and that $(h_n)_{n=1}^\infty$ is a sequence in M_α which converges uniformly to h. If $0 \le s < t \le 1$, then $\log(h_n(t)-h_n(s)) \to \log(h(t)-h(s))$ as $n \to \infty$, so that $\sup\{|\log(h(t)-h(s)) - \log(t-s)| : 0 \le s < t \le 1\} \le \alpha$. Consequently $|(h(t)-h(s)) - (t-s)| \le (e^\alpha - 1)(t-s)$, and so $h(t) - h(s) \ge (2 - e^\alpha)(t-s) > 0$; h is strictly increasing and continuous on $[0,1]$. Since $h(0) = 0$ and $h(1) = 1$, it follows that h is a homeomorphism of $[0,1]$ onto itself. Finally $\nu_B(h) \le \alpha$, so that $h \in M_\alpha$. □

On the other hand, (G_B, ρ_B) is not separable. Let $t_n = 1 - 1/2^n$, let $s_n = (t_n + t_{n+1})/2$ and let $b_n = s_n + 1/2^{n+2}$ for $n \in \mathbf{N}$. If $A \in P(\mathbf{N})$ let $f_A(t_n) = t_n$ and let

$$f_A(s_n) = \begin{cases} b_n & \text{if } n \in A \\ s_n & \text{if } n \notin A. \end{cases}$$

Let f_A be defined linearly between these values, and let $f_A(1) = 1$. Then $f_A \in G_B$, and $\rho_B(f_A, f_C) = \log 2$ if $A \ne C$.

We now define a new metric on $D[0,1]$ which is equivalent to d_S. Let

$$W_a = \{(f,g) \in D[0,1] \times D[0,1] : \inf\{\|f - g \circ h\|_\infty : h \in M_a\} \le a\},$$

and let $d_B(f,g) = \inf\{a : (f,g) \in W_a\}$.

Theorem 6.4.5 *d_B is a metric on $D[0,1]$.*

Proof The proof is exactly similar to the proof of Theorem 6.3.1. □

Theorem 6.4.6 *The metric space $(D[0,1], d_B)$ is complete.*

Proof Suppose that $(f_n)_{n=1}^\infty$ is a Cauchy sequence in $D[0,1]$. By extracting a subsequence if necessary, we may suppose that $d_B(f_{n-1}, f_n) < 1/2^n$, for $n > 1$. Thus for each $n > 1$ there exists $h_n \in M_{1/2^n}$ such that $\|f_{n-1} - f_n \circ h_n\|_\infty < 1/2^n$. If $m < n$, let $h_{m,n} = h_{m+1} \circ \cdots \circ h_n$. Then $h_{m,n} \in M_{1/2^m}$, and if $m < n < p$ then

$$\rho_B(h_{m,n}, h_{m,p}) = \nu_B(h_{n,p}) \leq 1/2^n.$$

Thus the sequence $(h_{m,n})_{n=m+1}^\infty$ is a ρ_B-Cauchy sequence, which converges to an element $h_{m,\infty}$.

If $m < n$, let $f_{m,n} = f_m \circ h_{m,n}$. Then

$$d_B(f_{m,n}, f_m) \leq 1/2^m \text{ and } \|f_{m,n} - f_n\|_\infty \leq 1/2^m.$$

Similarly, let $f_{m,\infty} = f_m \circ h_{m,\infty}$. Again, $d_B(f_{m,n}, f_m) \leq 1/2^m$ and

$$d_B(f_{m,\infty}, f_m) \leq 1/2^m \text{ and } \|f_{m,\infty} - f_\infty\|_\infty \leq 1/2^m.$$

Now $f_{m,\infty} = f_m \circ h_{m,n} \circ h_{n,\infty}$, so that

$$f_{m,\infty} - f_{n,\infty} = (f_m \circ h_{m,n} - f_n) \circ h_{n,\infty}.$$

Consequently

$$d_\infty(f_{m,\infty}, f_{n,\infty}) = d(f_m \circ h_{m,n}, f_n) \leq 1/2^m.$$

Thus the sequence $(f_{m,\infty})_{m=1}^\infty$ is a uniform Cauchy sequence, which by Theorem 6.2.1 converges uniformly to an element f_∞ of $D[0,1]$. Further, $d_B(f_{m,\infty}, f_\infty) \leq 1/2^m$, so that

$$d_B(f_m, f_\infty) \leq d_B(f_m, f_{m,\infty}) + d_B(f_{m,\infty}, f_\infty) \leq 2/2^m.$$

Thus $f_m \to f_\infty$ as $m \to \infty$. □

Theorem 6.4.7 *The metrics d_S and d_B on $D[0,1]$ are equivalent.*

Proof First we show that the identity mapping $(D[0,1], d_B) \to (D[0,1], d_S)$ is uniformly continuous. Suppose that $0 < \alpha < \frac{1}{2}$. If $d_B(f,g) < \alpha$ there exists $h \in M_\alpha$ such that $d_\infty(f, g \circ h) < \alpha$. By Proposition 6.4.3, $\nu_s(h) < 2\alpha$, and so $d_S(f,g) < 2\alpha$.

The identity mapping $(D[0,1], d_S) \to (D[0,1], d_B)$ cannot be uniformly continuous, since $(D[0,1], d_B)$ is complete, and $(D[0,1], d_s)$ is not. We show that it is continuous. Suppose that $f \in D[0,1]$ and that $\epsilon > 0$.

By Theorem 6.1.1, there exists a dissection $D = (0 = t_0 < \cdots < t_k = 1)$ such that $\omega_{(D)} < \epsilon/2$. Let $\delta = \min_{1 \le j \le k}(t_j - t_{j-1})$, and let $\eta = \delta\epsilon/4$. Suppose that $d_S(f, g) < \eta$, so that there exists $h \in \text{Homeo}^+[0, 1]$ with $\nu_S(h) < \eta$ for which $\|f - g \circ h\|_\infty < \eta$.

Now let $k(t_j) = h(t_j)$ for $1 \le j \le k$, and let k be linear between these values. Thus $h^{-1}k([t_{j-1}, t_j)) = [t_{j-1}, t_j)$, so that

$$
\begin{aligned}
|f(t) - g \circ k(t)| &\le |f(t) - f \circ h^{-1}k(t)| + |f \circ h^{-1}k(t) - g \circ k(t)| \\
&\le \omega_{(D)}(f) + d_S(f, g) \le \epsilon/2 + \eta < \epsilon,
\end{aligned}
$$

and so $\|f - g)\|_\infty < \epsilon$. Further, if $1 \le j \le k$ then

$$
\begin{aligned}
|(k(t_j) - k(t_{j-1})) - (t_j - t_{j-1})| &= |(h(t_j) - h(t_{j-1})) - (t_j - t_{j-1})| \\
&\le \eta \le \epsilon(t_j - t_{j-1})
\end{aligned}
$$

so that, since k is piecewise linear, $|(k(t) - k(s)) - (t - s)| \le \epsilon(t - s)/4$ for $0 < s < t < 1$, and so $|l(k)(s, t)| < \epsilon/2$. Thus $k \in M_\epsilon$, and so $d_B(f, g) \le \epsilon$. Thus the identity mapping $(D[0, 1], d_S) \to (D[0, 1], d_B)$ is continuous at f. □

Corollary 6.4.8 *$(D[0, 1], \tau_S)$ is a Polish space.*

7
Banach Spaces

7.1 Normed Spaces and Banach Spaces

Many of the metric spaces that we shall consider are real or complex vector spaces, or subsets of such spaces. Let us denote the underlying field **R** or **C** by K.

Suppose that E is a real or complex vector space. It is natural to consider metrics d on E which are

(i) *translation invariant* ($d(x + a, y + a) = d(x, y)$ for all $x, y, a \in E$);
(ii) *scaling homogeneous* ($d(\lambda x, \lambda y) = |\lambda| d(x, y)$ for all $x, y \in E$) and $\lambda \in K$.

Note that

(iii) d is *inversion invariant*; $d(-x, -y) = d(x, y)$ for all $x, y \in E$.

Suppose that d is a metric on E with these properties. If $x \in E$, let $\|x\| = d(x, 0)$. Then $d(x, y) = d(x - y, 0) = \|x - y\|$, so that $\|.\|$ determines d. The function $\|.\|$ then has the following properties:

(a) $\|x + y\| \leq \|x\| + \|y\|$, for $x, y \in E$ (*subadditivity*);
(b) $\|\lambda x\| = |\lambda|\,\|x\|$, for $\lambda \in K$ and $x \in E$ (*positive homogeneity*);
(c) $\|x\| = \|-x\|$ for $x \in E$ (*symmetry*); and
(d) $\|x\| = 0$ if and only if $x = 0$.

(a) follows, since

$$\|x + y\| = d(x + y, 0) \leq d(x + y, y) + d(y, 0) = d(x, 0) + d(y, 0) = \|x\| + \|y\|.$$

(b) is a consequence of (ii) and (iii), and (c) follows, since $d(x, 0) = 0$ if and only if $x = y$.

A function $\|.\|$ on E which satisfies (a), (b), (c) and (d) is called a *norm*, and $(E, \|.\|)$ is called a *normed space*. A function which satisfies (a), (b) and (c) is called a *seminorm*.

If $(E, \|.\|)$ is a normed space, then the function $d(x, y) = \|x - y\|$ is a metric on E which satisfies (i), (ii) and (iii), the metric *defined by* the norm $\|.\|$. The mappings $(x, y) \to x+y$ from $E \times E \to E$ and $(\lambda, x) \to \lambda x$ from $\mathbf{R} \times E$ to E are jointly continuous, and the mapping $x \to -x$ is an isometry of E onto itself.

A subset A of a normed space $(E, \|.\|)$ is *bounded in norm* if $\sup\{\|a\| : a \in A\} < \infty$. Since

$$\sup\{\|a\| : a \in A\} \leq \operatorname{diam}(A) \leq 2 \sup\{\|a\| : a \in A\},$$

A is bounded in norm if and only if it is bounded in the metric d. The *open unit ball* $U(E)$ is the set $U(E) = N_1(0) = \{x : \|x\| < 1\}$ and the *closed unit ball* $B(E)$ is the set $B(E) = M_1(E) = \{x : \|x\| \leq 1\}$. A set A is bounded if there exists $\lambda > 0$ such that $\lambda A \subseteq B(E)$.

A *Banach space* is a normed space which is complete under the metric defined by the norm.

Exercise 7.1.1 Suppose that $(E, \|.\|)$ is a normed space. Use the proof of Proposition 2.5.3 to show that $Ca(E)$ is a vector space, that if $p(x) = \phi(x, 0)$ then p is a seminorm on $Ca(E)$ which determines ϕ, and that $x \sim y$ if and only if $x - y \in N = \{z : z_n \to 0 \text{ as } n \to \infty\}$. Conclude that N is a linear subspace of $Ca(E)$ and q is the linear quotient mapping of $Ca(E)$ onto $\hat{E} = Ca(E)/N$; $\hat{E}$ is a vector space. Further, there is a norm $\|.\|^\wedge$ on $\hat{E}$ such that $\hat{d}(u, v) = \|u - v\|^\wedge$ and $(\hat{E}, \|.\|^\wedge)$ is a Banach space. Finally, j is a linear isometry of $(E, \|.\|)$ onto a dense subspace of $(\hat{E}\ \|.\|^\wedge)$: the completion of $(E, \|.\|)$ is a Banach space.

If a normed space is topologically complete, it must be a Banach space.

Theorem 7.1.2 (Klee) *Suppose that $(E, \|.\|)$ is a topologically complete normed space. Then $(E, \|.\|)$ is a Banach space.*

Proof Let $(\hat{E}, \|.\|\hat{\ })$ be the completion of $(E, \|.\|)$. By Alexandroff's theorem (Theorem 2.6.2), E is the intersection of a sequence $(O_n)_{n=1}^\infty$ of open dense subsets of $\hat{E}$. Suppose that $E \neq \hat{E}$, so that there exists $\hat{x} \in \hat{E} \setminus E$. Then $(O_n + \hat{x})_{n=1}^\infty$ is a sequence of open dense subsets of $\hat{E}$, and

$$(\cap_{n=1}^\infty O_n) \cap (\cap_{n=1}^\infty (O_n + \hat{x})) = E \cap (E + \hat{x}) = \emptyset,$$

contradicting Baire's category theorem. □

Let us give a few examples of Banach spaces that we shall need.

(i) If S is a non-empty set and $(E, \|.\|)$ is a normed space, then the space $B(S, E)$ of bounded functions taking values in E is a vector space with the operations of co-ordinate addition and scalar multiplication, and the function $\|f\|_\infty = \sup\{\|f(s)\| : s \in S\}$ is a norm on it, which defines the metric d_∞. It is a Banach space if and only if $(E, \|.\|)$ is complete (Proposition 2.3.1). In particular the spaces $B(S)$ and l_∞ are Banach spaces.

(ii) If (X, τ) is a Hausdorff topological space and $(E, \|.\|)$ is a normed space, then the space $C_b(X, E)$ of bounded continuous functions on X taking values in E is a closed linear subspace of $B(X, E)$; we give $C_b(X, E)$ the norm $\|.\|_\infty$. Then $(C_b(X, E), \|.\|_\infty)$ is a Banach space if and only if E is. In particular $(C_b(X), \|.\|_\infty)$ is a closed linear subspace of $(B(X), \|.\|_\infty)$, and is a Banach space. In particular, if (K, τ) is a compact Hausdorff space then $C(K) = C_b(K)$ is Banach space. Note that if $f \in C(K)$ then there exists $x \in K$ such that $|f(x)| = \|f\|$.

Exercise 7.1.3 Let $c = \{x \in K^{\mathbf{N}} : x_n \text{ tends to a limit as } n \to \infty\}$. If $x \in c$, let $\|x\|_\infty = \sup_{n=1}^\infty |x_n|$. Show that c is a linear subspace of $K^{\mathbf{N}}$, that $\|.\|_\infty$ is a norm on c and that $(c, \|.\|_\infty)$ is a Banach space. Let $c_0 = \{x \in c : x_n \to 0$ as $n \to \infty\}$. Show that c_0 is a closed hyperplane in c.

(iii) Suppose that (X, τ) is a compact topological space and that $(E, \|.\|)$ is a normed space. If f is a continuous function on X taking values in E, then $f(X)$ is a norm bounded subset of E, and so $C(X, E) = C_b(X, E)$. Again, we give $C(X, E)$ the norm $\|.\|_\infty$.

Exercise 7.1.4 Let $l^1 = \{x \in K^{\mathbf{N}} : \|x\|_1 = \sum_{i=1}^\infty |x_i| < \infty\}$. Show that l^1 is a linear subspace of $K^{\mathbf{N}}$, that $\|.\|_1$ is a norm on l^1 and that $(l^1, \|.\|_1)$ is a Banach space.

Exercise 7.1.5 If $x \in K^n$, let $\|x\|_1 = \sum_{i=1}^n |x_i|$. Show that $(K^n, \|.\|_1)$ is a locally compact Banach space.

Exercise 7.1.6 Show that any two norms on a finite-dimensional space E are Lipschitz equivalent, and make E a complete locally compact space.

Exercise 7.1.7 Suppose that F is a closed linear subspace of a normed space $(E, \|.\|)$, and that $q : E \to E/F$ is the quotient mapping. Let $\|q(x)\|_q = \inf\{\|x + f\| : f \in F\}$. Show that $\|.\|_q$ is a norm on E/F and that $q : (E, \|.\|) \to (E/F, \|.\|_q)$ is continuous. Show that if $(E, \|.\|)$ is complete, so is $(E/F, \|.\|_q)$. Show that if $(F, \|.\|)$ and $(E/F, \|.\|_q)$ are complete, then so is $(E, \|.\|)$.

Exercise 7.1.8 Show that a finite-dimensional subspace of a normed space is closed. Show that if F is a closed linear subspace of a normed space $(E, \|.\|)$ and D is a finite-dimensional subspace of E then $F + D$ is closed.

Exercise 7.1.9 Suppose that $(E, \|.\|)$ is a locally compact normed space with closed unit ball B_E. Show that there is a finite set F such that $B_E \subseteq F + \frac{1}{2}B_E$. Let $G = \operatorname{span}(F)$. Show inductively that $B_E \subseteq G + B_E/2^n$, and deduce that $E = G$, so that E is finite-dimensional.

Exercise 7.1.10 If (X, d) is a metric space, show that the space $U_b(X, E)$ of bounded uniformly continuous functions on X taking values in a normed space

$(E, \|.\|)$ is a closed linear subspace of $C_b(X, E)$. Show that $(U_b(X, E), \|.\|_\infty)$ is a Banach space if and only if E is.

Exercise 7.1.11 Let

$$C^{(1)}([0,1]) = \{f \in C([0,1] : f \text{ is continuously differentiable}\}.$$

Show that $C^{(1)}([0,1])$ is a meagre subset of $(C([0,1]), \|.\|_\infty)$. Show that $\|f\|^{(1)} = \sup_{x\in[0,1]}(|f(x)| + |f'(x)|)$ is a norm on $C^{(1)}([0,1])$ under which $C^{(1)}([0,1])$ is a Banach space.

7.2 The Space *BL(X)* of Bounded Lipschitz Functions

Spaces of Lipschitz functions provide further examples of Banach spaces. Suppose that (X, d) is a metric space. The space $L(X)$ of Lipschitz functions on X is a linear subspace of $C(X)$ and the space $BL(X)$ of bounded Lipschitz functions on X is a linear subspace of $C_b(X)$. If $f \in BL(X)$, we set $\|f\|_{BL} = p_L(f) + \|f\|_\infty$.

Theorem 7.2.1 *Suppose that (X, d) is a metric space. p_L is a seminorm on $L(X)$. $\|.\|_{BL}$ is a norm on $BL(X)$, and its unit ball B_{BL} is closed in $(C_b(X), \|.\|_\infty)$. $(BL(X), \|.\|_{BL})$ is a Banach space. Further, $BL(X)$ is a Banach algebra under pointwise multiplication: if $f, g \in BL(X)$, then $fg \in BL(X)$, and $\|fg\|_{BL} \leq \|f\|_{BL} \cdot \|g\|_{BL}$.*

Proof It is clear that p_L is a seminorm and that $\|.\|_{BL}$ is a norm on $BL(X)$.

Suppose that $(f_n)_{n=1}^\infty$ is a sequence in B_{BL} which converges uniformly to $f \in C_b(X)$. Then $p_L(f) \leq 1$, by Corollary 2.8.3, and so $f \in BL(X)$. Since $\|f_n\|_\infty \to \|f\|_\infty$ as $n \to \infty$, given $\epsilon > 0$ there exist n_0 such that $\|f_n\|_\infty > \|f\|_\infty - \epsilon$, for $n \geq n_0$. Hence $p_L(f_n) < 1 - \|f\|_\infty + \epsilon$ for $n \geq n_0$. By Corollary 2.8.3, $p_L(f) \leq 1 - \|f\|_\infty + \epsilon$, and so $\|f\|_{BL} \leq 1 + \epsilon$. Since ϵ is arbitrary, $f \in B_{BL}$. Thus B_{BL} is closed in $(C_b(X), \|.\|_\infty)$.

Completeness follows from Theorem 2.4.5.

If $f, g \in BL(X)$ and $x, y \in X$, then

$$\begin{aligned} |f(x)g(x) - f(y)g(y)| &\leq |f(x)(g(x) - g(y))| + |(f(x) - f(y))g(y)| \\ &\leq (\|f\|_\infty \cdot \|g\|_L + \|f\|_L \cdot \|g\|_\infty)d(x, y), \end{aligned}$$

so that $fg \in BL(X)$ and $\|fg\|_L \leq \|f\|_\infty \cdot \|g\|_L + \|f\|_L \cdot \|g\|_\infty$. Thus

$$\begin{aligned} \|fg\|_{BL} &\leq \|f\|_\infty \cdot \|g\|_\infty + \|f\|_\infty \cdot \|g\|_L + \|f\|_L \cdot \|g\|_\infty \\ &\leq \|f\|_{BL} \cdot \|g\|_{BL} \cdot \end{aligned}$$

□

Proposition 7.2.2 *Suppose that (X, d) is compact. Then the unit ball $B_{BL}(X)$ is a compact subset of $(C_b(X), \|.\|_\infty)$.*

Proof For B_{BL} is a bounded equicontinuous subset of $C_b(X)$, and so is a totally bounded subset of $(C_b(X), \|.\|_\infty)$, by the Arzelà–Ascoli theorem. It is also complete, and so it is compact. □

Theorem 7.2.3 *Let (X, d) be a compact metric space. There exists a countable subset G of $BL(X)$, with $\sup_{g \in G} p_L(g) \leq 1$, which is dense in $C(X)$.*

Proof For each $n \in \mathbf{N}$, there exists a finite $1/n$-net A_n in X. Let G_n be the countable set of rational-valued functions on A_n with $p_L(g) \leq 1$, and let $G = \cup_{n=1}^\infty G_n$. By the McShane–Whitney extension theorem, for each $g \in G_n$ there exists a Lipschitz function h_g on X with $p_L(h_g) = p_L(g) \leq 1$ which extends g. We show that the set $H = \{h_g : g \in G_n, n \in \mathbf{N}\}$ is $\|.\|_\infty$-dense in $C(X)$.

Suppose then that $f \in C(X)$ and that $\epsilon > 0$. Then f is uniformly continuous, and so there exists $n > 3/\epsilon$ such that if $d(x, y) < 1/n$ then $|f(x) - f(y)| < \epsilon/3$. There exists $g \in G_n$ such that $\max\{|g(a) - f(a)| : a \in A_n\} < \epsilon/3$. Suppose that $x \in X$. There exists $a \in A_n$ with $d(x, a) < 1/n$. Then

$$\begin{aligned} |f(x) - h_g(x)| &\leq |f(x) - f(a)| + |f(a) - h_g(a)| + |h_g(a) - h_g(x)| \\ &< \epsilon/3 + \epsilon/3 + 1/n < \epsilon. \end{aligned}$$

□

On the other hand, if (X, d) is an infinite compact metric space, then $(BL(X), \|.\|_{BL})$ is not separable. There exists a sequence $(x_n)_{n=1}^\infty$ of distinct points which converges to x_∞, say. By picking a subsequence, we can suppose that $d(x_n, x_\infty)$ is a decreasing sequence and that $d(x_n, x_m) > \frac{1}{2}d(x_n, x_\infty)$ for $m > n$. Let $Y = \{x_n : n \in \mathbf{N}\}$. Suppose that $A \subseteq \mathbf{N}$. Let $f_A(x_{2n}) = d(x_n, x_\infty)$ if $n \in A$, let $f_A(x_{2n}) = -d(x_n, x_\infty)$ if $n \notin A$, and let $f_A(x_{2n+1}) = 0$. Then $f_A \in BL(Y)$, and by the McShane–Whitney extension theorem (Theorem 2.8.6), we can extend each f_A to a Lipschitz function g_A on X. Then $\|g_A - g_B\|_{BL} \geq \|f_A - f_B\|_{BL} \geq 1$ if $A \neq B$, and so $(BL(X), \|.\|_{BL})$ is not separable.

Exercise 7.2.4 Suppose that (X, d) is an infinite metric space. Show that $BL(X)$ is a meagre subspace of $C_b(X)$.

7.3 Introduction to Convexity

If $(E, \|.\|)$ is a normed space, then the function $\|.\|$ is a convex function on E, and the closed unit ball $B(E)$ and the open unit ball $U(E)$ are convex sets.

These simple facts have a profound effect on analysis. Here we begin the study of convexity; we shall return to it several times.

A subset C of a real or complex vector space is *convex* if $(1-\lambda)x+\lambda y \in C$ whenever $x, y \in C$ and $0 \leq \lambda \leq 1$.

The intersection of a set of convex sets is convex. If S is any subset of a vector space E, then $\Gamma(S) = \cap\{C : C \text{ convex}, S \subseteq C\}$ is a convex set, the smallest convex set containing S: $\Gamma(S)$ is the *convex cover* of S.

Exercise 7.3.1 Show that if S is a non-empty subset of a vector space E then

$$\Gamma(S) = \left\{ \alpha_1 s_1 + \cdots + \alpha_n s_n : n \in \mathbf{N}, s_j \in S, \alpha_j \geq 0, \sum_{j=1}^{n} \alpha_j = 1 \right\}.$$

A convex set C is *absolutely convex* if $\alpha C \subseteq C$ for all $|\alpha| \leq 1$. If $(E, \|.\|)$ is a normed space, then the closed unit ball $B(E)$ and the open unit ball $U(E)$ are absolutely convex sets.

If C is a convex set and $\alpha \in K$ then αC is convex. If A is any subset of E, and $\alpha, \beta > 0$, then $(\alpha+\beta)A \subseteq \alpha A + \beta A$.

Exercise 7.3.2 Show that if C is a convex set and $\alpha, \beta > 0$ then $(\alpha+\beta)C = \alpha C + \beta C$.

We now turn to convex functions. A function f on a vector space E taking values in $(-\infty, \infty]$ is a *convex function* if

$$f((1-\lambda)x+\lambda y) \leq (1-\lambda)f(x) + \lambda f(y)$$

whenever $x, y \in C$ and $0 < \lambda < 1$.

Why do we allow the value ∞?

Proposition 7.3.3 *Suppose that F is a set of convex functions on a convex subset C of a real vector space E. Then $g = \sup\{f : f \in F\}$ is a convex function.*

Proof Suppose that $x, y \in C$ and that $0 < \lambda < 1$. If $t < g((1-\lambda)x+\lambda y)$, there exists $f \in F$ with $t < f((1-\lambda)x+\lambda y)$. Then

$$\begin{aligned} t < f((1-\lambda)x+\lambda y) &\leq (1-\lambda)f(x) + \lambda f(y) \\ &\leq (1-\lambda)g(x) + \lambda g(y). \end{aligned}$$

Since this holds for all $t < g((1-\lambda)x+\lambda y)$,

$$g((1-\lambda)x+\lambda y) \leq (1-\lambda)g(x) + \lambda g(y).$$

□

Since the supremum of finite-valued functions can be infinite, this explains why the value ∞ is allowed. Recall that if f is a function taking values in $[-\infty, \infty]$, its *effective domain* Φ_f is the set where f is finite, and that the function f is *proper* if its effective domain is non-empty.

Exercise 7.3.4 Show that the effective domain of a proper convex function is convex.

If f is a convex function on a convex subset C of a vector space E – that is, $f((1-\lambda)x+\lambda y) \leq (1-\lambda)f(x)+\lambda f(y)$ whenever $x, y \in C$ and $0 < \lambda < 1$ - it can be extended to a convex function on E by setting $f(x) = \infty$ for $x \in E \setminus C$. For example, if C is a convex subset of E, we define the function 0_C by setting

$$0_C(x) = 0 \text{ for } x \in C, \text{ and } 0_C(x) = \infty \text{ for } x \notin C.$$

A function on E taking values in $[-\infty, \infty)$ is *concave* if $-f$ is convex, and a real-valued function on E is *affine* if it is both convex and concave: that is, $f((1-\lambda)x+\lambda y) = (1-\lambda)f(x)+\lambda f(y)$ for $x, y \in E$ and $0 \leq \lambda \leq 1$. f is affine if and only if $f = \phi + c$, where $c = f(0)$ and ϕ is a linear functional on E.

Exercise 7.3.5 Suppose that f is a convex function on a convex subset C of a real vector space E. Show that the sets $\{x \in C : f(x) \leq \lambda\}$ and $\{x \in C : f(x) < \lambda\}$ are convex, for each $\lambda \in (-\infty, \infty]$.

A function p from a vector space E to $(-\infty, \infty]$ which satisfies

(a) $p(x+y) \leq p(x)+p(y)$, for $x, y \in E$ (*subadditivity*);
(b) $p(\lambda x) = \lambda p(x)$, for $\lambda > 0$ and $x \in E$ (*positive homogeneity*);
(c) $p(0) = 0$,

is called an *extended sublinear functional*. If it is real-valued, it is called a *sublinear functional*. A seminorm is an example of a sublinear functional. Extended sublinear functionals are proper convex functions. If p is an extended sublinear functional, then $p(x)+p(-x) \geq p(0) = 0$. If p is a seminorm, this implies that $p(x) \geq 0$ for all $x \in E$. This need not be the case when p is a sublinear functional; for example, a function p on $\mathbf{R}$ is a sublinear functional if and only if $p(x) = \alpha x$ for $x \leq 0$ and $p(x) = \beta x$ for $x \geq 0$, where $\alpha \leq \beta$.

Suppose that C is a convex subset of a real vector space E, and that $0 \in C$. If $x \in E$, then $\mathcal{I}_C(x) = \{\lambda \geq 0 : \lambda x \in C\}$ is an interval in $[0, \infty)$. If $\mathcal{I}_C(x) = \{0\}$, set $p_C(x) = \infty$, and if $\mathcal{I}_C(x) = [0, \infty)$ set $p_C(x) = 0$. Otherwise let $p_C(x) = \inf\{\alpha : \alpha > 0, x \in \alpha C\}$.

Suppose that $x, y \in E$. If $p_C(x) = \infty$ or $p_C(y) = \infty$, then trivially $p_C(x+y) \leq p_C(x)+p_C(y)$. Otherwise, if $\alpha > p_C(x)$ and $\beta > p_C(y)$, then $x \in \alpha C$ and $y \in \beta C$, so that $x+y \in \alpha C + \beta C = (\alpha+\beta)C$. Thus $p_C(x+y) \leq \alpha+\beta$, and so $p_C(x+y) \leq p_C(x) + p_C(y)$. Clearly, $p_C(\lambda x) = \lambda p_C(x)$ for $\lambda > 0$, and so p_C is an extended sublinear functional on E; it is the *gauge*, or *Minkowski functional* of C.

Exercise 7.3.6 Suppose that C is a convex subset of a vector space E, and that $0 \in C$. Show that

$$\{x : p_C(x) < 1\} \subseteq C \subseteq \{x : p_C(x) \leq 1\}.$$

We shall consider convex functions further in Chapters 10 and 12.

7.4 Convex Sets in a Normed Space

We now consider convex sets in a normed space.

Proposition 7.4.1 *Suppose that A is a convex subset of a normed space $(E, \|.\|)$. Then $\overline{A}$ and A^{int} are convex sets.*

Proof Suppose that $x, y \in \overline{A}$ and that $\epsilon > 0$. Let $z = (1-\lambda)x + \lambda y$. There exist $a, b \in A$ such that $\|x - a\| < \epsilon$ and $\|y - b\| < \epsilon$. Then $c = (1-\lambda)a + \lambda c \in A$ and

$$\begin{aligned} \|z - c\| &= \|(1-\lambda)(x-a) + \lambda(y-b)\| \leq (1-\lambda)\|x-a\| + \lambda\|y-b\| \\ &< (1-\lambda)\epsilon + \lambda\epsilon = \epsilon, \end{aligned}$$

so that $z \in \overline{A}$.

Suppose that $a, b \in A^{int}$, and that $0 \leq \lambda \leq 1$. Let $c = (1-\lambda)a + \lambda b$. There exists $\delta > 0$ such that if $\|x - a\| < \delta$ then $x \in A$ and if $\|y - b\| < \delta$ then $y \in A$. Suppose that $\|z - c\| < \delta$. Then $a + (z - c) \in A$ and $b + (z - c) \in A$, so that

$$z = (z - c) + (1-\lambda)a + \lambda b = (1-\lambda)(a + (z-c)) + \lambda(b + (z-c)) \in A;$$

thus $c \in A^{int}$. □

If S is any subset of a normed space $(E, \|.\|)$, then

$$\overline{\Gamma}(S) = \cap\{C : C \text{ convex and closed}, S \subseteq C\}$$

is a closed convex set, the smallest closed convex set containing S: $\overline{\Gamma}(S)$ is the *closed convex cover* of S.

Exercise 7.4.2 Show that, if S is any subset of a normed space $(E, \|.\|)$, then $\overline{\Gamma}(S) = \overline{\Gamma(S)}$.

Proposition 7.4.3 *If S is a totally bounded subset of a normed space $(E, \|.\|)$, then $\overline{\Gamma}(S)$ is totally bounded.*

Proof S is norm bounded: let $M = \sup\{\|x\| : x \in S\}$. Suppose that $\epsilon > 0$. There exist a finite $\epsilon/2$-net F in S. Let $\eta = \epsilon/2(M+1)|F|$. There exists a finite η-net B in $[0,1]$. Let G be the finite set $\{\sum_{y\in F}\beta_y y : \beta_y \in B\}$. We show that $\Gamma(S) \subseteq \cup_{g\in G}N_\epsilon(g)$. Suppose that $u = \sum_{j=1}^n \alpha_j x_j \in \Gamma(S)$, with $\alpha_j \geq 0$ and $\sum_{j=1}^n \alpha_j = 1$. For each j there exists $y_j \in F$ such that $\|x_j - y_j\| < \epsilon/2$. Let $v = \sum_{j=1}^n \alpha_j y_j$. Then $v \in \Gamma(S)$, and $\|u - v\| < \epsilon/2$. Gathering terms together, we can write $v = \sum_{y\in F}\gamma_y y$, where $\gamma_y = \sum\{\alpha_j : y_j = y\}$; thus $\gamma_y \geq 0$ and $\sum_{y\in F}\gamma_y = 1$. For each y there exists $\beta_y \in B$ with $|\beta_y - \gamma_y| \leq \eta$. Then $\|u - v\| \leq \sum_{y\in F}|\beta_y - \gamma_y|.\,\|y\| < \epsilon/2$. Thus $\Gamma(S) \subseteq \cup_{g\in G}N_\epsilon(g)$, and so $\Gamma(S)$ is totally bounded. Since the closure of a totally bounded set is totally bounded, $\overline{\Gamma}(S)$ is totally bounded. □

Exercise 7.4.4 Show that if S is a compact subset of a Banach space $(E, \|.\|)$ then $\overline{\Gamma}(S)$ is compact.Give an example of a compact subset S in a normed space $(E, \|.\|)$ for which $\overline{\Gamma}(S)$ is not compact.

Proposition 7.4.5 *Suppose that C is a closed convex subset of a normed space $(E, \|.\|)$, and that $0 \in C$. Then $C = \{x \in E : p_C(X) \leq 1\}$ (where p_C is the gauge of C).*

Proof By Exercise 7.3.6, $C \subseteq \{x \in E : p_C(X) \leq 1\}$. On the other hand, if $x \notin C$, then $x \neq 0$ and there exists $\epsilon > 0$ such that $(x + \epsilon B(E)) \cap C$ is empty. Thus $x \notin (1 - \epsilon/\|x\|)^{-1}C$, so that $p_C(x) > 1$. □

Proposition 7.4.6 *Suppose that p is a sublinear functional which is bounded above in a neighbourhood of 0. Then p is a Lipschitz function on E.*

Proof There exist $\epsilon > 0$ and $M > 0$ such that if $\|h\| \leq \epsilon$, then $p(h) \leq M$. Suppose that $x, y \in E$, and that $x \neq y$. Let $x' = \epsilon x/\|x - y\|$ and let $y' = \epsilon y/\|x - y\|$. Then $\|x' - y'\| = \epsilon$, so that

$$\begin{aligned} p(x') - p(y') &\leq p(x' - y') \leq M, \\ \text{and } p(y') - p(x') &\leq p(y' - x') \leq M, \end{aligned}$$

so that $|p(x') - p(y')| \leq M$. By positive homogeneity,

$$|p(x) - p(y)| = \frac{\|x - y\|}{\epsilon}|p(x') - p(y')| \leq \frac{M}{\epsilon}\|x - y\|.$$

Thus p is an (M/ϵ)-Lipschitz function on E. □

Proposition 7.4.7 *Suppose that C is an open convex subset of a normed space $(E, \|.\|)$ and that $0 \in C$. Then p_C is real-valued and continuous on E, $C = \{x \in E : p_C(x) < 1\}$ and $\overline{C} = \{x \in E : p_C(x) \leq 1\}$.*

Proof There exists $\epsilon > 0$ such that $\epsilon B(E) \subseteq C$. If $x \in E$, then $\epsilon x/(\|x\| + 1) \in C$, and so $p_C(x) \leq (\|x\| + 1)/\epsilon$, and p_C is finite-valued. Further, $p_C(x) \leq 1/\epsilon$ for $x \in B(E)$, and so p_C is continuous, by Proposition 7.4.6.

$\{x \in E : p_C(x) < 1\} \subseteq C$, by Exercise 7.3.6. If $x \in C \setminus \{0\}$, there exists $\epsilon > 0$ such that $x + \epsilon B(E) \subseteq C$. In particular, $(1 + \epsilon/\|x\|)x \in C$, so that $p_C(x) < 1$. Consequently, $C \subseteq \{x \in E : p_C(x) < 1\}$.

Since p_C is continuous, $\{x \in E : p_C(x) \leq 1\}$ is closed. Since $C \subseteq \{x \in E : p_C(x) \leq 1\}$, by Exercise 7.3.6, $\overline{C} \subseteq \{x \in E : p_C(x) \leq 1\}$. On the other hand, if $p_C(x) \leq 1$ then $p_C((1 - 1/n)x) < 1$, so that $(1 - 1/n)x \in C$. Since $(1 - 1/n)x \to x$ as $n \to \infty$, $x \in \overline{C}$. Thus $\{x \in E : p_C(x) \leq 1\} \subseteq \overline{C}$. □

Exercise 7.4.8 If A is a convex subset of a normed space $(E, \|.\|)$ and $A^{int} \neq \emptyset$ then $\overline{A} = \overline{A^{int}}$.

A subset B of a normed space $(E, \|.\|)$ is a *convex body* if it is convex, norm bounded and has a non-empty interior. If $b_0 \in B^{int}$ and $B_0 = B - b_0$, then p_{B_0} is a non-negative sublinear functional, and by positive homogeneity there exist positive constants m and M such that

$$m\|x\| \leq p_{B_0}(x) \leq M\|x\|, \text{ for } x \in E.$$

Exercise 7.4.9 A convex body in a normed space $(E, \|.\|)$ is *symmetric* if it is absolutely convex. Show that if B is a symmetric convex body then p_B is a norm, uniformly equivalent to $\|.\|$.

7.5 Linear Operators

We now return to the study of normed spaces. It is natural to consider continuous linear mappings between normed spaces. We prove some standard results.

Theorem 7.5.1 *Suppose that $(E_1, \|.\|_1)$ and $(E_2, \|.\|_2)$ are normed spaces and that T is a linear mapping from E_1 to E_2. The following are equivalent:*

(i) $K = \sup\{\|Tx\|_2 : \|x\|_1 \leq 1\} < \infty$;
(ii) there exists $C \in \mathbf{R}$ such that $\|Tx\|_2 \leq C\|x\|_1$, for all x in E_1;
(iii) T is Lipschitz;
(iv) T is uniformly continuous on E_1;
(v) T is continuous on E_1;
(vi) T is continuous at 0.

Proof (i) implies (ii): (ii) is trivially satisfied if $x = 0$. Otherwise, let $x_1 = x/\|x\|_1$. Then

$$\|T(x)\|_2 = \|T(\|x\|_1 x_1)\|_2 = \|\|x\|_1 T(x_1)\|_2 = \|x\|_1 \|T(x_1)\|_2 \leq K\|x\|_1.$$

(ii) implies (iii): $\|T(x_1) - T(x_2)\|_2 = \|T(x_1 - x_2)\|_2 \leq C\,\|x_1 - x_2\|_1$.
Obviously (iii) implies (iv), (iv) implies (v) and (v) implies (vi).
(vi) implies (i): there exists $\delta > 0$ such that if $\|x\|_1 \leq \delta$ then $\|T(x)\|_2 \leq 1$. If $\|x\|_1 \leq 1$ then $\|\delta x\|_1 \leq \delta$, so that

$$\|T(x)\|_2 = \|\delta^{-1}T(\delta x)\|_2 = \delta^{-1}\,\|T(\delta x)\|_2 \leq \delta^{-1}.$$

□

Two norms $\|.\|_1$ and $\|.\|_2$ on a vector space E are *equivalent* if they define the same topology on E.

Corollary 7.5.2 *Suppose that $\|.\|_1$ and $\|.\|_2$ are two norms on a vector space E. They are equivalent if and only if they are uniformly equivalent, and if and only if there exist positive c and C such that*

$$c\,\|x\|_1 \leq \|x\|_2 \leq C\,\|x\|_1$$

for each $x \in E$.

Corollary 7.5.3 *Suppose that $\|.\|_1$ and $\|.\|_2$ are two equivalent norms on a vector space E. Then $(E, \|.\|_1)$ is a Banach space if and only if $(E, \|.\|_2)$ is a Banach space.*

We denote the set of continuous linear mappings from E_1 to E_2 by $L(E_1, E_2)$. We write $L(E)$ for $L(E, E)$. A linear mapping from E_1 to E_2 is also called a *bounded linear mapping*, or a *linear operator*; a continuous linear mapping from E to itself is called an *operator on E*.

We have the following extension theorem.

Theorem 7.5.4 *Suppose that F is a dense linear subspace of a normed space $(E, \|.\|_E)$, and that T is a continuous linear mapping from F to a Banach space $(G, \|.\|_G)$. Then there is a unique continuous linear mapping $\tilde{T}$ from E to G which extends T: $\tilde{T}(y) = T(y)$ for $y \in F$. If T is an isometry then so is $\tilde{T}$.*

Proof By Theorem 7.5.1, T is uniformly continuous, and so by Theorem 2.4.4 there is a unique continuous extension $\tilde{T}$, which is an isometry if T is. We must show that $\tilde{T}$ is linear. Suppose that $x, y \in E$ and that α, β are scalars. There exist sequences $(x_n)_{n=1}^\infty$ and $(y_n)_{n=1}^\infty$ in F such that $x_n \to x$ and $y_n \to y$ as $n \to \infty$. Then $\alpha x_n + \beta y_n \to \alpha x + \beta y$ as $n \to \infty$, and so

$$\begin{aligned}
\tilde{T}(\alpha x + \beta y) &= \lim_{n\to\infty} \tilde{T}(\alpha x_n + \beta y_n) = \lim_{n\to\infty} T(\alpha x_n + \beta y_n) \\
&= \lim_{n\to\infty} (\alpha T(x_n) + \beta T(y_n)) = \alpha \lim_{n\to\infty} T(x_n) + \beta \lim_{n\to\infty} T(y_n) \\
&= \alpha \lim_{n\to\infty} \tilde{T}(x_n) + \beta \lim_{n\to\infty} \tilde{T}(y_n) = \alpha\tilde{T}(x) + \beta\tilde{T}(y).
\end{aligned}$$

□

Theorem 7.5.5 *(i) $L(E_1, E_2)$ is a linear subspace of the vector space of all linear mappings from E_1 to E_2.*

(ii) If $T \in L(E_1, E_2)$, set $\|T\| = \sup\{\|T(x)\|_2 : \|x\|_1 \leq 1\}$. Then $\|T\|$ is a norm on $L(E_1, E_2)$, the operator *norm.*

(iii) If $T \in L(E_1, E_2)$, and $x \in E_1$ then $\|T(x)\|_2 \leq \|T\| . \|x\|_1$.

Proof (i) We use condition (i) of Theorem 7.5.1. Suppose that $S, T \in L(E_1, E_2)$ and that α is a scalar. Then

$$\sup\{\|(\alpha T)(x)\|_2 : \|x\|_1 \leq 1\} = |\alpha| \sup\{\|T(x)\|_2 : \|x\|_1 \leq 1\},$$

so that $\alpha T \in L(E_1, E_2)$ and

$$\begin{aligned}&\sup\{\|(S+T)(x)\|_2 : \|x\|_1 \leq 1\} \leq \\ &\leq \sup\{\|S(x)\|_2 : \|x\|_1 \leq 1\} + \sup\{\|T(x)\|_2 : \|x\|_1 \leq 1\},\end{aligned}$$

so that $S + T \in L(E_1, E_2)$.

(ii) If $\|T\| = 0$, then $T(x) = 0$ for x with $\|x\| \leq 1$, and so $T(x) = 0$ for all x; thus $T = 0$. $\|\alpha T\| = |\alpha| \, \|T\|$ and $\|S + T\| \leq \|S\| + \|T\|$, by the equation and inequality that we have established to prove (i).

(iii) This is true if $x = 0$. Otherwise, let $y = x / \|x\|_1$. Then $\|y\|_1 = 1$, so that

$$\|T(x)\|_2 = \|T(\|x\|_1 \, y)\| = \|x\|_1 \, \|T(y)\|_2 \leq \|T\| \, \|x\|_1 .$$

□

Theorem 7.5.6 *If $(E_1, \|.\|_1)$ is a normed space and $(E_2, \|.\|_2)$ is a Banach space then $L(E_1, E_2)$ is a Banach space under the operator norm.*

Proof Let (T_n) be a Cauchy sequence in $L(E_1, E_2)$. First we identify what the limit must be. Since, for each $x \in E_1$, $\|T_n(x) - T_m(x)\|_2 \leq \|T_n - T_m\| \, \|x\|_1$, $(T_n(x))$ is a Cauchy sequence in E_2, which converges, by the completeness of E_2, to $T(x)$, say. Secondly, we show that T is a linear mapping from E_1 to E_2. This follows, since

$$T(\alpha x + \beta y) - \alpha T(x) - \beta T(y) = \lim_{n \to \infty} (T_n(\alpha x + \beta y) - \alpha T_n(x) - \beta T_n(y)) = 0,$$

for all $x, y \in E_1$ and all scalars α, β. Thirdly we show that T is continuous. There exists N such that $\|T_n - T_m\| \leq 1$, for $m, n \geq N$. Then

$$\|(T - T_N)(x)\| = \lim_{n \to \infty} \|(T_n - T_N)(x)\| \leq \|x\|_1 ,$$

for each $x \in E_1$, so that $T - T_N \in L(E_1, E_2)$. Since $L(E_1, E_2)$ is a vector space, $T = (T - T_N) + T_N \in L(E_1, E_2)$. Finally we show that $T_n \to T$. Given $\epsilon > 0$ there exists M such that $\|T_n - T_m\| \leq \epsilon$, for $m, n \geq M$. Then if $m \geq M$, and $x \in E_1$,

$$\|(T - T_m)(x)\| = \lim_{n\to\infty} \|(T_n - T_m)(x)\| \leq \epsilon \, \|x\|_1 ,$$

so that $\|T - T_m\| \leq \epsilon$. □

A *linear functional* on a vector space E is a linear mapping from E into K. The vector space of continuous linear functionals on a normed space $(E, \|.\|)$ is called the *dual space* E'; it is given the *dual norm* $\|\phi\|' = \{\sup |\phi(x)| : \|x\| \leq 1\}$. This is simply the operator norm from $(E, \|.\|)$ into the scalars.

Corollary 7.5.7 *The dual space $(E', \|.\|')$ of a normed space $(E, \|.\|)$ is a Banach space.*

Exercise 7.5.8 Show that the dual space of $(c_0, \|.\|_\infty)$ is naturally linearly isometrically isomorphic to $(l^1, \|.\|_1)$, and that the dual space of $(l^1, \|.\|_1)$ is naturally linearly isometrically isomorphic to $(l^\infty, \|.\|_\infty)$. What about the dual of $(c, \|.\|_\infty)$?

Exercise 7.5.9 Let $cs = \{x \in l^\infty : \sum_{i=1}^\infty x_i \text{ is convergent}\}$. If $x \in cs$ let $\|x\|_{cs} = \sup_n(|\sum_{i=1}^n x_i|)$. Show that $\|.\|_{cs}$ is a norm on cs, and that $(cs, \|.\|_{cs})$ is a Banach space. To what space is it naturally linearly isometrically isomorphic? Find natural representations of its dual and its bidual.

We can extend these results to multilinear mappings. It is clearly enough to consider bilinear mappings.

Exercise 7.5.10 Suppose that $(E, \|.\|_E)$, $(F, \|.\|_F)$ and $(G, \|.\|_G)$ are normed spaces and that B is a bilinear mapping from $E \times F$ into G. Show that B is continuous if and only if there exists $M \geq 0$ such that $\|B(x, y)\|_G \leq M \|x\|_E \|y\|_F$ for all $(x, y) \in E \times F$.

Exercise 7.5.11 Suppose that $(E, \|.\|_E)$, $(F, \|.\|_F)$ and $(G, \|.\|_G)$ are normed spaces. If $T \in L(E, L(F, G))$, and $x \in E$, $y \in F$, let $j(T)(x, y) = (T(x))(y)$. Show that $j(T)$ is a continuous bilinear mapping from $E \times F$ into G; show that j is a bijective linear mapping of $L(E, L(F, G))$ onto the vector space $B(E, F; G)$ of continuous bilinear mappings from $E \times F$ into G. If $b \in B(E, F; G)$, let

$$\|b\| = \sup\{\|b(x, y)\|_G : \|x\|_E \leq 1, \|y\|_F \leq 1\}.$$

Show that this is a norm on $B(E, F; G)$, and with this norm the mapping j is an isometry. Thus if G is a Banach space, then so is $B(E, F; G)$.

7.6 Five Fundamental Theorems

We now prove five fundamental theorems, each of which depends on Baire's category theorem.

Theorem 7.6.1 (The Principle of Uniform Boundedness) *Suppose that $(E, \|.\|_E)$ is a Banach space, that $(F, \|.\|_F)$ is a normed space and that $\mathcal{T} \subseteq L(E, F)$. Then $\mathcal{T}$ is norm bounded in $L(E, F)$ if and only if $\{T(x) : T \in \mathcal{T}\}$ is norm bounded in F, for each $x \in E$.*

Proof If $\mathcal{T}$ is norm bounded in $L(E, F)$ then certainly $\{T(x) : T \in \mathcal{T}\}$ is norm bounded in F, for each $x \in E$.

If $x \in E$ and $T \in \mathcal{T}$, let $p_T(x) = \|T(x)\|_F$. Then p_T is a continuous seminorm on $(E, \|.\|_E)$. If $\{T(x) : T \in \mathcal{T}\}$ is norm bounded in F, for each $x \in E$, then $P = \sup\{p_T : T \in \mathcal{T}\}$ is a lower semi-continuous seminorm on $(E, \|.\|_E)$, and $A_n = \{x : P(x) \leq n\}$ is closed, for each n. Since $E = \cup_{n=1}^{\infty} A_n$, it follows from Baire's category theorem that A_n^{int} is not empty, for some n. If $x \in A_n^{int}$, then $0 = \frac{1}{2}x + \frac{1}{2}(-x) \in A_n^{int}$. Thus there exists $\epsilon > 0$ such that $\epsilon B(E) \subseteq A_n$. Hence, if $\|x\| \leq 1$ then $T(x) \leq n/\epsilon$, for each $T \in \mathcal{T}$; that is, $\|T\| \leq n/\epsilon$, for each $T \in \mathcal{T}$. □

We shall prove a non-linear version of this theorem in Theorem 10.2.4.

Theorem 7.6.2 (The open mapping theorem) *Suppose that $(E, \|.\|_E)$ and $(F, \|.\|_F)$ are two Banach spaces, and that $T \in L(E, F)$ is surjective. Then T is an open mapping: if U is open in E then $T(U)$ is open in F.*

Proof Since translation and dilation are homeomorphisms, it is sufficient to show that there exists $r > 0$ such that if $y \in F$ and $\|y\|_F < r$ then there exists $x \in B(E)$ such that $T(x) = y$.

Let $F_n = \overline{T(nB(E))}$. Then $F = \cup_{n=1}^{\infty} F_n$. By Baire's category theorem, there exists $n \in \mathbf{N}$ such that $F_n^{int} \neq \emptyset$. By homogeneity, $F_1^{int} \neq \emptyset$. Thus there exists $y \in F_1$ and $r > 0$ such that $y+2rB(F) \subseteq F_1$. By symmetry, $-y+2rB(F) \subseteq F_1$, and by convexity,

$$2rB(F) \subseteq \tfrac{1}{2}((y + rB(F)) + (-y + rB(F))) \subseteq F_1.$$

Suppose now that $y \in F$ and that $\|y\| < r$. There exists $x_1 \in \frac{1}{2}B(E)$ such that $\|y - T(x_1)\| \leq r/2$. Arguing inductively, there exists a sequence $(x_n)_{n=1}^{\infty}$ in E with $\|x\|_n \leq 1/2^n$ such that

$$\left\| \left(y - \sum_{j=1}^{n} T(x_j) \right) - T(x_{n+1}) \right\| < r/2^{n+1}, \text{ for } n \in \mathbf{N}.$$

But then $\sum_{j=1}^{\infty} x_j$ converges in E to an element x of $B(E)$, and

$$T(x) = \lim_{n\to\infty} T\left(\sum_{j=1}^{n} x_j\right) = \lim_{n\to\infty} \sum_{j=1}^{n} T(x_j) = y.$$

□

Theorem 7.6.3 (The isomorphism theorem) *Suppose that $(E, \|.\|_E)$ and $(F, \|.\|_F)$ are two Banach spaces, and that $T \in L(E, F)$ is bijective. Then T is a homeomorphism: T^{-1} is continuous.*

Proof This is an immediate consequence of the open mapping theorem. □

Suppose that $(E, \|.\|)$ and $(F, \|.\|_F)$ are Banach spaces, and that T is a linear mapping from E to F. Then the graph $\Gamma_T = \{(x, T(x)) : x \in E\}$ is a linear subspace of $E \times F$. If we give $E \times F$ the product norm $\|(x, y)\| = \|x\|_E + \|y\|_F$, and if T is continuous, then Γ_T is closed in $E \times F$. The closed graph theorem says that the converse is true.

Theorem 7.6.4 (The closed graph theorem) *Suppose that $(E, \|.\|)$ and $(F, \|.\|_F)$ are Banach spaces, and that T is a linear mapping from E to F. If Γ_T is closed in $E \times F$, then T is continuous.*

Proof We give Γ_T the norm inherited from $E \times F$. Since Γ_T is a closed linear subspace of $E \times F$, it is a Banach space. If $(x, T(x)) \in \Gamma_T$, let $\pi_1(x, T(x)) = x$, and $\pi_2(x, T(x)) = T(x)$. Then π_1 is a norm-decreasing bijection of Γ_T onto E and π_2 is a norm-decreasing injective mapping of Γ_T into F. By the isomorphism theorem, it follows that π_1^{-1} is continuous, and so therefore is $T = \pi_2 \circ \pi_1^{-1}$. □

A sequence $(e_n)_{n=1}^\infty$ in a Banach space $(E, \|.\|)$ is a *Schauder basis* if each $x \in E$ can be written *uniquely* as $x = \sum_{n=1}^\infty x_n e_n$, where the co-efficients x_n are scalars, and convergence is in the norm topology. Let $\phi_n(x) = x_n$: then ϕ_n is a linear functional on E. Similarly, let $P_n(x) = \sum_{j=1}^n x_j e_j$: then P_n is a projection of E onto an n-dimensional subspace E_n of E, and $P_m P_n = P_n P_m = P_m$, for $m \leq n$.

Theorem 7.6.5 *Suppose that $(e_n)_{n=1}^\infty$ is a Schauder basis for a Banach space $(E, \|.\|)$. Let $!x! = \sup_{n \in \mathbf{N}} \|P_n(x)\|$, for $x \in E$. Then $!\,.\,!$ is a norm on E, $(E, !\,.\,!)$ is a Banach space, and $!\,.\,!$ is equivalent to the norm $\|.\|$. The co-ordinate functions ϕ_n and the projections P_n are continuous on $(E, \|.\|)$; $\sup \|\phi_n\|' < \infty$ and $\sup \|P_n\| < \infty$.*

Proof If $x \in E$, then $P_n(x) \to x$ as $n \to \infty$, and so $\sup_{n \in \mathbf{N}} \|P_n(x)\| < \infty$. It follows from this that $!\,.\,!$ is a norm on E, and $\|x\| \leq !\,x\,!$, for $x \in E$. We shall show that $!\,.\,!$ is a complete norm on E. It then follows from the isomorphism theorem that $!\,.\,!$ and $\|.\|$ are equivalent norms. The remaining results then follow from this.

Suppose then that $(x^{(k)})_{k=1}^\infty$ is a $!\,.\,!$-Cauchy sequence in E. For each $n \in \mathbf{N}$, $(P_n(x^{(k)}))_{k=1}^\infty$ is a Cauchy sequence in $P_n(E)$, which converges to an element

$x^{[n]} = \sum_{j=1}^{n} x_j^{[n]} e_j$ of $P_n(E)$. If $m \leq n$, then $x_j^{[m]} = x_j^{[n]}$ for $1 \leq j \leq m$, since $P_m P_n = P_n P_m = P_m$. Thus there exist $x_j \in E$ such that $x_j^{[n]} = x_j$ for $1 \leq j \leq n$. We shall show that $\sum_{j=1}^{\infty} x_j e_j$ converges to an element x of E, and that $x^{(k)} \to x$ in ! . !-norm as $k \to \infty$.

Suppose that $\epsilon > 0$. There exists K such that $!x^{(k)} - x^{l)}! < \epsilon/3$ for $k, l \geq K$, so that $\left\| P_n(x^{(k)}) - P_n(x^{(l)}) \right\| < \epsilon/3$ for all n, for $k, l \geq N$. Letting $l \to \infty$,

$$\left\| P_n(x^{(k)}) - x^{[n]} \right\| \leq \epsilon/3, \text{ for } k \geq K, \text{ and for all } n. \qquad (*)$$

There exists N such that $\left\| P_n(x^{(K)}) - P_m(x^{(K)}) \right\| < \epsilon/3$ for $m, n \geq N$. Consequently, if $N \leq m < n$ then

$$\begin{aligned} &\left\| x^{[n]} - x^{[m]} \right\| \\ &\leq \left\| x^{[n]} - P_n(x^{(K)}) \right\| + \left\| P_n(x^{(K)}) - P_m(x^{(K)}) \right\| + \left\| P_m(x^{(K)}) - x^{[m]} \right\| \\ &< \epsilon, \end{aligned}$$

so that $x^{[n]}$ converges in $(E, \|.\|)$ as $n \to \infty$ to an element x of E, and $x^{[n]} = P_n(x)$. It therefore follows from (*) that $!x^{(k)} - x! \leq \epsilon$, for $k \geq K$. □

If $(e_n)_{n=1}^{\infty}$ is a Schauder basis for a Banach space $(E, \|.\|)$, $\|.\|$ is a *monotone basis* if $\|P\|_n \leq 1$ for each n, and is a *bimonotone basis* if it is monotone, and if $\|P_n - P_m\| \leq 1$ for $m, n \in \mathbf{N}$.

Corollary 7.6.6 *There exists an equivalent bimonotone norm $|||.|||$ on E.*

Proof Let

$$|||x||| = \sup\{\max(P_n(x) - P_m(x), P_n(x)) : m, n \in \mathbf{N}\}, \text{ for } x \in E.$$

Then $|||.|||$ is a bimonotone norm and $!x! \leq |||x||| \leq 2!x!$, so that $|||.|||$ is equivalent to $\|.\|$. □

A Banach space with a Schauder basis is separable. The converse is not true, but is hard to prove. The separable Banach spaces that one meets in practice have Schauder bases, though it can sometimes be difficult to give an explicit construction.

Exercise 7.6.7 Let (ω, τ) denote the linear space $K^{\mathbf{N}}$ with the product topology, and let ϕ be the subspace of sequences with only finitely many non-zero terms. A *Banach sequence space* $(E, \|.\|)$ is a linear subspace E of ω which contains ϕ, with a complete norm $\|.\|$ for which the inclusion mapping $(E, \|.\|) \to (\omega, \tau)$ is continuous. Suppose that $(E, \|.\|)$ and $(F\, \|.\|)$ are two Banach sequence spaces and that (a_{ij}) is a doubly-infinite matrix for which

$(\sum_{j=1}^{\infty} a_{ij}x_j)_{i=1}^{\infty} \in F$ for each $x \in E$. Use the closed graph theorem to show that (a_{ij}) defines a continuous linear mapping from E into F.

Exercise 7.6.8 Suppose that $\|.\|$ is a norm on $C([0,1])$ with the property that if $\|f_n\| \to 0$ then $f_n(t) \to 0$ as $n \to \infty$, for each $t \in [0,1]$. Show that if $\|.\|$ is a complete norm then $\|.\|$ is equivalent to the uniform norm $\|.\|_\infty$, and give an example for which $\|.\|$ is not complete, and is therefore not equivalent to $\|.\|_\infty$.

7.7 The Petal Theorem and Daneš's Drop Theorem

We now give two applications of Ekeland's variational theorem. If A is a non-empty bounded closed subset of a Banach space, and $z \notin A$, then $d(z,A) = \inf_{a\in A} d(z,a)$ is not necessarily achieved. Both theorems show that there is a (weaker) alternative. We need two definitions. Suppose that a and b are distinct points of a Banach space $(E, \|.\|)$ and that $\gamma > 0$ and $0 < r < \|a-b\|$. Then the *petal* $P_\gamma(a,b)$ is the set

$$\{x \in E : \gamma \|x-a\| + \|x-b\| \leq \|a-b\|\}$$

and the *drop* $D_r(a,b)$ is the set

$$\{x \in E : x \in \Gamma(M_r(a) \cup b) = \{x = (1-t)c + tb : \|c-a\| \leq r\}.$$

Exercise 7.7.1 Draw some pictures to illustrate the shape of a petal and a drop (raindrop).

Theorem 7.7.2 (The petal theorem) *Suppose that Y is a non-empty closed subset of a Banach space $(E, \|.\|)$, that $y_0 \in Y$ and that $z \in E \setminus Y$. Suppose that $0 < r < d(z,Y)$ and that $\gamma > 0$. Then there exists $\tilde{y}$ in Y, with $\|\tilde{y} - y_0\| < (\|y_0 - z\| - r)/\gamma$ such that $\tilde{y} \in P_\gamma(y_0, z)$ and $Y \cap P_\gamma(\tilde{y}, b) = \{\tilde{y}\}$.*

Proof Consider the function $f(y) = \|y-z\|$ on Y, put $\alpha = \gamma$ and $\epsilon = f(z_0) - r$ in Ekeland's variational principle (Theorem 4.4.1). Then there exists $\tilde{y} \in Y$ satisfying $\|\tilde{y} - y_0\| < (\|y_0 - z\| - r)/\gamma$, (ii) and (iii). But (ii) implies that $\tilde{y} \in P_\gamma(y_0, z)$, and (iii) implies that $Y \cap P_\gamma(\tilde{y}, z) = \{\tilde{y}\}$. □

Theorem 7.7.3 (Daneš's drop theorem) *Suppose that Y is a non-empty closed subset of a Banach space $(E, \|.\|)$ and that $z \in E \setminus Y$. Suppose that $0 < r < d(z,Y) < \rho$. Then there exists a point $\tilde{y}$ of Y such that $\|\tilde{y} - z\| \leq \rho$, and if $y \in \Gamma(M_r(z), \tilde{y}) \cap Y$ then $y = \tilde{y}$.*

Proof We may assume that $z = 0$. Let $Y_\rho = Y \cap M_\rho(0)$, let $R = d(0,Y)$ and let $\alpha = (R-r)/2(\rho + r)$. Applying Ekeland's variational principle to the function $\|.\|$, the set Y_ρ and the parameter α, there exists $\tilde{y} \in Y_\rho$ such that if $y \in Y_\rho$ then

$$\|\tilde{y}\| < \|y\| + \alpha \|y - \tilde{y}\| .$$

Suppose now that $y \in Y_\rho$ belongs to the drop $D_r(0, \tilde{y}) = \Gamma(M_r(0), \tilde{y})$ and that $y \neq \tilde{y}$. There exist $w \in M_r(0)$ and $0 \leq t \leq 1$ such that $y = (1-t)\tilde{y} + tw$, so that $\tilde{y} - y = t(\tilde{y} - w)$. Note that in fact $0 < t < 1$ and that

$$\|y\| \leq (1-t)\|\tilde{y}\| + t\|w\| \leq (1-t)\rho + tr.$$

Thus $t(\rho - r) \leq \|\tilde{y}\| - \|y\|$. Now

$$\begin{aligned}\|\tilde{y}\| - \|y\| &< \alpha \|y - \tilde{y}\| = t\alpha \|\tilde{y} - w\| \\ &\leq \left(\frac{\|\tilde{y}\| + \|w\|}{2(\rho + r)}\right).(t(\rho - r)) \leq \tfrac{1}{2}(\|\tilde{y}\| - \|y\|)\end{aligned}$$

giving a contradiction. □

Corollary 7.7.4 *Suppose that f is a proper lower semi-continuous function on a Banach space $(E, \|.\|)$, that $z \in E$ and that $s < f(z)$. Suppose that*

$$0 < r < \inf\{\max(d(z, x), |s - t|) : x \in \Phi_f, t \geq f(x)\} < \rho.$$

Then there exists a point $\tilde{y}$ of Φ_f and $\tilde{t} \in \mathbf{R}$ such that $\tilde{t} \geq f(\tilde{y})$, with $\|\tilde{y} - z\| < \rho$ and $|\tilde{t} - s| < \rho$ such that if $y \in \Phi_f$ and $t \geq f(y)$ and $(y, t) \in \Gamma((M_r(z), |t-s| \leq r), (\tilde{y}, \tilde{z}))$ then $y = \tilde{y}$ and $t = \tilde{t}$. If f is continuous on E, then $\tilde{t} = f(\tilde{y})$.

Proof The first statement follows by considering the episum A_f when $E \times \mathbf{R}$ is given the norm $\|x, u\| = \max(\|x\|, |u|)$. If f is continuous, and $f(\tilde{y}) < \tilde{t}$ then the line segment $[(\tilde{y}, f(\tilde{t})), (z, s)]$ contains more than one point of the episum, giving a contradiction. □

8
Hilbert Spaces

In this chapter, we consider Hilbert spaces. These are Banach spaces with a great deal of symmetry, and important geometrical properties.

8.1 Inner-product Spaces

Suppose that E is a real or complex vector space. An *inner product* on E is a scalar-valued function $(x, y) \to \langle x, y\rangle$ on $E \times E$ which satisfies the following:

(i) (bilinearity)

$$\langle \alpha_1 x_1 + \alpha_2 x_2, y\rangle = \alpha_1 \langle x_1, y\rangle + \alpha_2 \langle x_2, y\rangle ,$$
$$\langle x, \beta_1 y_1 + \beta_2 y_2\rangle = \beta_1 \langle x, y_1\rangle + \beta_2 \langle x, y_2\rangle ,$$

for all x, x_1, x_2, y, y_1, y_2 in E and all real $\alpha_1, \alpha_2, \beta_1, \beta_2$;

(ii) (skew-symmetry)

$$\langle y, x\rangle = \overline{\langle x, y\rangle} \text{ for all } x, y \text{ in } E;$$

(iii) (positive definiteness)

$$\langle x, x\rangle > 0 \text{ for all non-zero } x \text{ in } E.$$

A function which satisfies (i) and (ii) is called a *skew-symmetric bilinear form.*

A vector space E equipped with an inner product is called an *inner-product space*. Note that if $(E, \langle ., .\rangle)$ is a complex inner-product space then $\langle ., .\rangle_{\mathbf{R}}$, its real part, is an inner product on $E_{\mathbf{R}}$, the underlying real space, and $\langle x, x\rangle = \langle x, x\rangle_{\mathbf{R}}$.

Here is a most important example of an inner-product space. Let l_2 denote the set of all sequences $(a_n)_{n=1}^{\infty}$ for which $\sum_{n=1}^{\infty} |a_n|^2$ is finite. Then l_2 is a vector space (with the algebraic operations defined pointwise), such that if $a, b \in l_2$ then $\sum_{n=1}^{\infty} a_n \overline{b_n}$ converges absolutely, and that the function $(a, b) \to \langle a, b \rangle = \sum_{n=1}^{\infty} a_n \overline{b_n}$ is an inner product on l_2.

Similarly, if $E = \mathbf{C}^d$, we define the *usual* inner product by setting $\langle z, w \rangle = \sum_{i=1}^{d} z_i \overline{w_i}$ for $z = (z_i), w = (w_i)$.

As another example, the space $C[a, b]$ of continuous real-valued functions on the closed interval $[a, b]$ is an inner-product space when the inner product is defined by

$$\langle f, g \rangle = \int_a^b f(x)\overline{g(x)}\,dx.$$

If x is a vector in E, we set $\|x\| = \langle x, x \rangle^{\frac{1}{2}}$. We shall show that $\|.\|$ is a norm on E, the *inner-product norm* on E. Note that, in the complex case, $\|x\| = \langle x, x \rangle_{\mathbf{R}}$. Thus, when we consider metric properties of E we can frequently suppose that we are dealing with a real inner-product space. Certainly $\|x\| = 0$ if and only if $x = 0$, and $\|\lambda x\| = |\lambda|\, \|x\|$.

We now establish some basic properties of inner product spaces.

Proposition 8.1.1 (The Cauchy–Schwarz inequality) *If x and y are vectors in an inner-product space E then*

$$|\langle x, y \rangle| \le \|x\| . \|y\|,$$

with equality if and only if x and y are linearly dependent.

Proof This depends upon the quadratic nature of the inner product. The inequality is trivially true if $\langle x, y \rangle = 0$. If $\|x\| = 0$, then $x = 0$ and $\langle x, y \rangle = 0$, so that the inequality is true, and the same holds if $\|y\| = 0$.

Otherwise, if λ is a scalar, then

$$\begin{aligned} 0 \le \|x + \lambda y\|^2 &= \langle x + \lambda y, x + \lambda y \rangle \\ &= \langle x, x \rangle + \overline{\lambda} \langle x, y \rangle + \lambda \langle y, x \rangle + |\lambda|^2 \langle y, y \rangle . \end{aligned}$$

Put

$$\lambda = -\frac{\langle x, y \rangle}{|\langle x, y \rangle|} . \frac{\|x\|}{\|y\|}, \text{ so that } |\lambda| = \frac{\|x\|}{\|y\|}.$$

It follows that

$$0 \le \|x\|^2 - 2\frac{|\langle x, y \rangle|^2}{|\langle x, y \rangle|} . \frac{\|x\|}{\|y\|} + \frac{\|x\|^2}{\|y\|^2} \|y\|^2 = 2\left(\|x\|^2 - |\langle x, y \rangle| . \frac{\|x\|}{\|y\|} \right),$$

so that $|\langle x, y \rangle| \le \|x\| . \|y\|$.

If $x = 0$ or $y = 0$, then equality holds, and x and y are linearly dependent.

Otherwise, if equality holds, then $\|x + \lambda y\| = 0$, so that $x + \lambda y = 0$, and x and y are linearly dependent. Conversely, if x and y are linearly dependent, then $x = \alpha y$ for some scalar α, and so

$$|\langle x, y\rangle| = |\alpha|\, \|y\|^2 = \|x\| \,.\, \|y\| \,.$$

$\square$

Corollary 8.1.2 $\|x + y\| \leq \|x\| + \|y\|$, *with equality if and only if* $y = 0$ *or* $x = \alpha y$, *with* α *real and non-negative.*

Proof We have

$$\begin{aligned}\|x + y\|^2 &= \|x\|^2 + \langle x, y\rangle + \langle y, x\rangle + \|y\|^2 \\ &\leq \|x\|^2 + 2\, \|x\| \,.\, \|y\| + \|y\|^2 = (\|x\| + \|y\|)^2.\end{aligned}$$

Equality holds if and only if $\langle x, y\rangle + \langle y, x\rangle = 2\, \|x\|^2 \,.\, \|y\|^2$, which is equivalent to the condition stated. $\square$

Thus $\|.\|$ is a norm on E.

Note also that the inner product is determined by the norm: we have the *polarization formulae*.

Exercise 8.1.3 Show that

$$\begin{aligned}\langle x, y\rangle &= \tfrac{1}{2}(\|x + y\|^2 - \|x\|^2 - \|y\|^2) \\ &= \tfrac{1}{4}(\|x + y\|^2 - \|x - y\|^2)\end{aligned}$$

(in the real case), and

$$\begin{aligned}\langle x, y\rangle &= \tfrac{1}{4}\sum_{k=1}^{4} i^k \left\|x + i^k y\right\|^2 - (\|x\|^2 + \|y\|^2) \\ &= \frac{1}{2\pi}\int_0^{2\pi} e^{it}\left\|x + e^{it} y\right\|^2 dt - (\|x\|^2 + \|y\|^2)\end{aligned}$$

(in the complex case).

Exercise 8.1.4 Suppose that T is a linear mapping of a complex inner-product space into itself. Establish a polarization formula for $\langle T(x), y\rangle$, and deduce that if $\langle T(x), x\rangle = 0$ for all x, then $T = 0$. Give a two-dimensional example to show that the same is not true for real inner-product spaces.

We also have the following.

Exercise 8.1.5 (The parallelogram law) By expanding the first two terms, show that if x and y are vectors in an inner-product space E, then

$$\|x + y\|^2 + \|x - y\|^2 = 2\|x\|^2 + 2\|y\|^2.$$

Proposition 8.1.6 *If $y \in E$, an inner-product space, then the mapping l_y defined by $l_y(x) = \langle x, y \rangle$ is a continuous linear functional on E, and $\|l_y\| = \|y\|$. l is a linear (in the real case) or antilinear (in the complex case) isometry of E into its dual space E'.*

Proof Since $|\langle x, y \rangle| \leq \|x\| . \|y\|$, $l_y \in E'$, and $\|l_y\| \leq \|y\|$. If $y \neq 0$, then $\langle y/\|y\|, y \rangle = \|y\|$ and so $\|l_y\| = \|y\|$. Trivially, $\|l_0\| = \|0\| = 0$. The antilinearity of l in the complex case follows from the sesquilinearity of the inner product. □

Many of the geometric and metric properties of inner-product spaces can be expressed in terms of orthogonality. Vectors x and y in an inner-product space E are said to be *orthogonal* if $\langle x, y \rangle = 0$; if so, we write $x \perp y$. In the complex case, x and y are orthogonal in the underlying real space if and only if $\langle x, y \rangle$ is pure imaginary; the notions of orthogonality are different.

Proposition 8.1.7 (Pythagoras' theorem) *If x and y are vectors in a real inner-product space E then $\|x + y\|^2 = \|x\|^2 + \|y\|^2$ if and only if $x \perp y$. If x and y are vectors in a complex inner-product space E then $\|x + y\|^2 = \|x\|^2 + \|y\|^2$ if and only if $\langle x, y \rangle$ is pure imaginary.*

Proof For $\|x + y\|^2 = \|x\|^2 + \|y\|^2 + \langle x, y \rangle + \langle y, x \rangle$. □

If A is a subset of an inner-product space E, we set

$$A^{\perp} = \{x \in E : \langle a, x \rangle = 0 \text{ for all } a \in A\}.$$

We write $x^{\perp}$ for $\{x\}^{\perp}$. $A^{\perp}$ is the *annihilator* of A.

Proposition 8.1.8 *If A is a subset of an inner-product space E, then $A^{\perp}$ is a closed linear subspace of E.*

Proof If $a \in A$ then $a^{\perp} = \{x : l_a(x) = 0\}$ is closed, since l_a is continuous, and is clearly a linear subspace of E. Then $A^{\perp} = \cap_{a \in A} a^{\perp}$ is closed, and a linear subspace of E. □

Exercise 8.1.9 Suppose that A and B are subsets of an inner-product space E. Show the following.

(i) $A^{\perp} = \{x \in E : \langle x, a \rangle = 0 \text{ for all } a \in A\}$.
(ii) If $A \subseteq B$ then $B^{\perp} \subseteq A^{\perp}$.
(iii) $A \subseteq A^{\perp\perp}$.
(iv) $A^{\perp} = A^{\perp\perp\perp}$.
(v) $A \cap A^{\perp} \subseteq \{0\}$.

Suppose that x is a unit vector in E, and that $z \in E$. Let $\lambda = \langle z, x \rangle$ and let $y = z - \lambda x$. Then $\langle y, x \rangle = \langle z, x \rangle - \langle z, x \rangle \langle x, x \rangle = 0$. Thus $z = \lambda x + y$, where

$\lambda x \in \text{span}(x)$ and $y \in x^{\perp}$. If $z = \mu x + w$, with $w \in x^{\perp}$, then $\langle z, x\rangle = \mu$, so that $\mu = \lambda$ and $w = y$; the decomposition is unique.

Here is an application.

Proposition 8.1.10 *Suppose that x, y and z are elements of an inner-product space E and that $\|z - x\| = \|y - x\| + \|z - y\|$. Then there exists $0 \le \lambda \le 1$ such that $y = (1 - \lambda)x + \lambda z$.*

Proof By translation, we can suppose that $x = 0$. Let $y = \lambda z + w$, where $w \perp z$. Then

$$\begin{aligned} \|z\| &= \|\lambda z + w\| + \|(1 - \lambda)z - w\| \\ &\ge \|\lambda z\| + \|(1 - \lambda)z\| \\ &= (|\lambda| + |1 - \lambda|)\,\|z\|, \end{aligned}$$

with equality throughout if and only if $w = 0$ and $0 \le \lambda \le 1$. □

A corresponding result does not hold for general Banach spaces; in $(\mathbf{R}^2, \|.\|_\infty)$, take $x = (0, 0)$, $y = (1, 1)$ and $z = (2, 0)$.

8.2 Hilbert Space; Nearest Points

An inner-product space which is complete under the inner-product norm is called a *Hilbert space*. A finite-dimensional real Hilbert space is called a *Euclidean space*, and a finite-dimensional complex Hilbert space is called a *Hermitian space*.

The following result is very important; the corresponding result for Banach spaces is not in general true.

Theorem 8.2.1 *Suppose that C is a non-empty closed convex subset of a real Hilbert space H and that $x \in H$. Then there exists a unique point $c(x)$ in C such that*

$$\|x - c(x)\| = d(x, C) \qquad (= \inf\{\|x - y\| : y \in C\}).$$

($c(x)$ is the nearest point *to x in C.)*

$$\langle x, c(x)\rangle = \sup\{\langle x, c\rangle : c \in C\} \text{ so that } C \subseteq H_- = \{z : \langle x, z\rangle \le c(x)\}.$$

The mapping $x \to c(x)$ is a retraction of H onto C.

Proof Let $d = d(x, C)$. For each n, there exists $c_n \in C$ with $\|x - c_n\| < d + 1/n$. By the parallelogram law,

$$4\|x-(c_m+c_n)/2\|^2+\|c_m-c_n\|^2$$
$$=2\|x-c_m\|^2+2\|x-c_n\|^2.$$

Since, by convexity, $\frac{1}{2}(c_m+c_n)\in C$,

$$\begin{aligned}4d^2+\|c_m-c_n\|^2 &\le 4\|x-(c_m+c_n)/2\|^2+\|c_m-c_n\|^2\\ &=2\|x-c_m\|^2+2\|x-c_n\|^2\\ &\le 2(d+1/m)^2+2(d+1/n)^2,\end{aligned}$$

so that

$$(c_m-c_n)^2\le\frac{4d}{m}+\frac{2}{m^2}+\frac{4d}{n}+\frac{2}{n^2}\to 0$$

as $m,n\to\infty$. Thus (c_m) is a Cauchy sequence in C. Since H is complete, (c_m) converges, to $c(x)$, say, and $c(x)\in C$, since C is closed. Then $d\le\|x-c(x)\|=\lim\|x-c_n\|\le d$, so that $\|x-c(x)\|=d$.

If c and c' are two nearest points,

$$\|2x-c-c'\|^2+\|c-c'\|^2=2\|x-c\|^2+2\|x-c'\|^2=4d^2 \text{ and}$$

$$\|2x-c-c'\|^2=4\left\|x-\frac{1}{2}(c+c')\right\|^2\ge 4d^2,$$

since $\frac{1}{2}(c+c')\in C$, so that $\|c-c'\|^2\le 0$, and $c=c'$.

By translation, and then scaling, we can suppose that $\|x\|=1$ and that $c(x)=0$. Suppose that $c\in C$. We can write $c=\alpha x+w$, where $\alpha\in\mathbf{R}$ and $w\perp x$. Suppose, if possible, that $\alpha>0$. Then if $\beta>0$, $\|x-\beta c\|^2=1-2\alpha\beta+\beta^2\|c\|^2<1$ for small values of β. But $\beta c\in C$, for $0<\beta\le 1$, and so we have a contradiction; thus $\alpha\le 0$, and C s contained in the half-space $H_-=\{z:\langle x,z\rangle\le 0=c(x)\}$.

We must show that the mapping $x\to c(x)$ is continuous. Suppose that $0<\epsilon<1$. Let $\delta=\epsilon/5$ and suppose that $\|y-x\|<\delta$. Let $a(y)=y-c(y)$. Then

$$\|a(y)\|\le\|y-c(x)\|=\|y\|\le 1+\delta.$$

Let $\lambda=\langle a(y),x\rangle$. Then

$$\lambda=\langle y,x\rangle-\langle c(y),x\rangle\ge 1+\langle y-x,x\rangle\ge 1-\delta,$$

so that $\|a(y)-\lambda x\|\le\|a(y)\|-\lambda\le 2\delta$. Now

$$a(y)-x=(a(y)-\lambda x)-(1-\lambda)x \text{ and } (a(y)-\lambda x)\perp x,$$

so that, using Pythagoras' theorem,

$$\|a(y)-x\|^2=\|a(y)-\lambda x\|^2+(1-\lambda)^2\le 5\delta^2\le\epsilon^2/4.$$

Thus

$$c(y) = \|a(y) - y\| \leq \|a(y) - x\| + \|x - y\| \leq \epsilon/2 + \epsilon/2 = \epsilon,$$

and so c is continuous. □

Suppose that V is a vector space. A *projection* P is a linear mapping: $V \to V$ such that $P^2 = P$.

Proposition 8.2.2 *If P is a projection, and $U = P(V), W = P^{-1}(0)$, then $V = U \oplus W$ ($U + W = V$ and $U \cap W = \{0\}$) and $U = \{x \in V : P(x) = x\}$, $W = \{x \in V : P(x) = 0\}$.*

Proof If $x = P(y) \in U$ then $P(x) = P^2(y) = P(y) = x$, while if $P(x) = x$ then $x \in U$; thus $U = \{x \in V : P(x) = x\}$. If $x \in V$, then $x = P(x) + (x - P(x))$: $P(x) \in U$ and $x - P(x) \in W$. If $x \in U \cap W$, $P(x) = x$, since $x \in U$, and $P(x) = 0$, since $x \in W$. Thus $U \cap W = \{0\}$, and $V = U \oplus W$. □

P is the *projection onto U along W*. $I - P$ is then the projection onto W along U.

Theorem 8.2.3 *Suppose that M is a closed linear subspace of a Hilbert space H. If $x \in H$, let $P_M(x)$ be the unique nearest point to x in M.*

(i) $x - P_M(x) \in M^\perp$, and $P_M(x)$ is the only point in M with this property.
(ii) P_M is linear and continuous, with $\|P_M\| = 1$ (unless $M = \{0\}$, when $P_M = 0$).
(iii) P_M is a projection onto M, along $M^\perp$.
(iv) $I - P_M = P_{M^\perp}$, so that $H = M \oplus M^\perp$. Thus $M = M^{\perp\perp}$.
(v) If $x, y \in H$, then $\langle P_M(x), y\rangle = \langle P_M(x), P_M(y)\rangle = \langle x, P_M(y)\rangle$.

Proof (i) Let $z = x - P_M(x)$. Suppose that $y \in M$ and that $\langle z, y\rangle \neq 0$. Then $y \neq 0$. Let $w = y/\|y\|$. Then $w \in M$, $\|w\| = 1$ and $\langle z, w\rangle = \alpha \neq 0$. Then $\langle z - \alpha w, \alpha w\rangle = |\alpha|^2 - |\alpha|^2 = 0$. By Pythagoras' theorem,

$$\|z\|^2 = \|z - \alpha w\|^2 + \|\alpha w\|^2 = \|z - \alpha w\|^2 + |\alpha|^2 > \|z - \alpha w\|^2.$$

But $z - \alpha w = x - (P_M(x) + \alpha w)$, and $P_M(x) + \alpha w \in M$, so that we get a contradiction. Thus $z \in M^\perp$. Suppose that $u \in M$, and that $x - u \in M^\perp$. Then $P_M(x) - u \in M$ and $P_M(x) - u = (x - u) - (x - P_M(x)) \in M^\perp$, so that $\|P_M(x) - u\|^2 = \langle P_M(x) - u, P_M(x) - u\rangle = 0$, and $P_M(x) = u$.

(ii) and (iii) If $x, y \in H$ and α, β are scalars,

$$(\alpha x + \beta y) - (\alpha P_M(x) + \beta P_M(y)) = \alpha(x - P_M(x)) + \beta(y - P_M(y)) \in M^\perp,$$

so that $\alpha P_M(x) + \beta P_M(y) = P_M(\alpha x + \beta y)$, by (i). Thus P_M is linear. Since $(x - P_M(x)) \perp P_M(x)$, $\|x\|^2 = \|P_M(x)\|^2 + \|x - P_M(x)\|^2$, so that $\|P_M\| \leq 1$

and P_M is continuous. Since $P_M(x) = x$ if and only if $x \in M$, $\|P_M\| = 1$ (unless $M = \{0\}$), and P_M is a projection onto M. By (i), $P_M(x) = 0$ if and only if $x \in M^\perp$, so that P_M is the projection onto M along $M^\perp$.

(iv) If $x \in H$, $x - (I - P_M)(x) = P_M(x) \in M \subseteq M^{\perp\perp}$, so that $(I - P_M)(x) = P_{M^\perp}(x)$, by (i).

(v) If $x, y \in H$, then $P_M(x) \in M$ and $y - P_M(y) \in M^\perp$, so that $\langle P_M(x), y - P_M(y)\rangle = 0$; thus $\langle P_M(x), y\rangle = \langle P_M(x), P_M(y)\rangle$. Similarly, $\langle x, P_M(y)\rangle = \langle P_M(x), P_M(y)\rangle$. □

Theorem 8.2.4 *Suppose that H is a Hilbert space.*

(i) $C \subseteq H$ is a closed linear subspace if and only if $C = C^{\perp\perp}$.
(ii) If A is a non-empty subset of H then $\overline{span}(A) = A^{\perp\perp}$.

Proof (i) The condition is sufficient, by Proposition 8.1.8. If C is a closed linear subspace of H, then $H = C \oplus C^\perp$. If $x \in C^{\perp\perp}$, we can write $x = y + z$, with $y \in C$ and $z \in C^\perp$. But then $\langle z, z\rangle = \langle x, z\rangle - \langle y, z\rangle = 0$, since $x \in C^{\perp\perp}$ and $z \in C^\perp$, so that $z = 0$ and $x = y \in C$. Thus $C^{\perp\perp} \subseteq C$. Since the reverse inclusion always holds, the condition is also necessary.

(ii) Let $M = \overline{span}(A)$. Then $A^{\perp\perp} \subseteq M^{\perp\perp} = M$, by (i). But $A^{\perp\perp}$ is a closed linear subspace of H containing A, and so $A^{\perp\perp} \supseteq M$. □

8.3 Orthonormal Sequences; Gram–Schmidt Orthonormalization

A finite or infinite sequence (x_n) in an inner-product space is an *orthogonal sequence* if $\langle x_m, x_n\rangle = 0$ if $m \neq n$; it is an *orthonormal sequence* if, in addition, $\langle x_m, x_m\rangle = 1$ for all m.

Exercise 8.3.1 Let $C_{\mathbf{C}}([0, 1])$ be given the inner product

$$\langle f, g\rangle = \int_0^1 f(t)\overline{g(t)}\, dt,$$

and let $\gamma(t) = e^{2\pi i t}$. Show that the sequence $(\gamma^n)_{n=-\infty}^\infty$ is an orthonormal sequence.

Exercise 8.3.2 Let $C_{\mathbf{R}}([a, b])$ be given the inner product

$$\langle f, g\rangle = \int_a^b f(x)g(x)w(x)\, dx,$$

where w is a positive continuous weight function. Let $(Q_n)_{n=1}^\infty$ be defined inductively by

$$Q_0 = 1, \quad Q_1 = x - c_1, \quad Q_n = (x - c_n)Q_{n-1} - d_n Q_{n-2}$$

where

$$c_n = \langle xQ_{n-1}, Q_{n-1}\rangle \,/\, \|Q_{n-1}\|^2 \quad \text{and} \quad d_n = \langle xQ_{n-1}, Q_{n-2}\rangle \,/\, \|Q_{n-2}\|^2 .$$

Show that $(Q_n)_{n=0}^{\infty}$ is an orthogonal sequence of monic polynomials, with each Q_n of degree n.

Exercise 8.3.3 In Exercise 8.3.2, show that

$$\|Q_n\| = \mathit{inf}\{\|Q\| : Q \text{ a monic polynomial of degree } n\}.$$

Exercise 8.3.4 In Exercise 8.3.2, show that Q_n has n distinct roots in $[a, b]$. (Consider $\langle Q_n, P\rangle$, where $P = \prod_{i=1}^{k}(x - r_i)$.)

Exercise 8.3.5 If $[a, b] = [-1, 1]$ and $w = 1$ in Exercise 8.3.2 the resulting polynomials are called the *Legendre polynomials* and are usually denoted by $(X_n)_{n=0}^{\infty}$. Calculate X_n for $0 \leq n \leq 4$.

Proposition 8.3.6 *Suppose that $(e_1, \ldots e_n)$ is a finite orthonormal sequence in an inner product space H. Let $A_n = \{e_1, \ldots, e_n\}$ and let $M_n = span(A_n)$. If $x \in H$ then $P_{M_n}(x) = \sum_{i=1}^{n} \langle x, e_i\rangle\, e_i$.*

Proof $\left\langle x - \sum_{i=1}^{n} \langle x, e_i\rangle\, e_i, e\right\rangle_j = 0$ for $1 \leq j \leq n$, so that

$$x - \sum_{i=1}^{n} \langle x, e_i\rangle\, e_i \in A_n^{\perp} = M_n^{\perp}.$$

Also $\sum_{i=1}^{n} \langle x, e_i\rangle\, e_i \in M_n$, so that the result follows from Theorem 8.2.3. □

Theorem 8.3.7 (Gram–Schmidt orthonormalization) *Let $(x_n)_{n=1}^{\infty}$ be a linearly independent sequence in an inner product space V and let $V_n = span(x_1, \ldots, x_n)$. Then there exists an orthonormal sequence $(e_n)_{n=1}^{\infty}$ in V such that*

(i) $span(e_1, \ldots, e_n) = V_n$, for $n = 1, 2, \ldots$, and
(ii) $e_n \perp V_{n-1}$, for $n = 2, 3, \ldots$.

The sequence $(e_n)_{n=1}^{\infty}$ is fairly unique: if (f_n) is another orthonormal sequence which satisfies (i) and (ii) then $f_n = \lambda_n e_n$ for each n, where $|\lambda_n| = 1$.

Proof Let $y_n = x_n - P_{V_{n-1}}(x_n)$. $y_n \neq 0$, since $x_n \notin V_{n-1}$. Let $e_n = y_n / \|y_n\|$. Then e_n is a unit vector in V_n orthogonal to V_{n-1}, and so $(e_n)_{n=1}^{\infty}$ is an orthonormal sequence. In particular, $(e_1, \ldots, e_n)$ is a linearly independent sequence in the n-dimensional space V_n, and so $span(e_1, \ldots, e_n) = V_n$.

If (f_n) is another orthonormal sequence satisfying (i) and (ii) then, since $f_n \in V_n$, we can write $f_n = \alpha_1 e_1 + \cdots + \alpha_n e_n$, and, since $f_n \perp V_{n-1}$, $\alpha_i = 0$ for $i < n$; thus $f_n = \alpha_n e_n$ and $|\alpha_n| = \|f_n\| = 1$. □

Theorem 8.3.8 (The Riesz–Fischer theorem) *Suppose that $(e_n)_{n=1}^\infty$ is an orthonormal sequence in a Hilbert space H. Then $\sum_{i=1}^\infty c_i e_i$ converges in norm if and only if $\sum_{i=1}^\infty |c_i|^2 < \infty$. If so, then $\left\|\sum_{i=1}^\infty c_i e_i\right\|^2 = \sum_{i=1}^\infty |c_i|^2$.*

Proof Let $s_n = \sum_{i=1}^n c_i e_i$. Suppose that s_n converges in norm, to c say. Then

$$\sum_{i=1}^n |c_i|^2 = \|s_n\|^2 \to \|c\|^2 \text{ as } n \to \infty$$

so that $\sum_{i=1}^\infty |c_i|^2 < \infty$.

Suppose conversely that $\sum_{i=1}^\infty |c_i|^2 < \infty$. Then if $m > n$, $\|s_m - s_n\| = \left\|\sum_{i=n+1}^m c_i e_i\right\| = (\sum_{i=n+1}^m |c_i|^2)^{1/2}$, so that (s_n) is a Cauchy sequence. By completeness, it converges. □

Theorem 8.3.9 *Suppose that $(e_n)_{n=1}^\infty$ is an orthonormal sequence in a Hilbert space H. Let $A = \operatorname{span}\{e_n : n \in \mathbf{N}\}$ and $M = \overline{A}$. If $x \in H$ then $P_M(x) = \sum_{i=1}^\infty \langle x, e_i\rangle e_i$.*

Proof $M = A^{\perp\perp}$ and $M^\perp = A^\perp$. Let $s_n = \sum_{i=1}^n \langle x, e_i\rangle e_i$. Then $\langle s_n, x\rangle = \sum_{i=1}^n |\langle x, e_i\rangle|^2 = \langle s_n, s_n\rangle$, so that $s_n \perp (x - s_n)$. Thus $\|s_n\|^2 = \|x\|^2 - \|x - s_n\|^2 \leq \|x\|^2$, and we can apply the Riesz–Fischer theorem. Let $s = \sum_{i=1}^\infty \langle x, e_i\rangle e_i$. Then $s \in M$, and $s_n \to s$. As $\langle x - s_n, e_m\rangle = 0$ for $n \geq m$, $\langle x - s, e_m\rangle = 0$, and so $x - s \in A^\perp = M^\perp$. The result now follows from Theorem 8.2.3. □

Corollary 8.3.10 $\sum_{n=1}^\infty |\langle x, e_n\rangle|^2 \leq \|x\|^2$ *(*Bessel's *inequality).*

Proof $\sum_{n=1}^\infty |\langle x, e_n\rangle|^2 = \|P_M(x)\|^2 \leq \|x\|^2$. □

When do we have equality in Bessel's inequality?

Corollary 8.3.11 *The following are equivalent:*

(i) $x \in M$;
(ii) $x = \sum_{i=1}^\infty \langle x, e_i\rangle e_i$ (Parseval's equation);
(iii) $\|x\|^2 = \sum_{i=1}^\infty |\langle x, e_i\rangle|^2$.

Proof (i) and (ii) are equivalent, and imply (iii), by the Riesz–Fischer theorem. If (iii) holds then

$$\left\|x - \sum_{i=1}^k \langle x, e_i\rangle e_i\right\|^2 = \|x\|^2 - \sum_{i=1}^k |\langle x, e_i\rangle|^2 \to 0,$$

so that (i) holds. □

8.4 Orthonormal Bases

An orthonormal sequence $(e_n)_{n=1}^{\infty}$ in a Hilbert space H is an *orthonormal basis* for H if H is the closed linear span of $(e_n)_{n=1}^{\infty}$. We consider the case where H is infinite-dimensional; the finite-dimensional case is simply a matter of linear algebra.

Proposition 8.4.1 *An orthonormal sequence $(e_n)_{n=1}^{\infty}$ in a Hilbert space H is an orthonormal basis for H if and only if $x = \sum_{n=1}^{\infty} \langle x, e_n \rangle e_n$ for each $x \in V$. If so, and if $x, y \in V$, then $\langle x, y \rangle = \sum_{i=1}^{\infty} \langle x, e_i \rangle \langle e_i, y \rangle$.*

Proof $(e_n)_{n=1}^{\infty}$ is an orthonormal basis for H if and only if $M = \overline{span}\{e_n : n \in \mathbf{N}\}$, and this happens if and only if $x = P_M(x)$ for each $x \in H$. Apply Theorem 8.3.9. If the condition is satisfied, and if $x, y \in H$, then

$$\langle x, y \rangle = l_y(x) = l_y\left(\sum_{i=1}^{\infty} \langle x, e_i \rangle e_i\right) = \sum_{i=1}^{\infty} \langle x, e_i \rangle l_y(e_i) = \sum_{i=1}^{\infty} \langle x, e_i \rangle \langle e_i, y \rangle .$$

□

Note that we use the continuity of l_y in an essential way.

Theorem 8.4.2 *A separable Hilbert space H has an orthonormal basis.*

Proof There exists a dense sequence (y_k) in H. Let

$$K = \{k : y_k \notin span(y_1, \ldots, y_{k-1})\} = \{k_1 < k_2 < \ldots\},$$

and let $x_n = y_{k_n}$. Then (x_n) is a linearly independent sequence; applying Gram–Schmidt orthonormalization, there exists an orthonormal sequence $(e_n)_{n=1}^{\infty}$ satisfying the conditions of Theorem 8.3.7. Then $\overline{\text{span}}\{e_n : n \in \mathbf{N}\} = \overline{\text{span}}\{x_n : n \in \mathbf{N}\} = \overline{\text{span}}\{y_n : n \in \mathbf{N}\} = H$, so that $(e_n)_{n=1}^{\infty}$ is an orthonormal basis for H. □

Corollary 8.4.3 *If H is a separable Hilbert space, there is a linear isometry of H onto l_2.*

Proof Let (f_n) be an orthonormal basis for H. Let $J(x) = (\langle x, f_n \rangle)_{n=1}^{\infty}$. Applying Corollary (iii), we see that J is an isometry into l_2. □

It is surjective, by the Riesz–Fischer theorem.

Unfortunately, at this stage we do not have examples of Hilbert spaces other than l_2. When we have defined Lebesgue measure in Part II, it will follow that $(\gamma^n)_{n=-\infty}^{\infty}$ is an orthonormal basis for $L^2_{\mathbf{C}}([0, 1])$, and the normalized Legendre polynomials $(X_n / \|X_n\|)_{n=0}^{\infty}$ form an orthonormal basis for $L^2_{\mathbf{R}}([-1, 1])$.

8.5 The Fréchet–Riesz Representation Theorem; Adjoints

Theorem 8.5.1 (The Fréchet–Riesz representation theorem) *Suppose that H is a Hilbert space. If $y \in H$, let $l_y(x) = \langle x, y \rangle$. Then l is an antilinear (in the complex case) or linear (in the real case) isometry of H onto the dual space H'.*

Proof We have seen (Proposition 8.1.6) that l is a linear isometry of H into H'. It remains to see that it is surjective. Suppose that ϕ is a non-zero element of H'. Let N be its null-space. Then N is a proper closed linear subspace of H, and so $H = N \oplus N^{\perp}$. Since ϕ is one-one on $N^{\perp}$, $\dim(N^{\perp}) = 1$. Thus if y is a unit vector in $N^{\perp}$, any element x of H can be written uniquely as $x = n + \alpha y$, with $n \in N$, and then $\phi(x) = \alpha\phi(y)$. But if $z = \phi(y)y$, then

$$l_z(x) = \langle n, z \rangle + \alpha \langle y, z \rangle = \alpha\phi(y) = \phi(x),$$

so that $\phi = l_z$. □

Theorem 8.5.2 *Suppose that H_1 and H_2 are Hilbert spaces and that $T \in L(H_1, H_2)$. There exists a unique $T^* \in L(H_2, H_1)$ such that*

$$\langle T(x), y \rangle = \left\langle x, T^*(y) \right\rangle \text{ for all } x \in H_1, y \in H_2.$$

In the complex case, the mapping $T \to T^*$ is an antilinear isometry of $L(H_1, H_2)$ onto $L(H_2, H_1)$, and in the real case it is a linear isometry. $T^{**} = T$. Further, $\|T\|^2 = \|T^*T\| = \|TT^*\|$.

Proof Suppose that $y \in H_2$. Let $\psi_y(x) = \langle T(x), y \rangle$, for $x \in H_1$. ψ_y is a linear functional on H_1, and

$$|\psi_y(x)| \leq \|T(x)\| \, \|y\| \leq \|T\| \, \|y\| \, \|x\| ,$$

so that $\psi_y \in H_1'$, and $\left\|\psi_y\right\| \leq \|T\| \, \|y\|$. By the Fréchet–Riesz representation theorem, there exists a unique element of H_1, $T^*(y)$ say, such that $\psi_y(x) = \langle x, T^*(y) \rangle$ and $\|T^*(y)\| = \left\|\psi_y\right\|$. It is easily verified that the mapping $T^* : H_2 \to H_1$ is linear, and so

$$\left\|T^*\right\| = \sup\left\{\left\|\psi_y\right\| : \|y\| \leq 1\right\} \leq \|T\| .$$

Since

$$\begin{aligned}\left\langle x, (\alpha S + \beta T)^* y \right\rangle &= \langle (\alpha S + \beta T)x, y \rangle = \alpha \langle S(x), y \rangle + \beta \langle T(x), y \rangle \\ &= \left\langle x, \bar{\alpha} S^*(y) \right\rangle + \left\langle x, \bar{\beta} T^*(y) \right\rangle = \left\langle x, (\bar{\alpha} S^* + \bar{\beta} T^*)(y) \right\rangle,\end{aligned}$$

the mapping $T \to T^*$ is antilinear. Also

$$\left\langle y, T^{**}(x) \right\rangle = \left\langle T^*(y), x \right\rangle = \left\langle x, T^*(y) \right\rangle = \langle T(x), y \rangle = \langle y, T(x) \rangle ,$$

so that by uniqueness, $T = T^{**}$. Thus $\|T\| \leq \|T^*\|$, and so $\|T\| = \|T^*\|$.

Finally, $\|T^*T\| \leq \|T^*\| . \|T\| = \|T\|^2$ and

$$\begin{aligned}\|T\|^2 &= \sup\{\|T(x)\|^2 : \|x\| \leq 1\} \\ &= \sup\{\langle T(x), T(x)\rangle : \|x\| \leq 1\} \\ &= \sup\{|\langle x, T^*T(x)\rangle| : \|x\| \leq 1\} \\ &= \sup\{|\langle T^*T(x), x\rangle| : \|x\| \leq 1\} \leq \|T^*T\|,\end{aligned}$$

so that $\|T^*T\| = \|T\|^2$. Applying this to T^*, we see that $\|TT^*\| = \|T\|^2$, as well. □

T^* is called the *adjoint* of T.

Exercise 8.5.3 Suppose that H is a Hilbert space, that S, T are linear mappings from H to itself which satisfy $\langle S(x), y\rangle = \langle x, T(y)\rangle$ for all $x, y \in H$. Show that S and T are continuous, and that $T = S^*$.

Exercise 8.5.4 Give an example where E is an inner-product space and S, T are discontinuous linear mappings from E to itself which satisfy $\langle S(x), y\rangle = \langle x, T(y)\rangle$ for all $x, y \in E$.

Exercise 8.5.5 Suppose that H_1, H_2 and H_3 are Hilbert spaces, that $S \in L(H_1, H_2)$ and $T \in L(H_2, H_3)$. Show that $(TS)^* = S^*T^*$.

Suppose that H is a Hilbert space and that $T \in L(H)$. T is *normal* if $TT^* = T^*T$.

Exercise 8.5.6 Suppose that H is a complex Hilbert space and that $T \in L(H)$. Show the following.

(i) T is normal if and only if $\|T(x)\| = \|T^*(x)\|$ for each $x \in H$.
(ii) If T is normal then $\|T\| = \|T^*\|$, and $\|T^n\| = \|T\|^n$ for each $n \in \mathbf{N}$.
(iii) If T is normal, then $T(x) = 0$ if and only if $x \in T(H)^\perp$, if and only if $T^*(x) = 0$, and if and only if $x \in T^*(H)^\perp$.

Suppose that H is a Hilbert space and that $T \in L(H)$. T is *self-adjoint* if $T = T^*$. A self-adjoint operator is clearly normal. In the complex case, a self-adjoint operator is also called a *Hermitian operator*, and in the real case is called a *symmetric operator*.

Exercise 8.5.7 Suppose that $T \in L(H)$, where H is a complex Hilbert space. By considering $T_1 = \frac{1}{2}(T + T^*)$ and $T_2 = -\frac{1}{2}i(T - T^*)$, show that T can be written as $T_1 + iT_2$, where T_1 and T_2 are Hermitian. Show that such a representation is unique.

Exercise 8.5.8 Suppose that $T \in L(H)$, where H is a real Hilbert space. T is *skew-symmetric* if $\langle T(x), y\rangle = -\langle x, T(y)\rangle$ for all $x, y \in H$. If $T \in L(H)$, show that T can be written uniquely as $T = T_1 + T_2$, where T_1 is symmetric, and T_2 is skew-symmetric.

Exercise 8.5.9 Suppose that H is a complex Hilbert space and that $T \in L(H)$.

(i) Show that TT^* and T^*T are Hermitian.
(ii) Show that if T is Hermitian then $\langle T(x), x\rangle \in \mathbf{R}$ for each $x \in H$.

A Hermitian operator is *positive* if $\langle T(x), x\rangle \geq 0$ for all $x \in H$.

Exercise 8.5.10 Show that the set of all Hermitian operators on a complex Hilbert space H is a closed real linear subspace of $L(H)$.

Suppose that H is a Hilbert space. Recall that an element P of $L(H)$ is a projection if $P^2 = P$. By long tradition, a projection which is self-adjoint is called an *orthogonal projection.*

Theorem 8.5.11 *If P is an orthogonal projection, then $M = P(H)$ is closed, $P = P_M$ and $H = P(H) \oplus (I - P)(H)$. Conversely if N is a closed linear subspace of H, then P_N is an orthogonal projection.*

Proof $M = \{x : (I - P)(x) = 0\}$ is closed (this is generally true for continuous projections on Banach spaces). If $x \in H$ and $y \in M$,

$$\langle x - P(x), y\rangle = \langle x, y\rangle - \langle P(x), y\rangle = \langle x, y\rangle - \langle x, P(y)\rangle = \langle x, y\rangle - \langle x, y\rangle = 0.$$

The results now follow from Exercise 8.1.9. □

Exercise 8.5.12 Suppose that P is a projection on a Hilbert space. Show that the following are equivalent.

(i) P is an orthogonal projection.
(ii) P is normal.
(iii) $\|P(x)\|^2 = \langle P(x), x\rangle$ for all $x \in H$.

An element $U \in L(H)$, where H is a complex Hilbert space is *unitary* if it is an invertible isometry of H. The unitary elements of $L(H)$ form a group under composition, which is closed in $L(H)$.

Exercise 8.5.13 Suppose that $U \in L(H)$, where H is a complex Hilbert space. Show that U is unitary if and only if it is invertible and $\langle U(x), U(y)\rangle = \langle x, y\rangle$ for all $x, y \in H$, and if and only if it is invertible and $U^* = U^{-1}$.

Exercise 8.5.14 Suppose that $T \in L(H)$, where H is a complex Hilbert space. By considering power series in T, show how to define $e^{iT} \in L(H)$. Show that if T is Hermitian then e^{iT} is unitary.

An element $O \in L(H)$, where H is a real Hilbert space is called an *orthogonal isometry* if it is an invertible isometry of H. Most of the properties of unitary mappings carry over to orthogonal isometries. The orthogonal isometries of $L(H)$ form a group under composition, which is closed in $L(H)$.

Theorem 8.5.15 (von Neumann's theorem) *Let T be an isometric linear mapping of a Hilbert space H into itself. Let $F = \{y : T(y) = y\}$ and let $S = \{y - T(y) : y \in H\}$. Then*

(i) *$F = S^{\perp}$ and $H = F \oplus \bar{S}$.*

Let $A_n(x) = \frac{1}{n}(x + T(x) + \cdots + T^{n-1}(x))$. Then

(ii) *$A_n(x) = x$ for $x \in F$;*
(iii) *$A_n(x) \to 0$ for $x \in S$;*
(iv) *$A_n(x) \to 0$ for $x \in \overline{S}$;*
(v) *$A_n(x) \to P_F(x)$ for $x \in H$ (where P_F is the orthogonal projection onto F).*

Proof (i) and (ii) are easy. For (iii) if $y - T(y) \in S$,

$$A_n(y) = \frac{1}{n}(y - T^{n+1}(y)) \to 0 \text{ as } n \to \infty.$$

(iv) Note that $\|A_n(x)\| \leq \|x\|$ for $x \in H$. If $z \in \overline{S}$ and $\epsilon > 0$, there exists $y \in S$ with $\|z - y\| < \epsilon$. Then

$$\|A_n(z)\| \leq \|A_n(z - y)\| + \|A_n(y)\| \leq (\epsilon + \frac{1}{n}(\|A\|_n (z) + \epsilon),$$

from which the result follows.

(v) If $x \in H$, by (i) we can write $x = y + z$, where $y \in F$ and $z \in \overline{S}$. Then $A_n(x) \to y \in F$, and $y \perp (x - y)$, so that $y = P_F(x)$. □

9

The Hahn–Banach Theorem

9.1 The Hahn–Banach Extension Theorem

In the remaining chapters, we consider convex sets and convex functions. The first and most important theorems here are the Hahn–Banach extension theorem, and its geometric equivalent, the separation theorem.

We begin by proving a one-dimensional extension theorem, which is at the heart of the matter. We then use the axiom of choice to prove a general extension theorem. (This, with Tychonoff's compactness theorem, is one of the first times that analysts meet the axiom of choice.) When we consider separable Banach spaces, it is however possible to use a version of the Hahn–Banach theorem which does not use the axiom of choice, and we also give an account of this.

Theorem 9.1.1 *Suppose that p is a convex function on a real vector space E, that F_0 is a linear subspace of E, that $w \in E \setminus F_0$ and that $F_1 = \operatorname{span}(F_0 \cup \{w\})$. Suppose that f is a linear functional on F_0 satisfying $f(x) \leq p(x)$ for all $x \in F_0$. Then there exists a linear functional g on F_1 such that $g(x) = f(x)$ for all $x \in E_0$ (g extends f) and such that $g(y) \leq p(y)$ for all $y \in F_1$ (control is maintained).*

Proof If $y \in F_1$, we can write $y = x + \theta w$ uniquely, with $x \in F_0$. Let $\alpha \in \mathbf{R}$ and let $g(y) = f(x) + \theta\alpha$. Then g is a linear functional on F_1 which extends f, and we need to choose α to maintain control.

We require that $f(x) + \theta\alpha \leq p(x + \theta w)$ for all $x \in E_n$ and all real θ. The inequality is certainly satisfied if $\theta = 0$. For $\theta > 0$ we require that

$$\alpha \leq \frac{p(x + \theta w) - f(x)}{\theta},$$

for all $x \in F_0$, and for $\theta < 0$, writing $\phi = -\theta$, we require

$$\alpha \geq \frac{f(y) - p(y - \phi w)}{\phi},$$

for all $y \in E_n$. If θ and ϕ are positive, then

$$\begin{aligned}
&\frac{p(x+\theta w)}{\theta} + \frac{p(y-\phi w)}{\phi} \\
&\quad = \frac{\theta+\phi}{\theta\phi}\left(\frac{\phi}{\theta+\phi}p(x+\theta w) + \frac{\theta}{\theta+\phi}p(y-\phi w)\right) \\
&\quad \geq \frac{\theta+\phi}{\theta\phi}p\left(\frac{\phi}{\theta+\phi}(x+\theta w) + \frac{\theta}{\theta+\phi}(y-\phi w)\right) \\
&\quad = \frac{\theta+\phi}{\theta\phi}p\left(\frac{\phi x+\theta y}{\theta+\phi}\right) \geq \frac{\theta+\phi}{\theta\phi}f\left(\frac{\phi x+\theta y}{\theta+\phi}\right) \\
&\quad = \frac{f(x)}{\theta} + \frac{f(y)}{\phi}.
\end{aligned}$$

Thus

$$\begin{aligned}
&\sup\left\{\frac{f(y)-p(y-\phi w)}{\phi} : \phi > 0, y \in F_1\right\} \\
&\quad \leq \inf\left\{\frac{p(x+\theta w)-f(x)}{\theta} : \theta > 0, x \in F_1\right\},
\end{aligned}$$

and we can indeed find an α which satisfies both requirements. □

We now use the axiom of choice, in the form of Zorn's lemma, to prove the general Hahn–Banach extension theorem.

Theorem 9.1.2 (The Hahn–Banach extension theorem) *Suppose that p is a convex function on a real vector space E and that F is a linear subspace of E. Suppose that f is a linear functional on F satisfying $f(x) \leq p(x)$ for all $x \in F$. Then there exists a linear functional g on E such that $g(x) = f(x)$ for all $x \in E$ (g extends f) and such that $g(y) \leq p(y)$ for all $y \in E$ (control is maintained).*

Proof Let P be the set of all pairs (G, h), where G is a linear subspace of E containing F, and h is a linear functional on G which extends f; $h(x) = f(x)$ for $x \in F$. Partially order P by setting $(G, h) \prec (G', h')$ if $G \subset G'$ and h' extends h; $h'(x) = h(x)$ for $x \in G$. Then it is immediate that if C is a chain in P then (G', h') is an upper bound for C, where $G' = \cup\{G : (G, h) \in C\}$ and $h'(x)$ is the common value of $h(x)$ for those (G, h) for which $x \in G$. Thus we can apply Zorn's lemma; there exists a maximal (G_{max}, h_{max}) in P.

We claim that $G_{max} = E$, which completes the proof. If not, there exists $w \in E \setminus G_{\max}$. But then, by Theorem 9.1.1, we can extend h_{max} to $\text{span}(G_{max} \cup \{w\})$, retaining control, contradicting the maximality of (G_{max}, h_{max}). □

Here is a classical application of the Hahn–Banach theorem.

Exercise 9.1.3 (Banach limits) If $x \in l^\infty_{\mathbf{R}}$, let $p(x) = \limsup((\sum_{i=1}^n x_i)/n)$. Show that p is a sublinear functional on $l^\infty_{\mathbf{R}}$. Show that there is a linear functional LIM on $l^\infty_{\mathbf{R}}$ (a *Banach limit*) such that $\liminf(x_n) \leq LIM(x) \leq \limsup(x_n)$, $LIM(x) = LIM(L(x))$, where $L(x)_n = x_{n+1}$, and $LIM(x) = \lim_{n\to\infty}(x_n)$ if $x \in c$.

Do we need the axiom of choice? When E is a separable normed space and p is continuous, we can avoid it. We need an easy preliminary result.

Proposition 9.1.4 *Suppose that E_0 is a linear subspace of a separable normed space $(E, \|.\|)$. Then either there exists a finite sequence $(x_1, \ldots, x_n)$ in E such that, setting $E_j = \operatorname{span}(E_0, x_1, \ldots, x_j)$, the sequence $(E_j)_{j=0}^n$ is a strictly increasing sequence of linear subspaces of E, with $E_n = E$, or there exists an infinite sequence $(x_n)_{n=1}^\infty$ in E such that, setting $E_j = \operatorname{span}(E_0, x_1, \ldots, x_j)$, the sequence $(E_j)_{j=0}^\infty$ is a strictly increasing sequence of linear subspaces of E, with $\cup_{n=1}^\infty E_n$ dense in E.*

Proof Let $(e_n)_{n=1}^\infty$ be a dense sequence in E. Let $n_1 = \inf\{n : e_n \notin E_0\}$, and inductively let $n_{j+1} = \inf\{n : e_n \notin E_j = \operatorname{span}(E_0, x_1, \ldots, x_j)\}$, let $x_{j+1} = e_{n_j}$, and let $E_{j+1} = \operatorname{span}(E_j, e_{n_j}) = \operatorname{span}(E_0, x_1, \ldots, x_{j+1})$. If the process terminates, we are in the first case. Otherwise $\{e_j : j \in \mathbf{N}\} \subseteq \cup_{n=1}^\infty E_n$, so that $\cup_{n=1}^\infty E_n$ is dense in $(E, \|.\|)$. □

Theorem 9.1.5 *Suppose that p is a convex function on a separable real normed space $(E, \|.\|)$ with $p(0) = 0$ and that p is continuous at 0. Suppose that E_0 is a linear subspace of E and that f is a linear functional on E_0 satisfying $f(x) \leq p(x)$ for all $x \in E_0$. Then there exists a continuous linear functional g on E such that $g(x) = f(x)$ for all $x \in E_0$ (g extends f) and such that $g(y) \leq p(y)$ for all $y \in E$ (control is maintained).*

Proof Let (E_n) and (x_n) be sequences satisfying the conclusions of Proposition 9.1.4. Using Theorem 9.1.1, we can inductively define a linear function h on $\cup_{n=1}^\infty E_n$ which extends f and is controlled by p. Let $q(x) = \max(p(x), p(-x))$; q is continuous at 0. Then $|h(x+y) - h(x)| = |h(y)| \leq q(y)$, so that h is continuous; it therefore extends to a continuous linear functional g on E. g certainly extends f, and, since g and p are both continuous, $g(x) \leq p(x)$ for all $x \in E$. □

In what follows, if a result is true in general, but can be proved in the separable case without using the axiom of choice, we shall include [separable] in its statement.

Since a sublinear functional is convex, the next theorem follows immediately.

Theorem 9.1.6 (The separable Hahn–Banach extension theorem) *Suppose that p is a continuous sublinear functional on a [separable] real normed space $(E, \|.\|)$, that F is a linear subspace of E and that f is a linear functional on E_0 satisfying $f(x) \leq p(x)$ for all $x \in F$. Then there exists a continuous linear functional g on E such that $g(x) = f(x)$ for all $x \in F$ (g extends f) and such that $g(y) \leq p(y)$ for all $x \in F$ (control is maintained).*

Corollary 9.1.7 *Suppose that p is a continuous sublinear functional on a [separable] real normed space $(E, \|.\|)$, and that $x \in E$. There exists a linear functional g on E with $g(x) = p(x)$, and with $g(y) \leq p(y)$ for all $y \in E$. Thus $p(x) = \sup\{g(x) : g$ linear, and $g(y) \leq p(y)$ for all $y \in E\}$, and the supremum is achieved.*

Proof Take $E_0 = \operatorname{span}(x)$, and let $f(\lambda x) = \lambda p(x)$. Then if $\lambda \geq 0$ we have $f(\lambda x) = p(\lambda x)$, while if $\lambda < 0$ then

$$f(x) = \lambda p(x) = -p(|\lambda|x) \leq p(\lambda x),$$

since $0 = p(0) \leq p(\lambda x) + p(-\lambda x)$. □

Recall that if $(E, \|.\|)$ is a normed space, then the dual space E' of continuous linear functionals on E is a Banach space under the norm $\|f\|' = \sup\{|f(x)| : x \in E,\ \|x\| \leq 1\}$.

Theorem 9.1.8 *Suppose that f is a continuous linear functional on a linear subspace F of a [separable] real normed space $(E, \|.\|)$. Then there exists a continuous linear functional g on E which extends f, with $\|g\|' = \|f\|'$.*

Proof Let $p(y) = \|f\|\,\|y\|$, and apply Theorems 9.1.2 and 9.1.6. Then $|g(y)| \leq p(y) = \|f\|\,\|y\|$, so that $\|g\|' \leq \|f\|'$. Of course, $\|g\|' \geq \|f\|'$. □

Corollary 9.1.9 *Suppose that x is a non-zero element of $(E, \|.\|)$. There exists a continuous linear functional g on E such that $g(x) = \|x\|$ and $\|g\|' = 1$. Thus $\|x\| = \sup\{|g(x)| : \|g\|' = 1\}$, and the supremum is attained.*

Proof Let $F = \operatorname{span}(x)$, and if $\lambda x \in F$ let $f(\lambda x) = \lambda \|x\|$. Then $f \in F'$ and $\|f\| = 1$. Let g be an extension with the same norm. □

We can also consider the dual E'' of the dual E': E' is the *bidual* of E. There is a natural linear map $j : E \to E''$, given by $j(x)(f) = f(x)$.

Corollary 9.1.10 *The mapping $j : E \to E''$ is an isometry of E into E''.*

For

$$\|j(x)\|'' = \sup\{|j(x)(f)| : \|f\|' \leq 1\} = \sup\{|f(x)| : \|f\|' \leq 1\} = \|x\|\,.$$

The mapping j need not be surjective. If it is, we say that $(E, \|.\|)$ is *reflexive*. Hilbert space is an example of a reflexive space. The bidual of the separable space $(c_0, \|.\|_\infty)$ of null sequences is isomorphic to the non-separable space $(l_\infty, \|.\|_\infty)$, and so $(c_0, \|.\|_\infty)$ is not reflexive.

9.2 The Separation Theorem

What does the Hahn–Banach theorem mean geometrically?

We need some definitions. A subset A of a real vector space E is *absorbent* if whenever $x \in E$ there exists $\lambda > 0$ such that $\alpha x \in E$ for $0 \leq \alpha < \lambda$. A subset A of a real vector space E is *radially open* if $A - a$ is absorbent for each $a \in A$. The collection of radially open subsets of E is a topology on E, the *radially open* topology.

It follows that if A is convex and absorbent then the gauge p_A is a real-valued non-negative sublinear functional on E and that

$$\{x : p_A(x) < 1\} \subseteq A \subseteq \{x : p_A(x) \leq 1\}.$$

Proposition 9.2.1 *Suppose that A is a radially open convex subset of a real vector space E and that $0 \in A$. Then p_A is continuous in the radially open topology, and $A = \{x : p_A(x) < 1\}$.*

Proof If $x \in E$ and $\epsilon > 0$ then $x + \epsilon A$ is radially open. If $y \in x + \epsilon A$ then $p_A(x + y) - p_A(x) \leq p_A(y) < \epsilon$; p_A is continuous. If $x \in A$, then there exists $\eta > 0$ such that $x + \eta x \in A$. Thus $(1 + \eta)x \in A$, so that $p_A(x) < 1$. □

Theorem 9.2.2 (The separation theorem I) *Suppose that A and B are non-empty disjoint convex subsets of a real vector space $(E, \|.\|)$ and that A is radially open. Then there exists a linear functional g on E and a scalar λ such that $g(a) < \lambda$ for $a \in A$ and $g(b) \geq \lambda$ for $b \in B$.*

Proof Let $C = A - B = \{a - b : a \in A, b \in B\}$. C is convex, and $0 \notin C$. Since $C = \cup_{b \in B}(A - b)$, C is radially open.

Pick $c_0 = a_0 - b_0 \in C$, and let $D = C - c_0$. Then D is convex and radially open, $0 \in D$ and $-c_0 \notin D$. Let p_D be the gauge of D. Then $p_D(-c_0) \geq 1$. Let $W = \text{span}(c_0)$, and set $f(\alpha c_0) = -\alpha$, for $\alpha c_0 \in W$. Then f is a linear functional on W, and $f(\alpha c_0) \leq p_D(\alpha c_0)$. By the Hahn–Banach theorem, there exists a linear functional g on B extending f and such that $g(x) \leq p_D(x)$ for all $x \in V$. Now if $a \in A$ and $b \in B$ then $a - b - c_0 \in D$, and so $g(a - b - c_0) = g(a) - g(b) + 1 < 1$, by Proposition 9.2.1; thus $g(a) < g(b)$. Let $\lambda = \inf\{g(b) : b \in B\}$; λ is finite, and $g(a) \leq \lambda$ for $a \in A$. Finally, if $a \in A$

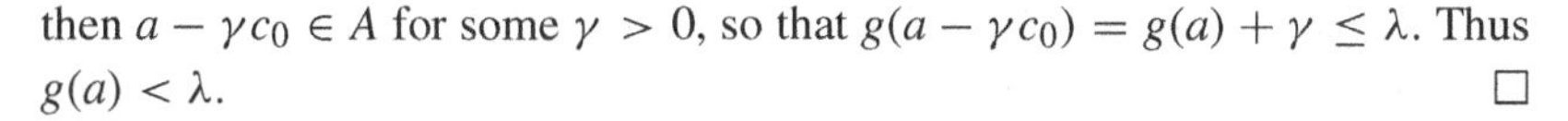

then $a - \gamma c_0 \in A$ for some $\gamma > 0$, so that $g(a - \gamma c_0) = g(a) + \gamma \leq \lambda$. Thus $g(a) < \lambda$. □

Since an open subset of a normed space $(E, \|.\|)$ is radially open, we obtain the more familiar separation theorem for normed spaces.

Corollary 9.2.3 (The separation theorem for normed spaces) *Suppose that A and B are non-empty disjoint convex subsets of a [separable] real normed space $(E, \|.\|)$ and that A is open. Then there exists a continuous linear functional g on E and a scalar λ such that $g(a) < \lambda$ for $a \in A$ and $g(b) \geq \lambda$ for $b \in B$.*

Proof For then p_D is continuous, and so therefore is g. □

It is also of interest to prove the separation theorem directly. We start with a slightly weaker result.

Theorem 9.2.4 *Suppose that A is a non-empty radially open convex set in a real vector space E and that F is a linear subspace of E disjoint from A. Then there exists a hyperplane G of E containing F and disjoint from A.*

Proof We use the axiom of choice by a simple application of Zorn's lemma. Partially order the linear subspaces of E which contain F and are disjoint from A by direct inclusion. By Zorn's lemma, there exists a maximal element G. We must show that G is a hyperplane.

Suppose not. Let $q : E \to E/G$ be the quotient mapping. Then $q(A)$ is radially open and convex. Let a_0 be any element of A. If G is not a hyperplane in E, there exists an element e_0 of E/G linearly independent of $q(a_0)$. Let us consider the two-dimensional space $P = \text{span}(q(a_0), e_0)$. Then $q(A) \cap P$ is a non-empty radially open subset of P. If $\theta \in [0, 2\pi]$, let $c(\theta) = q(a_0)\cos\theta + e_0 \sin\theta$ and let

$$J = \{\theta \in [0, 2\pi) : \text{there exists } r > 0 \text{ such that } rc(\theta) \in q(A)\}.$$

Then J is a non-empty open subset. Let $K = \{(j + \pi) \bmod (2\pi) : j \in J\}$. Then K is also open. But $J \cap K = \emptyset$, since if $rc(\theta) \in q(A)$ then $-sc(\theta) \notin q(A)$ for any $s > 0$, since $q(A)$ is convex and $0 \notin q(A)$. But $[0, 2\pi)$ is connected, and so there exists $m \notin J \cup K$. Thus $\text{span}(c(m) \cap q(A))$ is empty. Consequently $q^{-1}\text{span}(c(m))$ is a linear subspace of E which contains G strictly, which contains F and is disjoint from A. □

Thus the theorem depends upon the connectedness of $[0, 2\pi)$.

Corollary 9.2.5 (The separation theorem II) *Suppose that A and B are disjoint non-empty convex subsets of a real vector space E, and that A is radially open. Then there exists a linear functional ϕ on E and a constant λ such that*

$\phi(a) > \lambda$ for $a \in A$ and $\phi(b) \leq \lambda$ for $b \in B$. If $(E, \|.\|)$ is a normed space and A is open, then ϕ is continuous.

Proof Let $a_0 \in A, b_0 \in B$, and let $c_0 = a_0 - b_0$. Let $C = A - B$; then $0 \notin C$, so that there exists a hyperplane G such that $G \cap C = \emptyset$. Since $c_0 \in C$, $c_0 \notin G$; consequently $E = G \oplus \text{span}(c_0)$, and if $x \in E$, we can write x uniquely as $g + \alpha c_0$. Let $\phi(x) = \alpha$; then ϕ is a linear functional on E with null-space G and $\phi(c_0) = 1$. If $c \in C$, then $\phi(c) > 0$; for otherwise $\phi(c - \phi(c)c_0) = 0$. Thus if $a \in A$ and $b \in B$ then $\phi(a) > \phi(b)$; in particular, ϕ is bounded below on A. Let $\lambda = \inf_{a \in A} \phi(a)$, so that $\sup_{b \in B} \phi(b) \leq \lambda$. Since A is radially open, $\phi(A)$ is an open interval, and so $\phi(a) > \lambda$ for $a \in A$. □

We can deduce the analytic Hahn–Banach theorem from the geometric separation theorem.

Proof of Theorem 9.1.2. The set $S_p = \{(x, \lambda) : \lambda > p(x)\}$ is radially open in $V \times \mathbf{R}$. It is disjoint from the linear subspace $G(f) \subseteq F \times \mathbf{R} \subseteq V \times \mathbf{R}$. By the separation theorem, there is a hyperplane H in $E \times \mathbf{R}$ which contains $G(f)$ and is disjoint from S_p. If (h, k_1) and (h, k_2) are two elements of H, so is $(h, (1 - t)k_1 + tk_2)$ for all $t \in \mathbf{R}$, which is only possible if $k_1 = k_2$. Thus there exists a linear functional g on E such that $H = G(g)$. Then g extends f and $g(x) \leq p(x)$ for all $x \in E$. □

This corollary applies when p is a sublinear function, or semi-norm or norm.

9.3 Weak Topologies

Corollary 9.1.9 shows that a [separable] real normed space E and its dual E' form a dual pair; a *dual pair* (E, F) consists of two vector spaces E and F, over the same field K, together with a bilinear mapping $E \times F \to K$, written as $(e,f) \to \langle e,f \rangle$, with the properties that if $e \neq 0$ then there exists $f \in F$ with $\langle e,f \rangle \neq 0$ and that if $f \neq 0$ then there exists $e \in E$ with $\langle e,f \rangle \neq 0$. If (E, F) is a dual pair we can consider E as a vector space of functions on F – that is, E is a linear subspace of K^F, the vector space of all scalar-valued functions on F – and F as a vector space of functions on E.

When K is $\mathbf{R}$ or $\mathbf{C}$ we can give K^F the product topology – that is, the topology of pointwise convergence on the points of F. The subspace topology on E is then called the *weak topology* $\sigma(E, F)$. Similarly, we have the weak topology $\sigma(F, E)$ on F. Thus a basic neighbourhood of 0 is a set of the form

$$\{e \in E : |\langle e,f_i \rangle| < 1 f_i \in F, 1 \leq i \leq n\};$$

note that this is a convex set. A neighbourhood of a point e of E is a set of the form $e + N$, where N is a neighbourhood of 0.

Exercise 9.3.1 Show that $(E, \sigma(E,F))$ is a *topological vector space*; the mapping $(e_1, e_2) : (E, \sigma(E,F)) \times (E, \sigma(E,F)) \to (E, \sigma(E,F))$ is jointly continuous, and so is the mapping $(\lambda, e) \to \lambda e : K \times (E, \sigma(E,F)) \to (E, \sigma(E,F))$.

Proposition 9.3.2 *If (E,F) is a real or complex dual pair, then F can be identified with the space of $\sigma(E,F)$-continuous linear functionals on E.*

Proof Since $\sigma(E,F)$ is the topology of pointwise convergence, the evaluation functional $e \to \langle e,f \rangle$ is continuous, for each $f \in F$.

Suppose that ϕ is a continuous linear functional on $(E, \sigma(E,F))$. Since ϕ is continuous at 0, there exist $f_1, \ldots, f_r$ in F and $\epsilon > 0$ such that if $| \langle e,f_i \rangle | < \epsilon$ for $1 \leq i \leq r$ then $|\phi(e)| < 1$. By homogeneity, if $| \langle e,f_i \rangle | = 0$ for $1 \leq i \leq r$ then $|\phi(e)| = 0$.

Let $T : E \to \mathbf{R}^r$ be defined by $T(e) = (\langle e,f_1 \rangle, \ldots, \langle e,f_r \rangle)$. If $T(e_1) = T(e_2)$ then $\phi(e_1) = \phi(e_2)$, and so there exists a linear functional θ on $T(E)$ such that $\phi(e) = \theta(T(e))$ for all $e \in E$. By linear algebra, we can extend θ to a linear functional ψ on $\mathbf{R}^r$. There exist $\alpha_1, \ldots, \alpha_r$ such that $\psi(x_1, \ldots, x_r) = \alpha_1 x_1 + \cdots + \alpha_r x_r$. If $e \in E$, then

$$\begin{aligned} \phi(e) &= \theta(T(e)) = \psi(T(e)) \\ &= \alpha_1 \langle e,f_1 \rangle + \cdots + \alpha_r \langle e,f_r \rangle \\ &= \langle e, \alpha_1 f_1 + \cdots + \alpha_r f_r \rangle . \end{aligned}$$

Thus we can identify ϕ with $\alpha_1 f_1 + \cdots + \alpha_r f_r$.

Exercise 9.3.3 Show that if C is convex, then so is its $\sigma(E,F)$-closure. If $A \subseteq E$, let $\Gamma(A)$ denote the smallest convex set containing A; $\Gamma(A)$ is the convex cover of A. Show that $\overline{\Gamma(A)}$ is the smallest $\sigma(E,F)$-closed convex set containing A.

Exercise 9.3.4 Suppose that (E_1, F_1) and (E_2, F_2) are real or complex dual pairs and that $T : E_1 \to E_2$ is linear. Let $T'(f_2)(e_1) = \langle T(e_1), f_2 \rangle$ for $e_1 \in E_1$ and $f_2 \in F_2$. Then T is weakly continuous (that is, continuous from $(E_1, \sigma(E_1, F_1))$ to $(E_2, \sigma(E_2, F_2))$) if and only if $T'(F_2) \subseteq F_1$.

9.4 Polarity

When we have duality, it is natural to consider polarity. Here we restrict attention to the real case. Suppose that (E, F) is a real dual pair and that $A \subseteq E$. We define

$$A^o = \{f \in F : \langle a,f \rangle \leq 1 \text{ for all } a \in A\} = \cap_{a \in A} \{a\}^o .$$

Similarly if $B \subseteq F$ we define $B^o \subseteq E$.

Proposition 9.4.1 *Suppose that (E, F) is a real dual pair, and that $A \subseteq E$. Then A^o is $\sigma(F, E)$-closed, convex and contains* 0.

Proof For each of the sets $\{f \in F : \langle a,f\rangle \leq 1\}$ is $\sigma(F, E)$-closed, convex and contains 0. □

Exercise 9.4.2 Suppose that (E, F) is a real dual pair, and that $A, A_1, A_2 \subseteq E$. Show the following.

(i) If $A_1 \subseteq A_2$ then $A_1^o \supseteq A_2^o$.
(ii) $A \subseteq A^{oo}$ and $A^o = A^{ooo}$.
(iii) If λ is real and positive then $(\lambda A)^o = (1/\lambda)A^o$.

Theorem 9.4.3 (The theorem of bipolars) *Suppose that (E, F) is a real dual pair and that $A \subseteq E$. Then A^{oo} is the smallest $\sigma(E, F)$-closed convex set containing A and 0: that is, $A^{oo} = \overline{\Gamma(A \cup \{0\})}$.*

Proof Certainly $A^{oo} \supseteq \overline{\Gamma}(A \cup \{0\})$. Suppose that $x \notin \overline{\Gamma}(A \cup \{0\})$. Then there exists non-zero $f_1, \ldots, f_n$ such that if

$$M = \{y : |\langle x - y, f_i\rangle| < 1 \text{ for } 1 \leq i \leq n\},$$

then M is disjoint from $\Gamma(A \cup \{0\})$. Let $T(e) = (\langle e, f_i\rangle)_{i=1}^n$. T is a linear mapping from E into $\mathbf{R}^n$. Let $U = \{y \in \mathbf{R}^n : |y_i - T(x)_i| < 1 \text{ for } 1 \leq i \leq n\}$. Then U is a convex open subset of $\mathbf{R}^n$ disjoint from the convex set $B = T(\overline{\Gamma}(A \cup \{0\})$, and so by the separation theorem (applied to $\mathbf{R}^n$, with its usual topology) there exists a linear functional ϕ on $\mathbf{R}^n$ and $\lambda \in \mathbf{R}$ such that $\phi(y) > \lambda$ for $y \in U$ and $\phi(z) \leq \lambda$ for $z \in B$. Since $x \in B$, $\lambda \geq 0$. Let $\phi(e_i) = \phi_i$ and let $\psi = \sum_{i=1}^n \phi_i f_i \in F$. Then $\langle x, \psi\rangle > \lambda$ and $\langle y, \psi\rangle) \leq \lambda$ for $y \in \overline{\Gamma}(A \cup \{0\})$. Choose $\lambda < \mu < \psi(x)$, and let $\theta = \psi/\mu$. Then $\theta \in A^o$. Since $\langle x, \theta\rangle > 1$, $x \notin A^{oo}$. □

The following corollary is a convenient consequence of the theorem of bipolars.

Corollary 9.4.4 *If A is a non-empty $\sigma(E, F)$-closed convex subset of E and $x \notin A$ then there exists $f \in F$ such that $\langle x, f\rangle > \sup\{\langle a, f\rangle : a \in A\}$.*

Proof Choose $a_0 \in A$, and apply the theorem of bipolars to $A - a_0$. □

9.5 Weak and Weak* Topologies for Normed Spaces

When $(E, \|.\|)$ is a normed space, we can consider the weak topology $\sigma(E, E')$ on E. There are two weak topologies on E', the topology $\sigma(E', E)$, which is called the *weak*-topology* and the weak topology $\sigma(E', E'')$. They must not be confused. How are the weak and weak* topologies related to the corresponding norm topologies?

Proposition 9.5.1 *The weak topology $\sigma(E, E')$ on a real normed space $(E, \|.\|_E)$ is weaker than the norm topology on E. The following are equivalent.*

(i) The weak topology is the same as the norm topology on E.
(ii) The restriction of the weak topology to the unit ball B(E) is the same as the restriction of the norm topology on E.
(iii) E is finite-dimensional.

Proof Since the norm topology and the weak topology are translation invariant, it is sufficent to consider neighbourhoods of 0. Suppose that $U = \{x : |\langle x, \phi\rangle_i| < 1, 1 \leq i \leq n\}$ is a fundamental weak neighbourhood of 0. Let $M = \max_{1 \leq i \leq n} \|\phi_i\|$. If $\|x\| < 1/M$ then $x \in U$, and so the weak topology is weaker than the norm topology.

(i) certainly implies (ii). If (ii) holds, there exist $\phi_1, \ldots, \phi_n$ in E' such that if $\langle x, \phi_i\rangle | < 1$ then $\|x\| < 1$. Suppose that $\langle x, \phi_i\rangle = 0$ for $1 \leq i \leq n$. If $\alpha \in \mathbf{R}$ then $\langle \alpha x, \phi_i\rangle = 0$ for $1 \leq i \leq n$, so that $|\alpha|. \|x\| < 1$. Since this holds for all $\alpha \in \mathbf{R}$, $x = 0$. Thus the linear mapping $x \to (\langle x, \phi_i\rangle)_{i=1}^n : E \to \mathbf{R}^n$ is injective, and E is finite-dimensional; (ii) implies (iii).

Suppose that $(E, \|.\|)$ is finite-dimensional, and that $(e_1, \ldots, e_n)$ is a basis for E, with dual basis $(\phi_1, \ldots, \phi_n)$. Let

$$U = \{x : |\langle x, \phi_i\rangle| < \|e_i\| /n \text{ for } \leq i \leq n\}.$$

If $x \in U$ then $\|x\| \leq \sum_{i=1}^n |\langle x, \phi_i\rangle| \|e_i\| < 1$, and so the weak topology and the norm topology are the same on E; (iii) implies (i). □

Things can be different on the unit sphere. A real normed space $(E, \|.\|)$ is *uniformly convex* if whenever $\epsilon > 0$ there exists $\delta > 0$ such that whenever $\|x\| = \|y\| = 1$ and $\|x - y\| \geq \epsilon$ then $\left\|\frac{1}{2}(x + y)\right\| \leq (1 - \delta)$.

Proposition 9.5.2 *Suppose that $(E, \|.\|)$ is a uniformly convex Banach space. If $\epsilon > 0$ then there exists $\delta > 0$ such that if $\|x\| = 1$ and if x^* is a continuous linear functional on E with $\|x^*\|' = \langle x, x^*\rangle = 1$ then*

$$\left\{y : \langle y, x^*\rangle > 1 - \delta\right\} \subseteq \left\{y : \|y\| = 1, \|y - x\| \leq \epsilon\right\},$$

so that the weak topology and the norm topology coincide on the unit sphere $S(E) = \{x \in E : \|x\| = 1\}$.

Proof Suppose that $\epsilon > 0$. Let δ be the quantity assured by the definition of uniform convexity. If $\langle y, x^*\rangle \geq 1 - \delta$, then $\left\langle \frac{1}{2}(x + y), x^*\right\rangle \geq 1 - \delta/2$. Hence $\|x - y\| \leq \epsilon$. □

Theorem 9.5.3 *A Hilbert space H is uniformly convex.*

Proof Suppose that $\|x\| = \|y\| = 1$. Then $\|(x+y)/2\|^2 + \|x-y\|^2 = 1$, by the parallelogram law, and this gives the result. □

We shall meet more examples of uniformly convex spaces later; in Section 15.5 we shall define the L_p spaces, for $1 < p < \infty$, and show that they are uniformly convex.

If (E, F) is a dual pair and $B \subseteq E$, we say that B is *$\sigma(E,F)$-bounded* or *weakly bounded* if $\{\langle b,f\rangle : b \in B\}$ is bounded, for each $f \in F$. A set which is bounded for the weak* topology is said to be weak* bounded.

Proposition 9.5.4 *A subset B of the dual E' of a Banach space $(E, \|.\|)$ is weak* bounded if and only if it is norm bounded.*

Proof If B is norm bounded, then $\{b(x) : b \in B\}$ is bounded for each $x \in (E, \|.\|)$, since the mapping $\phi \to \phi(x)$ is a continuous linear functional on E'; thus B is weak* bounded. Conversely if B is weak* bounded, it is a pointwise bounded set of continuous linear functionals on $(E, \|.\|)$, and so it is norm bounded, by the principle of uniform boundedness (Theorem 7.6.1). □

Corollary 9.5.5 *A subset A of a [separable] normed space $(E, \|.\|)$ is weakly bounded if and only if it is norm bounded.*

Proof As before, A is weakly bounded if it is norm bounded. If A is weakly bounded then $j(A)$ is weak* bounded in E'', where j is the canonical isometric embedding of $(E, \|.\|)$ into E''. Thus $j(A)$ is norm bounded in E'', and so A is norm bounded. □

Exercise 9.5.6 Suppose that $(E_1, \|.\|_1)$ and $(E_2, \|.\|_2)$ are [separable] real Banach spaces and that T is a linear map from $(E_1, \|.\|_1)$ to $(E_2, \|.\|_2)$. Then T is norm continuous if and only if it is weakly continuous.

If (E, F) is a dual pair of vector spaces, a sequence $(x_n)_{n=1}^\infty$ is a $\sigma(E,F)$ *Cauchy sequence* if whenever N is a $\sigma(E,F)$ neighbourhood of 0 there exists n_0 such that $x_m - x_n \in N$ for $m, n \geq n_0$. That is to say, $(x_n)_{n=1}^\infty$ is a Cauchy sequence in the uniformity defined by $\sigma(E,F)$. A $\sigma(E,F)$-convergent sequence is clearly a $\sigma(E,F)$ Cauchy sequence. When appropriate, we use the terms 'weakly Cauchy' and 'weak* Cauchy'.

Although the norm topology and the weak topology on the Banach space l_1, and on its unit ball $B(l_1)$, are different, the two topologies have the same Cauchy sequences and convergent sequences.

Theorem 9.5.7 *A sequence $(y^{(n)})_{n=1}^\infty$ in l_1 is weakly Cauchy if and only if it is norm convergent.*

Proof A norm convergent sequence is certainly weakly Cauchy. Suppose, if possible, that $(x^{(n)})_{n=1}^{\infty}$ is weakly Cauchy, but not norm convergent. By choosing a subsequence if necessary, we can suppose that there exists $\epsilon > 0$ such that if $y^{(n)} = x^{2n} - x^{2n+1}$ then $(y^{(n)})_{n=1}^{\infty}$ converges weakly to 0, and $\|y^{(n)}\|_1 \geq \epsilon$ for all $n \in \mathbf{N}$. Note that $y_j^{(n)} \to 0$ as $n \to \infty$ for all $j \in \mathbf{N}$. We use a 'sliding hump' argument. Let $j_0 = 0$ and let $n_1 = 1$. There exists j_1 such that $\sum_{j=j_1+1}^{\infty} |y^{(n_1)}| < \epsilon/5$. We now show that there are strictly increasing sequences $(n_i)_{i=1}^{\infty}$ and $(j_i)_{i=1}^{\infty}$ in $\mathbf{N}$ such that

(i) $|y_j^{(n_i)}| < \epsilon/5j_{i-1}$ for $i > 1$ and $1 \leq j \leq j_{i-1}$, and
(ii) $\sum_{j=j_i+1}^{\infty} |y_j^{(n_i)}| < \epsilon/5$.

Suppose that we have found n_k and j_k satisfying (i) and (ii), for $1 \leq k \leq i$. By co-ordinatewise convergence, we can find n_{i+1} satisfying (i), and we can then find j_{i+1} satisfying (ii). For each i there exist ϕ_j for $j_{i-1} < j \leq j_1$, with $|\phi_j| = 1$ such that $\sum_{j=j_{i-1}+1}^{j_i} \phi_j(y_j^{(n_i)}) = \sum_{j=j_{i-1}+1}^{j_i} |y_j^{(n_i)}|$. Then $\phi = (\phi_j)_{j=1}^{\infty} \in l_{\infty}$, and $\|\phi\|_{\infty} = 1$. Now

$$\sum_{j=j_{i-1}+1}^{j_i} |y_j^{(n_i)}| \geq \left\|y^{(n_i)}\right\|_1 - \sum_{j=1}^{j_{i-1}} |y_j^{(n_i)}| - \sum_{j=j_i+1}^{\infty} |y_j^{(n_i)}| \geq 3\epsilon/5,$$

and so

$$\begin{aligned} |\phi(y^{(n_i)})| &\geq \sum_{j=j_{i-1}+1}^{j_i} \phi_j(y_j^{(n_i)}) - \sum_{j=1}^{j_{i-1}} \phi_j(y_j^{(n_i)}) - \sum_{j=j_i+1}^{\infty} \phi_j(x_j^{(n_i)}) \\ &\geq 3\epsilon/5 - \epsilon/5 - \epsilon/5 = \epsilon/5. \end{aligned}$$

Thus $y^{(n_i)}$ does not converge weakly to 0, giving a contradiction. □

Exercise 9.5.8 Deduce that the unit ball $B(l_1)$, with the weak topology $\sigma(l_1, l_{\infty})$, is not metrizable.

Proposition 9.5.9 *A subset L of l_1 is weakly compact (that is, $\sigma(l^1, l^{\infty})$-compact) if and only if it is norm compact.*

Proof A norm compact set is certainly weakly compact. Suppose that L is weakly compact. The metrizable topology of co-ordinatewise convergence is weaker than the weak topology on l_1, and so is the same as the weak topology on L. Consequently, L is a compact metrizable space in the weak topology, and so is weakly sequentially compact. Thus if $(x_n)_{n=1}^{\infty}$ is a sequence in L then there exists a weakly convergent subsequence $(x_{n_k})_{k=1}^{\infty}$. But this subsequence converges in norm, and so L is norm compact. □

Since the weak topology on a normed space is weaker than the norm topology, a weakly closed set is norm closed. For convex sets, the converse holds.

Theorem 9.5.10 *A norm closed convex subset B of a real [separable] normed space $(E, \|.\|)$ is weakly closed.*

Proof Since translation is a homeomorphism in either topology, we can suppose that $0 \in B$. Suppose that $x \notin B$. There exists $\epsilon > 0$ such that $U = \{y : \|y - x\| < \epsilon\}$ is disjoint from B. U is open and convex, and so, by the separation theorem, there exist a continuous linear functional g on $(E, \|.\|)$ and a real number λ such that $g(y) < \lambda$ for $y \in U$ and $g(b) \geq \lambda$ for $b \in B$. Since $0 \in B$, $\lambda \leq 0$. Choose $g(x) < \alpha < \lambda$, and let $h = g/\alpha$. Then $h(b) \leq 1$ for $b \in B$, so that $h \in B^o$, while $h(x) > 1$, so that $x \notin B^{oo}$. Thus $B = B^{oo}$, and B is weakly closed. □

9.6 Banach's Theorem and the Banach–Alaoglu Theorem

Proposition 9.6.1 *Suppose that G is a subset of a normed space $(E, \|.\|)$ and that* span(G) *is dense in E. Then the weak* topology on $B(E')$ is the same as the topology of pointwise convergence on the elements of G.*

Proof The topology of pointwise convergence on the elements of G is weaker than the weak* topology, and is the same as the topology of pointwise convergence on the elements of span(G).

Suppose that $\phi \in B(E')$ and that

$$N = \{\psi \in B(E')' : |\psi(y_i) - \phi(y_i)| < 1, 1 \leq i \leq j\}$$

is a basic $\sigma(E', E)$ neighbourhood of ϕ in $B(E')$. For each $1 \leq i \leq j$ there exists $c_i \in \text{span}(G)$ with $\|c_i - y_i\| < 1/3$. Let

$$M = \{\psi \in B(E') : |\psi(c_i) - \phi(c_i)| < 1/3, 1 \leq i \leq j\}.$$

M is a neighbourhood of ϕ in $B(E')$. If $\psi \in M$ and $1 \leq i \leq j$ then

$$\begin{aligned} |\psi(y_i) - \phi(y_i)| &\leq |\psi(y_i - c_i)| + |\psi(c_i) - \phi(c_i)| + |\phi(c_i - y_i)| \\ &< 1/3 + 1/3 + 1/3 = 1, \end{aligned}$$

so that $M \subseteq N$. Thus the two topologies are the same. □

Exercise 9.6.2 If $(E, \|.\|)$ is separable then $B(E')$ is metrizable in the weak* topology.

Theorem 9.6.3 (Banach's theorem) *The unit ball $B(E')$ of the dual of a separable real normed space $(E, \|.\|)$ is compact and metrizable in the weak* topology.*

Proof Let G be a countable dense subset of E. If $\phi \in B(E')$ and $g \in G$ let $j(\phi)_g = \phi(g)$. Then j is a homeomorphism of $B(E')$ into $K = \prod_{g\in G}([-\|g\|, \|g\|])$. K is compact, and so it is enough to show that $j(B(E'))$ is closed in K. An easy approximation argument shows that $u \in j(B(E'))$ if (and only if) $\alpha u_g + \beta u_h = u_{\alpha g+\beta h}$ for all $\alpha, \beta \in \mathbf{Q}$ and $g, h \in G$. Thus if $u \in K \setminus j(B')$, there exist $\alpha, \beta \in \mathbf{Q}$ and $g, h \in G$ such that

$$|(\alpha u_g + \beta u_h) - u_{\alpha g+\beta h}| = \gamma > 0.$$

Let $\epsilon = \gamma/(|\alpha| + |\beta| + 1)$ and let

$$V = \{v \in K : |v_g - u_g| < \epsilon, |v_h - u_h| < \epsilon, |v_{\alpha g+\beta g} - u_{\alpha g+\beta g}| < \epsilon\}.$$

Then V is a neighbourhood of u in K which is disjoint from $j(B(E'))$, and so $j(B(E'))$ is closed in K. □

If we accept the axiom of choice, so that we can use Tychonoff's theorem, we can improve Banach's theorem.

Theorem 9.6.4 (The Banach–Alaoglu theorem) *If $(E, \|.\|)$ is a normed space, and we accept the axiom of choice, then $B(E')$ is compact with the weak* topology $\sigma(E', E)$.*

Proof For then $\prod_{g\in E}([-\|g\|, \|g\|])$ is compact, and the argument of Theorem 9.6.3 goes through. □

Theorem 9.6.5 *A [separable] Banach space $(E, \|.\|)$ is reflexive if and only if its unit ball $B(E)$ is weakly compact.*

Proof The condition is necessary, by the Banach–Alaoglu theorem [or Banach's theorem]. Suppose that $B(E)$ is weakly compact. Let $j : E \to E''$ be the canonical embedding. Then $j(B(E))$ is $\sigma(E'', E')$-compact, and so $\sigma(E'', E')$-closed. By the theorem of bipolars, this means that $j(B(E)) = j(B(E))^{oo} = B(E'')$, the unit ball of E'', and so j is surjective. □

9.7 The Complex Hahn–Banach Theorem

The Hahn–Banach theorem is essentially a real theorem, but it can be usefully applied in the complex case. Usually this is done by considering the underlying real space $E_{\mathbf{R}}$. The complex linear functionals on a complex vector space E are closely related to the real linear functionals on the real space $E_{\mathbf{R}}$. If ψ is a complex linear functional on E then $\psi_{\mathbf{R}} = \Re(\psi)$ is a real linear functional on $E_{\mathbf{R}}$.

Proposition 9.7.1 *Suppose that E is a complex vector space, and that $E_{\mathbf{R}}$ is the underlying real space. If ϕ is a linear functional on $E_{\mathbf{R}}$ and $x \in E$ let $j(\phi)(x) = \phi(x) - i\phi(ix)$. Then $j(\phi)$ is a complex linear functional on E. If ψ is a complex linear functional on E, then $j(\psi_{\mathbf{R}}) = \psi$.*

Proof $j(\phi)(y_1) + j(\phi)(y_2) = j(\phi)(y_1 + y_2)$ and

$$j(\phi)(ix) = \phi(ix) - i\phi(-x) = i(\phi(x) - i(\phi((x)) = ij(\phi)(x),$$

so that $j(\phi)$ is a complex linear functional on E.

If $\psi(x) = a+ib$, then $\psi(ix) = -b+ia$, so that $\psi_{\mathbf{R}}(x) = a$ and $\psi_{\mathbf{R}}(ix) = -b$, so that $j(\psi_{\mathbf{R}})(x) = a - i(-b) = a + ib = \psi(x)$. □

Theorem 9.7.2 (The complex separation theorem) *Suppose that $(E, \|.\|)$ is a [separable] complex normed space, that A and B are disjoint convex sets, and that A is open. Then there exist a continuous linear functional f and a real number λ such that $\Re\langle a,f\rangle > \lambda$ for $a \in A$ and* $\sup_{b\in B} \Re\langle b,f\rangle \leq \lambda$.

Proof For we can separate A and B by a real linear functional k, and then set $f(x) = k(x) - ik(ix)$. □

The analytic Hahn–Banach theorem takes the following form.

Theorem 9.7.3 *Suppose that p is a continuous semi-norm on a [separable] complex normed space $(E, \|.\|)$, and that f is a linear functional on a subspace F of E for which $|f(x)| \leq p(x)$ for $x \in F$. Then there exists a linear functional g on E such that $f(x) = g(x)$ for $x \in F$ (g is an extension of f) and $|g(y)| \leq p(y)$ for $y \in E$ (control is maintained).*

Proof Let $E_{\mathbf{R}}$ and $F_{\mathbf{R}}$ be the underlying real spaces, and let $f_{\mathbf{R}}$ be the real part of f. Then $f_{\mathbf{R}}$ is a linear functional on $F_{\mathbf{R}}$ and $f_{\mathbf{R}}(x) \leq p(x)$ for $x \in F_{\mathbf{R}}$. By Theorem 9.1.6, $f_{\mathbf{R}}$ can be extended to a real linear functional k such that $k(y) \leq p(y)$ for $y \in E$. By Proposition 9.7.1, $g = j(k)$ is a complex linear functional on E. If $x \in F$ and $f(x) = re^{i\theta}$, then

$$f(x) = e^{i\theta} f(e^{-i\theta}x) = e^{i\theta} k(e^{-i\theta}x) = e^{i\theta} g(e^{-i\theta}x) = g(x),$$

so that g extends f. If $y \in E$ and $g(y) = re^{i\theta}$ then

$$|g(y)| = r = g(e^{-i\theta}y) = k(e^{-i\theta}y) \leq p(e^{-i\theta}y) = p(y),$$

so that control is maintained. □

If (E, F) is a complex dual pair and $A \subseteq E$, we set

$$A^o = \{f \in F : \Re \langle a,f \rangle \leq 1 \text{ for all } a \in A\}.$$

Then all the results corresponding to those of Sections 9.4–9.6 go through.

10
Convex Functions

Exercise 10.0.1 Suppose that f is a real-valued convex function on an open interval (a, b) of $\mathbf{R}$. Show the following.

(i) f is continuous.

(ii) If $x \in (a, b)$ then the *right derivative* $f'_+(x) = \lim_{t \searrow 0}(f(x+t) - f(x))/t$ and the *left derivative* $f'_-(x) = \lim_{t \searrow 0}(f(x) - f(x-t))/t$ exist and are finite.

(iii) The functions f'_+ and f'_- are increasing functions on (a, b). f'_+ is right continuous and f'_- is left continuous.

(iv) If $a < x < y < b$ then $f'_-(x) \leq f'_+(x) \leq f'_-(y) \leq f'_+(y)$.

(v) The set $J = \{x \in (a, b) : f'_+(x) \neq f'_-(x)\}$ is countable. If $x \in J$ then f'_+ and f'_- have a jump discontinuity at x, of size $f'_+(x) - f'_-(x)$. If $x \notin J$ then f'_+ and f'_- are continuous at x, and f is differentiable at x, with derivative $f'_+(x) \, (= f'_-(x))$.

Our aim is to see how far these results extend to convex functions on a vector space, and in particular to convex functions on a separable Banach space. We begin by considering a dual pair of real vector spaces.

10.1 Convex Envelopes

Suppose that (E, F) is a real dual pair. We denote the vector space of $\sigma(E, F)$-continuous affine functions on E (functions of the form $a(x) = \langle x, \phi \rangle + c$, where $\phi \in F$ and $c \in \mathbf{R}$) by $A(E)$. Suppose that f is a proper function on E taking values in $(-\infty, \infty]$. We set $L(f) = \{a \in A(E), a \leq f\}$. If $L(f) \neq \emptyset$, we set $\underline{f} = \sup\{a : a \in L(f)\}$. (If $L(f) = \emptyset$, we set $\underline{f} = -\infty$.) The function $\underline{f}$, which is convex, is the *lower convex envelope* of f. Note that if $L(f)$ is not empty and $a \in L(f)$ then $g = f - a \geq 0$ and $\underline{f} = \underline{g} + a$, so that in many circumstances, we can consider the case where $f \geq 0$.

Exercise 10.1.1 Suppose that (E, F) is a dual pair, and that f and g are functions on E taking values in $(-\infty, \infty]$. Show the following.

(i) $\underline{f}$ is convex and $\sigma(E, F)$ lower semi-continuous.
(ii) $\underline{f+g} \geq \underline{f} + \underline{g}$, with equality if f or g is affine, and $\underline{rf} = r\underline{f}$ for $r \geq 0$.
(iii) If f is bounded below then $\underline{f} \geq \inf_{x \in E} f(x)$.

Theorem 10.1.2 *Suppose that (E, F) is a dual pair, and that f and g are functions on E taking values in $(-\infty, \infty]$.*

(i) $\underline{f} \leq f$, and $f = \underline{f}$ if and only if f is convex and $\sigma(E, F)$ lower semi-continuous.
(ii) If $(E, \|.\|)$ is a normed space and $F = E'$ then $\underline{f}$ is lower semi-continuous in the norm topology. $f = \underline{f}$ if and only if f is convex and lower semi-continuous in the norm topology.
(iii) If f and g are bounded then $|\underline{f} - \underline{g}| \leq \|f - g\|_\infty$.

Proof (i) If $f = \underline{f}$ then f is convex and lower semi-continuous. Suppose that f is convex and lower semi-continuous. Then the epigraph $A_f = \{(f(x), r) : f(x) \leq r\}$ is a $\sigma(E, F)$-closed convex subset of $E \times \mathbf{R}$. Suppose that $s < f(x)$. By Corollary 9.4.4, there exists $(\phi, t) \in F \times \mathbf{R}^+$ such that $\phi(x) + ts < \lambda = \inf\{\phi(y) + tr : (y, r) \in A_f\}$. Let $\psi = \phi/t$ and let $\mu = \lambda/t$, so that $\psi(x) + s < \mu = \inf\{\psi(y) + r : (y, r) \in S\}$. Let $a(y) = \psi(x) - \psi(y) + s$:then $a \in A(K)$, and $a(x) = s$. Further,

$$a(y) \leq \psi(x) - (\psi(y) + f(y)) + f(y) + s \leq \psi(x) - \mu + f(y) + s \leq f(y),$$

and so $\underline{f}(x) \geq s$. Since this holds for all $s < f(x)$, $\underline{f}(x) \geq f(x)$.

(ii) Since $\underline{f}$ is the supremum of norm-continuous affine functions, it is lower semi-continuous in the norm topology. If f is convex and lower semi-continuous in the norm topology then A_f is a convex norm closed subset of $E \times \mathbf{R}$, and so it is weakly closed. Thus f is $\sigma(E, E')$-lower semi-continuous, and so $f = \underline{f}$ by (ii).

(iii) $\underline{f} = \underline{(f - g) + g} \geq \underline{f - g} + \underline{g}$, so that

$$\underline{g} - \underline{f} \leq -\underline{(f - g)} = \inf_{x \in E}(g(x) - f(x)) \leq \|f - g\|_\infty ,$$

and similarly $\underline{f} - \underline{g} \leq \|f - g\|_\infty$. Thus $|\underline{f} - \underline{g}| \leq \|f - g\|_\infty$. □

Corollary 10.1.3 $-\underline{f} \geq \underline{(-f)}$.

A proper $\sigma(E, F)$-lower semi-continuous convex function on E is called a *regular convex function*. This is unfortunate terminology, since the word 'regular' is much overused.

The *upper concave envelope* $\overline{f}$ of a function in E taking values in $[-\infty, \infty)$ is defined similarly. We set $U(f) = \{a \in A(E) : a \geq f\}$. If $U(f) \neq \emptyset$, we set $\overline{f} = \inf\{a : a \in U(f)\}$. (If $L(f) = \emptyset$, we set $\overline{f} = \infty$.) Then $\overline{f}$ is a $\sigma(E, F)$-upper semi-continuous convex function on E, which satisfies results corresponding to those of Theorem 10.1.2. Clearly, $\overline{f} = -\underline{(-f)}$.

We shall consider the lower convex envelope further in Chapter 11.

10.2 Continuous Convex Functions

We now consider convex functions on a normed space. When are they continuous? If f is a convex function on a normed space, we say that f is *very regular* if Φ_f^{int} is non-empty (f is finite on a non-empty open set) and f is continuous on Φ_f^{int}.

Theorem 10.2.1 *Suppose that f is a convex function on a normed space $(E, \|.\|)$. The following are equivalent.*

(i) f is very regular.
(ii) The graph $\Gamma_f \cap (\Phi_f^{int} \times \mathbf{R})$ of f in Φ_f^{int} is closed in $\Phi_f^{int} \times \mathbf{R}$.
(iii) Φ_f^{int} is non-empty, and there is a non-empty open subset U of Φ_f^{int} on which f is bounded above.
(iv) The strict epigraph S_f^{int} is non-empty and

$$S_f^{int} = \{(x, t) : x \in \Phi_f^{int}, t > f(x)\}.$$

(v) S_f^{int} is non-empty and $S_f \cap (\Phi_f^{int} \times \mathbf{R})$ is open in $\Phi_f^{int} \times \mathbf{R}$.

Proof If g is *any* continuous function, then its graph is closed and its strict epigraph is open, and so (i) implies (ii) and (v).

Suppose that $\Gamma_f \cap (\Phi_f^{int} \times \mathbf{R})$ is closed in $\Phi_f^{int} \times \mathbf{R}$ and that $x \in \Phi_f^{int}$. Suppose that $\epsilon > 0$. There exists a $\delta > 0$ such that $N_\delta(x) \subseteq \Phi_f^{int}$, and such that $(N_\delta(x) \times (\epsilon, \infty)) \cap \Gamma_f = \emptyset$. Thus f is bounded above on $N_\delta(x)$; (ii) implies (iii).

Suppose that (iii) holds. By translation, and addition of a constant if necessary, we can suppose that $0 \in \Phi_f^{int}$, and that $f(y) < 0$ for $y \in U$, so that S_f is a convex set containing the origin $(0, 0)$ of $E \times \mathbf{R}$, and we can consider its gauge p_{S_f}. Since $U \times (0, \infty)$ is an open subset of S_f, S_f^{int} is non-empty. Certainly, $S_f^{int} \subseteq \{(x, t) : x \in \Phi_f^{int}, t > f(x)\}$. We need to prove the converse inclusion. Suppose that $x \in \Phi_f^{int}$ and that $t > f(x)$. There exists $\lambda > 1$ such that $[0, \lambda x) \subseteq \Phi_f^{int}$. Let $f_x(s) = f(sx)$, for $s \in [0, \lambda]$; f_x is a convex function on $[0, \lambda]$, and $f_x(s) \leq (1 - s)f(0) + sf(x) < st$, for $s \in [0, 1]$. Since f_x is a continuous function on $[0, \lambda]$, there exists $1 < \mu \leq \lambda$ such that $f_x(s) < st$ for $s \in [0, \mu]$; that is, $(sx, st) \in S_f$ for $0 \leq s \leq \mu$. Thus $p_{S_f}(x, t) < 1$, and so $(x, t) \in S_f^{int}$. Thus $\{(x, t) : x \in \Phi_f^{int}, t > f(x)\} \subseteq S_f^{int}$, and (iv) holds.

Clearly, (iv) implies (v).

Suppose that (v) holds, and that $x \in \Phi_f^{int}$. We can suppose that $x = 0$ and that $f(x) = 0$. There exists $0 < \delta < 1$ such that

$$U_\delta = \{(y, \mu) : \|y\| < \delta, |\mu - 1| < \delta\} \subseteq S_f.$$

Suppose that $0 < \epsilon < 1$. Let $\eta = \epsilon\delta$. If $\|z\| < \eta$, let $y = z/\epsilon$, so that $\|y\| < \delta$ and $f(y) < 1$. Hence $f(z) \leq \epsilon f(y) < \epsilon$. Similarly $f(-z) < \epsilon$, so that $f(z) \geq -f(-z) > -\epsilon$; f is continuous at x, and (i) holds. □

Corollary 10.2.2 *Suppose that f is a convex function on a normed space $(E, \|.\|)$. Then Φ_f^{int} is non-empty and f is continuous on Φ_f^{int} if and only if f is upper semi-continuous at a point of Φ_f.*

If Φ_f^{int} is non-empty, and f is continuous at a point of Φ_f^{int} then f is continuous on Φ_f^{int}.

Proposition 10.2.3 *Suppose that f is a very regular convex function on a normed space $(E, \|.\|)$. Then f is locally Lipschitz on Φ_f^{int}; if $x \in \Phi_f^{int}$, there exists $\delta > 0$ such that f is $2/\delta$-Lipschitz on $x + \delta U(E)$.*

Proof There exists $\delta > 0$ such that $x + 2\delta U(E) \subseteq \Phi_f^{int}$ and $|f(y) - f(x)| \leq 1$ on $x + 2\delta U(E)$. Suppose that y and y' are distinct elements of $x + \delta U(E)$. If $\|y' - y\| \geq \delta$ then $f(y') - f(y) \leq 2 \leq (2/\epsilon)\|y' - y\|$. Otherwise, let $z = y + (\delta/\|y' - y\|)(y' - y)$. Then $\|z - y\| = \delta$, and $\|z - x\| < 2\delta$. By convexity,

$$\frac{f(y') - f(y)}{\|y' - y\|} \leq \frac{f(z) - f(y)}{\|z - y\|} \leq \frac{|f(z) - f(x)| + |f(y) - f(x)|}{\|z - y\|} \leq \frac{2}{\delta}.$$

Similarly, $f(y) - f(y') \leq (2R/\delta)\|y' - y\|$. □

When $(E, \|.\|)$ is a Banach space, things work well. The following result is related to the principle of uniform boundedness (Theorem 7.6.1).

Theorem 10.2.4 *Suppose that f is a convex function on a Banach space $(E, \|.\|)$ which is lower semi-continuous on a non-empty open convex subset U of Φ_f^{int}. Then f is very regular.*

Proof Suppose that $x \in U$. As before, we can suppose that $x = 0$ and that $f(x) = 0$. There exists $\delta > 0$ such that $V = \{y : \|y\| < \delta\} \subseteq U$. If $y \in V$ let $g(y) = \max(f(y), f(-y))$. Then g is a non-negative lower semi-continuous real-valued convex function on V, $g(0) = 0$ and $f(y) \leq g(y)$. If $n \in \mathbf{N}$, let $V_n = \{y \in V : g(y) \leq n\}$. Since g is lower semi-continuous, V_n is a closed subset of V. Since g is real-valued, $V = \cup_{n=1}^{\infty} V_n$. But V is topologically complete, and so, by Baire's category theorem, there exists $n \in \mathbf{N}$ such that V_n^{int} is not empty. But V_n^{int} is convex, and if $z \in V_n^{int}$ then $-z \in V_n^{int}$, and so $0 \in V_n^{int}$. Thus there exists $0 < \eta < \delta$ such that $\eta U(E) \subseteq V_n$; that is, if $\|y\| < \eta$ then $0 \leq g(y) \leq n$. Suppose that $\epsilon > 0$. Since $g(\lambda y) \leq \lambda g(y)$ for $0 \leq \lambda \leq 1$, if $\|y\| < \eta\epsilon/n$ then $0 \leq g(y) < \epsilon$; g is continuous at 0. Since $f(x) \leq g(x)$ for $x \in V, f$ is bounded above on an open subset W of V, and so f is continuous on Φ_f^{int}, by Theorem 10.2.1. □

Corollary 10.2.5 *Suppose that f is a convex function on a Banach space $(E, \|.\|)$ and that U is a non-empty open subset of Φ_f. Then f is very regular if and only if the epigraph $A_f \cap (U \times \overline{\mathbf{R}})$ of f in U is closed in $U \times \overline{\mathbf{R}}$.*

Corollary 10.2.6 *Suppose that F is a set of convex functions on a Banach space $(E, \|.\|)$ which are real-valued and continuous on an open convex subset U of E, and that $g(x) = \sup_{f \in F} f(x)$ is finite, for each $x \in U$. Then g is continuous on U.*

Proof For g is lower semi-continuous. □

11

Subdifferentials and the Legendre Transform

11.1 Differentials and Subdifferentials

We need several definitions, generalizing what happens in one dimension.

Suppose that f is a real-valued function defined on a radially open subset U of a real vector space E, that $x \in U$ and that h is a non-zero element of E. Then f has a *directional derivative* $d^+f_x(h)$ at x if

$$\frac{f(x+th)-f(x)}{t} \to d^+f_x(h) \text{ as } t \searrow 0.$$

f is *Gâteaux differentiable* at x, with *Gâteaux derivative* Df_x, if there is a linear functional Df_x on E for which

$$\frac{f(x+th)-f(x)}{t} \to Df_x(h), \text{ as } t \searrow 0 \text{ for each } h \in E \setminus \{0\}.$$

Clearly, if f is Gâteaux differentiable at x then it has directional derivatives in every direction, but the converse is not true.

Suppose now that f is a real-valued proper convex function on a vector space E, and that $x \in \Phi_f$. The *subdifferential* $\partial_\vee f(x)$ is the set of linear functionals on E for which

$$f(x+h) - f(x) - \phi(h) \geq 0 \text{ for all } h \in E.$$

So far, we have not considered the case where there is a topology on E. Suppose that $(E, \|.\|)$ is a [separable] normed space. Since the epigraph A_f of a convex function f is closed in the norm topology of $E \times \mathbf{R}$ if and only if it is weakly closed, a regular convex function is lower semi-continuous in the norm topology.

Exercise 11.1.1 Suppose that f is a regular convex function on a normed space $(E, \|.\|)$ and that $x \in \Phi_f^{int}$. Show that f has directional derivatives in all directions at x.

Proposition 11.1.2 *If f is a regular convex function on a normed space $(E, \|.\|)$ and $x \in \Phi_f^{int}$, then $\partial_\vee f(x)$ is a weak*-closed convex subset of E'.*

Proof Let $g(h) = f(x + h) - f(x)$. Then $g(0) = 0$, and g is lower semi-continuous, so there exists a neighbourhood N of 0, with $N \subseteq \Phi_f^{int}$ such that $g(h) > -1$ for $h \in N$. If ϕ is a discontinuous linear functional on E, then $\phi(N) = \mathbf{R}$, so that $\phi \notin \partial_\vee f(x)$. Thus $\partial_\vee f(x) \subseteq E'$. It is then clearly weak* closed and convex. □

Proposition 11.1.3 *Suppose that f is a regular convex function on a [separable] normed space $(E, \|.\|)$ and that $x \in \Phi_f^{int}$. f is Gâteaux differentiable at x if and only if $\partial_\vee f(x)$ is a singleton.*

Proof If f is Gâteaux differentiable at x, then $g(h) = f(x+h) - f(x) - Df_x(h)$ is a non-negative Gâteaux differentiable function on $\Phi_g^{int} = \Phi_f^{int} - x$, with $g(0) = 0$ and $Dg_0 = 0$. Clearly $\partial_\vee g(0) = \{0\}$, which shows that $\partial_\vee f(x)$ is a singleton. Conversely, suppose that $\partial_\vee f(x) = \{\phi\}$. Let $g(h) = f(x + h) - f(x) - \phi(h)$; g is a non-negative convex function, and $g(0) = 0$. Suppose that the directional derivative $d^+g(h_0) \neq 0$ for some h_0; then $d^+g(h_0) > 0$. Define a linear functional on span(h_0) by setting $\psi(\lambda h_0) = \lambda d^+g(h_0)$. Then $\psi(\lambda h_0) \leq g(\lambda h_0)$; extend by the Hahn–Banach theorem to a linear functional ψ dominated by g. But then $\partial_\vee g(0)$ is not a singleton, and neither is $\partial_\vee f(x)$. □

11.2 The Legendre Transform

Suppose that (E, F) is a real dual pair, and that f is a proper function on E taking values in $(-\infty, \infty]$, and with the property that there exists $a_0 \in A(E)$ such that $f \geq a_0$. Then the *Legendre transform* f^* is the function on F defined as

$$f^*(\phi) = \sup_{x \in E}(\langle x, \phi \rangle - f(x)).$$

Clearly $f^*(\phi) = \sup_{x \in \Phi_f}(\langle x, \phi \rangle - f(x))$, and if $x_0 \in \Phi_f$ then $f^*(\phi) \geq \langle x_0, \phi \rangle - f(x_0)$, so that f^* takes values in $(-\infty, \infty]$. Suppose that $a_0(x) = \langle x, \phi_0 \rangle + c_0$. Then $f^*(\phi_0) \leq -c_0$, so that f^* is a proper function. Since it is the supremum of a set of $\sigma(F, E)$-continuous affine functions, it is a regular convex function on F, when F is given the topology $\sigma(F, E)$.

Exercise 11.2.1 Suppose that (E, F) is a real dual pair. Establish the following.

(i) The Legendre transform is order reversing; if $f \leq g$, then $f^* \geq g^*$.
(ii) $0 \in \Phi_{f^*}$ if and only if f is bounded below.

(iii) If $g(x) = f(x) + c$ then $g^*(\phi) = f^*(\phi) - c$.
(iv) If $g(x) = f(x + c)$ then $g^*(\phi) = f^*(\phi) - \langle c, \phi \rangle$.
(v) If $g(x) = \alpha f(x)$, where $\alpha > 0$, then $g^*(\phi) = \alpha f^*(\phi/\alpha)$.
(vi) If $g(x) = f(\beta x)$, where $\beta > 0$, then $g^*(\phi) = f^*(\phi)/\beta$.
(vii) If $g(x) = f(-x)$, then $g^*(\phi) = f^*(-\phi)$; in particular, if f is an even function, then so is f^*.
(viii) If $(x, \phi) \in E \times F$, then $f^*(\phi) + f(x) \geq \langle x, \phi \rangle$.

In the same way, if g is a proper function on F taking values in $(-\infty, \infty]$, and with the property that there exists $a_0 \in A(F)$ such that $g \geq a_0$, we define the Legendre transform of g as

$$g^*(x) = \sup_{\phi \in F}(\langle x, \phi \rangle - g(\phi)).$$

We can therefore construct the function f^{**} on E.

Theorem 11.2.2 *Suppose that (E, F) is a real dual pair, and that f is a proper function on E taking values in $(-\infty, \infty]$, and with the property that there exists $a_0 \in A(E)$ such that $f \geq a_0$. Then $f^{**} = \underline{f}$, the lower convex envelope of f.*

Proof If $a \in A(E)$ and $a(x) = \langle x, \phi \rangle + c$ for $x \in E$, then $a \leq f$ if and only if $-c \geq \langle x, \phi \rangle - f(x)$ for all $x \in E$; that is, if and only if $-c \geq f^*(\phi)$. Thus

$$\begin{aligned} \underline{f}(x) &= \sup\{\langle x, \phi \rangle + c : -f^*(\phi) \geq c\} \\ &= \sup\{\langle x, \phi \rangle - f^*(\phi) : \phi \in F\} = f^{**}(x). \end{aligned}$$

□

Exercise 11.2.3 Show that f is a regular convex function on E if and only if $f = f^{**}$.

This clearly corresponds to the theorem of bipolars, for convex sets.

Theorem 11.2.4 *Suppose that (E, F) is a real dual pair, that f is a regular convex function on E, that $x \in \Phi_f$ and that $\phi \in \Phi_{f^*}$. Then the following are equivalent:*

(i) $\phi \in \partial_\vee f(x)$;
(ii) $x \in \partial_\vee f^*(\phi)$;
(iii) $f(x) + f^*(x) = \langle x, \phi \rangle$.

Proof

$$\begin{aligned} \phi \in \partial_\vee f(x) &\text{ if and only if } f(y) - f(x) \geq \langle y - x, \phi \rangle \text{ for all } y \in E \\ &\text{ if and only if } \langle y, \phi \rangle - f(y) \leq \langle x, \phi \rangle - f(x) \text{ for all } y \in E \\ &\text{ if and only if } f^*(\phi) + f(x) \leq \langle x, \phi \rangle. \end{aligned}$$

Since $f^*(y) + f(x) \geq \langle x, \phi \rangle$, (i) and (iii) are equivalent, and similarly (ii) and (iii) are equivalent. □

Exercise 11.2.5 Show that $\partial_\vee f^*$ is the transpose of $\partial_\vee f$: $(\phi, x) \in \partial_\vee f^*$ if and only if $(x, \phi) \in \partial_\vee f$.

It is convenient to consider concave functions in a similar way. If f is a proper function on E taking values in $[-\infty, \infty)$, and with the property that there exists $a_0 \in A(E)$ such that $f \leq a_0$. we define the *concave Legendre transform* $f^\dagger$ of f to be

$$f^\dagger(\phi) = \inf_{x \in E}(\langle x, \phi \rangle - f(x)).$$

Proposition 11.2.6 *Suppose that f is a proper function on E taking values in $[-\infty, \infty)$, and with the property that there exists $a_0 \in A(E)$ such that $f \leq a_0$, Let $S(f)(x) = -f(-x)$. Then $f^\dagger(\phi) = -(S(f))^*(\phi)$.*

Proof For

$$\begin{aligned} f^\dagger(\phi) &= -\sup_{x \in E}(-\langle x, \phi \rangle + f(x)) = -\sup_{x \in E}(\langle -x, \phi \rangle - (-f)(x)) \\ &= -\sup_{x \in E}(\langle x, \phi \rangle - (-f)(-x)) = -(S(f))^*(\phi). \end{aligned}$$

□

Thus $f^\dagger$ is a proper concave function, and $(S(f))^\dagger = -(S^2(f))^* = -f^*$. Arguing as in Theorem 11.2.2, if $a \in A(E)$ and $a(x) = \langle x, \phi \rangle + c$ for $x \in E$, then $a \geq f$ if and only if $-c \leq \langle x, \phi \rangle - f(x)$ for all $x \in E$; that is, if and only if $-c \leq f^\dagger(\phi)$. Thus

$$\begin{aligned} \bar{f}(x) &= \inf\{\langle x, \phi \rangle + c : -f^\dagger(\phi) \leq c\} \\ &= \inf\{\langle x, \phi \rangle - f^\dagger(\phi) : \phi \in F\} = f^{\dagger\dagger}(x), \end{aligned}$$

and other properties of $f^\dagger$ can be derived from corresponding properties of the Legendre transform. For example, we have the following.

Theorem 11.2.7 *Suppose that (E, F) is a real dual pair. If f is a regular concave function on E, if $x \in \Phi_f$ and if $\phi \in \Phi_{f^\dagger}$, then the following are equivalent:*

(i) $\phi \in \partial^\wedge f(x)$;
(ii) $x \in \partial^\wedge f^\dagger(\phi)$;
(iii) $f(x) + f^\dagger(x) = \langle x, \phi \rangle$.

11.3 Some Examples of Legendre Transforms

Let us give some examples. Suppose first that $f = 0_C$, where C is a $\sigma(E,F)$-closed convex subset of E; $f(x) = 0$ if $x \in C$, and $f(x) = \infty$ otherwise. Then $f^*(\phi) = \sup_{x \in C} \langle x, \phi \rangle$. This is a regular convex function on F, since $f^*(0) = 0$. If $0 \in C$ of E then

$$f^*(\phi) = \sup\{\langle x, \phi \rangle : x \in C\} = p_{C^\circ}(\phi,)$$

where p_{C° is the gauge of C°. In particular, if E is a normed space, $F = E'$ and $C = B(E)$ then $f^*(\phi) = \|\phi\|'$. Similarly if $C = B(E')$ then $f^*(x) = \|x\|$.

Exercise 11.3.1 Show directly that if $(E, \|.\|)$ is a normed space, and $n_1(x) = \|x\|$ then $n_1^* = 0_{B(E')}$.

Suppose that (E, F) is a real dual pair and that f is a regular convex function on E. How do we calculate the Legendre transform of f? The function $\langle x, \phi \rangle - f(x)$ is a regular concave function. If it attains its supremum at x, then Theorem 11.2.4 implies that $f^*(\phi) \in \partial_\vee f(x)$, and that $f^*(\phi) = \langle x, \phi \rangle - f(x)$.

We use this to calculate the Legendre transform of convex functions on $\mathbf{R}$, and concave Legendre transforms of concave functions. Here are some examples.

(a) Suppose that $1 < p < \infty$ and that $q = p/(p-1)$, so that $1/p + 1/q = 1$ and $(p-1)(q-1) = 1$. Let $f(x) = |x|^p/p$. If $s > 0$ then $sx - |x|^p/p$ attains its supremum at x_0, where $s = x_0^{p-1}$. Thus $x_0 = s^{q-1}$ and

$$f^*(s) = s.s^{q-1} - s^{(q-1)p}/p = s^q/q.$$

If $s < 0$ then $f^*(s) = |s|^q/q$.

(b) Suppose that $0 < p < 1$ and that $1/p + 1/q = 1$ (so that $q < 0$). Let

$$f(x) = \begin{cases} x^p/p & \text{for } x > 0, \\ -\infty & \text{for } x \leq 0. \end{cases}$$

Then f is a regular concave function on $\mathbf{R}$. If $t \leq 0$ then $f^\dagger(t) = -\infty$. If $t > 0$ then $tx - f(x)$ attains its infimum when $x = t^{q-1}$ and then $f^\dagger(t) = t^q/q = -t^q/|q|$. Thus

$$f^\dagger(t) = \begin{cases} -\infty & \text{for } t \leq 0, \\ t^q/q = -t^q/|q| & \text{for } t > 0. \end{cases}$$

(c) Suppose that $r > 0$ and that $1/s - 1/r = 1$ (so that $0 < s < 1$). Let

$$f(x) = \begin{cases} x^{-r}/r & \text{for } x > 0, \\ \infty & \text{for } x < 0. \end{cases}$$

Then it follows from the previous example that

$$f^*(t) = \begin{cases} \infty & \text{for } t \geq 0, \\ -|t|^s/s & \text{for } t < 0. \end{cases}$$

Exercise 11.3.2 Verify the details of the next three examples.

(d) Let $f(x) = e^{x-1}$. Then

$$f^*(t) = \begin{cases} t \log t & \text{for } t > 0, \\ 0 & \text{for } t = 0, \\ \infty & \text{for } t < 0. \end{cases}$$

(e) Let

$$f(x) = \begin{cases} x \log x & \text{for } x > 0, \\ 0 & \text{for } x = 0, \\ \infty & \text{for } x < 0. \end{cases}$$

Then $f^*(t) = e^{t-1}$.

(f) Let $f(x) = (|x| + 1)\log(|x| + 1)$. Then

$$f^*(t) = \begin{cases} 0 & \text{for } |t| \leq 1, \\ e^{|t|-1} - |t| & \text{for } |t| > 1. \end{cases}$$

We can apply results concerning convex functions on **R** to certain convex functions on a normed space.

Proposition 11.3.3 *Suppose that f is a non-negative convex function on $[0, \infty)$ for which $f(t) = 0$ if and only if $t = 0$, that $(E, \|.\|)$ is a normed space and that $F(x) = f(\|x\|)$ for $x \in E$. Then $F^*(\phi) = f^*(\|\phi\|')$ for $\phi \in E'$.*

Proof Since $\sup\{\phi(x) : \|x\| = \alpha\} = \alpha \, \|\phi\|'$,

$$\begin{aligned} F^*(\phi) &= \sup_{\alpha > 0}(\sup\{\phi(x) - f(\alpha) : \|x\| = \alpha\}) \\ &= \sup_{\alpha > 0}(\alpha \, \|\phi\|' - f(\alpha)) = f^*(\|\phi\|'). \end{aligned}$$

□

If $(E, \|.\|)$ is a normed space, and $1 \leq p < \infty$, we set $n_p(x) = \|x\|^p/p$ for $x \in E$ and $n'_p(\phi) = (\|\phi\|')^p/p$ for $\phi \in E'$; n_p is a continuous convex function on E and n'_p is a continuous convex function on E'.

Corollary 11.3.4 *If $p > 1$ then $n_p^* = n'_q$, where $1/p + 1/q = 1$.*

11.4 The Episum

What about the Legendre transform of a sum? Suppose that f and g are proper functions on E, each of which is bounded below. The *episum* or *inf-convolution* $f \odot g$ of f and g is defined as

$$f \odot g(x) = \inf\{f(y) + g(z) : y + z = x\} = \inf_{y \in E}\{f(y) + g(x - y)\}.$$

Note that $f \odot g = g \odot f$ and that $\Phi_{f \odot g} = \Phi_f + \Phi_g$.

Proposition 11.4.1 *If (E, F) is a real dual pair and f and g are regular convex functions on E, each of which is bounded below, then $f \odot g$ is convex.*

Proof Suppose that $x_0, x_1 \in \Phi_{f \odot g}$, that $0 \leq \lambda \leq 1$ and that $\epsilon > 0$. There exists $y_0, y_1 \in \Phi_f$ such that $f(y_0) + g(x_0 - y_0) < f \odot g(x_0) + \epsilon/2$ and $f(y_1) + g(x_1 - y_1) < f \odot g(x_1) + \epsilon/2$. Let $x_\lambda = (1 - \lambda)x_0 + \lambda x_1$ and $y_\lambda = (1 - \lambda)y_0 + \lambda y_1$. Then

$$\begin{aligned} &f(y_\lambda) + g(x_\lambda - y_\lambda) \\ &\leq ((1 - \lambda)f(y_0) + \lambda f(y_1)) + ((1 - \lambda)g(x_0 - y_0) + \lambda g(x_1 - y_1)) \\ &\leq (1 - \lambda)f \odot g(x_0) + \lambda f \odot g(x_1) + \epsilon. \end{aligned}$$

Since ϵ is arbitrary, the result follows. □

On the other hand, $f \odot g$ need not be lower semi-continuous. If f is a regular convex function on E which is bounded below, and if C is closed convex subset of E, then

$$f \odot 0_C(x) = \inf\{f(y) : y \in x - C\} \text{ for } x \in \Phi_f + C.$$

In particular, if C and D are closed convex subsets of E, then $0_C \odot 0_D = 0_{C+D}$. But $C + D$ need not be closed, as the following example shows; it shows that the sum of two closed linear subspaces of a Hilbert space need not be closed.

If $x \in l_2$, let $T(x) = (x_n/n)_{n=1}^\infty$. Then T is a bounded linear operator on l_2, whose image $T(l_2)$ is a dense subspace of l_2 which is not closed in l_2. Let $H = l_2 \oplus l_2$, let $F = \{(x, y) \in H : y = 0\}$ and let $G = \Gamma_T = \{(x, T(x)) : x \in l_2\}$. Then F and G are closed linear subspaces of H, and $F + G = l_2 \oplus T(l_2)$ is a proper dense subspace of H.

Why is the episum important?

Proposition 11.4.2 *Suppose that (E, F) is a real dual pair and that f and g are proper functions on E, each of which is bounded below. Then $(f \odot g)^* = f^* + g^*$.*

Proof If $\phi \in F$ then

$$\begin{aligned}(f \odot g)^*(\phi) &= \sup_{x\in E}\left(\langle x,\phi\rangle - \inf_{y\in E}(f(y)+g(x-y))\right)\\ &= \sup_{x,y\in E}(\langle x,\phi\rangle - f(y) - g(x-y))\\ &= \sup_{x,y\in E}((\langle y,\phi\rangle - f(y)) + (\langle x-y,\phi\rangle - g(x-y)))\\ &= f^*(\phi)+g^*(\phi).\end{aligned}$$

□

Corollary 11.4.3 *Suppose that f and g are regular convex functions on E, and that $0 \in \Phi_f \cap \Phi_g$. Then $f + g$ is a regular convex function on E and $(f+g)^* = (f^* \odot g^*)^{**}$.*

Proof The function $f + g$ is certainly a regular convex function on E. The condition implies that if $\phi \in F$ then $f^*(\phi) \geq \langle 0,\phi\rangle - f(0) = -f(0)$ and similarly $g^*(\phi) \geq -g(0)$, so that $f^* \odot g^*$ is defined. Then $(f^* \odot g^*)^* = f^{**} + g^{**} = f + g$, and $(f+g)^* = (f^{**}+g^{**})^* = (f^* \odot g^*)^{**}$. □

In fact, the episum is most useful when E is a normed space and g is a very regular convex function, as we shall see in the next section (Proposition 11.5.5).

11.5 The Subdifferential of a Very Regular Convex Function

Theorem 11.5.1 *Suppose that f is a very regular convex function on a [separable] normed space $(E, \|.\|)$ and that $x \in \Phi_f^{int}$. Then the subdifferential $\partial_\vee f_x$ at x is a non-empty convex weak*-compact subset of E'. If $x \in \Phi_f^{int}$ and $h \in E$ then $d^+f_x(h) = \sup\{\phi(h) : \phi \in \partial_\vee f_x\}$, and the supremum is attained.*

Proof By Proposition 10.2.3, the function f is locally Lipschitz, and so there exist $\delta > 0$ and $M > 0$ such that $|f(x+h) - f(x)| \leq M\|h\|$ for $\|h\| < \delta$. Consequently,

$$d^+f_x(h) = \lim_{t\searrow 0}(f(x+th) - f(x))/t \leq M\|h\| \text{ for } h \neq 0,$$

and so d^+f_x is a sublinear functional on E, and is continuous, by Proposition 7.4.6.

If $h \in E$, then by the Hahn–Banach theorem there exists a linear functional ϕ_h on E such that $\phi_h(h) = d^+f_x(h)$ and $\phi_h(k) \leq d^+f_x(k)$ for $k \in E$. Since d^+f_x

is continuous, so is ϕ_h, and so $\phi_h \in \partial_\vee f_x$; thus $\partial_\vee f_x$ is a non-empty convex set. Also, $d^+ f_x(h) = \sup\{\phi(h) : \phi \in \partial_\vee f_x\}$, and the supremum is attained.

If $\phi \in \partial_\vee f_x$ and $k \in E$, then $\phi(k) \leq d^+ f_x(k) \leq M \|k\|$ so that $\sup\{\|\phi\|' : \phi \in \partial_\vee f_x\} \leq M$: $\partial_\vee f_x$ is bounded in E'. Since it is weak* closed, it is therefore weak* compact, by Banach's theorem. □

Corollary 11.5.2 *The function f is determined, up to a constant, on Φ_f^{int} by $\partial_\vee f$.*

Proof Suppose that $x_0 \in \Phi_f^{int}$. If x_1 is another element of Φ_f^{int}, let $h = (x_1 - x_0)/\|x_1 - x_0\|$. Then $f(x_1) = f(x_0) + \int_0^{\|x_1 - x_0\|} \partial_\vee f^+(t)\, dt$. □

Suppose that A is a closed convex subset of a normed space $(E, \|.\|)$ with non-empty interior A^{int}. A is *strictly convex* if whenever x, y are distinct points of A and $0 < \lambda < 1$ then $(1-\lambda)x + \lambda y \in A^{int}$. Thus A is strictly convex if and only if the boundary ∂A contains no non-trivial line segments.

A Banach space $(E, \|.\|)$ is strictly convex if its unit ball $B(E)$ is strictly convex; it is easy to see that a uniformly convex Banach space is strictly convex. A very regular convex function f on a Banach space $(E, \|.\|)$ is said to be strictly convex if its epigraph A_f is strictly convex; that is, if x, y are distinct points of Φ_f and $0 < \lambda < 1$ then $f((1-\lambda)x + \lambda y) < (1-\lambda)f(x) + \lambda f(y)$.

Proposition 11.5.3 *Suppose that f is a strictly convex very regular function on a normed space $(E, \|.\|)$, and that x, y are distinct points of Φ_f. Then $\partial_\vee f(x)$ and $\partial_\vee f(y)$ are disjoint.*

Proof Suppose if possible that $\phi \in \partial_\vee f(x) \cap \partial_\vee f(y)$. If $0 < \lambda < 1$ then

$$\begin{aligned} f((1-\lambda)x + \lambda y) &\geq f(x) + \phi(\lambda(y - x)) \\ \text{and } f((1-\lambda)x + \lambda y) &\geq f(y) + \phi((1-\lambda)(x - y)), \end{aligned}$$

so that $f((1-\lambda)x + \lambda y) \geq (1-\lambda)f(x) + \lambda f(y)$, giving a contradiction. □

Proposition 11.5.4 *Suppose that f is a very regular convex function on a normed space $(E, \|.\|)$. Then the function $(x, h) \to d^+ f_x(h)$ is upper semi-continuous on $\Phi_f^{int} \times E$.*

Proof Suppose that $(x, h) \in \Phi_f^{int} \times E$, and that $\mu > d^+ f_x(h)$. There exists $t > 0$ such that $(f(x + th) - f(x))/t < \mu$. Since f is continuous there exists a neighbourhood N of (x, h) in $\Phi_f^{int} \times E$ such that

$$(f(y + tk) - f(y))/t < \mu \text{ for } (y, k) \in N.$$

Then $d^+ f_y(k) \leq (f(y + tk) - f(y))/t < \mu$ for $(y, k) \in N$. □

In fact, we can improve the result of Theorem 11.5.1. The episum of a regular convex function and a very regular function is well-behaved.

Proposition 11.5.5 *If f is a regular convex function on a normed space $(E, \|.\|)$ and g is a very regular function on E, each of which is bounded below, then $f \odot g$ is very regular. Thus $(f+g)^* = f^* \odot g^*$.*

Proof For $\Phi_{f \odot g}^{int} \neq \emptyset$, and $f \odot g$ is the infimum of continuous functions, and is therefore upper semi-continuous on $\Phi_{f \odot g}^{int}$. But it is convex, and so it is continuous on $\Phi_{f \odot g}^{int}$, by Corollary 10.2.2. Thus $f \odot g$ is very regular, and so $(f^* \odot g^*) = (f^* \odot g^*)^{**}$. □

Theorem 11.5.6 *If f and g are regular convex functions on a [separable] Banach space $(E, \|.\|)$ and $\partial_\vee f = \partial_\vee g$ then there is a constant c such that $f = g + c$.*

Proof Recall that $n_2(x) = \frac{1}{2}\|x\|^2$, for $x \in E$. Then n_2 is a very regular convex function on E, and $n_2^*(\phi) = n_2(\phi)$, for $\phi \in E'$, by Corollary 11.3.4. Let $f_1 = f + n_2$ and $g_1 = g + n_2$. Then

$$\partial_\vee f_1 = \partial_\vee f + \partial_\vee n_2 = \partial_\vee g + \partial_\vee n_2 = \partial_\vee g_1.$$

Thus $\partial_\vee f_1^* = \partial_\vee g^*$, by Theorem 11.2.4. But $f_1^* = f^* \odot n_2^*$ is continuous, and so is g_1^*. Thus it follows from Corollary 11.5.2 that there is a constant c such that $f_1^* = g_1^* - c$. But then $f_1 = f_1^{**} = g_1^{**} + c = g_1 + c$, and so $f = g + c$. □

A very regular convex function f on a separable Banach space is Gâteaux differentiable at many points of Φ_f^{int}.

Theorem 11.5.7 (Mazur) *Suppose that f is a very regular convex function on a real separable Banach space $(E, \|.\|)$. Let*

$$G = \{x \in \phi_f^{int} : f \text{ is Gâteaux differentiable at } x\}.$$

Then G is a dense G_δ subset of Φ_f^{int}.

Proof Let (x_n) be a dense subset of E. Let

$$G_{n,m} = \{x \in \Phi_f^{int} : \phi(x_n) - \psi(x_n) < 1/m \text{ for } \phi, \psi \in \partial_\vee f_x\},$$

and let $B_{n,m} = \Phi_f^{int} \setminus G_{n,m}$. Then $G \subseteq \cap_{m,n} G_{m,n}$. On the other hand, if $x \in \cap_{m,n} G_{m,n}$, then $\phi(x_n) = \psi(x_n)$ for $\phi, \psi \in \partial_\vee f_x$, and so $\phi = \psi$; thus $x \in G$.

We show that each $B_{n,m}$ is closed in Φ_f^{int} and that $G_{n,m}$ is dense in Φ_f^{int}. It then follows from Baire's category theorem that $G = \cap_{m,n} G_{m,n}$ is a dense G_δ subset of Φ_f^{int}.

Suppose that $y_k \in B_{n,m}$ and that $y_k \to y \in \Phi_f^{int}$ as $k \to \infty$. There exists a neighbourhood N of y such that f is a Lipschitz function on N, and so there exist k_0 and $M > 0$ such that $\partial_\vee f_{y_k} \subseteq MB(E')$ for $k \geq k_0$. For each k, there exist ϕ_k, ψ_k in $\partial_\vee f_{y_k}$ such that $\phi_k(x_n) - \psi_k(y_n) \geq 1/m$. Since $MB(E')$ is weak* compact and metrizable, by extracting a subsequence if necessary we can suppose that there exist ϕ, ψ such that $\phi_k \to \phi$ and $\psi_k \to \psi$ in the weak* topology.

Suppose that $z \in E$. Since the sequence $(\phi_k)_{k=1}^\infty$ is norm bounded,

$$\phi_k(z - y_k) - \phi_k(z - y) - \phi_k(y - y_k) \to 0 \text{ as } k \to \infty,$$

and so

$$\begin{aligned}\phi(z - y) = \lim_{k\to\infty} \phi_k(z - y) &= \lim_{k\to\infty} \phi_k(z - y_k)\\ &\leq \lim_{k\to\infty} (f(z) - f(y_k)) = f(z) - f(y),\end{aligned}$$

so that $\phi \in \partial_\vee f_y$; similarly $\psi \in \partial_\vee f_y$. Further,

$$\phi(x_n) - \psi(x_n) = \lim_{k\to\infty} (\phi_k(x_n) - \psi_k(x_n)) \geq 1/n,$$

so that $y \in B_{n,m}$. Thus $B_{n,m}$ is closed in Φ_f^{int}.

Next we show that $G_{n,m}$ is dense in Φ_f^{int}. Suppose that $x \in \Phi_f^{int}$ and that $\epsilon > 0$. Let $f_n(t) = f(x + tx_n)$. Then f_n is a convex function of t in a neighbourhood of 0, and so there exists $x' = x + t'x_n$ with $\|x' - x\| < \epsilon$ such that f_n is differentiable at t'. Suppose now that $\phi, \psi \in \partial_\vee f_{x'}$. Then $\phi(sx_n) = sf_n'(t') = \psi(sx_n)$ for $s \in \mathbf{R}$, and so $\phi(x_n) = \psi(x_n)$; thus $x' \in G_{n,m}$. □

11.6 Smoothness

Besides considering Gâteaux differentiability, we can also consider *Fréchet differentiability*. A real-valued function is Fréchet differentiable at a point x of Φ_f^{int}, with Fréchet derivative $\mathbf{D}f(x)$ if

$$f(x + h) = f(x) + \mathbf{D}f(x)(h) + r(h),$$

where $h \in \Phi_f$, $\mathbf{D}f(x) \in E'$ and $\|r(h)\| / \|h\| \to 0$ as $\|h\| \to 0$. If E is finite dimensional and f is convex (or concave), then f is Fréchet differentiable at x if and only if it is Gâteaux differentiable at x. In the case where E is a Hilbert space H, we denote the Fréchet derivative by ∇f_x, and call it the *gradient* or *grad* of f at x.

A very regular convex function f is *Gâteaux smooth* if it is Gâteaux differentiable at each point of Φ_f^{int}, and is *Fréchet smooth* if it is Fréchet differentiable at each point of Φ_f^{int}.

Theorem 11.6.1 *If f is a Gâteaux smooth very proper convex function on a normed space $(E, \|.\|)$, then the function $(x, h) \to Df_x(h)$ is continuous on $\Phi_f^{int} \times E$.*

Proof Since $Df_x(h) = d^+f_x(h)$, the function is upper semi-continuous, by Proposition 11.5.4. But $Df_x(h) = -Df_x^+(-h)$, and so it is also lower semi-continuous. □

Corollary 11.6.2 *If f is a Gâteaux smooth very proper convex function on a normed space $(E, \|.\|)$, then Df is a continuous function from Φ_f^{int}, with the norm topology, to E', with the weak* topology.*

Theorem 11.6.3 *If f is a Gâteaux smooth very proper convex function on a normed space $(E, \|.\|)$, then f is Fréchet smooth if and only if Df is a continuous function from Φ_f^{int}, with the norm topology, to E', with the dual norm topology.*

Proof Suppose first that Df is norm-to-norm continuous. Suppose that $x \in \phi_f^{int}$ and that $\epsilon > 0$. There exists $\delta > 0$ such that if $\|x - y\| < \delta$ then $y \in \Phi_f^{int}$ and $\|Df_y - Df_x\|' < \epsilon$. For such y,

$$Df_x(y - x) \leq f(y) - f(x) \text{ and } Df_y(x - y) \leq f(x) - f(y),$$

so that

$$\begin{aligned} 0 \leq f(y) - f(x) - Df_x(y - x) &\leq (Df_y - Df_x)(y - x) \\ &\leq \|Df_y - Df_x\|' . \|y - x\| \leq \epsilon \|y - x\|, \end{aligned}$$

and so f is Fréchet smooth.

Conversely suppose that f is Fréchet smooth, and suppose if possible that Df is not continuous at $x \in \Phi_f^{int}$. Thus there exists $0 < \alpha < 1$ such that $\lim_{r \searrow 0}(\sup_{y \in N_r(x)} \|Df_y - Df_x\|') > \alpha$. Since f is locally Lipschitz, Df is locally bounded, and so there exist $\delta > 0$ and $L \geq 1$ such that $N_\delta(x) \subseteq \Phi_f^{int}$ and $\sup_{y \in N_\delta(x)} \|Df_y\|' \leq L$. There exists $0 < \eta < \delta$ such that $|f(y) - f(x) - Df_x(y - x)| < (\alpha/4) \|y - x\|$ for $y \in N_\eta(x)$. There exists $0 < \theta < \eta$ such that $|f(y) - f(x)| < \alpha\eta/4$ for $y \in N_\theta(x)$, and there exists $z \in N_\theta(x)$ such that $\|z - x\| < \alpha\delta/4L$ and $\|Df_z - Df_x\|' > \alpha$. Consequently, there exists $h \in E$ with $\|h\| = 1$ such that $Df_z(h) - Df_x(h) > \alpha$. Let $k = \eta h$. Then $Df_z(x + k - z) \leq f(x + k) - f(z)$. Thus

$$\begin{aligned}\alpha\eta &\leq Df_z(k) - Df_x(k)\\ &\leq (f(x+k) - f(x) - Df_x(k)) + Df_z(x-z) + (f(x) - f(z))\\ &\leq \alpha\eta/4 + \alpha\eta/4 + \alpha\eta/4,\end{aligned}$$

giving the necessary contradiction. □

Exercise 11.6.4 Show that if f is a Gâteaux smooth very proper convex function on $\mathbf{R}^d$, then f is Fréchet differentiable on Φ_f^{int}, and Df is continuous on Φ_f^{int}.

As a simple and important example of a very regular function, we consider the norm of a [separable] normed space.

Theorem 11.6.5 *Suppose that $(E, \|.\|)$ is a [separable] normed space. Recall that $n_1(x) = \|x\|$. Then $(\partial_\vee n_1)_0 = B(E')$. If $x \neq 0$ then $\phi \in (\partial_\vee n_1)_x$ if and only if $\phi(x) = \|x\|$ and $\|\phi\|' = 1$.*

Proof $(\partial_\vee n_1)_0 = B(E')$, since

$$\begin{aligned}(\partial_\vee n_1)_0 &= \{\phi \in E' : \phi(h) \leq \|h\| \text{ for } h \in E\ )\}\\ &= \{\phi \in E' : |\phi(h)| \leq \|h\| \text{ for } h \in E\} = B(E').\end{aligned}$$

Suppose that $x \neq 0$, so that

$$(\partial_\vee n_1)_x = \{\phi \in E' : \phi(h) \leq \|x+h\| - \|x\| \text{ for } h \in E\}.$$

If $\phi \in (\partial_\vee n_1)_x$ then, since $\|x+h\| - \|x\| \leq \|h\|$, it follows that $\|\phi\|' \leq 1$. On the other hand if $0 < \alpha < 1$ then

$$\begin{aligned}\alpha\phi(x) = \phi(\alpha x) &\leq \|x + \alpha x\| - \|x\| = \alpha\,\|x\|\\ \text{and } -\alpha\phi(x) = \phi(-\alpha x) &\leq \|x - \alpha x\| - \|x\| = -\alpha\,\|x\|,\end{aligned}$$

so that $\phi(x) = \|x\|$ and $\|\phi\|' \geq 1$. Thus $\|\phi\|' = 1$.

Conversely, suppose that $\phi \in E'$, that $\|\phi\|' = 1$ and that $\phi(x) = \|x\|$. If $h \in E$ then

$$\phi(h) = \phi(x+h) - \phi(x) = \phi(x+h) - \|x\| \leq \|x+h\| - \|x\|,$$

so that $\phi \in (\partial_\vee n_1)_x$. □

Note that if $x \neq 0$ then $(\partial_\vee n_1)_x = (\partial_\vee n_1)_{x/\|x\|}$.

Exercise 11.6.6 Suppose that $(E, \|.\|)$ is a [separable] normed space. Determine $\partial_\vee n_p$ for $p > 1$.

A Banach space $(E, \|.\|)$ is said to be *smooth* if the function n_1 is Gâteaux smooth on $E \setminus \{0\}$.

Corollary 11.6.7 *A [separable] Banach space is smooth if and only if for each $x \in E$ with $\|x\| = 1$ there exists a unique $\phi \in E'$ for which $\phi(x) = \|\phi\|' = 1$.*

The following result is an immediate consequence of Mazur's theorem (Theorem 11.5.7).

Exercise 11.6.8 Use Mazur's theorem to show that if $(E, \|.\|)$ is a separable Banach space, then the set of smooth points of $S(E)$ is a dense G_δ subset of $S(E)$.

The norm of a Hilbert space H is smooth. For if $\|x\| = \|y\| = 1$ and $\langle x, y\rangle = 1$ then $\|x - y\|^2 = \|x\|^2 - 2\langle x, y\rangle + \|y\|^2 = 0$, so that $\partial_\vee n(x) = \{x\}$ and $Df_x = x$.

We shall give further examples in Chapter 15.

Smoothness and strict convexity of Banach spaces are related in the following way.

Theorem 11.6.9 *Suppose that $(E, \|.\|)$ is a [separable] reflexive Banach space.*

(i) If $(E', \|.\|')$ is strictly convex then $(E, \|.\|)$ is smooth.
(ii) If $(E', \|.\|')$ is smooth then $(E, \|.\|)$ is strictly convex.

Proof (i) Suppose that $(E, \|.\|)$ is not smooth. Then there exists x with $\|x\| = 1$, and distinct $\phi, \psi \in E'$ such that $\phi(x) = \|\phi\|' = \psi(x) = \|\psi\|' = 1$. If $0 < \lambda < 1$ then $((1 - \lambda)\phi + \lambda\psi)(x) = 1$, so that $\|(1 - \lambda)\phi + \lambda\psi\|' = 1$ and $[\phi, \psi]$ is a proper line segment in $S(E')$. Thus E' is not strictly convex.

(ii) Suppose that $(E, \|.\|)$ is not strictly convex, so that there exists a proper line segment $[x, y]$ in $S(E)$. By the Hahn–Banach theorem, there exists $\phi \in E'$ with $\phi(x) = \phi(y) = \|\phi\| = 1$. Thus $(E', \|.\|')$ is not smooth. □

Even if $(E, \|.\|)$ is strictly convex, the function n_1 is not strictly convex, since $\|(1 - \lambda)x + \lambda(2x)\| = (1 - \lambda)\|x\| + \lambda\|2x\| = 1 + \lambda$. What about the function n_p, for $p > 1$?

If $x \in l_2$, let $f(x) = \sum_{n=1}^{\infty} x_n^2/n$. Then f is Gâteaux smooth and strictly convex. Then $f^*(y) = \sum_{n=1}^{\infty} n x_n^2$, so that $\Phi_{f^*} = \{x : \sum_{n=1}^{\infty} n x_n^2 < \infty\}$, and f^* is not very regular.

We now consider the Legendre transform of certain strictly convex very regular functions.

Theorem 11.6.10 *Suppose that f is a non-negative finite-valued strictly convex very regular function on a [separable] reflexive Banach space $(E, \|.\|)$ for which*

- *$f(x) = 0$ if and only if $x = 0$;*
- *f is bounded on the bounded sets of E; and*
- *$f(x)/\|x\| \to \infty$ as $\|x\| \to \infty$.*

Then f^* is a non-negative finite-valued Gâteaux smooth function on E' with $f^*(0) = 0$. Df^* is a surjection of E' onto E.

Proof Suppose that $\phi \in E'$. Then $f^*(\phi) \geq \phi(0) - f(0) = 0$, and if $\phi(0) = 0$ then $f^*(\phi) = 0$. There exists $R > 0$ such that $f(x) \geq \|\phi\|' . \|x\|$ for $\|x\| \geq R$, so that $\phi(x) - f(x) \leq \|\phi\|' . \|x\| - f(x) \leq 0$, for $\|x\| \geq R$. Thus $f^*(\phi) = \sup_{\|x\| \leq R} \phi(x) - f(x) \leq R\,\|\phi\|$, and so $f^*(\phi)$ is finite, for all $\phi \in E'$.

Let $S = \sup_{\|x\| \leq R} f(x)$, and let

$$A = \{(x, t) : \|x\| \leq R, f(x) \leq t \leq S\}.$$

Then A is a bounded norm-closed convex subset of $E \times \mathbf{R}$, and so, since $E \times \mathbf{R}$ is reflexive, it is weakly compact. Thus the linear functional $(\phi, -1)$ attains its supremum on A at a point (x_0, t_0) of A, and, since $f(x_0) \leq t_0$, it follows that $t_0 = f(x_0)$. Thus $\phi \in \partial_\vee f_{x_0}$. Since f is strictly convex, x_0 is unique, and so f^* is Gâteaux differentiable at ϕ. If $x \in E$ then $\partial_\vee f_x \neq \emptyset$; if $\phi \in \partial_\vee f_x$, then $x = Df^*(\phi)$, by Theorem 11.2.4, and so Df^* is surjective. □

Corollary 11.6.11 *Suppose that f is also Gâteaux smooth. Then Df is a bijection of E onto E', with inverse Df^*, and if E is finite-dimensional, then ∇f is a homeomorphism of E onto E', with inverse ∇f^*.*

As an example, suppose that $(E, \|.\|)$ is a smooth and strictly convex reflexive Banach space, and that ψ is a strictly increasing, strictly convex differentiable function on $[0, \infty)$ for which $\psi(0) = \psi'(0) = 0$ and $\psi(t)/t \to \infty$ as $t \to \infty$ (for example, $\psi(t) = t^p/p$, where $p > 1$, or $\psi(t) = t\log(t + 1)$). Let us check that the function $f(x) = \psi \circ n_1$ satisfies the conditions of the theorem. Clearly $f(x)/\|x\| \to \infty$ as $\|x\| \to \infty$. It is straightforward to verify that $d^+f_x(h) = \psi'((Dn_1)_x(h))$ if $x \neq 0$, and that $Df_0(h) = 0$, so that f is Gâteaux smooth. Suppose that $x \neq x'$ and that $0 < \lambda < 1$. If $\|x\| \neq \|x'\|$ then

$$\begin{aligned} f((1 - \lambda)x + \lambda x') &= \psi(\|(1 - \lambda)x + \lambda x'\| \\ &\leq \psi((1 - \lambda)\|x\| + \lambda\|x'\|) \\ &\leq (1 - \lambda)\psi(\|x\|) + \lambda\psi(\|x'\|) \\ &= (1 - \lambda)f(x) + \lambda f(x'), \end{aligned}$$

while if $\|x\| = \|x'\| = r$ then $\|(1 - \lambda)x + \lambda x'\| < (1 - \lambda)\|x\| + \lambda\|x'\| = r$, so that

$$f((1 - \lambda)x + \lambda x') < \psi(r) = (1 - \lambda)f(x) + \lambda f(x').$$

Thus f is strictly convex.

11.7 The Fenchel–Rockafeller Duality Theorem

Theorem 11.7.1 (The Fenchel–Rockafeller duality theorem) *Suppose that $(E, \|.\|)$ is a [separable] normed space, that f is a regular convex function on E, that g is a very regular convex function on E, that $\Phi_f \cap \Phi_g^{int} \neq \emptyset$ and that $\inf_{x\in E}(f(x) + g(x)) = c > -\infty$.*

Then $f^*(-\phi) + g^*(\phi) \geq -c$ for $\phi \in E'$, and there exists $\phi_0 \in E'$ such that $f^*(-\phi_0) + g^*(\phi_0) = -c$.

Proof Setting $x = y$,

$$\begin{aligned} f^*(-\phi) + g^*(\phi) &= \sup_{x,y\in E} (\phi(x-y) - f(x) - g(y)) \\ &\geq -\inf_{x\in E}(f(x) + g(x)) = -c. \end{aligned}$$

Once again, we use the separation theorem. Let $h(x) = f(x) - c$, and let $A(h) = \{(x,t) : h(x) + t \leq 0\}$. The convex sets $A(h)$ and $S(g)$ are disjoint, since if $h(x) + t \leq 0$ and $t > g(x)$ then $f(x) + g(x) < c$. Since $S(g)$ is open in $E \times \mathbf{R}$, by the separation theorem there exists $(\psi_0, s) \in E' \times \mathbf{R}$ and $\lambda \in \mathbf{R}$ such that

$$\begin{aligned} \psi_0(x) + st > \lambda &\quad \text{for } (x,t) \in S(g), \\ \text{and } \psi_0(y) - sr \leq \lambda &\quad \text{for } (y,r) \in A(h). \end{aligned}$$

By considering $x_0 \in \Phi_f \cap \Phi_g^{int}$, it follows that $s \neq 0$, and, since if $(x,t) \in S_f$ and $t' > t$ then $(x,t') \in S_f$, it follows that $s > 0$. Let $\phi_0 = \psi_0/s$ and let $\mu = \lambda/s$. If $x \in \Phi_f$ and $y \in \Phi_g^{int}$ then

$$\begin{aligned} \phi_0(x) + t > \mu &\quad \text{if } t > f(x) \\ \text{and } \phi_0(y) - r \leq \mu &\quad \text{if } r \geq g(y) - c, \end{aligned}$$

so that

$$((-\phi_0)(x) - f(x)) + (\phi_0(y) - g(y)) \leq -c;$$

that is, $f^*(-\phi_0) + g^*(\phi_0) \leq -c$. □

Corollary 11.7.2 *If $x_0 \in \Phi_f \cap \Phi_g^{int}$, then $\partial_\vee(f+g)(x_0) = \partial_\vee f(x_0) + \partial_\vee g(x_0)$.*

Proof In general, $\partial_\vee f(x_0) + \partial_\vee g(x_0) \subseteq \partial_\vee(f+g)(x_0)$. Suppose that $\phi \in \partial_\vee(f+g)(x_0)$. By replacing f and g by

$$f(x + x_0) - f(x_0) - \phi(x) \text{ and } g(x + x_0) - g(x_0),$$

we may suppose that $x_0 = 0$, that $\phi = 0$ and that $f(0) = g(0) = 0$. Thus $\inf_{x\in E} f(x) + g(x) = 0 \geq 0$. By the theorem, there exists $\phi_0 \in E'$ such that $f^*(-\phi_0) = g^*(\phi_0) = 0$; that is,

$$\sup_{x\in E}(-\phi_0(x) - f(x)) + \sup_{y\in E}(\phi_0(y) - g(y)) = 0.$$

Setting $y = 0$, we see that $f(x) - f(0) = f(x) \geq -\phi_0(x)$ for all $x \in E$, and, setting $x = 0$, we see that $g(y) - g(0) = g(y) \geq \phi_0(y)$ for all $y \in E$; that is, $-\phi_0 \in \partial_\vee f(0)$ and $\phi_0 \in \partial_\vee g(0)$. Thus $\phi = 0 \in \partial_\vee f(x_0) + \partial_\vee g(x_0)$. □

Corollary 11.7.3 *If $x_0 \in \Phi_f^{int} \cap \Phi_g^{int}$, then $f + g$ is Gâteaux differentiable at x_0 if and only if f and g are Gâteaux differentiable at x_0.*

11.8 The Bishop–Phelps Theorem

Suppose that f is a regular convex function on a [separable] normed space $(E, \|.\|)$, that $x_0 \in \Phi_f$ and that $\epsilon > 0$. The *ϵ-subdifferential* $\partial_\epsilon(f)(x_0)$ of f at x is defined as

$$\partial_\epsilon(f)(x_0) = \{\phi \in E' : \phi(h) \leq f(x_0 + h) - f(x_0) + \epsilon \text{ for all } h \in E\}.$$

Exercise 11.8.1 Use the separation theorem to show that $\partial_\epsilon(f)(x_0)$ is not empty, for all $x \in \Phi_f$; use the definition of f^* to show that $\phi \in \partial_\epsilon(f)(x_0)$ if and only if $f(x_0) + f^*(\phi) \leq \epsilon$.

Theorem 11.8.2 *Suppose that f is a regular convex function on a [separable] Banach space $(E, \|.\|)$. Suppose that $x_0 \in \Phi_f$, that $\epsilon > 0$, that $\alpha > 0$ and that $\phi_0 \in \partial_\epsilon(f)(x_0)$. Then there exist $\tilde{x} \in \Phi_f$ and $\tilde{\phi} \in \partial_\vee f(\tilde{x})$ such that*

$$\|x_0 - \tilde{x}\| \leq \epsilon/\alpha \textit{ and } \left\|\phi_0 - \tilde{\phi}\right\| \leq \alpha.$$

Proof First, we simplify the problem. By considering $f(x + x_0) - f(x_0) - \phi_0(x) + \epsilon$, we can suppose that $x_0 = 0$, that $f(x) \geq 0$ and that $f(0) = \epsilon$. Let $k = \inf_{x\in E} f(x)$, and let $h = f - k$. Then h is a non-negative regular convex function on E, $\inf_{x\in E} h(x) = 0$, and $h(0) \leq \epsilon$. It is then sufficient to show that there exists $\tilde{x} \in \Phi_h$ with $\|\tilde{x}\| \leq \epsilon/\alpha$ and $\tilde{\phi} \in \partial_\vee h(\tilde{x})$ with $\left\|\tilde{\phi}\right\|' \leq \alpha$.

We apply Ekeland's variational principle (Theorem 4.4.1). There exists $\tilde{x} \in E$ with $\|\tilde{x}\| \leq \epsilon/\alpha$ such that $h(\tilde{x}) < h(x) + \alpha\|x - \tilde{x}\|$ for all $x \in E \setminus \{\tilde{x}\}$; $\tilde{x} \in \Phi_h$. Let $n(x) = \alpha\|x - \tilde{x}\|$. Then n is a very regular convex function on E, and $0 \in \partial_\vee(h + n)(\tilde{x})$. But $\partial_\vee(h + n)(\tilde{x}) = \partial_\vee h(\tilde{x}) + \partial_\vee n(\tilde{x})$, by Corollary 11.7.2, and so there exists $\tilde{\phi} \in E'$ such that $-\tilde{\phi} \in \{\phi \in E' : \|\phi\|' \leq \alpha\}$. Thus $\tilde{\phi} \in \partial_\vee h(\tilde{x})$ and $\left\|\tilde{\phi}\right\|' \leq \alpha$. □

Corollary 11.8.3 *The set $\{x \in \Phi_f : \partial_\vee f(x) \neq \emptyset\}$ is dense in Φ_f.*

Suppose that A is a non-empty subset of a normed space $(E, \|.\|)$. If $\phi \in E'$, we set

$$p_A(\phi) = \sup\{\phi(a) : a \in A\} \qquad b(A) = \{\phi \in E' : p_A(\phi) < \infty\}.$$

A non-zero element ϕ of $b(A)$ is a *support functional* of A if there exists $a_0 \in A$ such that $\phi(a_0) = p_A(\phi)$, and an element a of A is a *support point* of A if there exists $\phi \in b(A)$ such that $\phi(a) = p_A(\phi)$. We denote the set of support points of A by $s(A)$ and the set of support functionals of A by $s'(A)$.

Theorem 11.8.4 (The Bishop–Phelps theorem) *Suppose that C is a non-empty proper closed convex subset of a [separable] Banach space $(E, \|.\|)$.*

(i) $s(C)$ is dense in the boundary ∂C of C.
(ii) $s'(C)$ is norm dense in $b(C)$.

Proof (i) We apply Theorem 11.8.2 to the regular convex function 0_C. Suppose that $x_0 \in \partial C$ and that $0 < \epsilon < 1$. There exists $x_1 \in E \setminus C$ such that $\|x_0 - x_1\| < \epsilon/2$. By the separation theorem, there exists $\phi_0 \in E'$, with $\|\phi_0\|' = 1$, such that $\phi_0(x_1) > p_C(\phi)$. If $x \in C$ then

$$\begin{aligned} \phi_0(x - x_0) = \phi_0(x) - \phi_0(x_0) &< \phi_0(x_1) - \phi_0(x_0) = \phi_0(x_1 - x_0) \\ &\leq \epsilon/2 = \epsilon/2 + 0_c(x) - 0_c(x_0), \end{aligned}$$

so that $\phi_0 \in \partial_{\epsilon/2} 0_C(x_0)$. By Theorem 11.8.2, there exist $\tilde{x} \in C$ and $\tilde{\phi} \in \partial_\vee 0_C(\tilde{x})$ such that $\|x_0 - \tilde{x}\| \leq \epsilon$ and $\left\|\phi_0 - \tilde{\phi}\right\|' \leq \epsilon$. Thus $\tilde{\phi} \neq 0$. Since $\partial_\vee 0_C(x) = \{0\}$ for $x \in C^{int}$, $\tilde{x} \in \partial C$.

(ii) Suppose that $\phi_0 \in b(C)$. If $\phi_0 = 0$, and ψ is any element of $s'(C)$, then $\psi/n \to 0$ as $n \to \infty$, so that $\phi_0 \in \overline{s'(C)}$. Otherwise, since $\lambda\phi \in b(C)$ if $\phi \in b(C)$ and $\lambda > 0$, we can suppose that $\|\phi_0\|' = 1$. Suppose that $0 < \epsilon < 1$. There exists $x_0 \in C$ such that $\phi_0(x_0) > p_C(\phi_0) - \epsilon$. Thus if $x \in C$ then

$$\phi_0(x - x_0) < \epsilon = 0_C(x) - 0_C(x_0) + \epsilon,$$

and so $\phi_0 \in \partial_\epsilon 0_C(x_0)$. Take $\alpha = \epsilon$. Then by Theorem 11.8.2 there exist $\tilde{x} \in C = \phi_{0_C}$ and $\tilde{\phi} \in \partial_\vee 0_C(\tilde{x})$ such that $\|\tilde{x} - x_0\| \leq \epsilon$ and $\left\|\tilde{\phi} - \phi_0\right\| \leq \epsilon$. Thus $\tilde{\phi} \neq 0$, and if $x \in C$ then

$$\tilde{\phi}(x) \leq \tilde{\phi}(x) + (0_C(x) - 0_C(\tilde{x})) = \tilde{\phi}(\tilde{x}),$$

so that $\tilde{\phi}$ is a support functional of C. □

The following corollary is also often referred to as the Bishop–Phelps theorem.

Corollary 11.8.5 *The set* $\{\phi \in E' : \phi(x) = \|\phi\|'$ *for some* $x \in B(E)\}$ *is norm dense in* E'.

Proof Take C to be the unit ball $B(E)$ of E. □

A closed convex subset C of a separable Banach space is determined by its support functionals.

Proposition 11.8.6 *Suppose that* C *is a proper non-empty closed convex subset of a separable Banach space* $(E, \|.\|)$. *Let*

$$\tilde{C} = \cap\{x \in E : \phi(x) \leq p_C(x) \textit{ for } p \in s'(C)\}.$$

Then $C = \tilde{C}$.

Proof Certainly $C \subseteq \tilde{C}$. Suppose that $y_0 \notin C$. Let $d = d(y_0, C)$. By the separation theorem, there exists $\Phi_0 \in E'$ with $\|\phi\|' = 1$, and $\lambda \in \mathbf{R}$ such that $\phi_0(x) \leq \lambda$ if $x \in C$ and $\phi_0(y) > \lambda$ if $\|y - y_0\| < d$. Thus if $\|h\| < d$ then $\phi_0(y_0) - \phi_0(h) = \phi_0(y_0 - h) > p_C(\phi_0)$ and so

$$d = \sup\{\phi_0(h) : \|h\| < d\} \leq \phi_0(y_0) - p_C(y_0).$$

Suppose that $0 < \epsilon < d/4(d+1)$. Choose $x_0 \in C$ with $\|x_0 - y_0\| < d + \epsilon/2$. Note that $\phi_0(y_0 - x_0) < d + \epsilon/2$. If $x \in C$, then

$$\phi(x - x_0) = \phi_0(x - y_0) + \phi_0(y_0 - x_0) < -d + (d + \epsilon/2) = \epsilon/2,$$

so that $\phi_0 \in \partial_{\epsilon/2} 0_C(x_0)$. By Theorem 11.8.2, there exist $\tilde{x} \in C$ and $\tilde{\phi} \in \partial_\vee 0_C(\tilde{x})$ with $\|\tilde{x} - x_0\| \leq \epsilon$ and $\left\|\tilde{\phi} - \phi_0\right\|' \leq \epsilon$. In particular $\tilde{\phi} \neq 0$, so that $\tilde{x} \in \partial C$. Further, if $x \in C$ then

$$\begin{aligned}\tilde{\phi}(x - y_0) &\leq \tilde{\phi}(\tilde{x} - y_0) = (\tilde{\phi} - \phi_0)(\tilde{x} - y_0) + \phi_0(\tilde{x} - y_0) \\ &\leq \epsilon\,\|\tilde{x} - y_0\| + p_C(\phi) - \phi_0(y_0) \leq \epsilon(\epsilon + d + \epsilon/2) - d < -d/2.\end{aligned}$$

Thus $\tilde{\phi}(y_0) > p_C(\tilde{\phi})$, and so $y_0 \notin \tilde{C}$. □

Exercise 11.8.7 Suppose that f is a regular convex function on a [separable] Banach space $(E, \|.\|)$. By considering the epigraph A_f of f in $E \times \mathbf{R}$, show that if $x \in \Phi_f$, then

$$f(x) = \sup\{f(y) + \phi(x - y) : \phi \in \partial_\vee f(y) \text{ for some } y \in \Phi_f\}.$$

11.9 Monotone and Cyclically Monotone Sets

Suppose that $(E, \|.\|)$ is a normed space. A subset A of $E \times E'$ is said to be *monotone* if $\phi_1(x_1) + \phi_2(x_2) \geq \phi_1(x_2) + \phi_2(x_1)$ for all $(x_1, \phi_1), (x_2, \phi_2)$ in

A, and is said to be *cyclically monotone* if $\sum_{j=1}^n \phi_j(x_j) \geq \sum_{j=1}^n \phi_j(x_{j+1})$, for all $(x_1, \phi_1), \ldots, (x_n, \phi_n)$ in A, where we set $x_{n+1} = x_1$. Clearly a cyclically monotone set is monotone, and a set A is cyclically monotone if and only if $\sum_{j=1}^n \phi_j(x_j) \geq \sum_{j=1}^n \phi_j(x_{\sigma(j)})$ for all $(x_1, \phi_1), \ldots, (x_n, \phi_n)$ in A and for every permutation σ of $\{1, \ldots, n\}$.

These notions are frequently defined in terms of operators. A mapping T from a normed space $(E, \|.\|)$ into the subsets of its dual E' is said to be a *monotone operator* (*cyclically monotone operator*) if $\phi_1(x_1) + \phi_2(x_2) \geq \phi_1(x_2) + \phi_2(x_1)$ for all $x_1, x_2 \in E$ and $\phi_1 \in T(x_1), \phi_2 \in T(x_2)$ ($\sum_{j=1}^n \phi_j(x_j) \geq \sum_{j=1}^n \phi_j(x_{j+1})$, for all $x_1, \ldots, x_n \in E$ and $\phi_i \in T(x_i)$ for $2 \leq i \leq n$ and $\phi_{n+1} = \phi_1$).

Exercise 11.9.1 Show that a mapping T from a normed space $(E, \|.\|)$ into the subsets of its dual E' is a monotone operator (cyclically monotone operator) if and only if its graph

$$G(T) = \{(x, \phi) : x \in E, \phi \in T(x)\}$$

is a monotone set (cyclically monotone set).

Exercise 11.9.2 Suppose that f is a real-valued function on $\mathbf{R}$. Show that the graph of f is monotone if and only if f is non-decreasing.

Exercise 11.9.3 Suppose that $T \in L(H)$, where H is a Hilbert space. Show that the graph of T is monotone if and only if $\langle T(x), x\rangle \geq 0$ for all $x \in E$.

Suppose that f is a regular convex function on a Banach space $(E, \|.\|)$, with subderivative $\partial_\vee f$. Recall that

$$G(\partial_\vee f) = \{(x, \phi) : x \in \Phi_f, \phi \in \partial_\vee f_x\}.$$

Proposition 11.9.4 *Suppose that f is a regular convex function on a Banach space $(E, \|.\|)$. Then $G(\partial_\vee f)$ is cyclically monotone.*

Proof Suppose that $(x_1, \phi_1), \ldots, (x_n, \phi_n) \in G(\partial_\vee f)$, and that $x_{n+1} = x_1$. Then

$$\begin{aligned}\sum_{j=1}^n \phi_j(x_{j+1}) - \sum_{j=1}^n \phi_j(x_j) &= \sum_{j=1}^n \phi_j(x_{j+1} - x_j) \\ &\leq \sum_{j=1}^n (f(x_{j+1}) - f(x_j)) = 0.\end{aligned}$$

□

Exercise 11.9.5 Suppose that $T \in L(H)$ is positive and self-adjoint (where H is a real Hilbert space). Show that the graph of T is cyclically monotone.

Exercise 11.9.6 Suppose that R_θ is a rotation of $\mathbf{R}^2$ by θ, where $0 < \theta \leq \pi/2$. Show that the graph of R_θ is monotone, but not cyclically monotone.

We can use a cyclically monotone set to define a regular convex function.

Theorem 11.9.7 *Suppose that $(E, \|.\|)$ is a normed space and that A is a non-empty cyclically monotone subset of $E \times E'$. Then there exists a regular convex function f on E such that $A \subseteq G(\partial_\vee f)$.*

Proof Choose an element (x_0, ϕ_0) of A. If $x \in E$, set

$$f(x) = \sup\left\{\sum_{j=0}^{n-1} \phi_j(x_{j+1} - x_j) + \phi_n(x - x_n) : ((x_j, \phi_j))_{j=1}^n \in A^n, n \in \mathbf{N}\right\}.$$

Taking $x = x_0$, $n = 1$ and $(x_1, \phi_1) = (x_0, \phi_0)$, we see that $f(x_0) \geq 0$. But $f(x_0) \leq 0$, by cyclic monotonicity, and so $f(x_0) = 0$.

Since f is the supremum of a set of continuous affine functions on E it is a lower semi-continuous convex function on E; f is a regular convex function.

Suppose that $(x, \phi) \in A$; we must show that $f(x) < \infty$. Suppose that $\alpha < f(x)$ and that $y \in E$. There exist $(x_1, \phi_1), \ldots, (x_n, \phi_n)$ in A such that

$$\phi_n(x - x_n) + \phi_{n-1}(x_n - x_{n-1}) + \cdots + \phi_0(x_1 - x_0) > \alpha.$$

Set $(x_{n+1}, \phi_{n+1}) = (x, \phi)$. Then

$$\begin{aligned} f(y) &\geq \phi_{n+1}(y - x_{n+1}) + \phi_n(x_{n+1} - x_n) + \cdots + \phi_0(x_1 - x_0) \\ &= \phi(y - x) + \phi_n(x - x_n) + \cdots + \phi_0(x_1 - x_0) > \phi(y - x) + \alpha. \end{aligned}$$

In particular, $0 = f(x_0) \geq \phi(x_0 - x) + \alpha$, so that $f(x) \leq \phi(x - x_0) < \infty$, and $x \in \Phi_f$. Consequently, if $y \in E$ then $f(y) - f(x) \geq \phi(y - x)$. Thus $(x, \phi) \in G(\partial_\vee f)$. □

If $(E, \|.\|)$ is a normed space, the collection of monotone subsets of $E \times E'$ is ordered by inclusion.

Theorem 11.9.8 *If f is a regular convex function on a normed space $(E, \|.\|)$, then $G(\partial_\vee f)$ is a maximal monotone set.*

Proof Suppose that $(x_0, \phi_0) \in E \times E'$ and that $\phi(x) + \phi_0(x_0) \geq \phi(x_0) + \phi_0(x)$ for all $(x, \phi) \in G(\partial_\vee f)$. We must show that $(x_0, \phi_0) \in G(\partial_\vee f)$. First we simplify the problem. By replacing $f(x)$ by $f(x + x_0) - \phi_0(x)$, we can suppose that $x_0 = 0$ and that $\phi_0 = 0$. Thus we must show that $0 \in \partial_\vee f(0)$, or equivalently that $f(0) + f^*(0) = 0$.

We apply the Fenchel–Rockafeller theorem (Theorem 11.7.1), taking $g = n_2$. Thus there exists $\phi_1 \in E'$ such that

$$\inf\{f(x) + n_2(x) : x \in E\} = -f^*(\phi_1) - n_2(\phi_1).$$

First we show that $\phi_1 = 0$. Suppose not, and suppose that $0 < \epsilon < \|\phi_1\|'$. Then there exists $x_1 \in E$ such that

$$f(x_1) + n_2(x_1) + f^*(\phi_1) + n_2(\phi_1) \leq \epsilon^2/2.$$

Since $f(x_1) + f^*(\phi_1) \geq \phi_1(x_1)$ it follows that

$$0 \leq \tfrac{1}{2}(\|x_1\| - \|\phi_1\|')^2 \leq n_2(x_1) + \phi_1(x_1) + n_2(\phi_1) < \epsilon^2/2.$$

Thus $|\,\|x_1\| - \|\phi_1\|'\,| < \epsilon$ and $f(x_1) + f^*(\phi_1) - \phi_1(x_1) < \epsilon^2/2$. By Theorem 11.8.2, there exists $(x_2, \phi_2) \in G(\partial_\vee f)$ with $\|x_2 - x_1\| < \epsilon$ and $\|\phi_1 - \phi_2\|' < \epsilon$. In particular $|\,\|x_2\| - \|\phi_1\|'\,| < 2\epsilon$, so that $\|x_2\| \leq 3\,\|\phi_1\|'$.

By hypothesis $\phi_2(x_2) \geq 0$ and

$$\phi_2(x_2) \leq \phi_1(x_1) + \|x_2\| . \|\phi_1 - \phi_2\|' + \|x_1 - x_2\| . \|\phi_1\|' \leq \phi_1(x_1) + 4\epsilon\,\|\phi_1\|',$$

so that

$$\tfrac{1}{2}(\|\phi_1\|')^2 = n_2(\phi_1) \leq \epsilon^2/2 - \phi_1(x_1) \leq 4\epsilon\,\|\phi_1\|'.$$

Consequently, $\|\phi_1\|' \leq 8\epsilon$. Since ϵ can be chosen to be arbitrarily small, $\phi_1 = 0$.

Thus $\inf\{f(x) + f^*(0) + n_2(x) : x \in E\} = 0$. If $\epsilon > 0$ there exists x_1 such that $f(x_1) + f^*(0) + n_2(x) < \epsilon^2/2$. But then $\|x_1\| < \epsilon$ and $0 \leq f(x_1) + f^*(0) < \epsilon^2/2$. Since f is lower semi-continuous, it follows that $f(0) = 0$. □

Corollary 11.9.9 *If f is a regular convex function on a normed space $(E, \|.\|)$, $(x_0, \phi_0) \in G(\partial_\vee f)$ and $x \in E$ then*

$$f(x) - f(x_0) = \sup\left\{\sum_{j=0}^{n-1} \phi_j(x_{j+1} - x_j) + \phi_n(x - x_n) : ((x_j, \phi_j)) \in G(\partial_\vee f)^n, n \in \mathbf{N}\right\}.$$

Proof By Theorem 11.9.7, the expression in the formula defines a regular convex function g, and $G(\partial_\vee f) \subseteq G(\partial_\vee g)$. Since $G(\partial_\vee f)$ is maximal, $G(\partial_\vee f) = G(\partial_\vee g)$. Since $f(x_0) = g(x_0)$, it follows from Theorem 11.5.6 that $f = g$. □

12

Compact Convex Polish Spaces

12.1 Compact Polish Subsets of a Dual Pair

Suppose that (E, F) is a dual pair, and that K is a $\sigma(E, F)$-compact metrizable subset of E. We shall investigate the geometric and topological properties of K.

Theorem 12.1.1 *Suppose that (E, F) is a dual pair, and that K is a $\sigma(E, F)$-compact subset of E. The following are equivalent.*

(i) *K is metrizable.*

(ii) *K is separable.*

(iii) *There is an affine map $T : E \to \mathbf{R}^{\mathbf{N}}$, continuous for the weak topologiy $\sigma(E, F)$ and the product topology on $\mathbf{R}^{\mathbf{N}}$ such that $T : K \to T(K)$ is an affine homeomorphism of K onto a compact subset of the Hilbert cube.*

(iv) *There is a linear map $R : E \to \mathbf{R}^{\mathbf{N}}$, continuous for the weak topology $\sigma(E, F)$ and the product topology on $\mathbf{R}^{\mathbf{N}}$, such that $R : K \to S(K)$ is a homeomorphism of K onto a closed subset of the compact subset $L = \{x : |x_n| \leq 1/n \text{ for } n \in \mathbf{N}\}$ of l_2.*

If $(E, \|.\|)$ is a normed space and F is a closed linear subspace of $(E', \|.\|')$ then R can be chosen to be a compact linear operator from $(E, \|.\|)$ into $(l_2 \, \|.\|_2)$, which is also continuous for the weak topology $\sigma(E, F)$ on E and the norm topology on l_2.

Proof Clearly (i) implies (ii). Suppose that (ii) holds, and that $(x_n)_{n=1}^{\infty}$ is a $\sigma(E, F)$-dense sequence in K. Then there exists an array

$$\{(\phi_{m,n}) : 1 \leq m < \infty, 1 \leq n < \infty\}$$

in F such that $\langle x_m, \phi_{m,n} \rangle \neq \langle x_n, \phi_{m,n} \rangle$ for $x_m \neq x_n$. Let $M_{m,n} = \sup_{x \in K} |\langle x, \phi_{m,n} \rangle|$; $M_{m,n} < \infty$, since K is compact and $\phi_{m,n}$ is $\sigma(E, F)$-continuous. Let $j : \mathbf{N} \to \mathbf{N} \times \mathbf{N}$ be an enumeration of $\mathbf{N} \times \mathbf{N}$, and let

$$T(x)_n = \frac{\langle x, \phi_{j(n)} \rangle}{2M_{j(n)} + 1} + \tfrac{1}{2};$$

then T is an affine map of E into $\mathbf{R}^{\mathbf{N}}$. $T(K)$ is contained in the Hilbert cube, and the restriction of T to K is injective, so that it is a homeomorphism. Thus (ii) implies (iii). Similarly, if $S(x)_n = \langle x, \phi_{j(n)} \rangle / n(M_{j(n)} + 1)$, then S satisfies the conditions of (iv). Clearly, (iii) and (iv) each imply (i).

The final statement follows by taking $(\phi_{m,n})_{(m,n)=(1,1)}^{(\infty,\infty)}$ in the unit ball of F and $R(x)_n = \langle x, \phi_{j(n)} \rangle / n$, for then if $y \in l_2$ it follows that $T(x)_n y_n \in l_1$. □

Here are three circumstances in which we can apply the theorem.

1. K is a bounded weak*-closed subset of the dual of a separable Banach space $(G, \|.\|)$. Then K is weak* separable, and is weak* compact, by Banach's theorem. We take (E, F) as (G', G).
2. K is a weakly compact subset of a separable normed space $(E, \|.\|)$. By Banach's theorem, $B(E')$ is compact and metrizable in the weak* topology; therefore it is separable in the weak* topology. Let G be a countable weak*-dense subset of $B(E')$. Then $G^{\circ\circ} = B(E')$ and so G separates the points of K. Thus the topology of pointwise convergence on G is Hausdorff, and so the weak topology on K is the same as the topology of pointwise convergence on G, which is metrizable.
3. K is a compact subset of a normed space $(E, \|.\|)$. Then K is $\sigma(E, E')$-compact and metrizable.

The great advantage of this theorem is that we can apply simple Euclidean geometrical ideas to $S(K)$, and then transfer them back to K. Here is a simple example of this, which shows how we can avoid using the separation theorem.

Proposition 12.1.2 *Suppose that* $(E, \|.\|)$ *is a normed space and that* K_1 *and* K_2 *are disjoint separable weakly compact convex subsets of* E. *Then there exist* $\phi \in E'$ *and* $\alpha \in \mathbf{R}$ *such that* $\sup_{k \in K_1} \phi(k) < \alpha$ *and* $\inf_{k \in K_2} \phi(k) > \alpha$.

Proof Let $R : E \to l_2$ be a compact linear operator which satisfies the requirements of Theorem 12.1.1, and let $L_1 = R(K_1)$, $L_2 = R(K_2)$. Let

$$d = d(L_1, L_2) = \inf\{\|l_1 - l_2\|_2 : (l_1, l_2) \in L_1 \times L_2\};$$

by compactness, there exists $(m_1, m_2) \in L_1 \times L_2$ with $d = \|m_1 - m_2\|_2$. Let $y = m_2 - m_1$ and let $\alpha = \langle (m_1 + m_2)/2, y \rangle$. Then $\langle m_1, y \rangle < \alpha$ and $\langle m_2, y \rangle > \alpha$. Suppose, if possible, that there exists $l \in L_1$ such that $\langle l, y \rangle > \langle m_1, y \rangle$. Then $m - l_1 = \beta y + w$, where $\beta > 0$ and $\langle w, y \rangle = 0$. Let $m_\lambda = (1 - \lambda)m_1 + \lambda l$; by convexity, $m_\lambda \in L_1$. Then $l_2 - m_\lambda = (1 - \lambda\beta)y - \lambda w$, so that

$$\|l_2 - m_\lambda\|_2^2 = (1-\lambda\beta)^2 d^2 + \lambda^2 \|w\|_2^2 < d^2$$

for small positive λ, giving a contradiction. Thus $\langle m_1, y\rangle = \sup_{l\in L_1}\langle l, y\rangle$. Similarly, $\langle m_2, y\rangle = \inf_{l\in L_2}\langle l, y\rangle$. Then $\phi = R'(y)$ satisfies the requirements of the proposition. □

12.2 Extreme Points

We need some definitions. A *variety* V in a vector space E is a subset of E of the form $x + F$, where $x \in E$ and F is a linear subspace of E; a variety V is the translate of a linear subspace. If A is a non-empty subset of E, a *support variety* H of A is a variety in E for which $H \cap A \neq \emptyset$ and for which if $x \in H \cap A$, $y, z \in A$ and

$$x \in (y,z) = \{(1-\lambda)y + \lambda z : 0 < \lambda < 1\}$$

then $y, z \in H$. If so, then $H \cap A$ is called a *face* of A. If $H \cap A = \{x\}$, then x is an *extreme point* of A. A support variety is *proper* if it is a proper subset of E.

Exercise 12.2.1 If A is a radially open subset of a vector space E, then there is no proper support variety of A.

On the other hand, we have the following.

Proposition 12.2.2 *Suppose that A is a convex neighbourhood of 0 in a [separable] normed space $(E, \|.\|)$ and that H is a support variety of $\overline{A}$. Then there is a closed support hyperplane V of $\overline{A}$ which contains H.*

Proof Suppose that $H = x + F$, where $x \in \overline{A} \cap H$. Define a linear functional ϕ on span(H) by setting $\phi(\lambda(x+f)) = \lambda$. If $\lambda > 1$ then $x+f \in (0, \lambda(x+f))$, so that $\lambda(x+f) \notin A$. Therefore $\phi(y) < 1$ for $y \in A \cap \text{span}(H)$, and so $\phi(y) \leq p_A(y)$ for $y \in \text{span}(H)$, where p_A is the gauge of A. By the Hahn–Banach theorem, ϕ can be extended to a linear functional ψ on E satisfying $\psi(z) \leq p_A(z)$ for $z \in E$. Since A is a neighbourhood of 0, p_A and ψ are continuous. Then $V = \{z \in E : \psi(z) = 1\}$ is a closed support hyperplane V of $\overline{A}$ which contains H. □

Proposition 12.2.3 *Suppose that A is a subset of a real vector space E. If L is a face of A and M is a face of L, then M is a face of K.*

Proof Suppose that $x \in M$, and that $x = (1-\lambda)y + \lambda z$ with $y, z \in A$, and $0 < \lambda < 1$. Then $y, z \in L$, since $M \subseteq L$ and since L is a face of A. Consequently $y, z \in M$, since M is a face of L. □

Proposition 12.2.4 *Suppose that (E, F) is a real dual pair and that K is a [separable] $\sigma(E, F)$-compact convex subset of E and that L is a closed face of K. Then L contains an extreme point of K.*

Proof By the preceding proposition, it is enough to show that L has an extreme point. First suppose that K is separable. By Theorem 12.1.1, L is affinely homeomorphic to a compact subset of a Hilbert space H, and so we can suppose that L is a compact subset of a Hilbert space H. Let d be the diameter of L. If $L = \{l\}$ is a singleton, then l is an extreme point of L. Otherwise, $d > 0$. The function $(x, y) \to \|x - y\|$ is continuous on $L \times L$, and so there exist $x, y \in L$ such that $\|x - y\| = d$. We shall show that x and y are extreme points of K. Let $J = y + (y - x)^{\perp}$. If $z = y + w \in J$, then $\|z - x\|^2 = d^2 + \|w\|^2$, so that $J \cap L = \{y\}$, and y is an extreme point of L; similarly for x.

When K is not separable, we need the axiom of choice. Order the closed faces of L by inverse inclusion; $F \leq G$ if and only if $G \subset F$. If $\mathcal{C}$ is a chain of closed faces, then $\cap_{F \in \mathcal{C}} F$ is non-empty and closed, since L is compact, and it is easily verified that it is a face of L, which is an upper bound for $\mathcal{C}$. Thus we can apply Zorn's lemma; there is a maximal face G. We claim that $G = \{g\}$ is a singleton, so that g is an extreme point. If not, suppose that a and b are two distinct points of G. There exists $\phi \in F$ such that $\langle a, \phi \rangle > \langle b, \phi \rangle$. Let $\alpha = \sup_{c \in G} \langle c, \phi \rangle$. Then $G \cap \{x : \phi(x) = c\}$ is a closed face of L properly contained in G, contradicting the maximality of G. □

Theorem 12.2.5 (The Krein–Mil'man theorem) *Suppose that (E, F) is a dual pair of real vector spaces and that K is a non-empty [separable] $\sigma(E, F)$-compact subset of E. $K \subseteq \overline{\Gamma}(Ex(K))$, the closed convex cover of the set $Ex(K)$ of extreme points of K.*

Proof Suppose not. Then there exists $x \in K \setminus \overline{\Gamma}(Ex(K))$. It follows from Corollary 12.1.2, or the separation theorem, that there exists $f \in F$ with $\langle x, f \rangle > \sup\{\langle y, f \rangle : y \in \overline{\Gamma}(Ex(K))\}$. Let $M = \sup\{\langle k, f \rangle : k \in K\}$. Then $J = \{y \in E : \langle y, f \rangle = M\}$ is a closed support hyperplane of K disjoint from $\overline{\Gamma}(Ex(K))$, and $J \cap K$ is a face of K; by the preceding proposition, it contains an extreme point of K, giving a contradiction. □

Corollary 12.2.6 *Suppose that $(E, \|.\|)$ is a [separable] normed space, and that $x \in E$. Then there exists $\phi \in Ex(B(E'))$ for which $\phi(x) = \|x\|$.*

Proof The result is trivially true if $x = 0$. Otherwise, by Corollary 9.1.9 there exists $\psi \in B(E')$ for which $\psi(x) = \|x\|$. Then $H = \{\theta \in E' : \theta(x) = \|x\|\}$ is a weak*-closed support hyperplane of $B(E')$. If ϕ is an extreme point of $H \cap B(E')$ then ϕ is an extreme point of $B(E')$, and $\phi(x) = \|x\|$. □

Exercise 12.2.7 Suppose that (E, F) is a dual pair of real vector spaces and that K is a non-empty [separable] convex $\sigma(E, F)$-compact subset of E. Show that $K = \overline{\Gamma}(Ex(K))$.

Exercise 12.2.8 Suppose that $(E, \|.\|)$ is a [separable] normed space. Show that $B(E') = \overline{\Gamma}(Ex(B(E')))$ (the closure being taken in the weak* topology).

We have a partial converse to the Krein–Mil'man theorem. We need a lemma.

Lemma 12.2.9 *If $K_1, \ldots, K_n$ are compact convex sets, then $\Gamma(\cup_{j=1}^n K_j)$ is compact.*

Proof Let

$$\Delta = \left\{ \delta = (\delta_1, \ldots, \delta_n) \in \mathbf{R}^n : \delta_j \geq 0 \text{ for } 1 \leq j \leq n, \sum_{j=1}^n \delta_j = 1 \right\}.$$

Δ is compact. Let $T : (K_1 \times \cdots \times K_n \times \Delta) \to E$ be defined as

$$T(k_1, \ldots, k_n, \delta) = \sum_{j=1}^n \delta_j k_j.$$

Since $K_1 \times \cdots \times K_n \times \Delta$ is compact and T is continuous, the image, which is $\Gamma(\cup_{j=1}^n K_j)$, is compact. □

Theorem 12.2.10 (Mil'man's theorem) *Suppose that (E, F) is a dual pair of real vector spaces and that K is a non-empty metrizable $\sigma(E, F)$-compact convex subset of E. Suppose that D is a closed subset of K, and that $K = \overline{\Gamma}(D)$. Then $Ex(K) \subseteq D$.*

Proof Suppose that x is an extreme point of K. We show that $x \in \overline{D} = D$. Suppose that N is a closed convex $\sigma(E, F)$ neighbourhood of 0. The sets $\{N + d : d \in D\}$ cover D, and so there exists a finite subset D_0 of D such that $D \subseteq \cup_{d \in D_0}((N + d) \cap K)$. By the lemma, $K \subseteq \Gamma\left(\cup_{d \in D_0}((N + d) \cap K)\right)$. Thus we can write $x = \sum_{d \in D_0} \lambda_d y_d$, where $\lambda_d \geq 0$ and $\sum_{d \in D_0} \lambda_d = 1$. There exists $d \in D_0$ for which $\lambda_d > 0$. Then $x = \lambda_d y_d + (1 - \lambda_d)z$, with $z \in K$. Since x is an extreme point, $x = y_d = n + d$, so that $x - n \in D$. Thus $x \in \overline{D}$. □

Here is a finite-dimensional example, which illustrates the importance of extreme points. The set S_n of doubly stochastic matrices is the set of $n \times n$ matrices m with entries in $[0, 1]$ for which

$$\sum_{i=1}^n m_{ij} = \sum_{j=1}^n m_{ij} = 1 \text{ for all } i, j.$$

Theorem 12.2.11 (Birkhoff's theorem) *The set P_n of permutation matrices is the set of extreme points of the set S_n of doubly stochatic matrices.*

Proof It is easy to see that the permutation matrices are extreme points of S_n. Suppose that $m \in S_n \setminus P_n$. Let $V = \{(i,j) : 0 < m_{ij} < 1\}$. Choose $(i_0,j_0) \in V$. Then the i_0-th row must contain another element (i_0,j_1) of V, and the j_1-th column must contain another element (i_1,j_1) of V. Continuing in this way, there exists a sequence $(i_k,j_k)_{k=0}^{\infty}$ in V with $(i_k,j_{k+1}) \in V$. Since $\{1,\dots,n\} \times \{1,\dots,n\}$ is finite, there is a least k such that there exists $l < k$ such that either $(i_l,j_l) = (i_k,j_k)$ or $(i_l,j_{l+1}) = (i_k,j_{k+1})$. From this it follows that there exists a sequence $(i'_k,j'_k)_{k=0}^{l}$ of distinct elements in V such that, setting $j_{l+1} = j_0$, $(i'_k,j'_{k+1}) \in V$ for $0 \leq k \leq l$; thus we have a circuit with no two terms in common. Let $\epsilon_k = \min\{c_{i'_kj'_k}, c_{i'_kj'_{k+1}}, 1 - c_{i'_kj'_k}, 1 - c_{i'_kj'_{k+1}}\}$ and let $\epsilon = \min_{0 \leq k \leq l} \epsilon_k$. Then $\epsilon > 0$. We now define a matrix d by setting

$$d_{ij} = \begin{cases} \epsilon & \text{if } (i,j) = (i'_k,j'_k) \text{ for some } 0 \leq k \leq l, \\ -\epsilon & \text{if } (i,j) = (i'_k,j'_{k+1}) \text{ for some } 0 \leq k \leq l, \\ 0 & \text{otherwise.} \end{cases}$$

Then $m + d$ and $m - d$ are doubly stochastic matrices, and so m is not an extreme point of S_n. □

Exercise 12.2.12 Identify the extreme points of the unit balls of the following Banach spaces:

$$c_0; \quad l^1; \quad l^\infty; \quad C[0,1]; \text{Hilbert space};$$

and draw appropriate conclusions.

12.3 Dentability

Can we slice small bits off a convex set?

Theorem 12.3.1 *Suppose that (E,F) is a dual pair, that K is a non-empty metrizable convex $\sigma(E,F)$-compact subset of E, and that p is a seminorm on E with the property that $B = \{x : p(x) \leq 1\}$ is $\sigma(E,F)$-closed. If $\epsilon > 0$, there exists a closed convex proper subset C of K such that if $x, y \in A = K \setminus C$ then $p(x - y) \leq \epsilon$.*

Proof First we show that p is bounded on K. The closed sequence of sets $(nB)_{n=1}^{\infty}$ cover K. By Baire's category theorem there exists n such that $nB \cap K$ has a non-empty interior; there exist $x \in K$ and a $\sigma(E,F)$-neighbourhood U of 0 such that $(x+U) \cap K \subseteq nB \cap K$. Since $K - x$ is compact, there exists $R \geq 1$ such that $(K-x) \subseteq RU$. Thus if $y = x + h \in K$ then $x + h/R \in (x+U) \cap K \subseteq nB$, so that

$$\begin{aligned} p(y) &\leq p(x) + p(h) = p(x) + Rp(h/R) \\ &\leq p(x) + R(p(x+h/R) + p(x)) \leq (R+1)p(x) + nR. \end{aligned}$$

Let $D = \overline{Ex(K)}$ and let $d = \sup\{p(x-y) : x, y \in K\}$. If $d \leq \epsilon$ we can take C to be a singleton set. Otherwise, D is $\sigma(E,F)$-separable; let H be a countable dense subset of D. The weakly closed sets $\{(h + (\epsilon/4)B) \cap D : h \in H\}$ cover D. By Baire's category theorem, one of them has a non-empty interior W. Then $D \setminus W$ is a proper non-empty $\sigma(E,F)$-closed subset of D. Let $K_1 = \overline{\Gamma}(D \setminus W)$ and let $K_2 = \overline{\Gamma}(W)$. If $k, l \in K_2$ then $p(k-l) \leq \epsilon/2$. By Theorem 12.2.10, $K_1 \neq K$ and $K_2 \neq K$, and by Lemma 12.2.9, $K = \Gamma(K_1 \cup K_2)$. Suppose that $0 < r < \min(\epsilon/4d, 1)$. Let

$$C_r = \{(1-\lambda)k_1 + \lambda k_2 : k_1 \in K_1, k_2 \in K_2, \lambda \in [0, 1-r]\}.$$

As in Lemma 12.2.9, C_r is a convex weakly compact subset of K. If $x = (1-\lambda)k_1 + \lambda k_2$ is an extreme point of K in C_r, then $x = k_1 \in K_1$, so that $C_r \neq K$. If $y \in K \setminus C_r$, then $y = (1-\lambda)k_1 + \lambda k_2$, with $\lambda > 1-r$, so that $y - k_2 = (1-\lambda)(k_1 - k_2)$, and $p(y-k_2) \leq rd$. Thus if $y, y' \in K \setminus C_r$ then $p(y-y') \leq \epsilon/2 + 2rd < \epsilon$. □

If L is a $\sigma(E,F)$-bounded subset of a dual pair (E,F), a *slice* of L is a set of the form $\{y \in L : \langle y, f\rangle \geq \sup_{x\in L} \langle x, f\rangle - \alpha\}$, where $f \in F$ and $\alpha > 0$.

Corollary 12.3.2 *If $\epsilon > 0$, there exists a slice S of K such that*

$$p(x-y) \leq \epsilon \text{ for } x, y \in S.$$

Proof By Corollary 9.4.4, there exists $f \in E'$ such that if

$$r = \sup\{f(x) : x \in K\} \text{ and } s = \sup\{f(x) : x \in C\},$$

then $r > s$. Choose $0 < \alpha < r - s$. □

A bounded subset L of a normed space is *dentable* if whenever $\epsilon > 0$ there exists a slice of L with diameter less than ϵ.

Corollary 12.3.3 *A non-empty metrizable weakly compact convex subset of a Banach space is dentable. In particular, the unit ball of the dual of a separable Banach space is dentable, and the unit ball of a separable reflexive Banach space is dentable.*

13
Some Fixed Point Theorems

13.1 The Contraction Mapping Theorem

This is the most familiar fixed point theorem, included for completeness' sake. If f is a mapping of a set X into itself, then an element x of X is a *fixed point* of f if $f(x) = x$.

A mapping $f : (X, d) \to (X, d)$ of a metric space into itself is a *contraction mapping* of (X, d) if there exists $0 \leq K < 1$ such that $d(f(x), f(y)) \leq Kd(x, y)$ for all $x, y \in X$; that is, f is a Lipschitz mapping with constant strictly less than 1. The fact that the constant K is *strictly* less than 1 is of fundamental importance.

Theorem 13.1.1 (The contraction mapping theorem) *If f is a contraction mapping of a non-empty complete metric space (X, d) then f has a unique fixed point x_∞.*

Proof Let K be the Lipschitz constant of f. Let x_0 be any point of X. Define the sequence $(x_n)_{n=0}^\infty$ recursively by setting $x_{n+1} = f(x_n)$. Thus $x_n = f^n(x_0)$, and

$$d(x_n, x_{n+1}) \leq Kd(x_{n-1}, x_n) \leq K^2 d(x_{n-2}, x_{n-1}) \leq \cdots \leq K^n d(x_0, x_1).$$

We show that $(x_n)_{n=0}^\infty$ is a Cauchy sequence. Suppose that $\epsilon > 0$. There exists $n_0 \in \mathbf{N}$ such that $K^n < (1-K)\epsilon/(d(x_0, x_1)+1)$ for $n \geq n_0$. If $n > m \geq n_0$ then

$$\begin{aligned} d(x_m, x_n) &\leq d(x_m, x_{m+1}) + d(x_{m+1}, x_{m+2}) + \cdots + d(x_{n-1}, x_n) \\ &\leq K^m d(x_0, x_1) + K^{m+1} d(x_0, x_1) + \cdots + K^{n-1} d(x_0, x_1) \\ &\leq K^m d(x_0, x_1)/(1 - K) < \epsilon. \end{aligned}$$

Since (X, d) is complete, there exists $x_\infty \in X$ such that $x_n \to x_\infty$ as $n \to \infty$. Since f is continuous, $x_{n+1} = f(x_n) \to f(x_\infty)$ as $n \to \infty$, and so $x_\infty = f(x_\infty)$; x_∞ is a fixed point of f. If y is any fixed point of f then

$d(y,x_\infty) = d(f(y),f(x_\infty)) \leq Kd(y,x_\infty)$; hence $d(y,x_\infty) = 0$, and $y = x_\infty$; x_∞ is the unique fixed point of f. □

Corollary 13.1.2 *Suppose that g is a mapping from X to X which commutes with f: $f \circ g = g \circ f$. Then x_∞ is a fixed point of g.*

Proof $f(g(x_\infty)) = g(f(x_\infty)) = g(x_\infty)$; $g(x_\infty)$ is a fixed point of f, and so $g(x_\infty) = x_\infty$. □

We can strengthen the contraction mapping in the following way.

Corollary 13.1.3 *Suppose that $h : (X,d) \to (X_d)$ is a mapping of a complete metric space into itself, and suppose that h^k is a contraction mapping for some $k \in \mathbf{N}$. Then h has a unique fixed point.*

Proof Let $f = h^k$. Then f has a unique fixed point x_∞. As $f \circ h = h \circ f = f^{k+1}$, x_∞ is a fixed point of h. If y is a fixed point of h then $f(y) = h^k(y) = y$, so that $y = x_\infty$; x_∞ is the unique fixed point of f. □

Three points are worth making about this proof. First, we start with *any* point x_0 of X, and obtain a sequence which converges to the unique fixed point x_∞; further, $d(x_0,x_\infty) \leq d(x_0,x_1)/(1-K)$. Secondly, $d(x_{n+1},x_\infty) = d(f(x_n),f(x_\infty)) \leq Kd(x_n,x_\infty)$, so that $d(x_n,x_\infty) \leq K^n d(x_0,x_\infty)$; the convergence is exponentially fast. Thirdly, the condition that $d(f(x),f(y)) < d(x,y)$ for $x \neq y$ is not sufficient for f to have a fixed point. The function $f(x) = x + e^{-x} : [0,\infty) \to [0,\infty)$ does not have a fixed point, but satisfies the condition; if $0 \leq y < x < \infty$ then, by the mean-value theorem, $f(x) - f(y) = (1-e^{-c})(x-y)$ for some $x < c < y$, so that $|f(x)-f(y)| < |x-y|$.

The most familiar application of this theorem gives a proof of the existence and uniqueness of solutions of certain ordinary differential equations.

Theorem 13.1.4 *Suppose that $M \geq 0$ and that $L \geq 0$. Suppose that H is a continuous real-valued function on the triangle*

$$T = \{(x,y) \in \mathbf{R}^2 : 0 \leq x \leq b, |y| \leq Mx\},$$

that $|H(x,y)| \leq M$ and that $|H(x,y) - H(x,y')| \leq L|y-y'|$, for $(x,y) \in T$ and $(x,y') \in T$. Then there exists a unique continuously differentiable function f on $[0,b]$ such that $(x,f(x)) \in T$ for $x \in [0,b]$ and

$$\frac{df}{dx}(x) = H(x,f(x)) \text{ for all } x \in [0,b], \text{ and } f(0) = 0.$$

Proof If f is any solution, then

$$|f(x)| = |f(x) - f(0)| = |\int_0^x H(t,f(t))dt| \leq \int_0^x |H(t,f(t))|dt \leq Mx,$$

for $0 \le x \le b$, so that the graph of f is contained in T. The second condition is a Lipschitz condition, which is needed to enable us to use the contraction mapping theorem.

First let us observe that the fundamental theorem of calculus shows that solving this differential equation is equivalent to solving an integral equation. If f is a solution, then,

$$f(x) = f(x) - f(0) = \int_0^x f'(t)\,dt = \int_0^x H(t,f(t))\,dt \text{ for all } x \in [0,b].$$

Conversely, if f is a continuous function which satisfies this integral equation, then $f(0) = 0$ and the function $J(f)(x) = \int_0^x H(t,f(t))\,dt$ is differentiable, with continuous derivative $H(x,f(x))$. Thus $f'(x) = H(x,f(x))$ for $x \in [0,b]$.

Let $X = \{g \in C[0,b] : |g(x)| \le Mx \text{ for } x \in [0,b]\}$. X is a closed subset of the Banach space $(C[0,b], \|.\|_\infty)$, and so is a complete metric space under the metric defined by the norm. We define a mapping $J : X \to C[0,b]$ by setting

$$J(g)(x) = \int_0^x H(t,g(t))\,dt, \text{ for } x \in [0,b].$$

Then $J(g)$ is a continuous function on $[0,b]$ and

$$|J(g)(x)| \le \int_0^x M\,dt = Mx,$$

so that $J(g) \in X$. We now show by induction that, for each $n \in \mathbf{Z}^+$,

$$|J^n(g)(x) - J^n(h)(x)| \le \frac{L^n x^n}{n!} \|g-h\|_\infty, \text{ for } g,h \in X \text{ and } 0 \le x \le b.$$

The result is certainly true for $n = 0$. Suppose that it is true for n. Then

$$\begin{aligned} |J^{n+1}(g)(x) - J^{n+1}(h)(x)| &\le \int_0^x |H(t,J^n(g)(t)) - H(t,J^n(h)(t))|\,dt \\ &\le \int_0^x L|J^n(g)(t) - J^n(h)(t)|\,dt \\ &\le \int_0^x \frac{L^{n+1}}{n!} \|g-h\|_\infty t^n\,dt \\ &= \frac{L^{n+1}x^{n+1}}{(n+1)!} \|g-h\|_\infty . \end{aligned}$$

Thus

$$\left\| J^n(g) - J^n(h) \right\|_\infty \le \frac{L^n b^n}{n!} \|g-h\|_\infty .$$

Now $L^n b^n/n! \to 0$ as $n \to \infty$, and so there exists $k \in \mathbf{N}$ such that $L^k b^k/k! < 1$. Thus J^k is a contraction mapping of X. We apply Corollary 13.1.3. J has a unique fixed point f, and f is the unique solution of the integral equation. □

Suppose that (X, d) and (Y, ρ) are metric spaces and that $f : X \times Y \to Y$ is continuous. Can we solve the equation $y = f(x, y)$ for each $x \in X$? In other words, is there a function $\phi : X \to Y$ such that $\phi(x) = f(x, \phi(x))$ for each $x \in X$? If so, is it unique? Is it continuous?

Our next application of the contraction mapping theorem gives sufficient conditions for these questions to have a positive answer. It can be thought of as a contraction mapping theorem with a continuous parameter.

Theorem 13.1.5 (The Lipschitz implicit function theorem) *Suppose that (X, d) is a metric space, that (Y, ρ) is a complete metric space and that $f : X \times Y \to Y$ is continuous. If there exists $0 < K < 1$ such that $\rho(f(x,y), f(x,y')) \leq K\rho(y, y')$ for all $x \in X$ and $y, y' \in Y$ then there exists a unique mapping $\phi : X \to Y$ such that $\phi(x) = f(x, \phi(x))$ for each $x \in X$. Further, ϕ is continuous.*

Proof The proof of existence and uniqueness follows easily from the contraction mapping theorem. If $x \in X$ then the mapping $f_x : Y \to Y$ defined by $f_x(y) = f(x, y)$ is a contraction mapping, which has a unique fixed point $\phi(x)$. Then $f(x, \phi(x)) = f_x(\phi(x)) = \phi(x)$.

It remains to show that ϕ is continuous. Suppose that $x \in X$ and that $\epsilon > 0$. There exists $\delta > 0$ such that if $d(x, z) < \delta$ then

$$\rho(\phi(x), f(z, \phi(x))) = \rho(f(x, \phi(x)), f(z, \phi(x))) < (1 - K)\epsilon.$$

For such z, let $z_0 = \phi(x)$ and let $z_n = f_z(z_{n-1})$ for $n \in \mathbf{N}$. Then $z_n \to \phi(z)$, and

$$\rho(\phi(x), \phi(z)) \leq \rho(\phi(x), z_1) + \rho(z_1, \phi(z)) \leq \rho(\phi(x), z_1) + K\rho(\phi(x), \phi(z)),$$

so that

$$\rho(\phi(z), \phi(x)) \leq \frac{\rho(\phi(x), z_1)}{1 - K} = \frac{\rho(\phi(x), f(z, \phi(x)))}{1 - K} < \epsilon.$$

□

Exercise 13.1.6 Suppose that (x, d) is a compact metric space and that f is a mapping from X to X which satisfies $d(f(x), f(y)) < d(x, y)$ whenever $x, y \in X$ and $x \neq y$. Show that f has a unique fixed point.

Exercise 13.1.7 Suppose that f and g are contractions of a complete metric space (X, d). Show that there are $x, y \in X$ such that $f(x) = y$ and $g(y) = x$.

13.2 Fixed Point Theorems of Caristi and Clarke

We prove two fixed point theorems which depend upon Ekeland's variational principle.

Theorem 13.2.1 (Caristi's fixed point theorem) *Suppose that ϕ is a real-valued lower semi-continuous function on a complete metric space (X, d) which is bounded below and that f is a mapping from X into itself which satisfies*

$$d(x,f(x)) \leq \phi(x) - \phi(f(x)) \text{ for all } x \in X.$$

Then f has a fixed point.

Proof By adding a constant, we can suppose that $\inf_{x \in X} \phi(x) = 0$. By Ekeland's variational principle, there exists a point $\tilde{x}$ such that

$$\phi(x) \geq \phi(\tilde{x}) - \tfrac{1}{2}d(x, \tilde{x}) \text{ for all } x \in X.$$

Thus

$$d(\tilde{x},f(\tilde{x})) \leq \phi(\tilde{x}) - \phi(f(\tilde{x})) \leq \tfrac{1}{2}d(\tilde{x}, f(\tilde{x})),$$

so that $d(\tilde{x},f(\tilde{x})) = 0$, and $\tilde{x}$ is a fixed point. □

Note that no continuity conditions are placed on f.

Exercise 13.2.2 Suppose that (X, d) is a complete metric space, that ψ is a lower semi-continuous function on $X \times X$ and that f is a continuous mapping from X to X which satisfies $\psi(f(x),f(y)) \leq \psi(x, y) - d(x, y)$ for all $(x, y) \in X \times X$. Show that f has a unique fixed point.

Clarke's theorem is an extension of the contraction mapping theorem (Theorem 13.1.1).

Theorem 13.2.3 (Clarke's fixed point theorem) *Suppose that g is a continuous mapping of a complete metric space (X, d) into itself, with the property that there exists $0 \leq k < 1$ such that if $x \in X$ then there exists $0 < t \leq 1$ and x_t in X such that*

$$d(x, x_t) = td(x, g(x)) \text{ and } d(x_t, g(x)) = (1 - t)d(x, g(x)),$$

and $d(g(x), g(x_t)) \leq kd(x, x_t)$. Then g has a fixed point.

Proof Let $\alpha = (1 - k)/2$. Let $f(x) = d(x, g(x))$. Then f is continuous, and so by Ekeland's variational principle (Theorem 4.4.1) there exists $\tilde{x} \in X$ such that

$$d(\tilde{x}, g(\tilde{x})) \leq d(x, g(x)) + \alpha d(x, \tilde{x}) \text{ for all } x \in X.$$

Setting $x = \tilde{x}_{\tilde{t}}$ we see that

$$\begin{aligned} d(\tilde{x}, g(\tilde{x})) &\leq d(\tilde{x}_{\tilde{t}}, g(\tilde{x}_{\tilde{t}})) + \alpha d(\tilde{x}_{\tilde{t}}, \tilde{x}) \\ &\leq d(\tilde{x}, g(\tilde{x})) + d(g(\tilde{x}), g(\tilde{x}_{\tilde{t}})) + \alpha d(\tilde{x}_{\tilde{t}}, \tilde{x}) \\ &\leq (1 - \tilde{t})d(\tilde{x}, g(\tilde{x})) + k(d(\tilde{x}, \tilde{x}_{\tilde{t}}) + \alpha d(\tilde{x}_{\tilde{t}}, \tilde{x}) \end{aligned}$$

so that

$$\tilde{t}d(\tilde{x}, g(\tilde{x})) \leq (k + \alpha)\tilde{t}d(\tilde{x}, g(\tilde{x}))$$

which ensures that $d(\tilde{x}, g(\tilde{x})) = 0$, and $\tilde{x}$ is a fixed point of g. □

Exercise 13.2.4 Let g be a continuous mapping of a closed subset X of a Banach space $(E, \|.\|)$ into itself for which $[x, g(x)] \subseteq X$ for each x and for which there exists $0 < k < 1$ such that if $x \in X$ there exists $y \in (x, g(x)]$ such that $d(g(x), g(y)) \leq kd(x, y)$. Show that g has a fixed point.

13.3 Simplices

Our aim is to prove Brouwer's fixed point theorem; for this we need to consider simplices an triangulations. Suppose that $V = \{v_0, \ldots, v_n\}$ is a finite subset of a real normed vector space $(E, \|.\|)$, with the property that $\{v_i - v_0 : 1 \leq i \leq n\}$ is a linearly independent set. Then the convex cover $\Delta = \Delta(V) = \Gamma(V)$ is an *n-simplex* in E. The elements of V are the *vertices* of $\Delta(V)$. If $x \in \Delta$, then x can be written uniquely as $x = \sum_{i=0}^n x_i v_i$, where $x_i \geq 0$ for $1 \leq i \leq n$ and $\sum_{i=0}^n x_i = 1$. The numbers x_i are the *barycentric co-ordinates* of x. The set $s(x) = \{i : x_i \neq 0\}$ is the *support* of x.

As an example, let the basic unit vectors of $\mathbf{R}^{n+1}$ be denoted by $e_0, e_1, \ldots, e_n$, and let $E = \{e_0, e_1, \ldots, e_n\}$. Then $\Delta(E)$ is the *fundamental n-simplex*.

Suppose that F is a non-empty subset of $\{0, \ldots, n\}$. Then $\Delta_F = \{x \in \Delta : s(x) \subset F\}$ is a *face* of Δ; it is a $(|F| - 1)$-simplex. Thus the 0-faces of Δ are the singletons $\{v_i\}$, the 1-faces are line segments $[v_i, v_j]$, the 2- faces are triangles, the 3-faces are tetrahedra, and so on. An $(n - 1)$-face is called a *facet*. The *interior* Δ^{int} is the set $\{x \in \Delta : s(x) = \{0, \ldots, n\}\}$ and the *boundary* $\partial\Delta$ is the union of the $n + 1$ facets of Δ. For example the *barycentre* $b = b(V) = \sum_{i=0}^n v_i/(n + 1)$ is an element of Δ^{int}.

A *triangulation* $T(V)$ of $\Delta(V)$ is a finite set of n-simplices whose union is Δ, with the property that if Δ_1 and Δ_2 are two distinct elements of $T(V)$ then $\Delta_1^{int} \cap \Delta_2^{int} = \emptyset$. An important example is provided by the *barycentric triangulation*. This is defined inductively. If $n = 1$ then the barycentre $b = \frac{1}{2}(v_0 + v_1)$, and the barycentric triangulation is $[v_0, b]$, $[v_1, b]$. If the barycentric triangulation has been defined for $(n - 1)$-simplices, then the barycentric triangulation of an n-simplex $\Delta(V)$ is defined as follows. Each facet F of $\Delta(V)$ is an $(n - 1)$-simplex, and therefore there is a barycentric triangulation of $T(F)$. If $\Delta' \in F$ then $\Gamma(\Delta' \cup \{b(V)\})$ is an n-simplex. The collection

$$\{\Gamma(\Delta' \cup \{b(V)) : \Delta' \in T(F), F \text{ a facet of } \Delta\}$$

is then a triangulation of Δ, the *barycentric triangulation* $T_b(V)$ of Δ. The barycentric triangulation $T_b(V)$ has $(n+1)!$ elements.

Since the norm is a convex function,

$$\mathrm{diam}(\Delta(V)) = \max\left(\left\|v - v'\right\| : v, v' \in V\right\}.$$

It is an important fact that the diameters of the barycentric triangulation are smaller than the diameter of V.

Proposition 13.3.1 *If* $\Delta' = \Delta(V') \in T_b(V)$, *then*

$$\mathrm{diam}(\Delta') \leq \frac{n}{n+1}\mathrm{diam}(\Delta).$$

Proof Again, the proof is by induction. The result is true if $n = 1$. Suppose that it is true for $n-1$. Let b' be the barycentre of Δ', and let v be the vertex in V which does not belong to Δ'. Then

$$b = \frac{v}{n+1} + \frac{nb'}{n+1},$$

so that if v' is a vertex of Δ' then, using the inductive hypothesis,

$$\begin{aligned}
\left\|b - v'\right\| &\leq \left(\frac{1}{n+1}\right)\left\|v - v'\right\| + \left(\frac{n}{n+1}\right)\left\|v - b'\right\| \\
&\leq \left(\frac{1}{n+1} + \frac{n}{n+1}.\frac{n-1}{n}\right)\mathrm{diam}(\Delta) \\
&= \left(\frac{n}{n+1}\right)\mathrm{diam}(\Delta).
\end{aligned}$$

On the other hand, if $v_1, v_2 \in V'$ then

$$\|v_1 - v_2\| \leq \left(\frac{n-1}{n}\right)\mathrm{diam}(\Delta) \leq \left(\frac{n}{n+1}\right)\mathrm{diam}(\Delta).$$

□

Thus iterating the construction of barycentic triangulations, we can obtain a triangulation of Δ for which each simplex of the triangulation has small diameter.

13.4 Sperner's Lemma

Suppose that A is a subset of an n-simplex $\Delta(V)$. A mapping l from A into $\{0, \ldots, n\}$ is a *Sperner mapping* if $l(a) \in s(a)$, the support of a, for each $a \in A$. Thus if a is in a face F then $l(a)$ is a vertex of F.

Suppose that $T(V)$ is a triangulation of $\Delta(V)$. We denote the set of vertices of the n-simplices in $T(V)$ by $W(T(V))$. Suppose that l is a *Sperner mapping* on $W(T(V))$, and that $\Delta' \in T(V)$. Then Δ' is *completely labelled* if l maps the vertices of Δ' onto $\{0, \ldots, n\}$.

Theorem 13.4.1 (Sperner's Lemma) *Suppose that $T(V)$ is a triangulation of an n-simplex $\Delta(V)$, and that l is a Sperner mapping on $W(T(V))$. Then the number $N(T(V))$ of completely labelled n-simplices in $T(V)$ is odd, and so there exists at least one completely labelled n-simplex in $T(V)$.*

Proof The proof is by induction on n. First suppose that $n = 1$. Then $W(T(V)) = \{v_0 = q_0, \ldots, q_j = v_1\}$, where the vertices are listed in order on $[v_0, v_1]$. Let $d_j = l(q_j) - l(q_{j-1})$, let $A = \{j : d_j = 1\}$ and let $B = \{j : d_j = -1\}$. Then the number of completely numbered 1-simplices in $T(V)$ is $|A| + |B|$. But

$$1 = \sum_{i=1}^{j} d_i = \sum_{i \in A} 1 + \sum_{i \in B} (-1) = |A| - |B|,$$

so that $|A| + |B| = 1 + 2|B|$, and $|A| + |B|$ is odd.

Suppose now that the result holds for $n - 1$, that $T(V)$ is a triangulation of an n-simplex $\Delta(V)$, and that l is a Sperner mapping on $W(T(V))$. Suppose that $\Delta' = \Delta'(V') \in T(V)$, and that F is a facet of Δ'. We set $\phi(F) = 1$ if

$$\{l(v) : v \text{ a vertex of } F\} = \{0, \ldots, n - 1\},$$

and set $\phi(F) = 0$ otherwise. We set $\psi(\Delta') = \sum\{\phi(F) : F \text{ a facet of } \Delta'\}$.

We count $S = \sum\{\psi(\Delta') : \Delta' \in T(V)\}$ in two different ways. Let $R(\Delta') = \{l(v) : v \in V'\}$. If Δ' is completely labelled, then $F(\Delta') = 1$. If $R(\Delta') = \{1, \ldots, n - 1\}$ then just one value is taken twice, so that $\psi(\Delta') = 2$, and otherwise $\psi(\Delta') = 0$. Adding, $S - N(T(V))$ is even.

On the other hand, let $\mathcal{F} = \{F : F \text{ a facet of } \Delta', \text{ for some } \Delta' \in T(V)\}$. Let

$$\mathcal{F}_1 = \{F \in \mathcal{F} : F \subseteq \partial\Delta\}, \qquad \mathcal{F}_2 = \{F \in \mathcal{F} : F \cap \Delta^{int} \neq \emptyset\}.$$

If $F \in \mathcal{F}_1$, then F is a facet of exactly one Δ' in $T(V)$, and $\phi(F) = 1$ if and only if it is a completely numbered $n - 1$-simplex in the triangulation of the facet $\Gamma\{v_0, \ldots, v_{n-1}\}$. By the inductive hypothesis, there is an odd number of these. If $F \in \mathcal{F}_2$, then F is a facet of exactly two elements of $T(V)$, and so $\phi(F)$ is counted twice. Consequently, S is odd. Since $S - N(T(V))$ is even it follows that $N(T(V))$ is odd.

□

13.5 Brouwer's Fixed Point Theorem

Theorem 13.5.1 (Brouwer's fixed point theorem) *Suppose that K is a compact convex subset of $\mathbf{R}^n$ and that $f : K \to K$ is continuous. Then f has a fixed point; there exists $x \in K$ such that $f(x) = x$.*

Proof The result is a topological one. Since K is homeomorphic to a simplex, we may assume that $K = \Delta(E)$, the fundamental n-simplex. We give $\mathbf{R}^{n+1}$ the metric defined by the norm $\|x\| = \sum_{i=0}^n |x_i|$.

Suppose that the result is false. Let

$$g(x) = (g_0(x), \dots, g_n(x)) = x - f(x).$$

Then $\sum_{i=0}^n g_i(x) = 0$, for all $x \in \Delta(E)$. Since g is continuous and K is compact, $\epsilon = \inf_{x\in K} \|g(x)\| > 0$, and there exists $\delta > 0$ such that if $\|x - y\| < \delta$ then $\|g(x) - g(y)\| < \epsilon/2(n+1)$.

We now define a Sperner mapping on the set $\Delta(E)$. Suppose that $x \in \Delta(E)$. Since $\sum_{i=0}^n g_i(x) = 0$ and $\sum_{i=0}^n |g_i(x)| \geq \epsilon$, it follows that $m(x) = \max_{i=0}^n g_i(x) \geq \epsilon/2(n+1)$. We set $l(x) = \inf\{i : g_i(x) = m(x)$, so that $g_{l(x)}(x) = m(x)$. Then $x_{l(x)} = g_{l(x)}(x) + f_{l(x)}(x) \geq m(x) > 0$, and so l is a Sperner mapping.

There is a triangulation $T(E)$ such that all the simplices have diameter less than δ. It follows from Sperner's lemma that one of the n-simplices in $T(E)$, $\Delta'(V')$, say, is completely labelled, and we can label the vertices so that $l(v_i') = i$. Then $g_0(v_0') \geq \epsilon/2(n+1)$, and if $1 \leq i \leq n$ then

$$g_i(v_0') \geq g_i(v_i') - |g_i(v_i') - g_i(v_0')| \geq \epsilon/2(n+1) - \epsilon/2(n+1) = 0.$$

Thus $\sum_{i=0}^n g_i(v_0') > 0$, giving a contradiction. □

Here is another application of Sperner's lemma.

Theorem 13.5.2 *Suppose that K is a compact convex subset of $\mathbf{R}^m$, that $f : K \to K$ is continuous and that $f(x) = x$ for $x \in \partial K$. Then f is surjective: $f(K) = K$.*

Proof Once again, we can suppose that $K = \Delta(E)$. It is clearly sufficient to show that if $y \in \Delta(E)^{int}$ then there exists $x \in \Delta(E)$ with $f(x) = y$. If $x \in \Delta(E)$ let

$$l(x) = \inf\{i : f_i(x)/y_i = \max\{f_j(x)/y_j : 0 \leq j \leq n\}\}.$$

Then l is a Sperner mapping. Note that $f(x)_{l(x)} \geq y_{l(x)}$.

By Sperner's lemma, there exists a decreasing sequence $(\Delta_k(V_k))_{k=1}^{\infty}$ of completely labelled n-simplices, with $\mathrm{diam}(\Delta_k(V_k)) \to 0$. Then $\cap_{k=1}^{\infty}\Delta_k(V_k)$ is a singleton $\{x\}$. We show that $f(x) = y$.

We label $V_k' = \{v_0^{(k)}, \dots v_n^{(k)}\}$ in such a way that $l(v_i^{(k)}) = i$ for $0 \leq i \leq n$. If $0 \leq i \leq n$ then

$$f_i(x) = f_i(v_i^{(k)}) + (f_i(x) - f_i(v_i^{(k)})) \geq y_i - |f_i(x) - f_i(v_i^{(k)})|.$$

But $f_i(v_i^{(k)}) \to f_i(x)$ as $k \to \infty$, and so $f_i(x) \geq y_i$, for $0 \leq i \leq n$. But $\sum_{i=0}^n f_i(x) = 1 = \sum_{i=0}^n y_i$, and so $f_i(x) = y_i$ for $0 \leq i \leq n$. □

Corollary 13.5.3 *There is no retract of K onto ∂K.*

In fact, we can do better. Suppose that K is a compact convex subset of $\mathbf{R}^m$. Let $F = \mathrm{span}\{x - y : x, y \in K\}$, and let $\partial_{rel}K$ be the boundary of K in F; $\partial_{rel}K$ is the *relative boundary* of K.

Corollary 13.5.4 *Suppose that K is a compact convex subset of $\mathbf{R}^m$, that $f : K \to K$ is continuous and that $f(x) = x$ for $x \in \partial_{rel}K$. Then f is surjective: $f(K) = K$. Thus, unless K is a singleton, there is no retract of K onto $\partial_{rel}(K)$.*

We can use Theorem 13.5.2 to give another proof of Brouwer's fixed point theorem. This time, we can take K to be the unit ball $B(\mathbf{R}^n)$ of $\mathbf{R}^n$ with its Euclidean norm. Suppose that $f : K \to K$ does not have a fixed point. If $x \in K$, the ray $\{f(x) + t(x - f(x)) : t \geq 1\}$ meets ∂K. Let $t_x = \inf\{t \geq 0 : x + t(x - f(x)) \in \partial K\}$, and let $r(x) = x + t_x(x - f(x))$. Then r is a retract of K onto ∂K. It remains to show that r is continuous. It is easy to see that the function $x \to t_x$ is continuous when $x \in K^{int}$; let us consider the case where $x \in \partial K$. Suppose that $\epsilon > 0$. There exists $0 < \delta < \epsilon$ such that $|x - f(x)| > 2\delta$, and there exists $0 < \eta < \delta$ such that if $|x - y| < \eta$ then $|r(y) - x| = |r(y) - r(x)| < \delta < \epsilon$. Thus r is a continuous retract of K onto ∂K, giving a contradiction.

13.6 Schauder's Fixed Point Theorem

We now extend the Brouwer fixed point theorem to infinite-dimensional spaces.

Theorem 13.6.1 (Schauder's fixed point theorem) *Suppose that (E, F) is a dual pair, and that K is a $\sigma(E, F)$-compact convex metrizable subset of E. If $f : K \to K$ is continuous, then it has a fixed point; there exists $x \in K$ with $f(x) = x$.*

Proof By Theorem 12.1.1, we can suppose that K is a compact subset of L, where $L = \{x : x \in l_2, |x_n| \leq 1/n \text{ for } n \in \mathbf{N}\}$.

Let $(e_n)_{n=1}^{\infty}$ be the standard orthonormal basis for l_2, and let $E_n = \operatorname{span}(e_1, \ldots, e_n)$. Let $j_n : E_n \to l_2$ be the inclusion mapping, and let P_n be the orthogonal projection of l_2 onto E_n. Then $L_n = P_n(L) \subseteq L$. Let $f_n = P_n \circ f \circ j_n$. Then f_n is a continuous mapping from L_n to itself, and so, by Brouwer's fixed point theorem, it has a fixed point x_n. Note that

$$\|x_n - f(x_n)\|^2 = \|P_n w(f(x_n)) - f(x_n)\|^2 \leq \sum_{j=n+1}^{\infty} |f(x_n)|_j^2 \leq 1/n.$$

Since M is compact, there exists a convergent subsequence $(x_{n_j})_{j=1}^{\infty}$, convergent to x say. But then $f(x_{n_j}) \to f(x)$, so that $\|x - f(x)\| = \lim_{j\to\infty} \left\|x_{n_j} - g(x_{n_j})\right\| = 0$, and $f(x) = x$. □

Exercise 13.6.2 Give an example of a continuous mapping f from the unit sphere $S(H)$ of a Hilbert space H into itself which has no fixed point.

Exercise 13.6.3 Let T be a continuous mapping of a closed convex subset A of a Banach space $(E, \|.\|)$ into itself for which $\overline{T(A)}$ is compact. Show that T has a fixed point.

Theorem 13.6.4 *Suppose that $(E, \|.\|)$ is a Banach space, that F is a linear subspace of E' for which (E, F) is a dual pair and the unit ball $B(E)$ is $\sigma(E, F)$-compact and separable. If $f : B(E) \to B(E)$ is continuous for the topology $\sigma(E, F)$ and $f(x) = x$ for $x \in S(E) = \{x : \|x\| = 1\}$ then f maps $B(E)$ onto $B(E)$.*

Proof By Theorem 12.1.1, there exists a linear map $R : E \to l_2$ such that $K = R(B(E)) \subseteq L = \{x : |x_n| \leq 1/n \text{ for } n \in \mathbf{N}\}$, and such that R is a homeomorphism of $(B(E), \sigma(E, F))$ onto the compact set $(K, \|.\|_2)$. Let $M = R(S(E))$, and let $g = R \circ f \circ R^{-1} : K \to K$. Let $P_n : l_2 \to l_2$ be the projection $l_2 \to l_2$ described in Theorem 7.6.5, let $K_n = P_n(K)$, let $M_n = P_n(M)$ and let $g_n = P_n \circ g : K \to K_n$. Suppose that $x \in K$. Let $x_n = P_n(x)$. Note that $\partial_{rel} K_n \subseteq M_n$. Thus g_n maps K_n continuously into K_n, and $g_n(y) = y$ for $y \in \partial_{rel} K_n$. It therefore follows from Brouwer's theorem that there exists $w_n \in K$ such that if $z_n = P_n(w_n)$, then $g_n(z_n) = x_n$. Now the sequence $(w_n)_{n=1}^{\infty}$ is contained in the compact set K, and so there is a subsequence $(w_{n_j})_{j=1}^{\infty}$ which converges in norm to an element w of K; then $z_n \to w$ as well. If $n < n_j$ then $P_n(x_{n_j}) = x_n$, so that $g_n(z_{n_j}) = x_n$. Thus $g_n(w) = x_n$, and so $g(w) = x$; g maps K onto K, and so f maps $B(E)$ onto $B(E)$. □

Corollary 13.6.5 *There is no retract of $(B(E), \sigma(E,F))$ onto the unit sphere $S(E) = \{x \in E : \|x\| = 1\}$.*

This theorem applies to weak* duals of separable Banach spaces, and so to separable reflexive Banach spaces.

13.7 Fixed Point Theorems of Markov and Kakutani

Schauder's fixed point theorem shows that there is at least one fixed point, but gives no further information about the set of fixed points. The situation is different when the mapping is affine; here we do not need to suppose that K is metrizable.

Theorem 13.7.1 (The Markov–Kakutani fixed point theorem) *Suppose that (E,F) is a separated dual pair, that K is a non-empty $\sigma(E,F)$-compact convex subset of E and that T is a weakly continuous affine mapping of E into itself for which $T(K) \subseteq K$. Then the set of fixed points of T in K is a non-empty convex compact subset of K.*

Proof If $n \in \mathbf{N}$ and $T \in \mathcal{T}$, let $A_n(T) = (1/n)(I + T + T^2 + \cdots + T^{n-1})$. Then $A_n(T)(K)$ is a $\sigma(E,F)$-compact subset of K. If $T_1, \ldots, T_k \in \mathcal{T}$, $(n_1, \ldots, n_k) \in \mathbf{Z}^k$, and σ is a permutation of $\{1, \ldots, n\}$ with $\sigma(1) = j$ then

$$A_{n_j} T_j(K) \supseteq A_{n_{\sigma(1)}}(T_{\sigma(1)}) \ldots A_{n_{\sigma(k)}}(T_{\sigma(k)})(K) = A_{n_1}(T_1) \ldots A_{n_k}(T_k)(K),$$

so that

$$\cap_{j=1}^k A_{n_j}(T_j)(K) \supseteq A_{n_1}(T_1) \ldots A_{n_k}(T_k)(K) \neq \emptyset.$$

Using the finite intersection property,

$$C = \cap\{A_n(T)(K) : T \in \mathcal{T}, n \in \mathbf{N}\}$$

is a non-empty closed convex subset of K. We show that if $x \in C$ then $T(x) = x$ for all $T \in \mathcal{T}$. Suppose not, and that $T(x_0) \neq x_0$, for $x_0 \in C$. There exists $f \in F$ such that $\langle T(x_0) - x_0, f \rangle \neq 0$. f is bounded on K; let $M = \sup\{| \langle x - y, f \rangle | : x, y \in K\}$. Since $x_0 \in C$, if $n \in \mathbf{N}$, there exists $y_n \in K$ such that $A_n(T)(y_n) = x_0$. But then $x_0 - T(x_0) = (1/n)(y_n - T^n(y_n))$, so that

$$| \langle T(x) - x, f \rangle | = \frac{1}{n} | \left\langle y_n - T^n(y_n), f \right\rangle | \leq \frac{M}{n}.$$

Since this holds for all $n \in \mathbf{N}$, we have a contradiction. □

This is a start; but the non-commutative case remains. We need another fixed point theorem. First, we obtain some easy standard results about weak* topologies.

Exercise 13.7.2 Suppose that $(E, \|.\|)$ is a normed space, and that $\mathcal{A}$ is the collection of totally bounded subsets of E. For $\phi \in E'$ let $B_{tb}(\phi) = \{\phi + A^\circ : A \in \mathcal{A}\}$. There is a topology $\tau_{tb}(E')$ on E', such that B_{tb} is a base of neighbourhoods of ϕ, for each $\phi \in E'$.

The topology t_{tb} is finer than the weak* topology $\sigma(E', E)$. On bounded subsets of E' they are the same.

Proposition 13.7.3 *Suppose that $(E, \|.\|)$ is a normed space and that B is a norm-bounded subset of E'. Then the restrictions of τ_{tb} and $\sigma(E', E)$ to B are the same.*

Proof By scaling, we can assume that B is contained in the unit ball of E'. Suppose that $\phi \in B$ and that $\phi + A^\circ$ is a τ_{tb} neighbourhood of ϕ. We must show that there exists a finite set A_0 in E such that if $\psi \in B$ and $\max_{x \in A_0} |\langle x, \phi - \psi\rangle| < \frac{1}{2}$ then $\psi \in \phi + A^\circ$.

There exists a finite 1/4-net A_0 in A. Let

$$N = \{\psi \in B : |\langle a_0, \phi - \psi\rangle| < 1/2 \text{ for } a_0 \in A_0\}.$$

N is a weak* neighbourhood of ϕ in B. If $a \in A$, there exists $a_0 \in A_0$ such that $\|a - a_0\| \le 1/4$. Thus if $\psi \in N$

$$|\langle a, \phi - \psi\rangle| \le |\langle a_0, \phi - \psi\rangle| + |\langle a - a_0, \phi\rangle| + |\langle a - a_0, \psi\rangle| \le 1,$$

and so $N \subset \phi + A^\circ$. □

Theorem 13.7.4 (Kakutani's fixed point theorem) *Suppose that $(E, \|.\|)$ is a separable Banach space, that K is a non-empty bounded closed convex subset of the dual E', and that $\mathcal{S}$ is a bounded semigroup in $L(E)$ with the property that $\{T(x) : T \in \mathcal{S}\}$ is totally bounded, for each $x \in E$, and such that $T'(K) \subseteq K$ for each $T \in \mathcal{S}$. Then there exists $\phi \in K$ such that $T'(\phi) = \phi$, for each $T \in \mathcal{S}$.*

Proof We give a proof which avoids the use of the axiom of choice. We can suppose that $I \in \mathcal{S}$. Let $(x_n)_{n=1}^\infty$ be a dense sequence in $B(E)$. Then it is easy to see the set $A = \{T(x_n)/2^n : n \in \mathbf{N}, T \in \mathcal{S}\}$ is a totally bounded subset of E. Let $p_A(\phi) = \sup\{|\phi(a)| : a \in A\}$. Since $(x_n)_{n=1}^\infty$ is a dense sequence in $B(E)$, p_A is a norm on E'; applying Proposition 13.7.3, it defines the weak* topology on K. Let $\mathcal{K}$ be the set of non-empty weak*-closed (and so weak*-compact) convex $\mathcal{S}$ invariant subsets of K. We order $\mathcal{K}$ by inverse inclusion. If $\mathcal{C}$ is a chain in $\mathcal{K}$, then $\cap_{L \in \mathcal{C}} L$ is an upper bound for $\mathcal{C}$, and so we can apply the Brézis–Browder lemma (Theorem 4.3.1) to the function d, where $d(L) = \text{diam}(L)$; there exists $L \in \mathcal{K}$ such that if $M \in \mathcal{K}$ and $M \subseteq L$ then $\text{diam}(M) = \text{diam}(L)$.

We now show that L is a singleton set, which establishes the result. Suppose not, so that $\delta = \text{diam}(L) > 0$. Let $D = \{\phi : p_A(\phi) \leq \delta\}$. Thus $L - L \subseteq D$, and if $0 < \alpha < 1$ then $L - L \nsubseteq \alpha D$. Note that if $T \in \mathcal{S}$ then $T'(D) \subseteq D$: D is $\mathcal{S}$-invariant.

The sets $\{D/2 + \phi : \phi \in L\}$ cover L, and so there is a finite subcover $\{D/2 + \phi_j : 1 \leq j \leq k\}$. Let $\phi_0 = (1/k)(\phi_1 + \cdots + \phi_k)$. Then $p_0 \in L$. If $\phi \in K_0$, there exists $1 \leq j \leq k$ such that $\phi \in D/2 + \phi_j$. Thus

$$\phi_0 - \phi = \frac{1}{k}((\phi_1 - \phi) + \cdots + (\phi_k - \phi)) \in (1 - \frac{1}{2k})D,$$

and so $\phi_0 \in (1 - 1/2k)D + \phi$. Now let

$$L_1 = L \cap \left(\cap \{(1 - \frac{1}{2k})D + \phi : \phi \in K_0\} \right).$$

Then L_1 is a compact convex subset of L_0, and is non-empty, since $\phi_0 \in L_1$. Since D is $\mathcal{S}$-invariant, so is L_1. If $\phi, \psi \in L_1$ then $\phi \in (1-1/2k)D + \psi$, so that $L_1 - L_1 \subseteq (1-1/2k)D$, and $\text{diam}L_1 < \delta$, giving the required contradiction. □

We shall use this in Section 16.12 to show that if $(G, \times)$ is a compact metrizable topological group, there exist a left Haar measure and right Haar measure on G.

13.8 The Ryll–Nardzewski Fixed Point Theorem

We prove another fixed point theorem. First, we need a definition. Suppose that E is a Banach space, that $\mathcal{S}$ is a semigroup in $L(E)$ and that K is a convex $\mathcal{S}$-invariant subset of E. Then $\mathcal{S}$ is *non-contracting* on K if for each $k_1, k_2 \in K$, with $k_1 \neq k_2$, there exists $\delta > 0$ such that $\|T(k_1) - T(k_2)\| \geq \delta$, for each $T \in \mathcal{S}$.

Theorem 13.8.1 (The Ryll-Nardzewski fixed point theorem) *Suppose that E is a separable Banach space, that $\mathcal{S}$ is a semigroup in $L(E)$, that K is a non-empty weakly compact convex $\mathcal{S}$-invariant subset of E and that $\mathcal{S}$ is non-contracting on K. Then $\mathcal{S}$ has a fixed point in K.*

Proof First observe that it is enough to prove the result when $\mathcal{S}$ is finitely generated. If $\mathcal{S}_1$ is a finitely generated sub-semigroup, then $\{k \in K : T(k) = k \text{ for } T \in \mathcal{S}_1\}$ is a non-empty weakly closed convex subset of K, and the collection of all such sets has the finite intersection property. Any point in the intersection is then a fixed point for $\mathcal{S}$.

Suppose then that $\mathcal{S}$ is generated by $T_1, \ldots, T_k$. Let $A = (T_1 + \cdots + T_k)/k$. Then K is A-invariant, and so it has a fixed point x_0, by the Markov fixed point theorem. We claim that $T_i(x_0) = x_0$ for $1 \leq i \leq k$. Suppose not. By relabelling, we can suppose that $T_i(x_0) \neq x_0$ for $1 \leq i \leq j$ and that $T_i(x_0) = x_0$ for $j < i \leq k$. Let B be the average $(T_1 + \cdots + T_j)/j$. Then $kx_0 = kA(x_0) = jB(x_0) + (k - j)x_0$, so that $B(x_0) = x_0$. Let $\mathcal{S}_2$ be the semigroup generated by $T_1, \ldots, T_j$. There exists $\delta > 0$ such that $\|TT_i(x_0) - T(x_0)\| > \delta$ for all $T \in \mathcal{S}_1$ and for $1 \leq i \leq j$. Let $L = \overline{\Gamma}\{T(x_0) : T \in \mathcal{S}_2\}$: then L is a separable weakly compact convex subset of K, and $x_0 = B(x_0) \in L$. By Corollary 12.3.3 there exists a slice M of L of diameter less than δ, and there exists $T \in \mathcal{S}_1$ such that $T(x_0) \in M$. But

$$T(x_0) = TB(x_0) = (TT_1(x_0) + \cdots + TT_j(x_0))/j,$$

so that there exists $1 \leq i \leq j$ such that $TT_i(x_0) \in M$. Thus $\|T(x_0) - TT_i(x_0)\| < \delta$, giving a contradiction. □

The Ryll-Nardzewski theorem can be established in other settings.

Exercise 13.8.2 Suppose that $\mathcal{S}$ is a semigroup in $L(E)$, that $\{T(x) : T \in \mathcal{S}\}$ is totally bounded, for each $x \in E$, and that $\mathcal{S}' = \{T' : T \in \mathcal{S}\}$ and that K is a non-empty weak*-compact convex $\mathcal{S}'$-invariant subset of E', with the following non-contraction property: for each $k_1, k_2 \in K$, with $k_1 \neq k_2$ there exists a totally bounded subset A of E such that $p_A(T'(k_1) - T'(k_2)) \geq 1$, for each $T \in \mathcal{S}$. Let A be the totally bounded subset of E described in Theorem 13.7.4. Show that there exists a weak*-closed slice M in K for which $P_A(x - y) \leq 1$ for $x, y \in M$. Arguing as in Theorem 13.8.1, deduce that $\mathcal{S}'$ has a fixed point in K.

This again will give a proof of the existence of Haar measure on a compact Hausdorff topological group.

PART TWO

Measures on Polish Spaces

14
Abstract Measure Theory

In this chapter, we give a brief account of the theory of measures on abstract sets. It contains definitions and statements of results, and the notation that is used. Abstract measure theory proceeds by many simple steps; many of these are set as exercises, although some comments are made. Proofs can be found in [G III], [Bi II],[H] or [B].

14.1 Measurable Sets and Functions

First, we describe the notation that we shall use. If $f : X \to Y$ is a mapping, if $c \in Y$ and if $A \subseteq Y$, we write $(f = c)$ for the set $\{x \in X : f(x) = c\}$ and $(f \in A)$ for the set $\{x \in X : f(x) \in A\}$, and use similar notation for other such sets.

A non-empty collection Σ of subsets of a non-empty set X is called a *σ-ring* if

(i) if $(A_j)_{j=1}^{\infty}$ is a sequence of sets in Σ then $\cup_{j=1}^{\infty} A_j \in \Sigma$, and
(ii) if $A, B \in \Sigma$ then $A \setminus B \in \Sigma$.

If

(iii) $X \in \Sigma$,

then Σ is a *σ-field*.

Exercise 14.1.1 Suppose that Σ is a non-empty collection of subsets of a non-empty set X. The following are equivalent.

(i) Σ is a σ-field.
(ii) If $A \in \Sigma$ then $X \setminus A \in \Sigma$, and if $(A_n)_{n=1}^{\infty}$ is a sequence of disjoint elements of Σ then $\cup_{n=1}^{\infty} A_n \in \Sigma$.
(iii) If $A \in \Sigma$ then $X \setminus A \in \Sigma$, and if $(A_n)_{n=1}^{\infty}$ is an increasing sequence in Σ then $\cup_{n=1}^{\infty} A_n \in \Sigma$.

Exercise 14.1.2 Let $\underline{s}$ be a collection of σ-fields on a set X. Show that

$$\cap\{\Sigma : \Sigma \in \underline{s}\} = \{A : A \in \Sigma, \text{ for all } \Sigma \in \underline{s}\}$$

is a σ-field. Is the union of two σ-fields necessarily a σ-field?

A *measurable space* is a pair (X, Σ), where X is a set and Σ is a σ-field of subsets of X.

Proposition 14.1.3 *Suppose that $\mathcal{F}$ is a set of subsets of a set S. Then there is a smallest σ-field $\sigma(\mathcal{F})$ of subsets of S which contains $\mathcal{F}$.*

Proof Let $\underline{s}$ be the collection of those σ-fields of subsets of S which contain $\mathcal{F}$. It is non-empty, since the collection $P(S)$ of all subsets is in $\underline{s}$. Let

$$\sigma(\mathcal{F}) = \cap\{\Sigma : \Sigma \in \underline{s}\} = \{A : A \in \Sigma, \text{ for all } \Sigma \in \underline{s}\}.$$

Then $\sigma(\mathcal{F})$ is a σ-field which belongs to $\underline{s}$. It is clearly the smallest element of $\underline{s}$. □

The σ-field $\sigma(\mathcal{F})$ is called the *σ-field generated by $\mathcal{F}$*. An important feature of this proposition is that its proof is indirect, and gives no indication of the structure of sets in $\sigma(\mathcal{F})$. This fact gives a particular flavour to much of measure theory.

A mapping $f : (X_1, \Sigma_1) \to (X_2, \Sigma_2)$ from a measurable space (X_1, Σ_1) to a measurable space (X_2, Σ_2) is said to be *measurable* if $f^{-1}(A) \in \Sigma_1$ for each $A \in \Sigma_2$. If so, then $\Sigma_f = \{f^{-1}(A) : A \in \Sigma_2\}$ is a sub-σ-field of Σ_1. If $\Sigma_2 = \sigma(\mathcal{F})$, then f is measurable if and only if $f^{-1}(F) \in \Sigma_1$ for each $F \in \mathcal{F}$.

The σ-field $\mathcal{B}$ generated by the collection of open subsets of a topological space (X, τ) is called the *Borel σ-field* of (X, τ). In the preceding definition, if X_1 is a topological space, and Σ_1 is its Borel σ-field, then f is said to be *Borel measurable*. When f is real-valued, then we always assume that Σ_2 is the Borel σ-field; thus f is measurable if and only if $f^{-1}(c, \infty) \in \Sigma_1$ for each $c \in \mathbf{Q}$; that is, if and only if $(f > c) \in \Sigma_1$ for each $c \in \mathbf{Q}$.

If f is a continuous map from a topological space (X, τ) into a topological space (Y, σ), then f is a Borel measurable map of X into Y, with its Borel σ-field, but a continuous map does not necessarily map Borel sets to Borel sets. Lebesgue mistakenly claimed that if A is a Borel set in the plane $\mathbf{R}^2$, then $p_1(A)$ is a Borel subset of the line $\mathbf{R}$ (where p_1 is the projection onto the first co-ordinate). Suslin invented the notion of *analytic set* (which we shall not consider) to show that this is not so. Borel sets are complicated!

The composition of measurable mappings is measurable.

Exercise 14.1.4 Suppose that (X, Σ) is a measurable space, that Y is a set and that $f : X \to Y$ is a mapping. Show the following.

(i) $\{A \subseteq Y : f^{-1}(A) \in \Sigma\}$ is a σ-field.
(ii) If $\mathcal{F}$ is a collection of subsets of Y, and $f^{-1}(A) \in \Sigma$ for $A \in \mathcal{F}$, then $f^{-1}(A) \in \Sigma$ for $A \in \sigma(\mathcal{F})$; the mapping $f : (X, \Sigma) \to (Y, \sigma(\mathcal{F}))$ is measurable.

Exercise 14.1.5 Let $A_n = \{\omega \in \Omega(\mathbf{N}) : \omega_k = 1 \text{ for } k > n\}$, and let $B = \Omega(\mathbf{N}) \setminus \cup_{n=1}^{\infty} A_n$. Show that B is a Polish subspace of $\Omega(\mathbf{N})$, that if $j(\omega) = \sum_{n=1}^{\infty} 2^{-n}\omega_n$ then j is a bijection of B onto $[0, 1]$, and that $C \subset B$ is a Borel set if and only if $j(C)$ is. Is j^{-1} continuous?

Exercise 14.1.6 Show that if Σ is a σ-field in X and $A \subseteq X$, then the indicator function I_A of A is Σ-measurable if and only if $A \in \Sigma$.

Exercise 14.1.7 Suppose that f and g are real-valued measurable functions on (X, Σ) and that $\alpha \in \mathbf{R}$. Show that each of the functions

$$\alpha f,\ f^+,\ f^-,\ |f|,\ f + g,\ fg,\ f \vee g,\ \text{and } f \wedge g$$

is Σ-measurable. The sets $(f < g)$, $(f \leq g)$ and $(f = g)$ are in Σ.

For example, $(f \vee g > c) = (f > c) \cup (g > c)$, and $(f + g > c) = \cup_{r \in \mathbf{Q}}((f > r) \cap (g > c - r))$.

Thus the set $\mathcal{L}^0 = \mathcal{L}^0(X, \Sigma)$ of real-valued measurable functions on X is a real vector space, when addition and scalar multiplication are defined pointwise.

We also consider extended real-valued functions, taking values in $\overline{\mathbf{R}} = \{-\infty\} \cup \mathbf{R} \cup \{\infty\}$. We say that such a function f is measurable if $(f \in A) \in \Sigma$ for each Borel set A in $\mathbf{R}$ and both $(f = -\infty)$ and $(f = \infty)$ are in Σ.

Exercise 14.1.8 Suppose that $(f_n)_{n=1}^{\infty}$ is a sequence of extended real-valued Σ-measurable functions on X. Then each of the extended real-valued functions $\sup_{n \in \mathbf{N}} f_n$, $\inf_{n \in \mathbf{N}} f_n$, $\limsup_{n \in \mathbf{N}} f_n$, and $\liminf_{n \in \mathbf{N}} f_n$ is Σ-measurable.

The set $C^* = \{x : f_n(x) \text{ converges in } \overline{\mathbf{R}} \text{ as } n \to \infty\}$ is in Σ. Let $f(x) = \lim_{n\to\infty} f_n(x)$ if $x \in C^*$, and let $f(x) = 0$ otherwise. Then f is Σ-measurable. If each f_n is real-valued, then the set

$$C = \{x : f_n(x) \text{ converges in } \mathbf{R} \text{ as } n \to \infty\}$$

is in Σ. Let $f(x) = \lim_{n\to\infty} f_n(x)$ if $x \in C$, and let $f(x) = 0$ otherwise. Then f is Σ-measurable.

For example,

$$(\limsup_{n\to\infty} f_n > c) = \cup_{k=1}^{\infty} \cap_{n=1}^{\infty} \cup_{m=n}^{\infty} (f_m > c + 1/k).$$

14.2 Measure Spaces

A *finite measure space* is a triple (X, Σ, μ), where (X, Σ) is a measurable space and μ is a *countably additive*, or σ*-additive* mapping of Σ into $\mathbf{R}^+$: if (A_n) is a sequence of disjoint elements of Σ then $\mu(\cup_{n=1}^\infty A_n) = \sum_{n=1}^\infty \mu(A_n)$. (Note that, since all the summands are non-negative, the sum does not depend upon the order of summation.) The function μ is called a *measure*. A *probability space* is a finite measure space $(\Omega, \Sigma, \mathbf{P})$ for which $\mathbf{P}(\Omega) = 1$; $\mathbf{P}$ is called a *probability measure*, and a measurable function on a probability space is called a *random variable*.

Exercise 14.2.1 Suppose that (X, Σ, μ) is a finite measure space. Suppose that $A, B \in \Sigma$, and that $(A_n)_{n=1}^\infty$ is a sequence in Σ.

(i) $\mu(A \cup B) + \mu(A \cap B) = \mu(A) + \mu(B)$.
(ii) If $B \subseteq A$ then $\mu(B) \leq \mu(A)$.
(iii) *(Upwards continuity)* If $(A_n)_{n=1}^\infty$ is an increasing sequence then

$$\mu(\cup_{n=1}^\infty A_n) = \sup_{n\in\mathbf{N}} \mu(A_n).$$

(iv) *(Downwards continuity)* If $(A_n)_{n=1}^\infty$ is a decreasing sequence then

$$\mu(\cap_{n=1}^\infty A_n) = \inf_{n\in\mathbf{N}} \mu(A_n).$$

(v) $\sup_{n\in\mathbf{N}} \mu(A_n) \leq \mu(\cup_{n=1}^\infty A_n) \leq \sum_{n=1}^\infty \mu(A_n)$.
(vi) $\mu(\cap_{n=1}^\infty A_n) \leq \inf_{n=1}^\infty \mu(A_n)$.

For example, if $(A_n)_{n=1}^\infty$ is an increasing sequence, let $D_1 = A_1$ and $D_n = A_n \setminus A_{n-1}$ for $n > 1$. Then

$$\mu(\cup_{n=1}^\infty A_n) = \mu(\cup_{n=1}^\infty D_n) = \sum_{n=1}^\infty \mu(D_n) = \lim_{n\to\infty} \sum_{j=1}^n \mu(D_j) = \lim_{n\to\infty} \mu(A_n).$$

Exercise 14.2.2 Show that if f is a measurable real-valued function on X, then $\mu(|f| > n) \to 0$ as $n \to \infty$.

Recall that if $(A_n)_{n=1}^\infty$ is a sequence of subsets of a set X then

$$\limsup_{n\to\infty} A_n = \cap_{n=1}^\infty \left(\cup_{j=n}^\infty A_j\right) \text{ and } \liminf_{n\to\infty} A_n = \cup_{n=1}^\infty \left(\cap_{j=n}^\infty A_j\right).$$

Theorem 14.2.3 (The first Borel–Cantelli lemma) *If* $\sum_{n=1}^\infty \mu(A_n) < \infty$ *then* $\mu(\limsup_{n\to\infty} A_n) = 0$.

Proof For $\mu(\limsup_{n\to\infty} A_n) \leq \mu(\cup_{n=m}^\infty A_n) \leq \sum_{m=n}^\infty \mu(A_m) \to 0$ as $n \to \infty$. □

A *σ-finite measure space* is a measurable space (X, Σ), together with a sequence $(I_k)_{k=1}^{\infty}$ of disjoint elements of Σ whose union is X, and a function μ on Σ (a *σ-finite measure*), taking values in $[0, \infty]$, with the properties that

(i) μ is countably additive; if $(A_n)_{n=1}^{\infty}$ is a sequence of disjoint elements of Σ then $\mu(\cup_{n\in\mathbf{N}} A_n) = \sum_{n=1}^{\infty} \mu(A_n)$, and
(ii) $\mu(I_k) < \infty$ for $k \in \mathbf{N}$.

Thus $A \in \Sigma$ if and only if $A \cap I_k \in \Sigma$ for each $k \in \mathbf{N}$; if so, then $\mu(A) = \sum_{k=1}^{\infty} \mu(A \cap I_k)$. In future we shall use the term 'measure' to mean either a finite measure or a σ-finite measure, adding 'finite' when necessary.

Proposition 14.2.4 *Suppose that ϕ is a measurable mapping from a finite measure space (X, Σ, μ) into a measurable space (Y, T). If $A \in T$, let $\phi_*\mu(A) = \mu(\phi^{-1}(A))$. Then $\phi_*\mu$ is a finite measure on T.*

Suppose that ϕ is a measurable mapping from a σ-finite measure space (X, Σ, μ) into a measurable space (Y, T). Suppose also that there exists an increasing sequence $(A_n)_{n=1}^{\infty}$ in T with union Y such that $\mu(\phi^{-1}(A_n)) < \infty$ for each $n \in \mathbf{N}$. If $A \in T$, let $\phi_*\mu(A) = \mu(\phi^{-1}(A))$. Then $\phi_*\mu$ is a σ-finite measure on T.

Exercise 14.2.5 Prove Proposition 14.2.4.

The measure $\phi_*\mu$ is called the *image measure*, or *push-forward measure*. If ϕ is real-valued, then $\phi_*\mu$ is called the *distribution* of ϕ. In the case where μ is a probability measure, it is also called the *law* of ϕ.

Suppose that (X, Σ, μ) is a measure space. An element N of Σ is a *null set* if $\mu(N) = 0$. The collection of null sets is a σ-ring contained in Σ.

A real-valued measurable function f is a *simple function* if it only takes finitely many values, each on a measurable set of finite measure; $f = \sum_{i=1}^{n} \lambda_i I_{A_i}$, where $\lambda_i \in \mathbf{R}$ and A_i is a measurable set of finite measure, for each i.

Theorem 14.2.6 *Suppose that f is a real-valued measurable function on a σ-finite measurable space (X, Σ). There exists a sequence $(g_n)_{n=1}^{\infty}$ of simple functions which converges pointwise to f. If f is non-negative, then the sequence $(g_n)_{n=1}^{\infty}$ can be taken to be a pointwise increasing sequence of non-negative simple functions.*

Proof Let $(A_n)_{n=1}^{\infty}$ be a disjoint sequence of sets of finite measure which cover X, let $C_n = \cup_{j=1}^{n} A_j$ and let $B_{n,j} = (j/2^n < f \leq (j+1)/2^n)$ for $|j| \leq 2^n$. Then set $g_n = \sum_{j=-2^n}^{j=2^n-1} (j/2^n) I_{C_n \cap B_{n,j}}$. □

14.3 Convergence of Measurable Functions

If a property holds on a measure space, except possibly on a null set, we say that it holds *almost everywhere* (probabilists use the term *almost surely*). Thus if $(f_n)_{n=1}^{\infty}$ is a sequence in $\mathcal{L}^0(X, \Sigma, \mu)$ and if $f \in \mathcal{L}^0(X, \Sigma, \mu)$ then $f_n \to f$ almost surely, as $n \to \infty$ if there exists a null set N such that $f_n(x) \to f(x)$ for all $x \in X \setminus N$.

Suppose that (X, Σ, μ) is a finite measure space and that $(f_n)_{n=1}^{\infty}$ is a sequence in $\mathcal{L}^0(X, \Sigma, \mu)$. We say that $f_n \to f$ *almost uniformly* if for each $\epsilon > 0$ there exists $A \in \Sigma$ with $\mu(A) < \epsilon$ such that $f_n \to f$ uniformly on $X \setminus A$ as $n \to \infty$. This is bad terminology, but is well established.

Theorem 14.3.1 (Egorov's theorem) *Suppose that (X, Σ, μ) is a finite measure space, that $(f_n)_{n=1}^{\infty}$ is a sequence in $\mathcal{L}^0(X, \Sigma, \mu)$ and that $f \in \mathcal{L}^0(X, \Sigma, \mu)$. Then $f_n \to f$ almost everywhere as $n \to \infty$ if and only if $f_n \to f$ almost uniformly as $n \to \infty$.*

Proof Suppose that $f_n \to f$ almost everywhere as $n \to \infty$ and that $\epsilon > 0$. Let $A_{j,k} = \cap_{i \geq j}(|f_i - f| \leq 1/k)$ for $j, k \in \mathbf{N}$. Then $(A_{j,k})_{j=1}^{\infty}$ is an increasing sequence and $\mu(X \setminus \cup_{j=1}^{\infty} A_{n,j}) = 0$, so that for each $k \in \mathbf{N}$ there exists j_k such that $\mu(X \setminus A_{j_k,k}) < \epsilon/2^k$. Then if $A = \cup_{k=1}^{\infty} A_{j_k,k}$, $\mu(X \setminus A) < \epsilon$ and $f_n \to f$ uniformly on A as $n \to \infty$.

If $f_n \to f$ almost uniformly then for each $k \in \mathbf{N}$ there exists a set B_k with $\mu(X \setminus B_k) < 1/k$ such that $f_n \to f$ uniformly on B_k as $n \to \infty$. Let $B = \cup_{k=1}^{\infty} B_k$. Then $\mu(X \setminus B) = 0$ and $f_n \to f$ pointwise on B as $n \to \infty$. □

Suppose that (X, Σ, μ) is a finite measure space, that $(f_n)_{n=1}^{\infty}$ is a sequence in $\mathcal{L}^0$ and that $f \in \mathcal{L}^0$. We say that $f_n \to f$ *in measure* if, for each $c > 0$, $\mu((|f_n - f| > c) \to 0$ as $n \to \infty$. In the case where $(X, \Sigma, \mathbf{P})$ is a probability space, 'convergence in measure' is called *convergence in probability*.

Theorem 14.3.2 *Suppose that (X, Σ, μ) is a finite measure space, that $(f_n)_{n=1}^{\infty}$ is a sequence in $\mathcal{L}^0$ and that $f \in \mathcal{L}^0$. If $f_n \to f$ almost everywhere as $n \to \infty$, then $f_n \to f$ in measure. If $f_n \to f$ in measure as $n \to \infty$, there exists a subsequence $(f_{n_k})_{k=1}^{\infty}$ which converges almost everywhere to f as $k \to \infty$.*

Proof If $f_n \to f$ almost everywhere then $f_n \to f$ in measure, by Egorov's theorem. Conversely, suppose that $f_n \to f$ in measure. For each $k \in \mathbf{N}$ there exists n_k such that $\mu(|f_n - f| > 1/k) < 1/2^k$ for $n \geq n_k$, and we can assume that $(n_k)_{k=1}^{\infty}$ is an increasing sequence. Let $B_k = (|f_{n_k} - f| > 1/k)$, so that $\sum_{k=1}^{\infty} \mu(B_k) < \infty$. By the first Borel–Cantelli lemma, $\mu(\limsup_{k\to\infty} B_k) = 0$. But if $x \notin \limsup_{k\to\infty} B_k$ then $f_n(x) \to f(x)$ as $n \to \infty$. □

If $f \in \mathcal{L}^0(X, \Sigma, \mu)$, f is called a *null function* if $f = 0$ almost everywhere. The set $\mathcal{N}^0(X, \Sigma, \mu)$ of null functions is a linear subspace of $\mathcal{L}^0(X, \Sigma, \mu)$; the quotient is denoted by $L^0(X, \Sigma, \mu)$. As is customary, we identify a measurable function f with its equivalence class $[f]$. Although the elements of $L^0(X, \Sigma, \mu)$ are equivalence classes of functions, we shall consider them and treat them as functions, identifying functions that are equal almost everywhere. For example, if $(f_n)_{n=0}^\infty$ is a sequence in $L^0(X, \Sigma, \mu)$, we say that $f_n \to f_0$ almost surely if whenever g_n is an element of $\mathcal{L}^0(X, \Sigma, \mu)$ which is in the equivalence class f_n, then $g_n(x) \to g_0(x)$ for almost all x. It is easy to see that this holds for one sequence of representatives if and only if it holds for any sequence of representatives.

The next result lies at the heart of measure theory. If μ is a finite measure on (X, σ) then convergence in measure in $\mathcal{L}_0(X, \Sigma, \mu)$ can be characterized by a pseudometric, and the corresponding metric on $L^0(X, \Sigma, \mu)$ is complete.

Theorem 14.3.3 *Suppose that (X, Σ, μ) is a finite measure space. If $f, g \in \mathcal{L}^0(X, \Sigma, \mu)$, let*

$$\rho_0(f, g) = \inf\{\epsilon > 0 : \mu(|f - g| > \epsilon) \leq \epsilon\}.$$

Then ρ_0 is a pseudometric and $\rho(f, g) = 0$ if and only if $f = g$ almost everywhere. If $(f_n)_{n=1}^\infty$ is a sequence in $\mathcal{L}^0(X, \Sigma, \mu)$, then $\rho(f_n, f) \to 0$ if and only if $f_n \to f$ in measure. If d_0 is the corresponding metric on L^0, then $(L^0(X, \Sigma, \mu), d_0)$ is complete. The simple measurable functions are dense in $(L^0(X, \Sigma, \mu), d_0)$.

Proof Suppose that $f, g, h \in \mathcal{L}^0(X, \Sigma, \mu)$. Certainly $\rho_0(f, g) = \rho_0(g, f)$. Suppose that $\epsilon_1 > \rho_0(f, g)$ and that $\epsilon_2 > \rho(g, h)$. Then there exist $\rho(f, g) \leq \eta_1 \leq \epsilon_1$ and $\rho(g, h) \leq \eta_2 \leq \epsilon_2$ such that if $A_1 = \{x : |f(x) - g(x)| > \eta_1$ then $\mu(A_1) \leq \eta_1$, and similarly for A_2. Then $\{x : |f(x) - h(x)| > \epsilon_1 + \epsilon_2\} \subseteq A_1 \cup A_2$ so that $\mu(\{x : |f(x) - h(x)| > \epsilon_1 + \epsilon_2\}) \leq \eta_1 + \eta_2 \leq \epsilon_1 + \epsilon_2$ and so $\rho_0(f, h) \leq \rho_0(f, g) + \rho_0(g, h)$. Thus ρ_0 is a pseudometric. Clearly, $\rho(f_n, f) \to 0$ if and only if $f_n \to f$ in measure, and $\rho(f, g) = 0$ if and only if $f = g$ almost everywhere.

Suppose now that $(f_n)_{n=1}^\infty$ is a d_0-Cauchy sequence. Then there exists a subsequence $(f_{n_k})_{k=1}^\infty$ such that $d_0(f_{n_k}, f_{n_{k+1}}) \leq 2^{-k}$, for each k. Let $C_k = \{x : |f_{n_k}(x) - f_{n_{k+1}}(x)| > 2^{-k}$, and let $C = \limsup_{k=1}^\infty C_k$. Since $\sum_{k=1}^\infty \mu(C_k) < \infty$, it follows from the first Borel–Cantelli lemma that $\mu(C) = 0$.

Suppose that (X, Σ, μ) is a finite measure space. If $x \notin C$ then there exists k_0 such that $|f_{n_k}(x) - f_{n_{k+1}}(x)| \leq 2^{-k}$ for $k \geq k_0$. Thus $|f_{n_k}(x) - f_{n_l}(x)| \leq 2.2^{-k}$ for $l > k > k_0$, and $(f_{n_k}(x))_{k=1}^\infty$ is a real Cauchy sequence, which converges to $f(x)$, say.

Set $f(x) = 0$ for $x \in C$. Then f is measurable, and $f_{n_k} \to f$ almost everywhere. Thus $f_{n_k} \to f$ almost uniformly, by Egorov's theorem, and so $d_0(f_{n_k}, f) \to 0$. Consequently $d(f_n, f) \to 0$ as $n \to \infty$, and $(L_0(X, \Sigma, \mu), d_0)$ is complete.

Suppose that $f \in L^0(X, \Sigma, \mu)$. Let $A_{j,n} = (j/2^n \leq f < (j+1)/2^n)$, and let f_n be the simple function

$$f_n = \sum_{j=-2^n n}^{2^n n-1} (j/2^n) I_{A_{j,n}}.$$

Then $|f - f_n| > 1/n$ only if $|f| > n$, so that $d_0(f, f_n) \to 0$ as $n \to \infty$, since $\mu(|f| > n) \to 0$ as $n \to \infty$. □

As an immediate result of Theorem 14.3.2, we have the following.

Theorem 14.3.4 *A subset A of a finite measure space (X, Σ, μ) is d_0-compact if and only if every sequence in A has a subsequence which converges almost everywhere to an element of A.*

Proposition 14.3.5 *Suppose that (X, Σ, μ) is a finite measure space, that $(f_n)_{n=1}^\infty$ and $(g_n)_{n=1}^\infty$ are sequences in $\mathcal{L}^0(X, \Sigma, \mu)$ and that $f, g \in \mathcal{L}^0(X, \Sigma, \mu)$. If $f_n \to f$ and $g_n \to g$ almost everywhere, as $n \to \infty$, then $f_n + g_n \to f + g$ and $f_n g_n \to fg$ almost everywhere, as $n \to \infty$. If $f_n \to f$ and $g_n \to g$ in measure, as $n \to \infty$, then $f_n + g_n \to f + g$, $f_n^2 \to f^2$ and $f_n g_n \to fg$ in measure, as $n \to \infty$.*

Proof Addition is easy, and left to the reader. Given $\epsilon > 0$, there exists $C > 1$ such that $\mu(|f| > C) < \epsilon/2$, and there exists n_0 such that $\mu(|f_n - f|) > \epsilon/(2C + \epsilon) < \epsilon/2$ for $n \geq n_0$. Since $|f_n^2 - f^2| = |f_n - f|.|f_n + f|$, $\mu(|f_n^2 - f^2| \geq \epsilon) < \epsilon$ if $n \geq n_0$; $f_n^2 \to f^2$ in measure, as $n \to \infty$. It then follows from polarization that $f_n g_n \to fg$ in measure, as $n \to \infty$. □

We can extend these ideas to σ-finite measure spaces. Suppose that (X, Σ, μ) is a σ-finite measure space, with a sequence $(I_k)_{k=1}^\infty$ of disjoint elements of Σ of finite positive measure, whose union is X. Suppose that $(f_n)_{n=1}^\infty$ is a sequence in $L^0(X, \Sigma, \mu)$, and that $f \in L^0(X, \Sigma, \mu)$. Let $\pi_k(f) = f_{|I_k}$. If $(f_n)_{n=1}^\infty$ is a sequence in $L^0(X)$, then f_n is said to converge to f *locally in measure* if and only if $\pi_k(f_n) \to \pi_k(f)$ in measure, as $n \to \infty$, for each $k \in \mathbf{N}$. Local convergence in measure can be defined by a complete product metric, such as, for example,

$$d_{(0,loc)}(f, g) = \sum_{k=1}^\infty 2^{-k} \inf\{\epsilon > 0 : \mu((|f - g| > \epsilon) \cap I_k) < \epsilon\}.$$

14.4 Integration

Suppose that f is a non-negative extended-real-valued measurable function defined on a finite or σ-finite measure space (X, Σ, μ). The *tail distribution function* λ_f is defined as

$$\lambda_f(t) = \mu(f > t) = f_*(\mu)((t, \infty]), \text{ for } t \in [0, \infty).$$

If $\lambda_f(t) = \infty$ for some $t > 0$ we define $\int_X f\,d\mu = \infty$. Otherwise, the function λ_f is a decreasing right-continuous real-valued function on $[0, \infty)$. We define

$$\int_X f\,d\mu = \int_0^\infty \lambda_f(t)\,dt.$$

Here, the integral on the right is an improper Riemann integral, taking values in $[0, \infty]$.

Exercise 14.4.1 Suppose that $f = \sum_{i=1}^n v_i I_{A_i}$ is a non-negative simple measurable function on a measure space (X, Σ, μ). Show that $\int_X f\,d\mu = \sum_{i=1}^n v_i \mu(A_i)$, and that the integral does not depend on the representation of f. Show that if f and g are non-negative simple measurable functions on a measure space (X, Σ, μ) then $\int_X (f + g)\,dm = \int_X f\,d\mu + \int_X g\,d\mu$.

If $A \in \Sigma$, we set $\int_A f\,d\mu = \int_X f I_A, d\mu$, where I_A is the indicator function of A.

If $0 \le f \le g$ almost everywhere then $\lambda_f \le \lambda_g$, and so $\int_X f\,d\mu \le \int_X g\,d\mu$. In particular, if $f = g$ almost everywhere then $\int_X f\,d\mu = \int_X g\,d\mu$, and if $f = 0$ almost everywhere then $\int_X f\,d\mu = 0$.

We now come to the fundamental theorem of integration theory. First we need an elementary result about Riemann integrals.

Exercise 14.4.2 Suppose that $(\lambda_n)_{n=1}^\infty$ is an increasing sequence of decreasing real-valued functions on an interval $[a, b]$ which converges pointwise to the real-valued function λ. By considering upper and lower Riemann sums, show that $\int_a^b \lambda_n(t)dt \to \int_a^b \lambda(t)dt$. (Here the integrals are Riemann integrals.)

Theorem 14.4.3 (The monotone convergence theorem) *Suppose that $(f_n)_{n=1}^\infty$ is an increasing sequence of non-negative measurable functions on a finite or σ-finite measure space (X, Σ, μ) which converges pointwise almost everywhere to a function f. Then*

$$\int_X f_n\,d\mu \to \int_X f\,d\mu \text{ as } n \to \infty.$$

Proof After approximating at 0 and ∞, this follows from the exercise. □

Corollary 14.4.4 (Fatou's lemma) *Suppose that $(f_n)_{n=1}^\infty$ is a sequence of non-negative measurable functions on a finite or σ-finite measure space (X, Σ, μ). Then*

$$\int_X \liminf_{n\to\infty} f_n\, d\mu \leq \liminf_{n\to\infty} \int f_n\, d\mu.$$

In particular, if $f_n \to f$ almost everywhere then $\int_X f\, d\mu \leq \liminf_{n\to\infty} \int f_n\, d\mu$.

Proof Let $g_n = \inf_{j\geq n} f_j$. Then $g_n \leq f_n$ and $(g_n)_{n=1}^\infty$ increases pointwise to $\liminf f_n$. Thus

$$\int_X \liminf f_n\, d\mu = \lim_{n\to\infty} \int_X g_n\, d\mu \leq \liminf_{n\to\infty} \int_X f_n\, d\mu.$$

□

14.5 Integrable Functions

Suppose that f is a real-valued, but not necessarily non-negative, measurable function f on a finite or σ-finite measure space (X, Σ, μ). Then we set

- $\int_X f\, d\mu = \int_X f^+\, d\mu - \int_X f^-\, d\mu$ if $\int_X f^+\, d\mu < \infty$ and $\int_X f^-\, d\mu < \infty$;
- $\int_X f\, d\mu = \infty$ if $\int_X f^+\, d\mu = \infty$ and $\int_X f^-\, d\mu < \infty$;
- $\int_X f\, d\mu = -\infty$ if $\int_X f^+\, d\mu < \infty$ and $\int_X f^-\, d\mu = \infty$;
- $\int_X f\, d\mu$ is not defined if $\int_X f^+\, d\mu = \infty$ and $\int_X f^-\, d\mu = \infty$.

The function f is said to be *integrable* if $\int_X f^+\, d\mu < \infty$ and $\int_X f^-\, d\mu < \infty$; this is clearly the case if and only if $\int_X |f|\, d\mu < \infty$. If $(\Omega, \Sigma, \mathbf{P})$ is a probability space, we write $\mathbf{E}(f)$ for $\int_\Omega f\, d\mathbf{P}$; then $\mathbf{E}(f)$ is the *expectation* of f.

Proposition 14.5.1 *If f and g are integrable, if h is a bounded measurable function and if $\alpha \in \mathbf{R}$ then hf, αf and $f + g$ are integrable, and*

$$\int_X \alpha f\, d\mu = \alpha \int_X f\, d\mu \quad \text{and} \quad \int_X (f+g)\, d\mu = \int_X f\, d\mu + \int_X g\, d\mu.$$

If $f \in \mathcal{N}^0$ then f is integrable and $\int_X f\, d\mu = 0$.

Proof Scalar multiplication is easy. The addition formula holds for non-negative simple functions, then by the monotone convergence theorem for non-negative integrable functions, and finally for arbitrary functions by considering positive and negative parts. □

Exercise 14.5.2 Suppose that f is a real-valued measurable function f on a finite or σ-finite measure space (X, Σ, μ). Show that $\int_X |f|\, d\mu = 0$ if and only if f is a null function.

We denote the set of integrable functions on (X, Σ, μ) by $\mathcal{L}^1(X, \Sigma, \mu)$. The proposition shows that $\mathcal{L}^1(X, \Sigma, \mu)$ is a vector space and the mapping $f \to \int_X f\, d\mu$ is a linear functional on it. We denote the quotient space $\mathcal{L}^1/\mathcal{N}^0$ by $L^1 = L^1(X, \Sigma, \mu)$. Again, we identify f with its equivalence class $[f]$. If f and g are equivalent functions then $\int_X f\, d\mu = \int_X g\, d\mu$, and so if $f \in L^1(X, \Sigma, \mu)$ then $\int_X f\, d\mu$ is a well-defined linear functional on L^1. The function $\|f\|_1 = \int_X |f|\, d\mu$ is then a norm on $L^1(X, \Sigma, \mu)$.

Theorem 14.5.3 *Suppose that (X, Σ, μ) is a σ-finite measure space. Then $(L^1(X, \Sigma, \mu), \|.\|_1)$ is complete and the simple measurable functions are dense in L^1.*

Proof $L^1(X, \Sigma, \mu) \subseteq L^0(X, \Sigma, \mu)$, and if $(y_n)_{n=1}^\infty$ is a $\|.\|_1$-Cauchy sequence in $(L^1(X, \Sigma, \mu), \|.\|_1)$, then it is a Cauchy sequence in the complete metric space $(L^0(X, \Sigma, \mu), d_{0,loc})$. Suppose that $(y_n)_{n=1}^\infty$ is a sequence in $M_\eta(y_0)$ and that $\|y_n - y\|_1 = \int_X |y_n - y|\, d\mu \to 0$ as $n \to \infty$. By extracting a subsequence if necessary, we can suppose that $y_n \to y$ almost everywhere. But then, by Fatou's lemma,

$$\int_X |y - y_0|\, d\mu \leq \liminf_{n\to\infty} \int_X |y_n - y_0|\, d\mu \leq \eta,$$

so that $M_\eta(y_0)$ is d_0-closed. The result therefore follows from Theorem 2.4.5.

The proof that the simple measurable functions are dense in L^1 is just the same as that of Theorem 14.3.3. □

Although this result can be proved directly just as easily, the proof underlines the importance of the completeness of $(L^0(X, \Sigma, \mu), d_{0,loc})$.

Theorem 14.5.4 (The dominated convergence theorem) *Suppose that $(f_n)_{n=1}^\infty$ is a sequence of measurable functions which converges pointwise almost everywhere to f, and that g is an integrable function such that $|f_n| \leq |g|$, for each n. Then*

$$\int_X f_n\, d\mu \to \int_X f\, d\mu \text{ as } n \to \infty.$$

Further, $\int_X |f_n - f|\, d\mu \to 0$ as $n \to \infty$.

Proof The functions $|g| - f_n$ and $|g| - f$ are non-negative and integrable. Appling Fatou's lemma,

$$\int_X (|g| - f)\, d\mu \leq \liminf_{n\to\infty} \int_X (|g| - f_n)\, d\mu,$$

so that

$$\int_X f\, d\mu \geq \limsup_{n\to\infty} \int_X f_n\, d\mu \geq \liminf_{n\to\infty} \int_X f_n\, d\mu \geq \int_X f\, d\mu,$$

so that all the terms are equal, and the first result follows. Since $|f_n - f| \le 2|g|$ and $|f_n - f| \to 0$ as $n \to \infty$, so does the second. □

Corollary 14.5.5 (The bounded convergence theorem) *Suppose that $(f_n)_{n=1}^\infty$ is a uniformly bounded sequence of measurable functions on a finite measure space (X, Σ, μ) which converges pointwise almost everywhere to f. Then*

$$\int_X f_n \, d\mu \to \int_X f \, d\mu \text{ and } \int_X |f_n - f| \, d\mu \to 0 \text{ as } n \to \infty.$$

Exercise 14.5.6 Suppose that (X, Σ, μ) is a measure space, and that $(A_n)_{n=1}^\infty$ is a sequence in Σ with $\mu(A_n) > 0$ for each n and $\sum_{n=1}^\infty \mu(A_n) < \infty$. Let $f_n = I_{A_n}/\mu(A_n) - 1$, and let $a_n = (f_1 + \cdots + f_n)/n$. Show that $\int_X a_n = 0$ for each n, but that $a_n \to -1$ almost everywhere.

Exercise 14.5.7 Suppose that (X, Σ, μ) is a measure space, that (Y, T) is a measurable space, and that ϕ is a measurable map from X to Y. Suppose that $f \in L^1(Y, T, \phi_*\mu)$. Show that $f \circ \phi \in L^1(\mu)$, and that $\int_X (f \circ \phi) \, d\mu = \int_Y f \, d(\phi_*\mu)$.

15
Further Measure Theory

In this chapter, we consider further properties of measures on abstract sets. This material may be less familiar, and so fuller details are given than in the previous chapter.

15.1 Riesz Spaces

Many of the Banach spaces that we consider have a natural partial order, which interacts with the topological, metric and geometric properties of the space. For example, a partial order on a $C(X)$ space is given by setting $f \leq g$ if and only if $f(x) \leq g(x)$ for all $x \in X$, and a partial order on L^0 is given by setting $f \leq g$ if and only if $f(x) \leq g(x)$ for almost all x.

A *partially ordered* vector space $(E, \leq)$ is a real vector space E and a partial order $\leq$ on E which satisfies

(i) if $x \leq y$ then $x + z \leq y + z$ and
(ii) if $x \leq y$ then $ax \leq ay$

for all $x, y, z \in E$ and all $a \geq 0$.

Thus $x \leq y$ if and only if $y - x \geq 0$. The set $\{x \in E : x \geq 0\}$ is the *positive cone* in E; it is a convex subset of E. An element of E^+ is called a *positive* element.

A partially ordered vector space $(E, \leq)$ is *Archimedean* if whenever $x, y \in E$ and $nx \leq y$ for all $n \in \mathbf{N}$ then $x \leq 0$. We shall assume, without saying so, that all the partially ordered spaces that we shall consider are Archimedean.

If x, y, z are elements of a partially ordered set, then z is the *least upper bound* of x and y if $x \leq z$ and $y \leq z$ and if whenever $x \leq z'$ and $y \leq z'$ then $z \leq z'$. If a least upper bound exists, it is denoted by $x \vee y$. A *greatest lower bound* is defined in the same way; when it exists, it is denoted by $x \wedge y$. A *lattice* is a partially ordered set in which each pair of elements has a least upper bound and a greatest lower bound.

A *Riesz space* is a partially ordered vector space which is a lattice. $C(X)$, $L^0(X, \Sigma, \mu)$ and $L^1(X, \Sigma, \mu)$ are Riesz spaces under their natural orderings.

Exercise 15.1.1 Suppose that $x, y, z \in E$, where $(E, \leq)$ is a Riesz space. Show that $(x + z) \vee (y + z) = (x \vee y) + z$.

Exercise 15.1.2 Suppose that $(E, \leq)$ is a partially ordered vector space. Show that E is a Riesz space if and only if x and 0 have a least upper bound, for each $x \in E$.

If $(E, \leq)$ is a Riesz space then $x \vee 0$ is denoted by x^+, and $(-x) \vee 0$ is denoted by x^-.

Proposition 15.1.3 *Suppose that $(E, \leq)$ is a Riesz space and that $x \in E$. Then $x = x^+ - x^-$ and $|x| = x^+ + x^-$.*

Proof Since $x^+ - x \geq 0$ and $x^+ - x \geq -x$, $x^+ - x \geq x^-$, and so $x^+ - x^- \geq x$. Similarly, $x^- + x \geq x$ and $x^- + x \geq 0$, so that $x^- + x \geq x^+$, and $x^+ - x^- \leq x$. Consequently, $x = x^+ - x^-$.

Further, $x \vee (-x) = (2x) \vee 0 - x = 2x^+ - x$ and $x \vee (-x) = 0 \vee (-2x) + x = 2x^- - x$; averaging, $|x| = x^+ + x^-$. □

Consequently, $E = E^+ - E^+$.

Proposition 15.1.4 *Suppose that $(E, \leq)$ is a Riesz space and that f is a real-valued function on E^+ which satisfies*

$$f(x + y) = f(x) + f(y) \text{ and } f(ax) = af(x) \text{ for } x, y \in E^+ \text{ and } a \geq 0.$$

Then there exists a unique linear functional ϕ on E which extends f; that is $\phi(x) = f(x)$ for $x \in E^+$.

Proof If $x \in E$ let $\phi(x) = f(x^+) - f(x^-)$. Then $\phi(-x) = -\phi(x)$, so that $\phi(ax) = a\phi(x)$ for $a \in \mathbf{R}$. Since $(x + y)^+ + x^- + y^- = (x + y)^- + x^+ + y^+$,

$$f((x + y)^+) + f(x^-) + f(y^-) = f((x + y)^-) + f(x^+) + f(y^+)$$

so that $\phi(x + y) = \phi(x) + \phi(y)$; ϕ is a linear functional on E which extends f, and ϕ is unique, since $E = E^+ - E^+$. □

The following decomposition result has important consequences.

Proposition 15.1.5 *Suppose that $(E, \leq)$ is a Riesz space, and that $x_1, x_2, y \in E^+$ and that $0 \leq y \leq x_1 + x_2$. Then there exist y_1, y_2 in E^+ such that $0 \leq y_1 \leq x_1$, $0 \leq y_2 \leq x_2$ and $y = y_1 + y_2$.*

Proof Let $y_1 = y \wedge x_1$, so that $0 \leq y_1 \leq x_1$ and let $y_2 = y - y_1$. Then $y_2 \geq 0$ and

$$x_2 - y_2 = (x_2 - y) + (y \wedge x_1) = x_2 \wedge (x_1 + x_2 - y) \geq 0.$$

□

Suppose that $(E, \leq)$ is a partially ordered vector space. If $x \leq y$, the *order interval* $[x, y]$ is the set $\{z : x \leq z \leq y\}$. A subset of E is *order bounded* if it is contained in an order interval. A linear functional ϕ on E is *positive* if $\phi(x) \geq 0$ whenever $x \geq 0$, and is *order bounded* if it is bounded on each order interval. If ϕ is positive and $[x, y]$ is an order interval, then $\phi(x) \leq \phi(z) \leq \phi(y)$ for $z \in [x, y]$, so that a positive linear functional is order bounded. We denote the vector space of order bounded linear functionals by $E^{\sim}$, and define a partial order on it by setting $\phi \leq \psi$ if $\psi - \phi$ is positive.

Theorem 15.1.6 *If $(E, \leq 0)$ is a Riesz space, then $(E^{\sim}, \leq)$ is a Riesz space. If $\phi \in E^{\sim}$ and $x \geq 0$ then $\phi^+(x) = \sup\{\phi(z) : z \in [0, x]\}$.*

Proof If $\phi \in E^{\sim}$ and $x \in E^+$, let $f(x) = \sup\{\phi(z) : z \in [0, x]\}$. Since ϕ is order bounded, $f(x)$ is finite. Clearly $f(ax) = af(x)$ for $a \geq 0$. Suppose that $x_1, x_2 \in E^+$ and that $0 \leq z \leq x_1 + x_2$. By Proposition 15.1.5 there exist $z_1 \in [0, x_1]$ and $z_2 \in [0, x_2]$ with $z_1 + z_2 = z$. Thus

$$f(x_1 + x_2) = \sup\{\phi(z_1 + z_2) : z_1 \in [0, x_1],\ z_2 \in [0, x_2]\} = f(x_1) + f(x_2).$$

By Proposition 15.1.4, there exists a linear function ψ which extends f, and ψ is positive. Suppose that $\theta \in E^{\sim}$ and that $\theta \geq \phi$. If $x \in E^+$ then $\theta(z) \geq \phi(z)$ for $z \in [0, x]$, so that $\theta \geq \psi$. Consequently, $\psi = \phi^+$, and $(E^{\sim}, \leq)$ is a Riesz space. □

Exercise 15.1.7 Suppose that x is a positive element of a Riesz space $(E, \leq)$. Show that $B_x = \{y \in E : |y| \leq x\}$ is a convex set.

A positive element e of a Riesz space $(E, \leq)$ is an *order unit* if whenever $y \in E$ then there exists $\lambda > 0$ such that $\lambda y \in B_x$; in other words, B_e is absorbent, and the gauge of B_e is a norm $\|.\|_e$ on E.

Exercise 15.1.8 Suppose that e is an order unit in a Riesz space $(E, \leq)$. Show that $E^{\sim} = (E, \|.\|_e)'$.

The function 1 is an order unit in the Riesz space $C_b(X)$, where (X, τ) is a topological space; the corresponding norm is simply $\|.\|_\infty$.

15.2 Signed Measures

So far we have been concerned with measures which take non-negative values. We now drop this requirement. A *signed measure* σ on a measurable space (X, Σ) is a real-valued function on Σ which is σ-additive; if $(A_n)_{n=1}^\infty$ is a sequence of disjoint elements of Σ, then

$$\sigma(\cup_{n=1}^\infty A_n) = \sum_{n=1}^\infty \sigma(A_n).$$

An important feature of this definition is that infinite values are not allowed. A finite measure is a signed measure; in this setting, we call such a measure a *positive* measure.

Proposition 15.2.1 *Suppose that σ is a signed measure on a measurable space (X, Σ).*

(i) $\sigma(\emptyset) = 0$.
(ii) If $(A_n)_{n=1}^\infty$ is a sequence of disjoint elements of Σ, then $\sum_{n=1}^\infty \sigma(A_n)$ converges absolutely.
(iii) If $(A_n)_{n=1}^\infty$ is an increasing sequence in Σ with union A then $\sigma(A) = \lim_{n\to\infty} \sigma(A_n)$.
(iv) If $(B_n)_{n=1}^\infty$ is a decreasing sequence in Σ with intersection B then $\sigma(B) = \lim_{n\to\infty} \sigma(B_n)$.
(v) $\sigma(\Sigma) = \{\sigma(A) : A \in \Sigma\}$ is a bounded subset of **R**.

Proof (i) Take $A_n = \emptyset$ for $n \in \mathbf{N}$. Then $\sum_{n=1}^\infty \sigma(A_n)$ converges, so that $\sigma(\emptyset) = 0$.
(ii) If τ is any permutation of **N** then $\sum_{n=1}^\infty \sigma(A_{\tau(n)})$ converges, so that the sum is absolutely convergent.
(iii) Let $C_1 = A_1$ and let $C_n = A_n \setminus A_{n-1}$ for $n > 1$. Then A is the disjoint union of the sequence $(C_n)_{n=1}^\infty$, so that

$$\sigma(A) = \sum_{n=1}^\infty \sigma(C_n) = \lim_{n\to\infty} \sum_{j=1}^n \sigma(C_j) = \lim_{n\to\infty} \sigma(A_n).$$

(iv) Since $(X \setminus B_n)_{n=1}^\infty$ increases to $X \setminus B$,

$$\begin{aligned}\sigma(B) &= \sigma(X) - \sigma(X \setminus B) = \sigma(X) - \lim_{n\to\infty} \sigma(X \setminus B_n) \\ &= \lim_{n\to\infty} (\sigma(X) - \sigma(X \setminus B_n)) = \lim_{n\to\infty} \sigma(B_n).\end{aligned}$$

(v) We need a lemma.

Lemma 15.2.2 *Let*

$$\mathcal{H} = \{H \in \Sigma : \{\sigma(C) : C \in \Sigma, C \subseteq H\} \text{ is unbounded}\}.$$

If $H \in \mathcal{H}$, then there exist $H' \in \mathcal{H}$ and $C \in \Sigma$ such that H is the disjoint union $H' \cup C$ and $|\sigma(C)| \geq 1$.

Proof There exists $D \in \Sigma$ such that $D \subseteq H$ and $|\sigma(D)| \geq |\sigma(H)| + 1$. Then $|\sigma(H \setminus D)| \geq 1$. If $D \in \mathcal{H}$, take $H' = D$ and $C = H \setminus D$. Otherwise, $H \setminus D$ must be in $\mathcal{H}$; take $H' = H \setminus D$ and $C = D$. □

Suppose that $X \in \mathcal{H}$. Let $A_0 = X$. Applying the lemma repeatedly, there exists a decreasing sequence $(A_n)_{n=1}^\infty$ in $\mathcal{H}$, such that if $C_n = A_{n-1} \setminus A_n$ then $|\sigma(C_n)| \geq 1$. But $(C_n)_{n=1}^\infty$ is a sequence of disjoint elements of Σ, and so $\sum_{n=1}^\infty \sigma(C_n)$ converges. This gives a contradiction. Thus $X \notin \mathcal{H}$, and so $\sigma(\Sigma)$ is bounded. □

The set $M(X, \Sigma)$ of signed measures on a measure space (X, Σ) contains the finite measures, and is a linear subspace of the Riesz space of all real-valued functions on Σ. Thus if π and ν are positive measures then $\pi - \nu$ is a signed measure. We can decompose a signed measure σ as the difference of two positive measures, in a canonical way.

Theorem 15.2.3 *If σ is a signed measure on a measurable space (X, Σ), then there exist disjoint P and N in Σ, with $X = P \cup N$, such that $\sigma(A) \geq 0$ for $A \subseteq P$ and $\sigma(A) \leq 0$ for $A \subseteq N$. Let*

$$\sigma^+(A) = \sigma(A \cap P), \ \sigma^-(A) = -\sigma(A \cap N).$$

Then σ^+ and σ^- are positive measures on Σ, and $\sigma = \sigma^+ - \sigma^-$.

Further, the decomposition is essentially unique; if $X = P' \cup N'$, where P' and N' are disjoint elements of Σ for which

$$\pi'(A) = \sigma(A \cap P') \geq 0, \ \nu'(A) = -\sigma(A \cap N) \geq 0 \text{ for } A \in \Sigma,$$

then $\pi = \sigma^+$ and $\nu = \sigma^-$.

Finally, $\sigma^+ = \sigma \vee 0$ and $\sigma^- = \sigma \wedge 0$.

Proof Say that A is strictly non-negative if $\sigma(B) \geq 0$ for all $B \subseteq A$. First we show that if $A \in \Sigma$ then there exists strictly non-negative $C \subseteq A$ with $\sigma(C) \geq \sigma(A)$. Suppose not. If $\sigma(A) \leq 0$ then we can take $C = \emptyset$. Suppose that $\sigma(A) > 0$. Let $l_0 = \inf\{\sigma(B) : B \subseteq A\}$: $-\infty < l_0 < 0$. Choose $B_1 \subseteq A$ such that $\sigma(B_1) < l_0/2$, and let $A_1 = A \setminus B_1$. Then $\sigma(A_1) > \sigma(A)$, and if $l_1 = \inf\{\sigma(B) : B \subseteq A_1\}$ then $l_0/2 < l_1 < 0$.

Repeating the process, we obtain a decreasing sequence (A_n) such that $\sigma(A_n)$ is increasing, and $l_n = \inf\{\sigma(B) : B \subseteq A_n\} \to 0$. Then $\sigma(\cap_n(A_n)) \geq \sigma(A)$ and $\cap_n(A_n)$ is strictly non-negative.

It follows that

$$M = \sup\{\sigma(A) : A \in \Sigma\} = \sup\{\sigma(A) : A \text{ strictly non-negative}\}.$$

There exist strictly non-negative P_n such that $\sigma(P_n) > M - 1/n$. Then $P = \cup_n P_n$ is strictly non-negative, and $\sigma(P) = M$. It follows that if $A \cap P = \emptyset$ then $\sigma(A) \leq 0$, so that we can take $N = X \setminus P$.

It is then immediate that σ^+ and σ^- are positive measures on (X, Σ), and that $\sigma = \sigma^+ - \sigma^-$.

Finally, suppose that P', N', π and ν satisfy the conditions of the theorem, and that $A \in \Sigma$. If $B \subseteq A \cap P'$ then $\sigma(B) \geq 0$, so that $\sigma^+(A \cap P') = \sigma(A \cap P') = \pi(A)$. Similarly, if $B \subseteq A \cap N'$ then $\sigma(B) \leq 0$, so that $\sigma^+(A \cap N') = 0$. Consequently,

$$\sigma^+(A) = \sigma(A \cap P') = \pi(A \cap P') = \pi(A),$$

so that $\pi = \sigma^+$. Similarly, $\nu = \sigma^-$. Finally, it is clear that $\sigma^+ = \sigma \vee 0$ and $\sigma^- = -(\sigma \wedge 0)$. □

Thus $M(X, \Sigma)$ is a Riesz space, and the notation is consistent; $\sigma^+ = \sigma \vee 0$, $\sigma^- = (-\sigma) \vee 0$ and $|\sigma| = \sigma \vee (-\sigma)$. The decomposition $\sigma = \sigma^+ - \sigma^-$ of this theorem is called the *Jordan decomposition* of σ.

Proposition 15.2.4 *If $\sigma \in M(X, \Sigma)$ and $A \in \Sigma$ then $|\sigma(A)| \leq |\sigma|(A)$.*

Proof

$$\begin{aligned} |\sigma(A)| &= |\sigma(A \cap P) + \sigma(A \cap N)| \leq |\sigma(A \cap P)| + |\sigma(A \cap N)| \\ &= |\sigma|(A \cap P) + |\sigma|(A \cap N) = |\sigma|(A). \end{aligned}$$

□

Exercise 15.2.5 Show that if $\sigma \in M(X, \Sigma)$ and $(A_n)_{n=1}^{\infty}$ is a sequence of disjoint elements of Σ with union A then $\sum_{n=1}^{\infty} |\sigma(A_n)| \leq |\sigma|(A)$.

15.3 $M(X)$, L^1 and L^∞

The space $M(X, \Sigma)$ can be given a norm $\|.\|_{TV}$ under which it is a Banach space.

Theorem 15.3.1 *If (X, Σ) is a measurable space and $\sigma \in M(X, \Sigma)$, let $\|\sigma\|_{TV} = |\sigma|(X)$. Then $\|.\|_{TV}$ is a norm on the vector space $M(X, \Sigma)$ of signed measures on (X, Σ) under which $M(X, \Sigma)$ is complete.*

Proof Let $\sigma = \sigma^+ - \sigma^-$ be the Jordan decomposition of σ. If $\lambda \geq 0$ then $\lambda\sigma = \lambda\sigma^+ - \lambda\sigma^-$ is the Jordan decomposition of $\lambda\sigma$, so that $\|\lambda\sigma\| = \lambda\|\sigma\|$. If $\lambda < 0$ then $\lambda\sigma = |\lambda|\sigma^- - |\lambda|\sigma^+$ is the Jordan decomposition of $\lambda\sigma$, so that $\|\lambda\sigma\|_{TV} = |\lambda|\sigma^-(X) + |\lambda|\sigma^+(X) = |\lambda|\,\|\sigma\|_{TV}$.

If σ_1, σ_2 are signed measures then there exists a positive measure ν such that

$$(\sigma_1 + \sigma_2)^+ = \sigma_1^+ + \sigma_2^+ - \nu \text{ and } (\sigma_1 + \sigma_2)^- = \sigma_1^- + \sigma_2^- - \nu$$

so that

$$\|\sigma_1 + \sigma_2\|_{TV} = \|\sigma_1\|_{TV} + \|\sigma_2\|_{TV} - 2\nu(X) \leq \|\sigma_1\|_{TV} + \|\sigma_2\|_{TV}.$$

Thus $\|.\|_{TV}$ is a norm on $M(X, \Sigma)$.

Suppose that $(\sigma_k)_{k=1}^\infty$ is a Cauchy sequence in $M(X, \Sigma)$. If $\epsilon > 0$ there exists K such that $\left\|\sigma_j - \sigma_k\right\|_{TV} \leq \epsilon/2$ for $j, k \geq K$.

If $A \in \Sigma$ then, by Proposition 15.2.4,

$$|\sigma_j(A) - \sigma_k(A)| \leq |\sigma_j - \sigma_k|(A) \leq \left\|\sigma_j - \sigma_k\right\|_{TV},$$

so that $(\sigma_k(A))_{k=1}^\infty$ is a Cauchy sequence in $\mathbf{R}$, which converges to $\sigma(A)$, say. Note that

$$|\sigma(A) - \sigma_k(A)| = \lim_{j\to\infty} |\sigma_j(A) - \sigma_k(A)| \leq \sup_{j\geq K} \left\|\sigma_j - \sigma_k\right\|_{TV} \leq \epsilon/2$$

for $k \geq K$. We shall show that σ is a signed measure and that $\sigma_k \to \sigma$ in norm as $k \to \infty$.

Suppose that $(A_n)_{n=1}^\infty$ is a sequence of disjoint elements of Σ with union A, and that $\epsilon > 0$. There exists $K \in \mathbf{N}$ such that $\left\|\sigma_j - \sigma_k\right\|_{TV} < \epsilon/2$ for $j, k \geq K$. By Exercise 15.2.5,

$$\sum_{n=1}^\infty |\sigma_j(A_n) - \sigma_K(A_n)| \leq |\sigma_j - \sigma_K|(A) \leq \left\|\sigma_j - \sigma_K\right\|_{TV} < \epsilon/2,$$

for $j \geq K$. Letting $j \to \infty$, it follows that $\sum_{n=1}^\infty |\sigma(A_n) - \sigma_K(A_n)| \leq \epsilon/2$, and similarly $|\sigma(A) - \sigma_K(A)| \leq \epsilon/2$. Thus

$$\begin{aligned}\left|\sum_{n=1}^\infty \sigma(A_n) - \sigma(A)\right| &= \left|\left(\sum_{n=1}^\infty (\sigma(A_n) - \sigma_K(A_n))\right) - (\sigma(A) - \sigma_K(A))\right| \\ &\leq \left(\sum_{n=1}^\infty (|\sigma(A_n) - \sigma_K(A_n)|)\right) + |\sigma(A) - \sigma_K(A)| \leq \epsilon.\end{aligned}$$

Since ϵ is arbitrary, it follows that σ is σ-additive.

Finally, if $k \geq K$ and $X = P_k \cup N_k$ is the Jordan decomposition for $\sigma - \sigma_k$ then

$$\|\sigma - \sigma_k\|_{TV} = (\sigma(P_k) - \sigma_k(P_k)) - (\sigma(N_k) - \sigma_k(N_k)) < \epsilon;$$

thus $\sigma_k \to \sigma$ as $k \to \infty$. □

The norm $\|.\|_{TV}$ is called the *total variation norm.*

Exercise 15.3.2 Suppose that $\mu \in M(X, \Sigma)$. Show that $\|\mu\|_{TV} = \sup_{A\in\Sigma} \{\mu(A) - \mu(X \setminus A)\}$.

If μ is a finite or σ-finite measure, then elements of $L^1(X, \Sigma, \mu)$ can be considered as signed measures on (X, Σ). First observe that $\|f\|_1 = \int_X |f| d\mu$ is a norm on $L^1(X, \Sigma, \mu)$ which defines the metric d_1, so that $(L_1(X, \Sigma, \mu), \|.\|_1)$ is a Banach space.

Exercise 15.3.3 Suppose that $(f_n)_{n=1}^\infty$ is a sequence in $L^1(X, \Sigma, \mu)$ which converges almost everywhere to $f \in L^1$, and that $\|f_n\|_1 \to \|f\|_1$ as $n \to \infty$. Show that $f_n \to f$ in $(L_1, \|.\|_1)$.

Theorem 15.3.4 *Suppose that $f \in L^1(X, \Sigma, \mu)$. If $A \in \Sigma$, let $f.d\mu(A) = \int_A f\, d\mu$. Then $f.d\mu$ is a signed measure, and the mapping $f \to f.d\mu$ is an isometric linear mapping of $L^1(X, \Sigma, \mu)$ onto a closed subspace of the space $M(X, \Sigma)$ of signed measures on (X, Σ).*

Proof Suppose that $(B_n)_{n=1}^\infty$ is an increasing sequence in Σ, with union B. Then $|fI_{B_n}| \leq |f|$ for each $n \in \mathbf{N}$, and $fI_{B_n} \to fI_B$ pointwise as $n \to \infty$. By the theorem of dominated convergence,

$$f.d\mu(B_n) = \int_{B_n} f\, d\mu \to \int_B f\, d\mu = f.d\mu(B),$$

so that $f.d\mu$ is a signed measure.

Clearly $f.d\mu = f^+.d\mu - f^-.d\mu$, so that

$$\begin{aligned}\|f.d\mu\|_{TV} &= \|f^+.d\mu\|_{TV} + \|f^-.d\mu\|_{TV} \\ &= \int_X f^+\, d\mu + \int_X f^-\, d\mu = \int_X |f|\, d\mu = \|f\|_1.\end{aligned}$$

Thus the mapping is an isometry. Since $(L^1(X, \Sigma, \mu), \|.\|_1)$ is complete, the image is closed. □

Suppose that (X, Σ, μ) is a finite or σ-finite measure space. A function f in $\mathcal{L}^0(X, \Sigma, \mu)$ is *essentially bounded* if there exists M such that $|f| < M$ almost everywhere; that is, there exists M such that $\lambda_{|f|}(M) = \mu(|f| > M) = 0$. We define $\mathcal{L}^\infty = \mathcal{L}^\infty(X, \Sigma, \mu)$ to be $\{f \in \mathcal{L}^0 : f \text{ is essentially bounded}\}$. $\mathcal{L}^\infty$ is a linear subspace of $\mathcal{L}^0$.

Theorem 15.3.5 *The function* $p(f) = \text{ess sup}(f)$ *is a seminorm on* $\mathcal{L}^\infty(X, \Sigma, \mu)$*, and*

$$\{f : p(f) = 0\} = \mathcal{N}^\infty(X, \Sigma, \mu) = \{f \in \mathcal{L}^\infty : f = 0 \text{ almost everywhere}\}.$$

Let $\|.\|_\infty$ *be the corresponding norm on the quotient space* $L^\infty(X, \Sigma, \mu)$*. Then* $(L^\infty(X, \Sigma, \mu), \|.\|_\infty)$ *is a Riesz space and a Banach space, and the inclusion*

$$(L^\infty(X, \Sigma, \mu), \|.\|_\infty) \to (L^0(X, \Sigma, \mu), \rho_0)$$

is continuous. If (X, Σ, μ) *is a finite measure space, the subspace* S *of simple measurable functions is dense in* $(L^\infty(X, \Sigma, \mu), \|.\|_\infty)$*.*

Proof It is easily seen that $\|.\|_\infty$ is a norm on L^∞, that L^∞ is a Riesz space and that $(L^\infty, \|.\|_\infty)$ is complete.

If $\|f_n - f\|_\infty \to 0$ as $n \to \infty$ then $f_n(x) \to f(x)$ almost everywhere, and so $\rho_0(f_n, f) \to 0$ as $n \to \infty$.

Suppose that $f \in L^\infty(X, \Sigma, \mu)$ and that $\|f\|_\infty < m \in \mathbf{N}$. Let $A_{n,k} = \{x : k/n < f(x) \leq (k+1)/n\}$ for $|k| \leq mn$, and let $f_n = \sum_{k=-mn}^{mn} (k/n) I_{A_{n,k}}$. Then $f_n \in S$ and $\|f - f_n\|_\infty \leq 1/n$. □

Norm convergence in $L^\infty(X, \Sigma, \mu)$ is called *uniform convergence almost everywhere.*

Unless it is finite-dimensional, $((L^\infty, \Sigma, \mu), \|.\|_\infty)$ is not separable. For if $(A_n)_{n=1}^\infty$ is a disjoint sequence of sets in Σ of positive measure, then the set $\{\sum_{n=1}^\infty \omega_n I_{A_n} : \omega \in \Omega(\mathbf{N})\}$ is an uncountable set for which the distance between any two distinct points is 2.

15.4 The Radon–Nikodym Theorem

Suppose that (X, Σ, μ) is a measure space and that ν is a finite measure on Σ. When does there exist $f \in L^1(X, \Sigma, \mu)$ such that $\nu = f.d\mu$? This is answered by the Radon–Nikodym theorem, which is a fundamental theorem of measure theory. But first we must introduce the Hilbert space $L^2(X, \Sigma, \mu)$. This is defined as

$$L^2 = L^2(X, \Sigma, \mu) = \{f \in L^0 : f^2 \in L^1(X, \Sigma, \mu)\}.$$

Theorem 15.4.1 *Suppose that* (X, σ, μ) *is a measure space. If* $f, g \in L^2(X, \Sigma, \mu)$ *then* $fg \in L^1(X, \Sigma, \mu)$*, and the function* $(f, g) \to \int_X fg\, d\mu$ *is an inner product on* $L^2(X, \Sigma, \mu)$*, under which* $L^2(X, \Sigma, \mu)$ *is a Hilbert space, whose unit ball is closed in* $L^0(X, \Sigma, \mu)$*.*

Proof Since $|fg| \leq \frac{1}{2}(f^2 + g^2)$, $fg \in L^1(X, \Sigma, \mu)$, and it then follows that $(f, g) \to \int_X fg\, d\mu$ is an inner product on $L^2(X, \Sigma, \mu)$. If $\|f_n\|_2 \leq 1$ for $n \in \mathbf{N}$

and f_n converges to f locally in measure, then there exists a subsequence $(f_{n_k})_{k=1}^{\infty}$ which converges almost everywhere to f. Then

$$\|f\|_2^2 = \int_X f^2 \, d\mu \leq \liminf_{k\to\infty} \int f_{n_k}^2 \, d\mu \leq 1,$$

by Fatou's lemma, so that the unit ball is closed in $L^0(X, \Sigma, \mu)$. It therefore follows from Theorem 2.4.5 that $(L^2(X, \Sigma, \mu), \|.\|_2)$ is complete. □

Corollary 15.4.2 *If $f, g \in L^2(X, \Sigma, \mu)$, let $l_g(f) = \int_X fg \, d\mu$. Then $g \to l_g$ is an isometric isomorphism of L^2 onto its dual $(L^2)'$.*

Proof This is simply a special case of the Fréchet–Riesz theorem. □

Theorem 15.4.3 (The Lebesgue decomposition theorem) *Suppose that (X, Σ, μ) is a finite or σ-finite measure space, and that ν is a finite measure on Σ. Then there exists a non-negative $f \in L^1(\mu)$ and a set $B \in \Sigma$ with $\mu(B) = 0$ such that $\nu(A) = \int_A f \, d\mu + \nu(A \cap B)$ for each $A \in \Sigma$.*

If we define $\nu_B(A) = \nu(A \cap B)$ for $A \in \Sigma$, then ν_B is a measure. The measures μ and ν_B are mutually singular; we decompose X as $B \cup (X \setminus B)$, where $\mu(B) = 0$ and $\nu_B(X \setminus B) = 0$; μ and ν_B live on disjoint sets.

Proof Let $\pi(A) = \mu(A) + \nu(A)$; π is a measure on Σ. Suppose that $g \in L^2_{\mathbf{R}}(\pi)$. Let $L(g) = \int g \, d\nu$. Then, by the Cauchy–Schwarz inequality,

$$|L(g)| \leq (\nu(X))^{1/2} (\int_X |g|^2 \, d\nu)^{1/2} \leq (\nu(X))^{1/2} \|g\|_{L^2(\pi)},$$

so that L is a continuous linear functional on $L^2_{\mathbf{R}}(\pi)$. By Corollary 15.4.2, there exists an element $h \in L^2_{\mathbf{R}}(\pi)$ such that $L(g) = \langle g, h \rangle$, for each $g \in L^2(\pi)$; that is, $\int_X g \, d\nu = \int_X gh \, d\mu + \int_X gh \, d\nu$, so that

$$\int_X g(1 - h) \, d\nu = \int_X gh \, d\mu. \qquad (*)$$

Taking g as an indicator function I_A, we see that

$$\nu(A) = L(I_A) = \int_A h \, d\pi = \int_A h \, d\mu + \int_A h \, d\nu$$

for each $A \in \Sigma$.

Now let $N = (h < 0)$, $G_n = (0 \leq h \leq 1 - 1/n)$, $G = (0 \leq h < 1)$ and $B = (h \geq 1)$. Then

$$\nu(N) = \int_N h \, d\mu + \int_N h \, d\nu \leq 0, \text{ so that } \mu(N) = \nu(N) = 0,$$

$$\text{and } \nu(B) = \int_B h \, d\mu + \int_B h \, d\nu \geq \nu(B) + \mu(B), \text{ so that } \mu(B) = 0.$$

Let $f(x) = h(x)/(1 - h(x))$ for $x \in G$, and let $f(x) = 0$ otherwise. Note that if $x \in G_n$ then $0 \leq f(x) \leq 1/(1 - h(x)) \leq n$. If $A \in \Sigma$, then, using $(*)$,

$$\nu(A \cap G_n) = \int_X \frac{1-h}{1-h} I_{A\cap G_n}\, d\nu = \int_X f I_{A\cap G_n}\, d\mu = \int_{A\cap G_n} f\, d\mu.$$

Applying the monotone convergence theorem, we see that $\nu(A \cap G) = \int_{A\cap G} f\, d\mu = \int_A f\, d\mu$. Thus

$$\nu(A) = \nu(A \cap G) + \nu(A \cap B) + \nu(A \cap N) = \int_A f\, d\mu + \nu(A \cap B).$$

Taking $A = X$, we see that $\int_X f\, d\mu < \infty$, so that $f \in L^1(\mu)$. □

This beautiful proof is due to von Neumann.

Exercise 15.4.4 Suppose that μ and $(\nu_n)_{n=1}^\infty$ are finite measures on a measurable space (X, Σ), that $\sum_{n=1}^\infty \nu_n = \nu$ and that each ν_n has Lebesgue decomposition $\nu_n = f_n.d\mu + \pi_n$ with respect to μ. Suppose that ν has Lebesgue decomposition $\nu = f.d\mu + \pi$. Show that $f = \sum_{n=1}^\infty f_n$ μ-almost everywhere, and that $\pi = \sum_{n=1}^\infty \pi_n$.

We can now identify the image of the inclusion mapping $L^1(X, \Sigma, \mu) \to M(X, \Sigma)$.

Suppose that (X, Σ, μ) is a measure space and that ψ is a real-valued function on Σ. We say that ψ is *absolutely continuous* with respect to μ if, given $\epsilon > 0$, there exists $\delta > 0$ such that if $\mu(A) < \delta$ then $|\psi(A)| < \epsilon$.

Corollary 15.4.5 (The Radon–Nikodym theorem) *Suppose that (X, Σ, μ) is a finite measure space and that ν is a finite measure on Σ. Then the following are equivalent.*

(i) *ν is absolutely continuous with respect to μ.*
(ii) *If $A \in \Sigma$ and $\mu(A) = 0$ then $\nu(A) = 0$.*
(iii) *There exists a non-negative $f \in L^1(\mu)$ such that $\nu(A) = \int_A f\, d\mu$ for each $A \in \Sigma$.*

Proof If ν is absolutely continuous with respect to μ, then (ii) holds. If (ii) holds, and B is the subset of the theorem, then $\nu(B) = 0$, and so (iii) holds. Finally, suppose that (iii) holds, and that $\epsilon > 0$. Let $B_n = (f > n)$. Then by the dominated convergence theorem, $\nu(B_n) = \int_{B_n} f\, d\mu \to 0$ as $n \to \infty$, and so there exists n such that $\nu(B_n) < \epsilon/2$. Let $\delta = \epsilon/2n$. Then if $\mu(A) < \delta$,

$$\nu(A) = \nu(A \cap B_n) + \int_{A\cap(0\leq f\leq n)} f\, d\mu < \epsilon/2 + n\delta = \epsilon,$$

so that (i) holds. □

There is also a 'signed' version of this corollary.

Theorem 15.4.6 *Suppose that (X, Σ, μ) is a finite measure space and that ψ is a bounded absolutely continuous real-valued function on Σ which is additive: if A, B are disjoint sets in Σ then $\psi(A \cup B) = \psi(A) + \psi(B)$. Then there exists $f \in L^1(\mu)$ such that $\psi(A) = \int_A f\,d\mu$, for each $A \in \Sigma$.*

Proof If $A \in \Sigma$, let $\psi^+(A) = \sup\{\psi(B) : B \subseteq A\}$. ψ^+ is a bounded additive non-negative function on Σ. We shall show that ψ^+ is countably additive. Suppose that A is the disjoint union of (A_i). Let $R_j = \cup_{i>j} A_i$. Then $R_j \searrow \emptyset$, and so $\mu(R_j) \to 0$ as $j \to \infty$. By absolute continuity, $\sup\{|\psi(B)| : B \subseteq R_j\} \to 0$ as $j \to \infty$, and so $\psi^+(R_j) \to 0$ as $j \to \infty$. This implies that ψ^+ is countably additive. Thus ψ^+ is a measure on Σ, which is absolutely continuous with respect to μ, and so it is represented by some $f^+ \in L^1(\mu)$. But now $\psi^+ - \psi$ is additive, non-negative and absolutely continuous with respect to μ, and so is represented by a function f^-. Let $f = f^+ - f^-$. Then $f \in L^1(\mu)$ and

$$\psi(A) = \psi^+(A) - (\psi^+(A) - \psi(A)) = \int_A f^+\,d\mu - \int_A f^-\,d\mu = \int_A f\,d\mu.$$

□

We can use this result to represent the dual of L^1.

Theorem 15.4.7 *Suppose that (X, Σ, μ) is a finite measure space. If $f \in L^\infty(\mu)$ and $g \in L^1(\mu)$, let $j_f(g) = \int_X fg\,d\mu$. Then j is an isometric isomorphism of $(L^\infty(\mu), \|.\|_\infty)$ onto the dual space $((L^1)', \|.\|_1)$.*

Proof The map j is clearly a norm-decreasing linear map of $(L^\infty(\mu), \|.\|_\infty)$ into the dual $((L^1)', \|.\|_1)$. If $f \in L^\infty(\mu)$ and $\epsilon > 0$, let

$$A_\epsilon = \{x : |f(x)| > \|f\|_\infty - \epsilon, \text{ and } g = \frac{\operatorname{sgn}(f)}{\mu(A_\epsilon)}.I_{A_\epsilon}.$$

Then $\|g\|_1 = 1$ and $\int fg\,d\mu \geq \|f\|_\infty - \epsilon$, so that j is an isometry. Finally, if $\phi \in (L^1)'$ and $A \in \Sigma$, let $\nu(A) = \phi(I_A)$. Then ν is a signed measure on Σ which is absolutely continuous with respect to μ, and so is represented by some $f \in L^1(\mu)$. But if $A \subseteq \{x : |f| > \|\phi\|_1'\}$, then $|\nu(A)| \leq \mu(A)$, from which it follows that $\mu(A) = 0$. Thus $f \in L^\infty$ and $j(f) = \phi$. □

All these results extend easily to σ-finite measure spaces. In particular, we have the following result.

Proposition 15.4.8 *Suppose that μ and ν are σ-finite measures on a measurable space (X, Σ), and that $0 \leq \nu(A) \leq \mu(A)$ for each $A \in \Sigma$. Then there exists $f \in L^\infty(\mu)$ such that $0 \leq f \leq 1$ and $\nu(A) = \int_A f\,d\mu$ for each $A \in \Sigma$.*

Proof There exists an increasing sequence $(B_n)_{n=1}^\infty$ in Σ such that $X = \cup_{n=1}^\infty B_n$ and $\mu(B_n) < \infty$ for each $n \in \mathbf{N}$. Let $\Sigma_n = \{A \in \Sigma : A \subseteq B_n\}$, and let μ_n and ν_n be the restrictions of μ and ν to Σ_n. By the Radon–Nikodym theorem, for each n there exists $f_n \in L^1(\mu_n)$ such that $\mu_n(A) = \int_A f_n \, d\mu_n$ for each $A \in \Sigma_n$. Then $0 \leq f_n \leq 1$, since $0 \leq \nu_n \leq \mu_n$. If $m > n$ and $A \in \Sigma_n$ then $\nu_n(A) = \int_A f_n \, d\mu_n = \int_A f_m \, d\mu_m$, from which it follows that $f_{m|B_n} = f_n$. There therefore exists $f \in L^\infty(\mu)$ such that $f_{|B_n} = f_n$. Thus if $A \in \Sigma$ then

$$\nu(A) = \lim_{n\to\infty} \nu(A \cap B_n) = \lim_{n\to\infty} \int_{A\cap B_n} f \, d\mu = \int_A f \, d\mu.$$

□

Exercise 15.4.9 Show how the results of this section can be extended to complex-valued functions and measures.

15.5 Orlicz Spaces and L^p Spaces

We now use the Legendre transform to define a large class of function spaces.

An *N-function* Φ on $\mathbf{R}$ is an even (that is, $\Phi(s) = \Phi(-s)$) convex non-negative real-valued function for which

(i) $\Phi(s) = 0$ if and only if $s = 0$,
(ii) Φ is differentiable at 0, with $\Phi'(0) = 0$, and
(iii) $\Phi(s)/|s| \to \infty$ as $|s| \to \infty$.

Exercise 15.5.1 Suppose that Φ is an N-function. Let $p(s) = \Phi'_+(s) = \lim_{u\searrow s}(\Phi(u) - \Phi(s))/(u - s)$ be the right-hand derivative of Φ, so that $\Phi(s) = \int_0^s p(u)\, du$. Let $q(t) = \inf_{v>t} p(v)$ be the right-continuous inverse of p. Then p and q are continuous increasing functions on $\mathbf{R}$; show that if $\Psi(t) = \int_0^t q(v)\, dv$, then Ψ is an N-function. Show that Ψ is the Legendre transform of Φ. Deduce that $uv \leq \Phi(u) + \Psi(v)$, with equality if and only if $p(u) = v$, (or equivalently $q(v) = u$).

Ψ is called the *complementary N-function* of Φ. The inequality $uv \leq \Phi(u) + \Psi(v)$, with equality if and only if $p(u) = v$ (or equivalently $q(v) = u$), is known as *Young's inequality*.

Suppose now that (X, Σ, μ) is a measure space. We define $B(L_\Phi(X))$ to be the set

$$B(L_\Phi(X)) = \{f \in L_0(X) : \int_X \Phi(f(x)) \leq 1\}.$$

Proposition 15.5.2 *$B(L_\Phi(X))$ is a convex subset of $L_0(X)$, which is closed in $(L_0(X), d_0)$.*

Proof Since Φ is a convex function, $B(L_\Phi(X))$ is a convex subset of $L_0(X)$. Suppose that $(f_n)_{n=1}^\infty$ is a d_0-Cauchy sequence in $B(L_\Phi)$. Then $(f_n)_{n=1}^\infty$ has a subsequence $(f_{n_k})_{k=1}^\infty$ which converges almost everywhere to a function $f \in L_0(X)$. Then $\Phi(f_{n_k})$ converges almost surely to $\Phi(f)$, and so, by Fatou's lemma, $\int_X \Phi(f)\,d\mu \leq 1; f \in B(L_\phi))$, $d_0(f_{n_k},f) \to 0$ as $k \to \infty$, and so $d_0(f_n,f) \to 0$ as $n \to \infty$. □

We define the *Orlicz space* $L_\Phi(X)$ to be the span of $B(L_\Phi(X))$, and define $\|.\|_\Phi$ to be the gauge of $B(L_\Phi(X))$. Thus $B(L_\Phi(X))$ is the unit ball of $(L_\Phi(X), \|.\|_\Phi)$, and $\|f\|_\Phi = \inf\{\alpha \geq 0 : \int_X \Phi(f(x)/\alpha \leq 1)\}$. $L_\Phi(X)$ is a Riesz space, and $\|.\|_\Phi$ is known as the *Luxemburg norm* on $L_\phi(X)$.

Proposition 15.5.3 *Suppose that Φ is an N-function.*

(i) If $\|f\|_\Phi \leq 1$ then $\int_X \Phi(f)\,d\mu \leq \|f\|_\Phi$. It follows from Proposition 15.5.2 that $(L_\Phi, \|.\|_\Phi)$ is complete.
(ii) If $1 \leq \int_X \Phi(f)\,d\mu = k \leq \infty$ then $\|f\|_\Phi \leq \int_X \Phi(f)\,d\mu$.

Proof (i) By convexity,

$$1 = \int_X \Phi\left(\frac{f}{\|f\|_\Phi}\right) \geq \int_X \frac{1}{\|f\|_\Phi}\Phi(f)\,d\mu = \frac{1}{\|f\|_\Phi}\int_X \Phi(f)\,d\mu.$$

(ii) By convexity again,

$$\int_X \Phi\left(\frac{f}{k}\right)\,d\mu \leq \frac{1}{k}\int_X \Phi(f)\,d\mu = 1,$$

so that $\|f\|_\Phi \leq k$. □

Exercise 15.5.4 (i) Show that $\|f\|_\Phi = 1$ if and only if $\int_X \Phi(f)\,d\mu = 1$.
(ii) Show that if $\|f_n\|_\Phi \to 0$ as $n \to \infty$ then $\int_X \Phi(f_n)\,d\mu \to 0$ as $n \to \infty$.

Proposition 15.5.5 *If μ is a finite measure then $L_\Phi \subseteq L_1$, and the inclusion mapping is continuous.*

Proof There exists t_0 such that $\Phi(t)/t \geq 1$ for $t \geq t_0$. Suppose that $f \in L_\Phi$ and that $\|f\|_\Phi = 1$. Then

$$\begin{aligned}\int_X |f|\,d\mu &\leq \int_{(|f|<t_0)} |f|\,d\mu + \int_{(|f|\geq t_0)} |f|\,d\mu \\ &\leq t_0\mu(X) + \int_{(|f|\geq t_0)} \Phi(f)\,d\mu \leq t_0\mu(X) + 1.\end{aligned}$$

□

We define the space $\check{L}_\Phi(X)$ to be

$$\check{L}_\Phi(X) = \{f \in L_0(X) : fg \in L_1(X) \text{ for each } g \in L_\Psi(X)\},$$

where Ψ is the complementary N-function.

Theorem 15.5.6

$$\check{L}_\phi(X) = \{f \in L_0(X) : \sup\{\|fg\|_1 : g \in B(L_\Psi(X))\} < \infty\}.$$

Proof Suppose not. Then there exists non-negative $f \in \check{L}_\Phi(X)$ and an increasing sequence $(g_n)_{n=1}^\infty$ of non-negative elements of $L_\Psi(X)$ such that $\|g_n - g_m\|_\Psi < 1/n$ for $m > n$, while $\|fg_n\|_1 > n$. Then $g_n \to g \in L_\Psi(X)$, and $\|fg\|_1 = \infty$, giving a contradiction. □

The quantity $\|f\|^\Phi = \sup_{g \in B(L_\Psi(X))} \|fg\|_1$ is a norm on $\check{L}_\Phi(X)$. It is called the *Orlicz norm* on $\check{L}_\Phi(X)$.

Theorem 15.5.7 *$\check{L}_\Phi(X) = L_\phi(X)$, and if $f \in L_\Phi(X)$ then $\|f\|_\Phi \le \|f\|^\Phi \le 2\,\|f\|_\Phi$.*

Proof It follows from Young's inequality that if $f \in L_\Phi(X)$ and $g \in B(L_\Psi(X))$ and $f \neq 0$ then

$$\int_X \frac{|fg|}{\|f\|_\phi}\,d\mu \le \int_X \Phi(\frac{f}{\|f\|_\phi})\,d\mu + \int_X g\,d\mu \le 2,$$

so that $\|f\|^\Phi \le 2\,\|f\|_\Phi$, and $L_\Phi(X) \subseteq \check{L}_\Phi(X)$.

On the other hand, it follows from the convexity of Ψ that if $\|h\|_\Psi \ge 1$ then

$$\|h\|_\Psi = \|h\|_\Psi \int_X \Psi\frac{h}{\|h\|_\Psi}\,d\mu \le \int_X \Psi(h)\,d\mu,$$

so that in general $\|h\|_\Psi \le 1 + \int_X \Psi(h)\,d\mu$. Further, if $\|g\|^\Psi \le 1$ then

$$\int_X |hg|\,d\mu = \|h\|_\Psi \int_X \frac{|hg|}{\|h\|_\Psi}\,d\mu \le \|h\|_\Psi\,.$$

Suppose now that $f \in \check{L}_\Phi(X)$, and $\|f\|^\Phi \le 1$. Let $g(x) = p(f(x))$. Then

$$\int_X \Phi(f)\,d\mu + \int_X \Psi(g)\,d\mu = \int_X |fg|\,d\mu \le \|g\|_\Psi \le 1 + \int_X \Psi(g)\,d\mu,$$

so that $\int_X \Phi(f)\,d\mu \le 1$ and $\|f\|_\Phi \le 1$, Thus $\check{L}_\Phi(X) \subset L_\phi(X)$, and $\|f\|_\Phi \le \|f\|^\Phi$. □

Suppose that Φ is an N-function and that $\alpha > 0$. We set $\Phi_\alpha(x) = \Phi(\alpha x)$. Then Φ_α is also an N-function, and $\{\Phi_\alpha : \alpha > 0\}$ is an increasing family of N-functions. The function conjugate to Φ_α is $\Psi_{1/\alpha}$, and $\{\Psi_\alpha : \alpha > 0\}$ is also an increasing family of N functions.

An N-function Φ is said to satisfy the Δ_2 *condition* if there exists $L > 0$ such that $\Phi_2 \le L\Phi$ for all $t > 0$. If Φ satisfies the Δ_2 condition and $2^{k-1} \le \alpha \le 2^k$, then $\Phi_\alpha \le \Phi^{2^k} \le L^k\Phi \le 2\alpha L\Phi$.

Exercise 15.5.8 Show that if Φ satisfies the Δ_2 condition with constant L then the conjugate function Ψ also satisfies the Δ_2 condition, with the same constant L.

Theorem 15.5.9 *Suppose that the N-function Φ satisfies the Δ_2 condition.*

(i) If $f \in L_0$, then $f \in L_\Phi$ if and only if $\int_X \Phi(f)\,d\mu < \infty$.
(ii) If $(f_n)_{n=1}^\infty$ is a sequence in L_Φ then $\int_X \Phi(f_n)\,d\mu \to 0$ as $n \to \infty$ if and only if $\|f_n\|_\Phi \to 0$ as $n \to \infty$.

Proof (i) Certainly, if $\int_X \Phi(f)\,d\mu < \infty$ then $f \in L_\Phi$, and if $f \in B(L_\Phi)$ then $\int_X \Phi(f)\,d\mu \le 1 < \infty$. If $f \in L_\Phi$ and $\|f\|_\Phi = \alpha > 1$, let $\alpha f = g$. Then

$$\int_X \Phi(f)\,d\mu = \int_X \Phi(g/\alpha)\,d\mu \le 2L\alpha \int_X g\,d\mu = 2L\alpha < \infty.$$

(ii) By Exercise 15.5.4, if $\|f_n\|_\Phi \to 0$ as $n \to \infty$, then $\int_X \Phi(f_n)\,d\mu \to 0$. Suppose that $0 < \epsilon < 1$, so that $\alpha = 1/\epsilon > 1$, and that $\int_X \Phi(f_n)\,d\mu \to 0$. Thus there exists N such that $\int_X \Phi(f_n)\,d\mu < \epsilon/2L$ for $n \ge N$. If $n \ge N$ then

$$\int_X \Phi(\alpha f_n)\,d\mu \le 2L\alpha \int_X \Phi(f_n)\,d\mu \le 1,$$

so that $\|f_n\|_\Phi \le 1/\alpha = \epsilon$. □

Exercise 15.5.10 Suppose that the N-function Φ satisfies the Δ_2 condition. Show that the simple measurable functions are dense in L_Φ.

Theorem 15.5.11 *Suppose that (X, Σ, μ) is a finite or σ-finite measure space, and that Φ is an N-function, with conjugate function Ψ, which satisfies the Δ_2 condition. If $f \in L_\Phi$ and $g \in L_\Psi$ let $j(g)(f) = \int_X fg\,d\mu$. Then the mapping j is an isometry of $(L_\Psi, \|.\|^\Psi)$ onto the dual space $((L_\Phi)', (\|.\|'_\Phi))$.*

Proof It follows from the definitions that j is an isometric injection. We need to show that it is surjective. By a standard approximation argument, we can suppose that (X, Σ, μ) is a finite measure space. Suppose that θ in $(L_\Phi)'$, and that $\|\theta\|'_\Phi = 1$. If $A \in \Sigma$, let $\tilde{\theta}(A) = \|\theta\|_\Phi (I_A)$. If $\mu(A_n) \to 0$ as $n \to \infty$, then $\int_{A_n} \Phi\,d\mu \to 0$ as $n \to \infty$, and so $\tilde{\theta}(A_n) = \left\|I_{A_n}\right\|_\Phi \to 0$ as $n \to \infty$. It therefore follows that $\tilde{\theta}$ is a signed measure on Σ, and that it is absolutely continuous with respect to μ. By the Radon–Nikodym theorem, there exists g in $L^1(X, \Sigma, \mu)$ such that $\tilde{\mu} = g\,d\mu$. But then $\theta(f) = \int_X fg\,d\mu$ for each $f \in L_\phi$, and so $g = j(\theta) \in L_\Psi$. □

Corollary 15.5.12 *If Φ satisfies the Δ_2 condition then L_Φ is reflexive.*

We now consider a more familiar special case.

If $1 < p < \infty$ then the function $\Phi(t) = |t|^p$ is an N-function which satisfies the Δ_2 condition, with conjugate function $\Psi(t) = |t|^q$, where $1/p + 1/q = 1$. We denote the corresponding Orlicz function by $L^p(X, \Sigma, \mu)$. The function $\|f\|_p = \left(\int_X |f|^p \, d\mu\right)^{1/p}$ is positive homogeneous, and is therefore the gauge of $B(L^p)$. Thus $\|f\|_\Phi = \|f\|_p$.

Exercise 15.5.13 Suppose that (X, Σ, μ) is a finite or σ-finite measure space, that $1 < p < \infty$ and that $1/p + 1/q = 1$. Show the following.

(i) L^p is a linear subspace of L^0, and the function $\|.\|_p = (\int_X |f|^p \, d\mu)^{1/p}$ is a norm on L^p under which L^p is a reflexive Banach space, with dual L^q.
(ii) Show that the space $(L^2, \|.\|_2)$ is a Hilbert space, with inner product $\langle f, g\rangle = \int_X fg \, d\mu$.
(iii) Suppose that $f \in L^p$ and $g \in L^q$. Then $fg \in L^1$, and

$$\left|\int fg \, d\mu\right| \leq \int |fg| \, d\mu \leq \|f\|_p \, \|g\|_q \, .$$

Equality holds throughout if and only if either $\|f\|_p \, \|g\|_{p'} = 0$, or $g = \lambda \mathrm{sgn}(f).|f|^{p-1}$ almost everywhere, where $\lambda \neq 0$.

The L^p spaces are uniformly convex, for $1 < p < \infty$.

Proposition 15.5.14 *Suppose that (X, Σ, μ) is a measure space and that $1 < p < \infty$. Then $(L^p(X, \Sigma, \mu), \|.\|_p)$ is uniformly convex.*

Proof Let $\delta(\epsilon) = 1 - 2(1 - \epsilon/2)^p/(1 + (1 - \epsilon)^p)$, for $0 < \epsilon < 2$. Then

$$\begin{aligned} \delta'(\epsilon) &= \frac{(1 + (1 - \epsilon)^p)(1 - \epsilon/2)^{p-1} - 2(1 - \epsilon/2)^p(1 - \epsilon)^{p-1}}{(1 + (1 - \epsilon)^p)^2} \\ &= \frac{(1 - \epsilon/2)^{p-1}}{(1 + (1 - \epsilon)^p)^2}(1 + (1 - \epsilon)^p - 2(1 - \epsilon/2)(1 - \epsilon)^{p-1}) > 0 \end{aligned}$$

so that δ is a strictly increasing non-negative function. It follows by homogeneity that if $|x| \geq |y|$ and $|x - y| \geq \epsilon |x|$ then

$$\left(\frac{x + y}{2}\right)^p \leq (1 - \delta(\epsilon))(|x|^p + |y|^p).$$

Now let $M = \{x \in X : |f(x) - g(x)|^p \geq \epsilon^p(|f(x)|^p + |g(x)|^p)/4\}$, let $f_M = f.I_M$, let $G_M = g.I_M$ and let $N = X \setminus M$. Then

$$\int_N |f(x) - g(x)|^p d\mu(x) \leq \frac{\epsilon^p}{4} \int_N |f(x)|^p + |g(x)|^p \, d\mu(x) = \frac{\epsilon^p}{2}$$

so that $\int_M |f(x) - g(x)|^p \, d\mu(x) \geq \epsilon^p/2$. Thus $\|f_M - g_M\| \geq \epsilon/2^{1/p}$, and $max(\|f_M\|_p, \|g_M\|_p) \geq \epsilon/2(1/p+1)$. But then

$$\begin{aligned} 1 - \|(f+g)/2\|^p &= \int_X \tfrac{1}{2}(|f(x)|^p + |g(x)|^p) - \frac{|f(x)+g(x)|^p}{2^p}\, dx \\ &\geq \int_M \tfrac{1}{2}(|f(x)|^p + |g(x)|^p) - \frac{|f(x)+g(x)|^p}{2^p}\, dx \\ &\geq \int_M \delta(\epsilon/4^{1/p})(\tfrac{1}{2}|f(x)|^p + \ |g(x)|^p)\, dx \\ &\geq \delta(\frac{\epsilon}{4^{1/p}})\frac{\epsilon^p}{2^{p+2}}, \end{aligned}$$

which gives the result. □

Corollary 15.5.15 *$(L^p(X, \Sigma, \mu), \|.\|_p)$ is smooth.*

Proof This follows from Theorem 11.6.9. □

Corollary 15.5.16 *The weak topology and norm topology coincide on the unit sphere of L^p.*

On the other hand, the space $(L_1(X, \Sigma, \mu), \|.\|_1)$ is not uniformly convex (consider $f = I_A/\mu(A)$ and $g = I_B/\mu(B)$, where A and B are two disjoint measurable subsets of positive measure; then $\|f\|_1 = \|g\|_1 = 1$ and $\|f - g\|_1 = \|f + g\|_1 = 2$). In particular, l_1 is not uniformly convex.

Exercise 15.5.17 If $f \in S(L^1(X, \Sigma, \mu))$, then the norm is smooth at f if and only if $\mu(f = 0) = 0$.

We can relate Orlicz spaces to L^p spaces.

Proposition 15.5.18 *Suppose that Φ is an N-function which satisfies the Δ_2 condition, with $\Phi(2t) \leq L\Phi(t) = 2^\alpha L_\Phi(t)$ for $t > 0$. Then $\Phi(t) \leq L\Phi(1)t^\alpha$ for $t > 1$ and $\Phi(t) \geq (\Phi(1)/L)t^\alpha$ for $0 < t < 1$. Thus if (X, Σ, μ) is a finite measure space then $L_\phi(X, \Sigma, \mu) \subseteq L_\alpha(X, \Sigma, \mu)$, and the inclusion is continuous. Similarly, if $\mathbf{N}$ is given counting measure then $l^\alpha \subseteq l_\Phi$, and the inclusion is continuous.*

Proof First observe that $\Phi(2^n) \leq L^n = 2^{\alpha n}$. Suppose that $t > 1$ and that $2^n < t \leq 2^{n+1}$. Then

$$\Phi(t) \leq \Phi(2^{n+1}) \leq L\Phi(2^n) \leq L(2^{\alpha n}) \leq Lt^\alpha.$$

Thus if $f \in L^\alpha(X, \Sigma, \mu)$ then

$$\int_X \Phi(f(x))\, d\mu(x) \leq \mu(X) + \int_{(|f|>1)} |f|^\alpha(x)\, d\mu(x),$$

and so $f \in L_\Phi(X, \Sigma, \mu)$. Since the inclusion mapping has a closed graph, the inclusion is continuous. The case where **N** is given counting measure is proved similarly. □

We can also define L^p spaces for $0 < p < 1$; we again set

$$L^p(X, \Sigma, \mu) = \{f \in L_0 : d_p(f) = \int_X |f|^p \, d\mu < \infty\}.$$

In this case, the function $t \to t^p$ is a concave function on $[0, \infty)$.

Exercise 15.5.19 If $f, g \in L^p(X, \Sigma, \mu)$, where $0 < p < 1$, show that the function $d_p(f, g) = \int_X |f - g|^p \, d\mu$ is a complete translation-invariant metric on L^p.

Theorem 15.5.20 *Suppose that (X, Σ, μ) is a finite measure space and $0 < p < r < \infty$. Then $L^r \subseteq L^p$, and the inclusion mapping is uniformly continuous. If $p \geq 1$ and μ is a probability measure, then the inclusion mapping is norm-decreasing.*

Proof Suppose that $f \in L_r$. Let $t = r/(r - p)$, so that $p/r + 1/t = 1$. Applying Hölder's inequality, with exponents t and r/p , to I_X and $|f|^p$, we find that

$$\int_X |f|^p \, d\mu \leq (\mu(X))^{1/t} . \left(\int_X |f|^r \, d\mu\right)^{p/r},$$

from which the theorem follows. □

Theorem 15.5.21 *Suppose that (X, Σ, μ) is a finite measure space, and that Φ is an N-function which satisfies the Δ_2 condition. Then $L^\infty(X, \Sigma, \mu) \subseteq L_\Phi(X, \Sigma, \mu)$, and the inclusion is continuous.*

Proof For if $\|f\|_\infty \to 0$ as $n \to \infty$, then $\int_X \Phi(f_n) \to 0$. □

16
Borel Measures

16.1 Borel Measures, Regularity and Tightness

Recall if (X, τ) is a topological space, then the *Borel σ-field $\mathcal{B}$* of X is the σ-field generated by the open sets of X. If (X, τ) and (Y, σ) are topological spaces and $f : X \to Y$ is continuous, then it follows from Exercise 14.1.4 that if B is a Borel set in Y then $f^{-1}(B)$ is a Borel set in X.

Proposition 16.1.1 *Suppose that Σ is σ-field of subsets of a metrizable space (X, d). The following are equivalent.*

(i) $\Sigma = \mathcal{B}$, *the Borel σ-field of X.*

(ii) $\Sigma = \Sigma_1$, *the smallest σ-field for which every continuous real-valued function on X is measurable.*

(iii) $\Sigma = \Sigma_2$, *the smallest σ-field for which every lower semi-continuous real-valued function on X is measurable.*

Proof Clearly $\Sigma_1 \subseteq \Sigma_2$. If f is lower semi-continuous, then $(f > c)$ is open, for each $c \in \mathbf{R}$, and so f is Borel measurable; hence $\Sigma_2 \subseteq \mathcal{B}$. Finally, if A is closed, then $A = \{x \in X : d(x, A) = 0\}$, so that $A \in \Sigma_1$, and $\mathcal{B} \subseteq \Sigma_1$. □

This proposition depends on the fact that a closed subset of a metrizable space is a G_δ set. This is not necessarily the case for more general topological spaces. Here it is necessary to consider the *Baire* σ-field – the σ-field generated by the closed G_δ sets – and the theory is consequently more complicated. In applications, the space is usually metrizable, and indeed is usually a Polish space, where the results are stronger, and easier to prove, and so we shall restrict our attention to such spaces.

For example, if μ is a Borel measure on a *separable* metrizable space (X, τ), we can define its support supp(μ). If μ is a finite Borel measure on a topological space (X, τ), a *closed* subset C of X is the *support* of μ, if $\mu(X \setminus C) = 0$ and C is the smallest closed subset of X with this property.

Proposition 16.1.2 *Suppose that μ is a Borel measure on a separable metrizable space (X, τ). Then μ has a support.*

Proof (X, τ) is second countable; let $(U_n)_{n=1}^\infty$ be a base of open sets, and let $K = \{n \in \mathbf{N} : \mu(U_n) = 0\}$. Let $U = \cup_{n\in K} U_n$. Then $\mu(U) = 0$, and U is the largest open subset of X with this property. Thus $X \setminus U$ is the support of μ. □

Exercise 16.1.3 Suppose that μ is a Borel measure on a metrizable space (X, τ) and that $x \in X$. Show that $x \in \operatorname{supp}(\mu)$ if and only if $\mu(N) > 0$ for each open neighbourhood of x.

A Finite Borel measure on a metrizable space has good regularity properties.

Theorem 16.1.4 *Suppose that μ is a finite Borel measure on a metrizable space (X, τ). Then μ is* closed-regular; *that is, if $A \in \mathbf{B}$ then*

$$\mu(A) = \sup\{\mu(B) : B \text{ closed}, B \subseteq A\}$$
$$= \inf\{\mu(C) : C \text{ open}, C \supseteq A\}.$$

Proof Let d be a metric on X which defines the topology, and let T be the collection of Borel sets for which the result holds. Suppose first that A is closed, and that $U_j = \{x \in X : d(x, A) < 1/j\}$. Then $(U_j)_{j=1}^\infty$ is a decreasing sequence of open sets with intersection A. Then $\mu(A) = \lim_{j\to\infty} \mu(U_j) = \inf_{j\in\mathbf{N}} \mu(U_j)$, and so $A \in T$.

It is therefore enough to show that T is a σ-field. Since $A \in T$ if and only if $X \setminus A \in T$, it is enough to show that if $(A_n)_{n=1}^\infty$ is a sequence in T then $A = \cup_{n=1}^\infty A_n \in T$. Suppose that $\epsilon > 0$. Then for each n there exist $F_n \subseteq A_n \subseteq U_n$ (F_n closed, U_n open) with $\mu(A_n \setminus F_n) < \epsilon/2^{n+1}$ and $\mu(U_n \setminus A_n) < \epsilon/2^n$. Then $U = \cup_n U_n$ is open, and $U \setminus A \subset \cup_n (U_n \setminus A_n)$, so that $\mu(U \setminus A) \leq \sum_n \mu(U_n \setminus A_n) < \epsilon$. Let $B_n = \cup_{i=1}^n A_i$. Then $B_n \nearrow A$, and so there exists N such that $\mu(A \setminus B_N) < \epsilon/2$. Then $G_N = \cup_{i=1}^N F_j$ is closed, and $B_N \setminus G_N \subseteq \cup_{i=1}^N (A_i \setminus F_i)$, so that

$$\mu(B_N \setminus G_N) \leq \sum_{i=1}^N \mu(A_i \setminus F_i) < \epsilon/2.$$

Thus $\mu(A \setminus G_N) < \epsilon$. □

Exercise 16.1.5 Suppose that U is an open subset of X and that F is a closed subset of X. Then

$$\mu(U) = \sup\{\int_X f\,d\mu : f \in C(X),\ 0 \leq f \leq I_U\}$$

$$\mu(F) = \inf\{\int_X f\,d\mu : f \in C(X),\ f \geq I_F\}.$$

For the next few exercises, F is the linear span of the indicator functions of closed sets, and G is the linear span of the indicator functions of open sets.

Exercise 16.1.6 Show that if μ is a finite Borel measure on a metrizable space (X, τ) then $C_b(X)$, F and G are each dense in $(L^1(\mu), \|.\|_1)$. If (X, τ) is separable, then $(L^1(\mu), \|.\|_1)$ is separable.

Exercise 16.1.7 Show that if μ is a finite Borel measure on a metrizable space (X, τ) then $C_b(X)$, F and G are each dense in $(L^0(\mu), d_0)$. If (X, τ) is separable, then $(L^0(\mu), d_0)$ is separable.

Exercise 16.1.8 Show that if μ is a finite Borel measure on a metrizable space (X, τ) and Φ is an N-function which satisfies the Δ_2 condition then $C_b(X)$, F and G are each dense in $(L_\Phi(\mu), \|.\|_\Phi)$. If (X, τ) is separable, then $(L_\Phi(\mu), \|.\|_\Phi)$ is separable.

In particular, this result holds for the L^p spaces, for $1 < p < \infty$.

Exercise 16.1.9 Show that if μ is a finite Borel measure on a metrizable space (X, τ) then $C_b(X)$ is closed in $(L^\infty, \|.\|_\infty)$, and is dense if and only if it is finite-dimensional.

We consider an even stronger property. A mapping f from the Borel sets of a metrizable space (X, τ) to $[0, \infty]$ is *tight* if $f(K) < \infty$ for each compact K in X and

$$f(A) = \sup\{f(K) : K \text{ compact }, K \subseteq A\}, \text{ for each } A \in \mathcal{B}(X).$$

Tightness is very powerful, as the next result shows. We consider non-negative functions on the Borel subsets of a metric space (X, τ) which can take infinite values. As before, a mapping $f : \mathcal{B}(X) \to [0, \infty]$ is *additive* if $f(A \cup B) = f(A) + f(B)$ whenever A and B are disjoint, and is σ*-additive* if $f(\cup_{n=1}^\infty A_n) = \sum_{n=1}^\infty f(A_n)$ for each sequence $(A_n)_{n=1}^\infty$ of disjoint Borel sets.

Proposition 16.1.10 *Suppose that f is a non-negative additive tight function on the Borel sets of a metrizable space (X, τ). Then f is σ-additive, and so it is a tight Borel measure on X.*

Proof Suppose that $(A_n)_{n=1}^\infty$ is a sequence of disjoint Borel sets whose union is A. First we consider the case where $f(A) < \infty$. Let $B_n = \cup_{j=1}^n A_j$. Since $B_n \subseteq A$, $\sum_{j=1}^n f(A_j) \leq f(A)$, and so $\sum_{j=1}^\infty f(A_j) \leq f(A)$.

Suppose, if possible that $\sum_{j=1}^\infty f(A_j) = s < f(A)$. Let $\epsilon = (f(A) - s)/2$ and let $C_n = A \setminus B_n$. Thus $f(C_n) \geq 2\epsilon$, for $n \in \mathbf{N}$. By combining blocks of terms, we can suppose that $f(B_n) > s - \epsilon/2^{n+1}$, for $n \in \mathbf{N}$. For each $n \in \mathbf{N}$ there exists a subset K_n of C_n with $f(K_n) > f(C_n) - \epsilon/2^{n+1}$. Let $L_n = \cap_{j=1}^n K_j$.

We show by induction that $f(L_n) \geq (1 + 1/2^n)\epsilon$ (so that $f(C_n \setminus L_n) \leq (1 - 1/2^n)\epsilon$). The result is true when $n = 1$; suppose that it is true for n. Now

$$\begin{aligned} f(C_n \setminus C_{n+1}) &= f(B_{n+1} \setminus B_n) = f(B_{n+1}) - f(B_n) \\ &\leq s - (s - \epsilon/2^{n+1}) = \epsilon/2^{n+1}; \end{aligned}$$

since $C_n \setminus K_{n+1} \subseteq (C_n \setminus C_{n+1}) \cup (C_{n+1} \setminus K_{n+1})$, it follows that

$$f(C_n \setminus K_{n+1}) \leq f(C_n \setminus C_{n+1}) + f(C_{n+1} \setminus K_{n+1}) < \epsilon/2^n.$$

Since $C_n = (C_n \setminus K_{n+1}) \cup (C_n \setminus L_n) \cup (L_n \cap K_{n+1})$

$$\begin{aligned} 2\epsilon &\leq f(C_n) \leq f(C_n \setminus K_{n+1}) + f(C_n \setminus L_n) + f(L_n \cap K_{n+1}) \\ &\leq \epsilon/2^{n+1} + (1 - 1/2^n)\epsilon + f(L_{n+1}), \end{aligned}$$

so that $f(L_{n+1}) \geq (1 + 1/2^{n+1})\epsilon$. This establishes the induction.

But $\bigcap_{n=1}^{\infty} L_n \subseteq \bigcap_{n=1}^{\infty} C_n = \emptyset$; since the sets L_n are compact, it follows that there exists $N \in \mathbf{N}$ for which $L_N = \emptyset$, so that $f(L_n) = 0$, giving a contradiction.

Finally, suppose that $f(A) = \infty$. If $M < \infty$, there exists a compact subset K of A with $f(K) > M$. Then

$$\sum_{n=1}^{\infty} f(A_n) \geq \sum_{n=1}^{\infty} f(A_n \cap K) = f(K) > M,$$

so that $\sum_{n=1}^{\infty} f(A_n) = \infty$. □

Proposition 16.1.11 *A finite Borel measure μ on a metric space (X, d) is* tight *if*

$$\sup\{\mu(K) : K \textit{ compact}, K \subseteq X\} = \mu(X).$$

Proof The condition is certainly necessary. Suppose that it is satisfied. Suppose that A is a Borel set, and that $\epsilon > 0$. There exists a closed set $B \subseteq A$ such that $\mu(B) \geq \mu(A) - \epsilon/2$ and there exists a compact K such that $\mu(K) > \mu(X) - \epsilon/2$. Then $B \cap K$ is a compact subset of A, and $\mu(B \cap K) \geq \mu(A) - \epsilon$. □

Exercise 16.1.12 Suppose that μ is a tight Borel measure on a metrizable space (X, τ). Show that there exists a σ-compact, and therefore separable Borel subset X_0 of X, for which $\mu(X_0) = \mu(X)$.

If μ is a tight measure on a metrizable space, we can prove a result whch goes beyond the monotone convergence theorem. We need a definition. A subset A of a partially ordered set is *directed upwards* if whenever $a_1, a_2 \in A$ there exists $b \in A$ such that $b \geq a_1$ and $b \geq a_2$.

Theorem 16.1.13 *Suppose that μ is a tight Borel measure on a metrizable space (X, τ), and that A is a set of lower semi-continuous function on X which is directed upwards. Let $s(x) = \sup_{a \in A} a(x)$ and let $J = \int_X s\, d\mu$. Then $J = \sup_{a\in A} \int_X a\, d\mu$.*

Proof First observe that s is lower semi-continuous, and so Borel measurable, so that J exists. Let $I = \sup_{a\in A} \int_X a\, d\mu$. Clearly $I \leq J$; we must show that equality holds. An easy induction shows that there exists an increasing sequence $(a_n)_{n=1}^\infty$ in A such that $\int_X a_n\, d\mu \to I$ as $n \to \infty$. Let $h = \sup_n a_n$. By the monotone convergence theorem, $\int_X h\, d\mu = I$. Suppose now that $I < J$, so that, in particular, $I < \infty$. Then there exists a Borel set B and $u < v$ such that $\mu(B) > 0$ and $h(y) < u < v < s(y)$ for $y \in B$. If C is a Borel set, let $\mu_B(C) - \mu(B \cap C)$; μ_B is a tight Borel measure. Let $y \in \operatorname{supp}(\mu_B)$. There exists $a' \in A$ such that $a'(y) > v$, and there exists an open neighbourhood N of y such that $a'(x) > v$ for $x \in N$. By induction, we can find an increasing sequence $(a'_n)_{n=1}^\infty$ in A such that $a'_n \geq a'$ and $a'_n \geq a_n$ for all n. Let $h' = \sup_n a'_n$. Then

$$\begin{aligned} I = \int_X h'\, d\mu &= \int_{B\cap N} h'\, d\mu + \int_{X\setminus(B\cap N)} h'\, d\mu \\ &\geq \int_{B\cap N} h + (v-u)\, d\mu + \int_{X\setminus(B\cap N)} h\, d\mu \\ &= \int_X h\, d\mu + (v-u)\mu(B\cap N) = I + (v-u)\mu(B\cap N), \end{aligned}$$

giving a contradiction, since $\mu(B \cap N) > 0$.

□

16.2 Radon Measures

Suppose that f is a function on the Borel subsets of a metric space X taking values in $[0, \infty]$. f is *locally finite* if for each $x \in X$ there exists a neighbourhood N of x with $f(N) < \infty$.

Proposition 16.2.1 *If $f : \mathcal{B}(X) \to [0, \infty]$ is a locally finite additive function on a metrizable space (X, τ) and K is a compact subset of X then $f(K) < \infty$.*

Proof For each $x \in K$ there exists a neighbourhood N_x of x with $f(N_x) < \infty$. The sets $\{N_x : x \in K\}$ cover K, and so there is a finite subcover. Additivity then ensures that $f(K) < \infty$. □

A *Radon measure* μ on a metrizable space (X, τ) is a tight additive function from $\mathcal{B}(X)$ to $[0, \infty]$ which is locally finite. By Proposition 16.1.10, a Radon measure μ is σ-additive, and, by the preceding property, $\mu(K) < \infty$ if K is compact.

Proposition 16.2.2 *If μ is a Radon measure on a separable metrizable space, then μ is σ-finite; there exists a countable set $\mathcal{W}$ of open sets for which $X = \bigcup_{W\in\mathcal{W}} W$ and $f(W) < \infty$ for all $W \in \mathcal{W}$.*

Proof Let d be a metric on X which defines the topology τ. For each $n \in \mathbf{N}$, let

$$U_n = \{x \in X : \text{there exists } r_x > 1/n \text{ such that } f(N_{r_x}(x)) < \infty\}.$$

If $x \in U_n$ and $d(x,y) < r_x - 1/n$, let $s_y = r_x - d(x,y)$. Then $s_y > 1/n$ and $N_{s_y}(y) \subseteq N_{r_x}(x)$, so that $y \in U_n$: U_n is open. Let C_n be a countable dense subset of U_n, and let $W_n(c) = N_{1/n}(c)$, for $c \in C_n$. Then $f(W_n(c)) < \infty$ for each $c \in C_n$. If $x \in U_n$ then there exists $c \in C_n$ with $d(x,c) < 1/n$, so that $U_n \subseteq \bigcup_{c\in C_n} W_n(c)$. Let $\mathcal{W} = \{W_n(c) : n \in \mathbf{N}, c \in C_n\}$. Then $X = \bigcup_{W\in\mathcal{W}} W$. □

Suppose that X and Y are metric spaces, that $f : X \to Y$ is continuous and that μ is a Radon measure on X. Then the push-forward measure $f_*(\mu)$ need *not* be a Radon measure on Y; let μ be counting measure on $\mathbf{N}$, and let $f : \mathbf{N} \to [0,\infty]$ be the inclusion mapping. Then $f_*(\mu)$ is not locally finite at ∞.

16.3 Borel Measures on Polish Spaces

A Borel measure on a compact metric space is tight. More generally, if (X,d) is a σ-compact metric space, then every Borel measure on X is tight. We can say more.

Theorem 16.3.1 (Ulam's theorem) *A finite Borel measure on a Polish space (X,τ) is tight.*

Proof We give two proofs of this important theorem. By Theorem 3.1.1 there is a homeomorphism j of X onto a G_δ subset Y of the Hilbert cube $\mathcal{H}$. Let $j_*(\mu)$ be the push-forward measure on $\mathcal{H}$. Since $\mathcal{H}$ is compact, $j_\star(\mu)$ is tight. Thus

$$j_*(\mu)(Y) = \sup\{j_*(\mu)(K) : K \text{ compact}, K \subseteq Y\};$$

since j is a homeomorphism,

$$\mu(X) = \sup\{\mu(K) : K \text{ compact}, K \subseteq X\}.$$

Here is the direct proof, given by Ulam. Let d be a complete metric on X which defines the topology τ. Let $(c_j)_{j=1}^\infty$ be a dense sequence in X, and let

$M_{j,n} = \{x \in X : d(x, c_j) \leq 1/n\}$. Let $A_{j,n} = \cup_{i=1}^{j} M_{i,n}$. Suppose that $\epsilon > 0$. If $n \in \mathbf{N}$, each $A_{j,n}$ is closed, and $A_{j,n} \nearrow X$ as $j \to \infty$; thus there exists J_n such that if $C_n = A_{J_n,n}$ then $\mu(C_n) > (1 - \epsilon/2^n)\mu(X)$. Further, $(M_{j,n})_{j=1}^{J_n}$ is a finite $1/n$-net which covers the closed set C_n. Let $D_n = \cap_{j=1}^{n} C_j$. Then $(D_n)_{n=1}^{\infty}$ is a decreasing sequence of closed sets, and $\mu(D_n) > (1 - (1 - 1/2^n)\epsilon)\mu(X) > (1 - \epsilon)\mu(X)$, for each $n \in \mathbf{N}$. Let $D = \cap_{n=1}^{\infty} D_n$. Then $\mu(D) \geq (1 - \epsilon)\mu(X)$ and D is closed and totally bounded, and is therefore compact. □

Corollary 16.3.2 *If μ is a σ-finite Borel measure on a Polish space (X, τ), then there exists an increasing sequence $(K_n)_{n=1}^{\infty}$ of compact sets such that $\mu(X \setminus \cup_{n=1}^{\infty} K_n) = 0$, and μ is tight.*

Proof Let (A_n) be an increasing sequence of sets of finite measure such that $\mu(X \setminus \cup_{n=1}^{\infty} A_n) = 0$. Consider the finite measures $I_{A_n}.d\mu$; there exists an increasing sequence $(K_n)_{n=1}^{\infty}$ of compact sets such that $\mu(X \setminus \cup_{n=1}^{\infty} K_n) = 0$. If A is a Borel set, then $\mu(A) = \lim_{n\to\infty} \mu(A \cap K_n)$. Since $\mu(A \cap K_n) = \sup\{\mu(K) : K \text{ compact } K \subseteq A \cap K_n\}$, the result follows. □

If (X, τ) is a Polish space, then Radon measures can be defined locally.

Theorem 16.3.3 *Suppose that (X, τ) is a Polish space, and that $(U_i)_{i=1}^{\infty}$ is a sequence of open sets in X which covers X; $\cup_{i=1}^{\infty} U_i = X$. Suppose that for each i, μ_i is a finite measure on the Borel sets of U_i and that these measures are compatible; if A is a Borel set of $U_i \cap U_j$ then $\mu_i(A) = \mu_j(A)$. Then there exists a unique Radon measure π on X for which $\pi(A) = \mu_i(A)$ for each Borel set A in of U_i, for each $i \in \mathbf{N}$.*

Proof Let $V_j = \cup_{i=1}^{j} U_i$. The compatiblity condition ensures that we can define a finite positive Borel measure ν_j on V_j such that $\nu_i(A) = \mu_i(A)$ if $1 \leq i \leq j$ and $A \subseteq U_i$. Further, if A is a Borel subset of V_j and $j \leq k$ then $\nu_j(A) = \nu_k(A)$. If A is a Borel subset of X, $(\nu_j(A \cap V_j))_{j=1}^{\infty}$ is an increasing sequence; let $\pi(A) = \lim_{n\to\infty} \nu_j(A \cap V_j)$; then it is easily verified that π is tight, locally finite and additive. □

16.4 Lusin's Theorem

Borel measurable functions on a Polish space are well-behaved.

Theorem 16.4.1 (Lusin's theorem) *Suppose that X and Y are Polish spaces, that μ is a finite Borel measure on X, that $f : X \to Y$ is Borel measurable and that $\epsilon > 0$. Then there exists a compact subset K of X, with $\mu(X \setminus K) < \epsilon$, such that the restriction of f to K is continuous.*

Proof Let d be a metric on Y which defines the topology of Y, and let $(y_n)_{n=1}^\infty$ be a dense sequence in Y. Suppose that $j \in \mathbf{N}$. Let

$$A_{n,j} = \{x \in X : d(f(x), y_n) < 1/j\},$$
$$B_{n,j} = A_{n,j} \setminus (\cup_{m=1}^{n-1} A_{m,j}),$$
$$C_{n,j} = \cup_{m=1}^{n} A_{m,j} = \cup_{m=1}^{n} B_{m,j}.$$

Then $(C_{n,j})_{n=1}^\infty$ is an increasing sequence of Borel subsets of X whose union is X, and so there exists N_j such that $\mu(X \setminus C_{N_j}) < \epsilon/2^{j+1}$. For each $1 \leq n \leq N_j$ there exists a compact subset $K_{n,j}$ of $B_{n,j}$ such that $\mu(B_{n,j} \setminus K_{n,j}) < \epsilon/2^{j+1}N_j$. Let $K_j = \cup_{n=1}^{N_j} K_{n,j}$. Then $\mu(X \setminus K_j) < \epsilon/2^j$. If $x \in K_{n,j}$, let $f_j(x) = y_n$. Since the sets $K_{n,j}$ are disjoint and closed, f_j is a continuous function on K_j. Further, $d(f_j(x), f(x)) < 1/j$ for $x \in K_j$.

Now let $K = \cap_{j=1}^\infty K_j$. Then K is a compact subset of X and $\mu(X \setminus K) \leq \sum_{j=1}^\infty \mu(X \setminus K_j) < \epsilon$. The restriction of each f_j to K is continuous, and $f_j \to f$ uniformly on K as $j \to \infty$, and so f is continuous on K. □

In these circumstances, we can improve on Egorov's theorem.

Corollary 16.4.2 *Suppose that $(f_i)_{i=1}^\infty$ is a sequence of Borel measurable functions from $X \to Y$, that $f_i \to f$ almost everywhere as $i \to \infty$ and that $\epsilon > 0$. Then there exists a compact subset K of X with $\mu(X \setminus K) < \epsilon$ such that each f_i is continuous on K and such that $f_i \to f$ uniformly on K as $i \to \infty$.*

Proof By Egorov's theorem, there exists a Borel subset E of X with $\mu(X \setminus E) < \epsilon/2$, such that $f_i \to f$ uniformly on E, and there exists a compact subset L of E with $\mu(X \setminus L) < \epsilon/2$. For each i there exists a compact subset K_i of L with $\mu(L \setminus K_i) < \epsilon/2^{i+1}$ such that f_i is continuous on K_i. Then $K = \cap_{i=1}^\infty K_i$ has the required properties. □

Corollary 16.4.3 *Suppose that f is a non-negative real-valued Borel measurable function on X. Then*

$$\int_X f\,d\mu = \sup\{\int_K f\,d\mu : K \text{ compact, } K \subseteq X, f_{|K} \text{ continuous}\}.$$

Proof There exists an increasing sequence $(K_j)_{j=1}^\infty$ of compact sets of X, with $\mu(X \setminus K_j) \to 0$ as $j \to \infty$, such that $f_{|K_j}$ is continuous. Let $f_j(x) = f(x)$ for $x \in K_j$ and $f_j(x) = 0$ otherwise. Then

$$\int_X f\,d\mu = \lim_{j\to\infty} \int_X f_j\,d\mu = \lim_{j\to\infty} \int_{K_j} f\,d\mu,$$

by monotone convergence. The result follows from this. □

Exercise 16.4.4 Suppose that A is a closed subset of $[0,1]$, and that μ is a Borel measure on $[0,1]$. Using the fact that ∂C is countable if C is closed, prove Lusin's theorem for I_A. Use this, and tightness, to prove Lusin's theorem for a Borel function on $[0,1]$.

16.5 Measures on the Bernoulli Sequence Space $\Omega(\mathbf{N})$

So far, we have not constructed any measures. Here we show how to construct Borel measures on $\Omega(\mathbf{N})$. This is rather simpler than constructing Borel measures on $\mathbf{R}$, because the geometry and topology is technically simpler. We can then consider measures on other spaces, such as $\mathbf{R}$, as push-forward measures of measures on $\Omega(\mathbf{N})$. We begin by considering the set $\mathrm{Cyl}(\Omega)$ of cylinder sets in Ω. Recall that the cylinder set $C_{\epsilon,m}$ of rank m is the set

$$C_{\epsilon,m} = \{\omega \in \Omega(\mathbf{N}) : \omega_j = \epsilon_j \text{ for } 1 \le j \le m\}.$$

Recall also that each cylinder set is open and closed, and that $\mathrm{Cyl}(\Omega)$ is a countable base for the topology of Ω.

Proposition 16.5.1 *If $C_{\epsilon,m}$ and $C_{\eta,n}$ are cylinder sets, and $m \le n$, then either $C_{\eta,n} \subseteq C_{\epsilon,m}$ or $C_{\eta,n} \cap C_{\epsilon,n} = \emptyset$.*

Proof If $\eta_j = \epsilon_j$ for $1 \le j \le m$, then $C_{\eta,n} \subseteq C_{\epsilon,m}$. Otherwise, $C_{\eta,n} \cap C_{\epsilon,m} = \emptyset$. □

Corollary 16.5.2 *If U is an open subset of Ω, then U is union of a disjoint sequence of cylinder sets.*

Corollary 16.5.3 *If U is an open and closed subset of Ω, and C is a set of disjoint cylinder sets whose union is U, then C is finite.*

Proof For C is an open cover of the compact set U. □

A cylinder set C of rank n is the disjoint union of two cylinder sets

$$C^{(0)} = \{x \in C : x_{n+1} = 0\} \text{ and } C^{(1)} = \{x \in C : x_{n+1} = 1\}$$

of rank $n+1$. We denote by $A(\Omega)$ the set of real-valued functions α on $\mathrm{Cyl}(\Omega)$ which satisfy

$$\alpha(C) = \alpha(C^{(0)}) + \alpha(C^{(1)})$$

for each $C \in \mathrm{Cyl}(\Omega)$. Elements of $A(\Omega)$ are called *dyadic martingales*. We also set $\alpha(\emptyset) = 0$, and set $A^+(\Omega) = \{\alpha \in A(\Omega) : \alpha \ge 0\}$.

Proposition 16.5.4 *Suppose that $\mathcal{C}$ is a set of disjoint cylinder sets whose union is the cylinder set D. If $\alpha \in A^+(\Omega)$ then $\alpha(D) = \sum_{C\in\mathcal{C}} \alpha(C)$. Consequently, $0 \le \alpha(D) \le \alpha(\Omega)$.*

Proof $|\mathcal{C}|$ is finite. The proof is by induction on $|\mathcal{C}|$. The result is true if $|\mathcal{C}| = 1$ or 2. Suppose that it holds when $|\mathcal{C}| = j$, and that $|\mathcal{C}| = j + 1$. There exists $C' = C_{\epsilon,n} \in \mathcal{C}$ with n maximal. Then $C_{\epsilon,n-1} = C^{(0)}_{\epsilon,n-1} \cup C^{(1)}_{\epsilon,n-1} = C' \cup C''$; then $C'' \subseteq D$, so that $C'' \in \mathcal{C}$. Replace C' and C'' by $C_{\epsilon,n-1}$ to obtain a new set $\mathcal{C}'$. Since $\alpha(C_{\epsilon,n-1}) = \alpha(C') + \alpha(C'')$, $\sum_{C\in\mathcal{C}} \alpha(C) = \sum_{C\in\mathcal{C}'} \alpha(C) = \alpha(D)$. □

If $\mu \in M^+(\Omega) = M^+(\Omega, \mathcal{B})$, let $j(\mu)(C) = \mu(C)$. Then j maps $M^+(\Omega)$ into $A^+(\Omega)$. Since $\mathcal{B} = \sigma(\mathrm{Cyl}(\Omega))$, j is injective. More importantly, it is also surjective.

Theorem 16.5.5 *The mapping $j : M^+(\Omega) \to A^+(\Omega)$ is bijective.*

Proof The proof is similar to, but simpler than, the proof of the existence of Lebesgue measure, and Lebesgue–Stieltjes measures.

Suppose that $\alpha \in A^+(\Omega)$. If U is an open subset of Ω, $U = \cup_i C_i$, where (C_i) is a finite or infinite sequence of disjoint cylinder sets. We set $l_\alpha(U) = \sum_j \alpha(C_j)$. We must show that this is well-defined. Suppose that (D_j) is another finite or infinite sequence of disjoint cylinder sets whose union is U. Then $U = \cup_{i,j}(C_i \cap D_j) = \cup_k E_k$, where (E_k) is a finite or infinite sequence of disjoint cylinder sets. For each i, $C_i = \cup_{k\in F_i} E_k$, where F_i is a finite set of indices. But then $l_\alpha(C_i) = \sum_{k\in F_i} l_\alpha(E_k)$, so that

$$\sum_i l_\alpha(C_j) = \sum_i \Big(\sum_{k\in F_i} l_\alpha(E_k) \Big) = \sum_k l_\alpha(E_k).$$

Similarly $\sum_j l_\alpha(D_j) = \sum_k l_\alpha(E_k)$, and so $l(U)$ is well-defined.

Here are the basic properties of the function l_α.

Lemma 16.5.6 *Suppose that U, $(U_n)_{n=1}^\infty$ and V are open subsets of Ω, and that $U = \cup_{n=1}^\infty U_n$.*

(i) *If $U_n \cap U_m = \emptyset$ for $m \ne n$ then $l_\alpha(U) = \sum_{n=1}^\infty l_\alpha(U_n)$.*
(ii) *If $V \subseteq U$ then $l_\alpha(V) \le l_\alpha(U)$.*
(iii) *If $(U_n)_{n=1}^\infty$ is an increasing sequence, then $l_\alpha(U) = \lim_{n\to\infty} l_\alpha(U_n)$.*
(iv) *$l_\alpha(U \cup V) + l_\alpha(U \cap V) = l_\alpha(U) + l_\alpha(V)$.*
(v) *$l_\alpha(U) \le \sum_{n=1}^\infty l_\alpha(U_n)$.*

Proof Let $U = \cup_i C_i$, where (C_i) is a sequence of disjoint cylinder sets, and let $V = \cup_j D_j$, where (D_j) is a sequence of disjoint cylinder sets.

(i) For each n, $U_n = \cup_{i=1}^\infty C_{n,i}$, where $(C_{n,i})$ is a sequence of disjoint cylinder sets. Then $U = \cup_{n,i} C_{n,i}$ so that

$$l_\alpha(U) = \sum_{n=1}^{\infty}\left(\sum_i l_\alpha(C_{n,i})\right) = \sum_{n=1}^{\infty} l_\alpha(U_n).$$

(ii) $V = \cup_{i,j}(D_j \cap C_i)$, so that

$$l_\alpha(V) = \sum_i \left(\sum_j \alpha(D_j \cap C_i)\right) \leq \sum_i \alpha(C_i) = l_\alpha(U).$$

(iii) Since $U_n \subseteq U$ for each n, $\lim_{n\to\infty} l_\alpha(U_n) \leq l_\alpha(U)$. For each i, $\{U_n\}$ is an open cover of C_i, and so there exists n_i such that $C_i \subseteq U_{n_i}$. Let $m_j = \max_{i\leq j} n_i$. Then $\cup_{1\leq j\leq i} C_j \subseteq U_{m_j}$, so that $\sum_{i=1}^{j} \alpha(C_i) \leq l_\alpha(U_{m_j})$. Thus $l_\alpha(U) \leq \lim_{n\to\infty} l_\alpha(U_n)$.

(iv) By (iii), it is enough to prove the result when $U = \cup_{i=1}^{m} C_i$ and $V = \cup_{j=1}^{n} D_j$ are unions of finitely many disjoint cylinder sets. We prove this by induction on $m+n$. The result is true when $m+n=2$. Suppose that it is true when $m+n=k$, and that $m+n=k+1$. Without loss of generality, we can suppose that $m \geq 2$. Let $U' = \cup_{i=1}^{m-1} C_i$. We consider three possibilities.

First, suppose that $C_m \cap V = \emptyset$. Then

$$U \cup V = (U' \cup V) \cup C_m \text{ and } U \cap V = U' \cap V,$$

so that

$$l_\alpha(U \cup V) + l_\alpha(U \cap V) = l_\alpha(U') + l_\alpha(V) + l_\alpha(C_m) = l_\alpha(U) + l_\alpha(V).$$

Secondly, suppose that $C_m \subseteq D_j$ for some j. Then

$$U \cup V = U' \cup V \text{ and } U \cap V = (U' \cap V) \cup C_m,$$

so that

$$l_\alpha(U \cup V) + l_\alpha(U \cap V) = l_\alpha(U') + l_\alpha(V) + l_\alpha(C_m) = l_\alpha(U) + l_\alpha(V).$$

Thirdly, suppose that $C_m \cap V \neq \emptyset$, but that $C_m \nsubseteq D_j$, for any j. Let $V' = V \setminus C_m$. Then

$$U \cup V = U \cup V' \text{ and } U \cap V = (U \cap V') \cup (V \cap C_m),$$

so that

$$l_\alpha(U \cup V) + l_\alpha(U \cap V) = l_\alpha(U) + l_\alpha(V') + l_\alpha(V \cap C_m) = l_\alpha(U) + l_\alpha(V).$$

(v) Let $W_n = \cup_{i=1}^{n} U_i$. Then, using (iii) and (iv),

$$l_\alpha(U) = \lim_{n\to\infty} l_\alpha(W_n) \leq \lim_{n\to\infty} \sum_{i=1}^{n} l_\alpha(U_i) = \sum_{n=1}^{\infty} l_\alpha(U_n).$$

□

Suppose now that A is a subset of Ω. We define the *α-outer measure* $\mu_\alpha^*(A)$ to be

$$\mu_\alpha^*(A) = \inf\{l_\alpha(U) : U \text{ open}, A \subseteq U\}.$$

The function μ_α^* has the following properties.

Theorem 16.5.7 *Suppose that A, $\{A_n : n \in \mathbf{N}\}$ and B are subsets of Ω, that $A = \cup_{n\in\mathbf{N}} A_n$ and that U is an open subset of Ω.*

(i) $\mu_\alpha^*(U) = l_\alpha(U)$.
(ii) *If $A \subseteq B$ then* $\mu_\alpha^*(A) \leq \mu_\alpha^*(B)$.
(iii) $\mu_\alpha^*(A) \leq \sum_{n=1}^\infty \mu_\alpha^*(A_n)$.
(iv) $\mu_\alpha^*(A \cup B) + \mu_\alpha^*(A \cap B) \leq \mu_\alpha^*(A) + \mu_\alpha^*(B)$.

Proof (i) and (ii) follow immediately from Lemma 16.5.6 (ii).

(iii) Suppose that $\epsilon > 0$. For each $n \in \mathbf{N}$ there exists an open U_n with $A_n \subseteq U_n$ for which $l_\alpha(U_n) < \mu_\alpha^*(A_n) + \epsilon/2^n$. Then

$$\mu_\alpha^* A) \leq l_\alpha(\cup_{n=1}^\infty U_n) \leq \sum_{n=1}^\infty l_\alpha(U_n) \leq \sum_{n=1}^\infty \mu_\alpha^*(A_n) - \epsilon,$$

by Lemma 16.5.6 (v). Since ϵ is arbitrary, the result follows.

(iv) Suppose that $\epsilon > 0$. There exist open sets U and V with $A \subseteq U$, $B \subseteq V$ and $l_\alpha(U) \leq \mu_\alpha^*(A) + \epsilon/2$, $l_\alpha(V) \leq \mu_\alpha^*(B) + \epsilon/2$. Then

$$\begin{aligned}\mu_\alpha^*(A) + \mu_\alpha^*(B) &\geq l_\alpha(U) + l_\alpha(V) - \epsilon = l_\alpha(U \cup V) + l_\alpha(U \cap V) - \epsilon \\ &\geq \mu_\alpha^*(A \cup B) + \mu_\alpha^*(A \cap B) - \epsilon,\end{aligned}$$

by Lemma 16.5.6 (iv). Since ϵ is arbitrary, the result follows. □

In particular, if $A \subseteq \Omega$ then $\mu_\alpha^*(A) + \mu_\alpha^*(\Omega \setminus A) \geq \mu_\alpha^*(\Omega) = l_\alpha(\Omega)$. We say that A is *α-measurable* if $\mu_\alpha^*(A) + \mu_\alpha^*(\Omega \setminus A) = \mu_\alpha^*(\Omega) = l_\alpha(\Omega)$, and denote the set of α-measurable sets by Σ_α. If $A \in \Sigma$, we set $\mu_\alpha(A) = \mu_\alpha^*(A)$. If C is a cylinder set then $C \in \Sigma_\alpha$. If $A, B \in \Sigma$, let $A' = \Omega \setminus A$ and $B' = \Omega \setminus B$. Then $A', B' \in \Sigma_\alpha$ and

$$\begin{aligned}\mu_\alpha^*(A \cup B) + \mu_\alpha^*(A \cap B) &\geq 2l_\alpha(\Omega) - \mu_\alpha^*(A' \cap B') - \mu_\alpha^*(A' \cup B') \\ &\geq 2l_\alpha(\Omega) - \mu_\alpha(A') - \mu_\alpha(B') \\ &= \mu_\alpha(A) + \mu_\alpha(B) \geq \mu_\alpha^*(A \cup B) + \mu_\alpha^*(A \cap B),\end{aligned}$$

and so there is equality throughout. Thus

$$\begin{aligned}&\mu_\alpha^*(A \cup B) + \mu_\alpha^*(\Omega \setminus (A \cup B)) = l_\alpha(\Omega) \\ \text{and } &\mu_\alpha^*(A \cap B) + \mu_\alpha^*(\Omega \setminus (A \cap B)) = l_\alpha(\Omega),\end{aligned}$$

and so $A \cup B$ and $A \cap B$ are in Σ_α. Consequently $A \setminus B \in \Sigma_\alpha$, and $\mu_\alpha(A) = \mu_\alpha(A \setminus B) + \mu_\alpha(A \cap B)$. Suppose that $(A_n)_{n=1}^\infty$ is a sequence of disjoint elements of Σ_α, and that $A = \cup_{n=1}^\infty A_n$. Let $A'_n = \Omega \setminus A_n$. Then

$$\begin{aligned}\mu_\alpha^*(A) &\leq \sum_{n=1}^\infty \mu_\alpha(A_n) = \sum_{n=1}^\infty (l_\alpha(\Omega) - \mu_\alpha(A')) \\ &\leq l_\alpha(\Omega) - \mu_\alpha^*(A) \leq \mu_\alpha^*(A),\end{aligned}$$

and so there is equality throughout; $A \in \Sigma_\alpha$, and $\mu_\alpha(A) = \sum_{n=1}^\infty \mu_\alpha(A_n)$.

Consequently, Σ_α is a σ-field, and μ_α is a finite measure on it. Since the cylinder sets are in Σ_α, Σ_α contains the Borel σ-field $\mathcal{B}(\Omega)$ of Ω. If we also denote the restriction of μ_α to $\mathcal{B}(\Omega)$ by μ_α, then $j(\mu_\alpha) = \alpha$; the mapping j is a bijection of $M^+(\Omega)$ onto $A(\Omega)$. □

Note that $(\Omega, \Sigma_\alpha, \mu_\alpha)$ is a complete measure space; it is the completion of $(\Omega, \mathcal{B}(\Omega), \mu_\alpha)$.

We can now push forward measures on $\Omega(\mathbf{N})$ to construct measures on compact metrizable spaces. Here are some examples.

Exercise 16.5.8 Let $\alpha(C_{\epsilon,m} = 2^{-m})$ for each ϵ. Verify that $\alpha \in A(\Omega)$. Let γ be the corresponding Borel measure. $\Omega(\mathbf{N})$ is a compact Abelian topological group, under co-ordinatewise addition mod 2. Show that $\gamma(B + \epsilon) = \gamma(B)$ for every Borel set B and $\epsilon \in \Omega(\mathbf{N})$.

Exercise 16.5.9 If $\epsilon \in \Omega(\mathbf{N})$ let $f(\epsilon) = \sum_{n=1}^\infty \epsilon_n/2^n$. Show that f maps $\Omega(\mathbf{N})$ continuously onto $[0, 1]$. Let $\lambda = f_*(\gamma)$. Show that λ is a Borel measure *(Lebesgue measure) on* $[0, 1]$ *for which* $\lambda([a, b]) = b - a$ *for each interval* $[a, b]$.

We can extend Lebesgue measure to the whole real line; if A is a Borel set in $\mathbf{R}$, we set $\lambda(A)$ to be $\sum_{n=-\infty}^\infty \lambda((A - n) \cap [0, 1])$.

Exercise 16.5.10 If $\epsilon \in \Omega(\mathbf{N})$ let $g(\epsilon) = 2\sum_{n=1}^\infty \epsilon_n/3^n$. Show that g maps $\Omega(\mathbf{N})$ continuously onto the Cantor set. Let $\mu = g_*(\gamma)$. Show that λ and μ are mutually singular.

16.6 The Riesz Representation Theorem

Suppose that (X, τ) is a Polish space. If $\mu = \mu^+ - \mu^-$ is a signed Borel measure on X and $f \in C_b(X)$, we set $j(\mu)(f) = \int_X f\,d\mu^+ - \int_X f\,d\mu^-$. We write $\int_X f\,d\mu$ for $\int_X f\,d\mu^+ - \int_X f\,d\mu^-$.

Proposition 16.6.1 *Suppose that (X,τ) is a Polish space. The mapping*

$$j : (M(X,\mathcal{B}), \|.\|_{TV}) \to (C_b(X)', \|.\|')$$

is a linear isometry of $(M(X,\mathcal{B}), \|.\|_{TV})$ into $(C(X)', \|.\|')$, and $j(M(X,\mathcal{B})$ is weak dense in $C_b(X)'$.*

Proof It is clear that $j(\alpha\mu) = \alpha\mu$ for $\alpha \geq 0$ and that $j(-\mu) = -\mu$. If $\mu_1, \mu_2 \in M(X)$, then, since

$$\mu_1^+ + \mu_2^+ = (\mu_1 + \mu_2)^+ + \nu \text{ and } \mu_1^+ - \mu_2^- = (\mu_1 + \mu_2)^- - \nu,$$

for some $\nu \in M^+(X)$, it follows from the definition that $j(\mu_1 + \mu_2) = j(\mu_1) + j(\mu_2)$; j is linear. Since $|j(\mu(f))| \leq \int_X |f|\,d\mu \leq \|f\|_\infty . \|\mu\|_{TV}$, j is norm-decreasing. If $\mu \in M(X)$, there exist disjoint Borel sets P and N, such that $\mu^+(P) = \mu^+(X)$ and $\mu^-(P) = \mu^-(X)$. If $\epsilon > 0$, there exist a closed subset K of P and a closed subset L of N such that $\mu^+(K) > \mu^+(P) - \epsilon/4$ and $\mu^-(L) > \mu^-(N) - \epsilon/4$. There exists $f \in C(X)$ with $\|f\|_\infty \leq 1$ for which $f(x) = 1$ for $x \in K$ and $f(x) = -1$ for $x \in L$. Then $j(\mu)(f)) \geq \|\mu\|_{TV} - \epsilon$; since ϵ is arbitrary $\|j(\mu)\|' \geq \|\mu\|_{TV}$. Thus j is an isometry.

Since $\|f\|_\infty = \sup\{\int_X f\,d\delta_x : x \in X\} = \sup\{\int_X f\,d\mu : \mu \in M_1(X,\mathcal{B})\}$, it follows from the theorem of bipolars that $j(M_1(X,\mathcal{B}))$ is weak* dense in $B(E')$. Consequently, $j(M(X,\mathcal{B}))$ is weak* dense in $C_b(X)'$. □

When (X,τ) is compact, the mapping j is surjective; every continuous linear function is represented by a signed Borel measure on X.

Theorem 16.6.2 (The Riesz representation theorem for compact metrizable spaces) *Suppose that (X,τ) is a compact metrizable topological space. The mapping $j : (M(X,\mathcal{B}), \|.\|_{TV}) \to (C(X)', \|.\|')$ is a linear isometry of $(M(X,\mathcal{B}), \|.\|_{TV})$* **onto** *$(C(X)', \|.\|')$.*

Proof First we consider the case where $X = \Omega$. Suppose that $\phi \in C(\Omega)'$. If C is a cylinder set, let $\alpha(C) = \phi(I_C)$. Then $\alpha \in A(\Omega)$ and $|\alpha(C)| \leq \|\phi\|'$. Let $\alpha^+(C) = \sup\{\phi(f) : 0 \leq f \leq I_C\}$. Then $0 \leq \alpha^+(C) \leq \|\phi\|'$. Since $C^{(0)}$ and $C^{(1)}$ are disjoint, it follows that $\alpha(C) = \alpha(C^{(0)}) + \alpha(C^{(1)})$; thus $\alpha^+ \in A^+(\Omega)$. By Theorem 16.5.5, there exists a positive Borel measure μ^+ on Ω such that if C is a cylinder set then $\alpha^+(C) = \mu^+(C)$. The function $\alpha^- = \alpha^+ - \alpha$ is also in $A^+(\Omega)$, and so there exists a positive Borel measure μ^- on Ω such that if C is a cylinder set then $\alpha^-(C) = \mu^-(C)$. Then $\mu = \mu^+ - \mu^-$ is a signed Borel measure on Ω, and $j(\mu) = \phi$.

Now we consider the general case. By Theorem 3.3.11, there exists a continuous surjection ψ of Ω onto X. If $f \in C(X)$, let $T_\psi(f) = f \circ \psi$. Then T_ψ is an isometric isomorphism of $C(X)$ into $C(\Omega)$, and so, by the Hahn–Banach theorem, T'_ψ is a surjection of $C(\Omega)'$ onto $C(X)'$. Thus if $\phi \in C(X)'$, there exists $\theta \in C(\Omega)'$ with $T'_\psi(\theta) = \phi$. There exists a signed Borel measure $\mu = \mu^+ - \mu^-$ on Ω which represents θ. Let $\nu = \psi_{\mu^+}$ and $\pi = \psi_{\mu^-}$ be the push-forward Borel measures on X. If $f \in C(X)$ then

$$\phi(f) = T'_\psi(\theta)(f) = \theta(f \circ \phi) = \int_\Omega (f \circ \phi)\, d\mu^+ - \int_\Omega (f \circ \phi)\, d\mu^-$$
$$= \int_X f\, d\nu - \int_X f\, d\pi = j(\nu - \pi)(f),$$

so that $\phi = j(U - \pi)$. Let $\rho = \nu - \pi$. Of course π need not equal ρ^+, nor need π equal ρ^-. □

Corollary 16.6.3 *$(M(X), \|.\|_{TV})$ is a Banach space.*

The weak* topology $\sigma^*(M(X, \mathcal{B}), C(X))$ is denoted by w, and called the w topology; it is also, misleadingly, called the *weak topology* on $M(X, \mathcal{B})$. As an immediate consequence of Banach's theorem, we have the following corollary.

Corollary 16.6.4 *$M_1(X, \mathcal{B})$ is w-compact and metrizable, and is therefore w-separable.*

Exercise 16.6.5 Suppose that $\phi : (K, \tau) \to (L, \sigma)$ is a continuous mapping from a compact metrizable space (K, τ) to a compact metrizable space (L, σ). If $f \in C(L)$, let $T_\phi(f)(x) = f(\phi(x))$, for $x \in K$. Show that $T_\phi(f) \in C(K)$, and $\|T_\phi(f)\|_\infty \leq \|f\|_\infty$. If $\mu \in M(K)$, show that $T'_\phi(\mu) = \phi_*(\mu)$, the push-forward measure of μ. If ϕ is surjective, show that T_ϕ is an isometry, and T'_ϕ maps $M(K)$ onto $M(L)$.

In fact, it is convenient to have a more explicit description of separability.

Proposition 16.6.6 *Suppose that (X, τ) is a compact metrizable space. Let C be a countable dense subset of X. Then the countable set*

$$A_C = \left\{ \sum_{i=1}^{n} \lambda_i \delta_{c_i} : n \in \mathbf{N}, \lambda_i \in \mathbf{Q}, \sum_{i=1}^{n} |\lambda_i| \leq 1, c_i \in C \right\}$$

is w-dense in $M_1(X)$.

Proof Suppose that $\mu \in M(X)$, that F is a finite subset of $C(X)$ and that $\epsilon > 0$. Let d be a metric on X which defines the topology of X, and let $M = \max\{\|f\|_\infty : f \in F\}$. Since each f is uniformly continuous on X, there exists $\delta > 0$ such that if $d(x, y) < \delta$ then $|f(x) - f(y)| < \epsilon/2$, for $f \in F$. Since X is totally bounded, there exists a finite partition $\mathcal{D}$ of X into non-empty Borel sets, each of diameter at most $\delta/2$. For each $D \in \mathcal{D}$ there exists $c_D \in C$ such that $d(c_D, y) < \delta$ for $y \in D$. Now let $\nu = \sum_{D \in \mathcal{D}} \mu(D)\delta_{c_D}$. If $f \in F$ then

$$\begin{aligned}\left|\int_X f\,d\mu - \int_X f\,d\nu\right| &= \left|\sum_{D\in\mathcal{D}}\left(\int_D f\,d\mu - \mu(D)f(c_D)\right)\right| \\ &= \left|\sum_{D\in\mathcal{D}}\int_D (f - f_D)\,d\mu\right| \\ &\le (\epsilon/2)\sum_{D\in\mathcal{D}} |\mu(D)| \le \epsilon/2.\end{aligned}$$

For each $D \in \mathcal{D}$ there exists $\lambda_D \in \mathbf{Q}$ such that $|\lambda_D - \mu(D)| < \epsilon/2(M+1)$ and such that $\sum_{D\in\mathcal{D}} |\lambda_D| \le 1$. Let $\pi = \sum_{D\in\mathcal{D}} \lambda_D\delta_{c_D}$. If $f \in F$ then

$$\left|\int_X f\,d\mu - \int_X f\,d\pi\right| = \left|\sum_{D\in\mathcal{D}} (\mu(D) - \lambda_D)f(c_D)\right| \le \epsilon/2,$$

so that $|\int_X f\,d\mu - \int_X f\,d\pi| < \epsilon$. □

Exercise 16.6.7 Suppose that (X,τ) is a compact metrizable space. Bearing in mind that $(C(X), \le)$ is a Riesz space with order unit 1, establish the following.

(i) If ϕ is a positive linear functional on $C(X)$ then there exists a unique positive Borel measure μ on Ω such that $\phi(f) = \int_X f\,d\mu$, for all $f \in C(X)$.
(ii) If $\phi \in C(X)'$ there exists a Borel probability measure μ on $C(X)$ such that $\phi(f) = \int_X f\,d\mu$, for all $f \in C(X)$ if and only if $\phi(1) = 1 = \|\phi\|'$.

The set $P(X) = \{\mu \in M_1(X), \mu(X) = 1\}$ is a w-closed convex subset of $M_1(X)$ and is therefore w-compact.

Exercise 16.6.8 Suppose that (X,τ) is a compact metrizable space. Let C be a countable dense subset of X. Show that the countable set

$$A_C = \left\{\sum_{i=1}^n \lambda_i\delta_{c_i} : n \in \mathbf{N}, \lambda_i \in \mathbf{Q}, \lambda_i \ge 0, \sum_{i=1}^n \lambda_i = 1, c_i \in C\right\}$$

is w-dense in $P(X)$.

16.7 The Locally Compact Riesz Representation Theorem

We now extend the Riesz representation theorem in a straightforward way to a locally compact Polish space (X,τ) which is not compact. By Theorem 3.4.1, (X,τ) is σ-compact, and there exists an increasing sequence $(K_n)_{n=1}^\infty$ of compact subsets of X such that $K_n \subseteq K_{n+1}^{int}$ for each $n \in \mathbf{N}$, and such that $X = \cup_{n=1}^\infty K_n$.

Recall that if f is a continuous real-valued function on a topological space (X,τ), the *support* of f is the closure of the set $\{x \in X : f(x) \ne 0\}$. If $s(f)$ is the support of f, then $X \setminus s(f)$ is the largest open subset of X on which f is equal to 0. If (X, τ) is a locally compact Polish space, we set $C_c(X)$ to be the set of all continuous real-valued functions on X with compact support. $C_c(X)$ is a Riesz space, under the usual pointwise ordering. If μ is a Radon measure on X, and $f \in C_c(X)$, let $j(\mu)(f) = \int_X f\,d\mu$. Then $j(\mu)$ is a positive linear functional on $C_c(X)$. The Riesz representation theorem for locally compact Polish spaces says that the converse is true.

Theorem 16.7.1 (The Riesz representation theorem for locally compact Polish spaces) *Suppose that (X, τ) is a locally compact Polish space which is not compact, and that ϕ is a positive linear functional on $C_c(X)$. Then there exists a unique Radon measure μ on X such that $\phi(f) = \int_X f\,d\mu$, for all $f \in C_c(X)$.*

Proof There exists an increasing sequence $(K_n)_{n=1}^\infty$ of compact subsets of X such that $K_n \subseteq K_{n+1}^{int}$ for each $n \in \mathbf{N}$, and such that $X = \cup_{n=1}^\infty K_n$. Let us set $E_n = \{f \in C_c(X) : \text{supp}(f) \subseteq K_n\}$, so that $(E_n)_{n=1}^\infty$ is an increasing sequence of subspaces of $C_c(X)$ whose union is $C_c(X)$. If $f \in E_{n+1}$ and $x \in K_n$ let $T_n(f)(x) = f(x)$. Then T_n is positive linear mapping of E_{n+1} onto $C(K_n)$. Using Tietze's extension theorem, it follows that there exists a sequence $(g_{n.r})_{r=1}^\infty$ in E_{n+1} which decreases pointwise to I_{K_n}. If $h \in C(K_{n+1})$, let $\psi_n(h) = \lim_{r\to\infty} \phi(g_{n,r}h)$; ψ_n is a positive linear functional on $C(K_{n+1})$, and if $T_n(h_1) = T_n(h_2)$ then $\psi(h_1) = \psi(h_2)$. Thus if $f = T_n(h) \in C(K_n)$ and we set $\theta_n(f) = \psi_n(h)$ then θ_n is a properly defined positive linear functional on $C(K_n)$, which is represented by a finite Borel measure μ_n, by the Riesz representation theorem. If A is a Borel set in X, we set $\mu(A) = \lim_{n\to\infty} \mu_n(A \cap K_n)$. It then follows from Theorem 16.3.3 that μ is a Radon measure on X; further, if $f \in C_c(X)$ then $\phi(f) = \int_X f\,d\mu$. □

16.8 The Stone–Weierstrass Theorem

We now use the Riesz representation theorem and the Krein–Mil'man theorem to prove the Stone–Weierstrass theorem, at least for compact metrizable spaces.

Theorem 16.8.1 (The Stone–Weierstrass theorem) *Suppose that K is a compact metrizable space and that A is a linear subspace of $C(K)$ which is an algebra under pointwise multiplication, which contains the constant functions and which separates points; if $x_1 \neq x_2$, there exists $g \in A$ such that $g(x_1) \neq g(x_2)$. Then A is dense in $(C(K), \|.\|_\infty)$.*

Proof Suppose that $\overline{A} \neq C(K)$. It then follows from the Hahn–Banach theorem that $A^\circ = \{\phi \in C(K)' : \phi(a) = 0 \text{ for } a \in \overline{A}\}$ is a non-zero weak*-closed linear subspace of $C(K)'$, and $B(A^\circ) = \{\phi \in A^\circ; \|\phi\|' \leq 1\}$ is a non-zero weak*-compact convex subset of $C(K)'$. By the Krein–Mil'man theorem, there exists an extreme point ϕ of $B(A^\circ)$, and this is represented by a non-zero signed measure μ, by the Riesz representation theorem. Let $\mu = \mu^+ - \mu^-$. Since $\mu(X) = 0$ and $|\mu|(X) = 1$, $\mu^+(X) = \mu^-(X) = \frac{1}{2}$.

Since $\mu(K) = \phi(1) = 0$, $\text{supp}(|\mu|)$ is not a singleton. Let x_1 and x_2 be two distinct points of $S = \text{supp}(|\mu|)$. By hypothesis there exists $g \in A$ with $g(x_1) < g(x_2)$. By adding a constant, we can assume that $g \geq 0$, and by scaling we can suppose that $\int_K g\, d|\mu| = 1$. Since A is an algebra, $\int_K g\, d\mu = 0$; so $\int_K g\, d\mu^+ = \int_K g\, d\mu^- = \frac{1}{2}$. Thus $\sup\{g(x) : x \in \text{supp}(|\mu|)\} = \alpha \geq 1$. There exist $x_0 \in \text{supp}(|\mu|)$ such that $g(x_0) = \alpha$. We consider two cases.

If $\alpha = 1$ then

$$\int_K |1 - g|\, d|\mu| = \int_K 1 - g\, d|\mu| = |\mu|(K) - \int_K g\, d|\mu| = 0,$$

so that $g = 1$ μ almost everywhere on $\text{supp}(|\mu|)$. But g is continuous, and so $g = 1$ on $\text{supp}(|\mu|)$. But $g(x_1) \neq g(x_2)$, giving a contradiction.

If $\alpha > 1$, let $\mu_2 = g.d\mu$. There is an open neighbourhood U of x_0 such that $g(x) > (1 + \alpha)/2$ for $x \in U$. Then there exists a Borel subset B of U such that $\mu(B) \neq \mu_2(B)$, so that $\mu \neq \mu_2$. Let $\lambda = 1/\alpha$, and let

$$\mu_1 = \left(\frac{1 - \lambda g}{1 - \lambda}\right) d\mu \text{ so that } \mu = (1 - \lambda)\mu_1 + \lambda\mu_2.$$

Then μ_1 and μ_2 are in $B(A^\circ)$. Since $\mu_2 \neq \mu$, again we have a contradiction.

Thus A is dense in $C(K)$. □

Corollary 16.8.2 *Suppose that K_1 and K_2 are compact metrizable spaces. Then the set*

$$\left\{\sum_{j=1}^n f_j \otimes g_j : n \in \mathbf{N}, f_j \in C(K_1), g_j \in C(K_2)\right\}$$

is dense in $C(K_1 \times K_2)$.

Corollary 16.8.3 *The space $BL(K)$ is dense in $C(K)$.*

Proof For $BL(K)$ is an algebra which contains the constants and separates points. □

Here is the usual way to prove the general Stone–Weierstrass theorem.

Exercise 16.8.4 Define a sequence of polynomials $(p_n)_{n=0}^{\infty}$ by setting $p_0 = 0$ and

$$p_{n+1}(x) = p_n(x) + \tfrac{1}{2}(x^2 - (p_n(x))^2).$$

Show that if $|x| \leq 1$ then $0 \leq p_n(x) \leq p_{n+1}(x) \leq |x|$, and that $p_n(x)$ tends uniformly to $|x|$.

Exercise 16.8.5 Suppose that A is a subalgebra of $C(X)$, where (X, τ) is a compact Hausdorff space. Show that $\overline{A}$ is a lattice.

Exercise 16.8.6 Suppose that L is a linear subspace of $C(X)$, where (X, τ) is a compact Hausdorff space, and that L is a lattice, that L contains the constants and that L separates points. Show that L is dense in $C(X)$.

16.9 Product Measures

We now use the Riesz representation theorem to develop the theory of product measures on Polish spaces. This is rather simpler than the abstract theory.

Theorem 16.9.1 *Suppose that X_1 and X_2 are two compact metrizable spaces, that $\mu_1 \in P(K_1)$ and $\mu_2 \in P(X_2)$. Then there exists a unique $\mu_1 \otimes \mu_2$ in $P(X_1 \times X_2)$ such that $(\mu_1 \otimes \mu_2)(B_1 \times B_2) = \mu_1(B_1).\mu_2(B_2)$ for B_1, B_2 Borel sets in X_1, X_2.*

Proof X_1 and X_2 are homeomorphic to G_δ-subsets of compact metrizable spaces K_1 and K_2. Push the measures forward to K_1 and K_2; it is enough to prove the result for K_1 and K_2. Let A be the algebra

$$\left\{\sum_{j=1}^{n} f_j \otimes g_j : n \in \mathbf{N}, f_j \in C(K_1), g_j \in C(K_2)\right\}.$$

If $h = \sum_{j=1}^{n} f_j \otimes g_j$, let

$$\phi(h) = \sum_{j=1}^{n} \int_{K_1} f_j \, d\mu_1 . \int_{K_2} g_j \, d\mu_2.$$

Now $(\int_{K_1} f_j \, d\mu_1) g_j \in C(K_2)$, so that

$$\phi(h) = \int_{K_2} \left(\int_{K_1} h(x, y) \, d\mu_1(x)\right) d\,\mu_2, (y)$$

and $\phi(h)$ does not depend upon the representation of h, and is properly defined, and unique. Further, ϕ is positive, so that ϕ is a continuous linear functional on A. Since A is dense in $C(K_1 \times K_2)$ (Corollary 16.8.2), ϕ extends uniquely to a positive linear functional on $C(K_1 \times K_2)$, which, by the Riesz representation theorem, is represented by an element $\mu_1 \otimes \mu_2$ of $P(K_1 \times K_2)$. Since ϕ is unique, so is $\mu_1 \otimes \mu_2$. Approximation arguments then show first that $(\mu_1 \otimes \mu_2)(C_1 \times C_2)$ for C_1, C_2 compact sets, and then $(\mu_1 \otimes \mu_2)(B_1 \times B_2)$ for B_1, B_2 Borel sets. □

We can extend this result to finite products, and to σ-finite measures. For example, we denote by λ_d the product $\lambda \otimes \cdots \otimes \lambda$ the product of d copies of Lebesgue measure λ; λ_d is *Lebesgue measure* on $\mathbf{R}^d$.

How can we evaluate $\int_{K_1 \times K_2} f \, d(\mu_1 \times \mu_2)$? We can do this by repeated integration.

Theorem 16.9.2 (Tonelli's theorem) *Suppose that X_1 and X_2 are two Polish spaces, that $\mu_1 \in P(X_1)$ and $\mu_2 \in P(X_2)$. If f is a non-negative Borel measurable function on $X_1 \times X_2$ then $\int_{X_1} f(x,y) \, d\mu_1(x)$ is a Borel measurable function on X_2, and*

$$\int_{X_1 \otimes X_2} f \, d(\mu_1 \otimes \mu_2) = \int_{X_2} (f(x,y) \, d\mu_1(x)) \, d\mu_2(y).$$

Proof Once again, it is enough to prove the result for compact metrizable spaces K_1 and K_2. First, suppose that f is continuous. Since $P(K_1) \subseteq C(K_1)'$, the function $y \to \int_{K_1} f(x,y) \, d\mu_1(x)$ is continuous. Since

$$\int_{K_1 \otimes K_2} f \, d(\mu_1 \otimes \mu_2) = \int_{K_2} (f(x,y) \, d\mu_1(x)) \, d\mu_2(y)$$

for $f \in A$, and A is dense in $C(K_1 \times K_2)$, the equation holds for continuous f.

If f is lower semi-continuous, there exists an increasing sequence $(f_n)_{n=1}^\infty$ in $C(K_1 \times K_2)$ which converges pointwise to f. Then $\int_{K_1} f_n(x,y) \, d\mu_1(x)$ converges pointwise to $\int_{K_1} f(x,y) \, d\mu_1(x)$, so that $\int_{K_1} f(x,y) \, d\mu_1(x)$ is lower semi-continuous. Further,

$$\begin{aligned}
\int_{K_1 \times K_2} f \, d(\mu_1 \otimes \mu_2) &= \lim_{n\to\infty} \int_{K_1 \times K_2} f_n \, d(\mu_1 \otimes \mu_2) \\
&= \lim_{n\to\infty} \int_{K_2} \left(\int_{K_1} f_n(x,y) \, d\mu_1(x) \right) d\mu_2(y) \\
&= \int_{K_2} \left(\lim_{n\to\infty} \int_{K_1} f_n(x,y) \, d\mu_1(x) \right) d\mu_2(y) \\
&= \int_{K_2} \left(\int_{K_1} f(x,y) \, d\mu_1(x) \right) d\mu_2(y).
\end{aligned}$$

In particular, the result holds for indicator functions of open sets. Now let $\mathcal{S}$ be the collection of Borel sets E whose indicator functions satisfy the theorem.

Then $\mathcal{S}$ is a σ-field which contains the open sets, and so is equal to the Borel σ-field B. Finally, a standard approximation argument shows that it holds for non-negative Borel measurable functions. □

This theorem can also be extended to σ-finite measures, by summing over a disjoint sequence of sets of finite measure, in the usual way.

Exercise 16.9.3 Suppose that f is a positive integrable Borel measurable function on $(X, \mathcal{B}, \mu)$, where μ is a σ-finite measure on a Polish space X. Let $S = \{(x, t) : x \in X, 0 \leq t \leq f(x)\}$. Show that $\int_X f\, d\mu = \mu(S)$. (The integral is the 'area under the curve'.)

By considering positive and negative parts, we obtain the following.

Theorem 16.9.4 (Fubini's theorem) *Suppose that μ_1 and μ_2 are Borel probability measures on Polish spaces X_1 and X_2, and that f is a Borel measurable function on $X_1 \times X_2$. Then the following are equivalent:*

(i) $f \in \mathcal{L}^1(\mu_1 \otimes \mu_2)$;

(ii) the function $f_y(x) = f(x, y)$ is in $\mathcal{L}^1(\mu_1)$ for μ_2-almost all y, and the function $y \to \int_{X_1} f_y\, d\mu_1$ is μ_2-measurable, and in $\mathcal{L}^1(\mu_2)$.

If so, then

$$\int_{X_1 \times X_2} f\, d(\mu_1 \otimes \mu_2) = \int_{X_2} \left(\int_{X_1} f_y(x)\, d\mu_1(x) \right) d\mu_2(y).$$

It may however happen that $\int_{X_1} f_y\, d\mu_1$ is infinite, or undefined, on a set of μ_2-measure 0.

What about infinite products? Suppose now that for each $n \in \mathbf{N}$, π_n is a Borel probability measure on a Polish space (X_n, τ_n). Let $(X, \tau) = \prod_{n=1}^{\infty}(X_n, \tau_n)$. Can we define an infinite product measure on (X, τ)? Again, we use the Riesz representation theorem to show that we can.

Theorem 16.9.5 *Suppose that, for each $n \in \mathbf{N}$, (X_n, τ_n) is a Polish space and $\pi_n \in P(X_n)$. Let $X = \prod_{n=1}^{\infty} X_n$, with the product topology, and let $p_n(x) = (x_1, \ldots, x_n)$, for each $n \in \mathbf{N}$ and $x \in X$. Then there exists a unique $\pi \in P(X)$ such that $p_{n*}(\pi) = \pi_1 \otimes \cdots \otimes \pi_n$.*

Proof For each n, there is a homeomorphism i_n of X_n onto a dense G_δ subset of a compact metrizable space $(\tilde{X}_n, \tilde{\tau}_n)$; then $\prod_{n=1}^{\infty} i_n(X_n)$ is a G_δ subset of $\tilde{X} = \prod_{n=1}^{\infty} \tilde{X}_n$. By considering the push-forward measures $(i_n)_* \pi_n$ we can therefore suppose that each X_n is compact.

If $x \in X$, let $p_n(x) = (x_1, \ldots, x_n)$. Let

$$F_n = \{\mu \in P(X) : p_{n*}(\mu) = \pi_1 \otimes \cdots \otimes \pi_n\}.$$

Then it follows from Exercise 16.6.5 that F_n is a non-empty w-compact subset of $P(X)$, and the sequence $(f_n)_{n=1}^{\infty}$ is decreasing, so that $F = \cap_{n=1}^{\infty} F_n$ is not empty. The set of functions

$$\{f \circ p_n : f \in C(X_1 \times \cdots \times X), n \in \mathbf{N}\}$$

is dense in $C(X)$, from which it follows that F is a singleton $\{\pi\}$.

Finally, π is unique, since the sets

$$\{p_n^{-1}(A) : A \text{ a Borel set in } X_1 \times \cdots \times X_n, n \in \mathbf{N}\}$$

generate the Borel sets in X. □

16.10 Disintegration of Measures

An important feature of the theorems of Tonelli and Fubini is that integrals with repect to product measures can be evaluated by repeated integration. We shall show in Theorem 16.10.2 that this extends to more general measures on products. For this, we need the idea of the disintegration of a measure.

Suppose that X and Y are Polish spaces, that $\mu \in P(X)$ and that T is a Borel measurable mapping from X into Y. Let $\nu = T_*(\mu)$ be the *push-forward* measure: $\nu(A) = \mu(T^{-1}(A))$. A *T-disintegration* of μ is a family $\{\lambda_y : y \in T(X)\}$ of Borel probability measures on X such that

(i) if $y \in T(X)$ then $\lambda_y(T^{-1}\{y\}) = 1$ for ν-almost all y, and
(ii) if $f \in L^1(X, \mu)$ then

  (a) $f \in L^1(\lambda_y)$ for almost all $y \in T(X)$,
  (b) the function $y \to \int_X f \, d\lambda_y$ is ν-measurable, and
  (c) $\int_X f \, d\mu = \int_Y \left(\int_X f \, d\lambda_y\right) d\nu$.

Theorem 16.10.1 *Suppose that X and Y are Polish spaces, that $\mu \in P(X)$ and that T is a Borel measurable mapping from X into Y. Then a T-disintegration of μ exists, and is essentially unique; if $\{\lambda_y : y \in T(X)\}$ and $\{\lambda'_y : y \in T(X)\}$ are two T-disintegrations, then $\lambda_y = \lambda'_y$ for almost all $y \in T(X)$.*

Proof Again, let $\nu = T_*(\mu)$. First we consider the case where X and Y are compact metric spaces and T is a continuous surjection of X onto Y.

We consider the product space $X \times Y$, and denote the projection of $X \times Y$ onto Y by p. If f and g are Borel measurable functions on X and Y respectively, we set $(f \otimes g)(x, y) = f(x)g(y)$; then $f \otimes g$ is a Borel measurable function on $X \times Y$, which is continuous if f and g are.

If $x \in X$, let $\gamma(x) = (x, T(x))$. γ is a homeomorphism of X onto Γ_T, the graph of T. Let $\pi = \gamma_*(\mu)$: $\pi \in P(X \times Y)$, and the support of π is contained in Γ_T. Further, $\nu = p_*(\pi)$.

Suppose now that $f \in L^1(X, \mu)$. Then $\gamma_*(f.d\mu) = (f \otimes 1).d\pi$. Let $\nu_f = T_*(f.d\mu)$; then $T_*(f.d\mu) = p_*((f \otimes 1).d\pi)$. If $\epsilon > 0$, there exists $\delta > 0$ such that if $\mu(E) < \delta$ then $\int_E |f|\, d\mu < \epsilon$. Thus if $\nu(f) < \delta$ then

$$|\nu_f(f)| = \left| \int_{T^{-1}(f)} f\, d\mu \right| \leq \int_{T^{-1}(f)} |f|\, d\mu < \epsilon.$$

Thus ν_f is absolutely continuous with respect to ν. By the Radon–Nikodym theorem, there exists $l_f \in L^1(\nu)$ such that $\nu_f = l_f.\, d\nu$.

We now need to restrict attention to a suitable countable dense subset of $C(X)$. Let C be a countable dense subset of $C(X)$ which is a vector space over $\mathbf{Q}$, and which contains 1. Then there exists a Borel subset Y_C of Y, with $\nu(Y \setminus Y_C) = 0$, such that if $y \in Y_C, f, g \in C$ and $\alpha \in \mathbf{Q}$ then

$$\begin{aligned} l_{\alpha f}(y) &= \alpha l_f(y), & l_{f+g}(y) &= l_f(y) + l_g(y), \\ l_1(y) &= 1, & l_f(y) &\geq 0 \text{ if } f \geq 0. \end{aligned}$$

It follows that if $y \in Y_C$ and $f, g \in C$ then

$$-\|f - g\|_\infty .1 \leq f(y) - g(y) \leq \|f - g\|_\infty .1,$$

and so $l_.(y)$ extends by continuity to a positive linear functional m_y on $C(X)$, with $m_y(1) = 1$. By the Riesz representation theorem, there exists $\lambda_y \in P(X)$ such that $m_y(f) = \int_X f\, d\lambda_y$, for $f \in C(X)$.

If $f \in C$ and $y \in Y_C$ then $\int_X f\, d\lambda_y = l_f(y)$, so that $\int_X f\, d\lambda_y$ is measurable, and

$$\int_X f\, d\mu = \int_Y d\nu_f = \int_Y l_f\, d\nu = \int_Y \left(\int_X f d\lambda_y(x) \right) d\nu(y).$$

This extends by bounded convergence to all f in $C(X)$. By Exercise 16.1.5, it extends to indicator functions of open sets and closed sets, and by regularity to indicator functions of Borel sets. It then extends by monotone convergence to functions in $L^1(\mu)$ (though if $f \in L^1(\mu)$, there may be a subset of Y_C of ν-measure 0 on which $\int_X f\, d\lambda_y$ is infinite, or undefined).

If $f \in C(X)$ and $g \in C(Y)$ then

$$\int_{X \times Y} f \otimes g\, d\pi = \int_Y g\, d\nu_f = \int_Y \left(\int_X (f \otimes g)(x, y)\, d\lambda_y \right) d\nu(y).$$

Again, this extends by linearity and continuity to all $h \in C(X \times Y)$, to give

$$\int_{X \times Y} h\, d\pi = \int_Y \left(\int_X h(x, y)\, d\lambda_y \right) d\nu(y).$$

By Exercise 16.1.5, this extends to indicator functions of closed subsets of $X \times Y$. Thus

$$1 = \int_{X\times Y} I_\Gamma \, d\pi = \int_Y \left(\int_X I_\Gamma(x, y) \, d\lambda_y(x) \right) d\nu(y),$$

so that

$$\int_X I_\Gamma(x, y) \, d\lambda_y(x) = \lambda_y(T^{-1}\{y\}) = 1$$

for ν almost all y.

Next, we establish uniqueness. Suppose that $\{\lambda_y : y \in T(X)\}$ and $\{\lambda'_y : y \in T(X)\}$ are two T-disintegrations. If $f \in C$ and A is a Borel subset in Y then

$$\begin{aligned}\int_A \left(\int_X f \, d\lambda_y \right) d\nu(y) &= \int_X f(x) I_A(T(x)) \, d\mu(x) \\ &= \int_A \left(\int_X f \, d\lambda'_y \right) d\nu(y),\end{aligned}$$

so that $\int_X f \, d\lambda_y = \int_X f \, d\lambda'_y$, for all $f \in C$, for almost all y, and so $\lambda_y = \lambda'_y$ for almost all y.

We now turn to the general case, where X and Y are Polish spaces, and T is Borel measurable. There exist a disjoint sequence $(K_n)_{n=1}^\infty$ of compact subsets of X with $\mu(\cup_{n=1}^\infty K_n) = 1$, such that the restriction of T to K_n is continuous, for each n. We can disintegrate μ on each K_n, and add the disintegrations together, to obtain the result. Again, uniqueness follows from the uniqueness on each K_n. □

The fact that we consider the measure π on the product $X \times Y$ suggests that we should consider more general measures on products.

Theorem 16.10.2 *Suppose that X, Y are Polish spaces, and that $\pi \in P(X \times Y)$. Let $\nu = p_*(\pi)$, where p is the projection of $X \times Y$ onto Y. (ν is the Y-*marginal distribution *of π.) Then there exists a Borel subset Y_C of Y, with $\nu(Y \setminus Y_C) = 0$, and a family $\{\mu_y : y \in Y_C\}$ in $P(X)$ such that if $f \in L^1(\pi)$ then there is a Borel subset Y_f of Y_C with $\nu(Y \setminus Y_f) = 0$ such that*

(i) $f(., y) \in L^1(\mu_y)$ for $y \in Y_f$,
(ii) $\int_X f(x, y) \, d\mu_y(x)$ is a measurable function on Y_f, and
(iii) $\int_{X\times Y} f \, d\pi = \int_{Y_f} \left(\int_X f(x, y) \, d\mu_y(x) \right) d\nu(y)$.

Proof In the theorem, we replace X by $X \times Y$, μ by π and T by p. Since the support of λ_y is contained in $X \times \{y\}$, there exists $\mu_y \in P(X)$ such that $\mu_y(A) = \lambda_y(A \times \{y\})$ for each Borel set A in X. The family $\{\mu_y : y \in Y_C\}$ clearly satisfies the corollary. □

Let us reconsider the formula in Theorem 16.10.1. Let us set $\mathcal{B}_T = T^{-1}\mathcal{B}(T(X))$. Then $\mathcal{B}_T$ is a sub σ-field of $\mathcal{B}$. Since $\nu = T_*(\mu)$, let us set $\mu_z = \lambda_{T(z)}$, for $z \in X$. Note that if $T(z) = T(z')$ then $\mu_z = \mu_{z'}$. We then see that if $f \in L^1(X, \mu)$ then

(a) $f \in L^1(\mu_z)$ for almost all $z \in X$,
(b) the function $z \to \int_X f\, d\mu_z$ is $\mathcal{B}_T$-measurable, and
(c) $\int_X f\, d\mu = \int_X \left(\int_X f(x)\, d\mu_z(x)\right) d\mu(z)$.

The important feature is (b). The function $\mathbf{E}(f|\mathcal{B}_T)(z) = \int_X f\, d\mu_z$ is the *conditional expectation* of f with respect to $\mathcal{B}_T$ and the function $z \to \mu_z$ is the *regular conditional probability* of μ, with respect to $\mathcal{B}_T$.

As an example, let $X = C([0, S])$, and let $\mu \in P(X)$. Thus μ is a probability measure on the continuous paths on $[0, S]$. Suppose that $0 < R < S$. Let $Y = C([0, R])$, and let T be the restriction mapping. Then if $y \in Y$, λ_y is the conditional probability of a path $g \in C([0, S])$ for which $g(x) = f(x)$ for $0 \le x \le R$.

16.11 The Gluing Lemma

We shall need a fundamental result about measures on products. Suppose that X and Y are Polish spaces, that $\mu \in P(X)$ and $\nu \in P(Y)$. If $(x, y) \in X \times Y$, let $p_1(x, y) = x$ and $p_2(x, y) = y$. We set

$$\Pi_{\mu,\nu} = \{\pi \in P(X \times Y) : (p_1)_*\pi = \mu, (p_2)_*\pi = \nu\}.$$

$\Pi_{\mu,\nu}$ is the set of Borel probability measures on $X \times Y$ with marginals μ and ν.

Theorem 16.11.1 (The gluing lemma) *Suppose that X, Y, Z are Polish spaces, that $\mu \in P(X)$, $\nu \in P(Y)$, $\pi \in P(Z)$ and that $\alpha \in \Pi_{\mu,\nu}$ and $\beta \in \Pi_{\nu,\pi}$. Then there exists $\gamma \in P(X \times Y \times Z)$ with marginals α on $X \times Y$ and β on $Y \times Z$.*

Proof We can embed X, Y and Z as G_δ subsets of compact metric spaces $\tilde{X}$, $\tilde{Y}$ and $\tilde{Z}$, push μ, ν and π forward to elements $\tilde{\mu}$, $\tilde{\nu}$ and $\tilde{\pi}$ of $P(\tilde{X})$, $P(\tilde{Y})$, $P(\tilde{Z})$, and α, β to elements $\tilde{\alpha}$, $\tilde{\beta}$ of $\Pi_{\tilde{\mu},\tilde{\nu}}$ and $\Pi_{\tilde{\nu},\tilde{\pi}}$. Thus we can suppose that X, Y and Z are compact metric spaces.

Let

$$V = \{f + g : f \in C(X \times Y), g \in C(Y \times Z)\};$$

V is a linear subspace of $C(X \times Y \times Z)$. If $f + g \in V$, we set $\phi(f + g) = \int_{X\times Y} f\, d\alpha + \int_{Y\times Z} g\, d\beta$.

First we show that ϕ is well-defined. If $f+g=f'+g'$ then $f-f'=g'-g$, so that $f-f'$ and $g'-g$ are functions of y only. Hence

$$\int_{X\times Y}(f-f')\,d\alpha=\int_Y(f-f')\,d\nu=\int_Y(g'-g)\,d\nu=\int_{Y\times Z}(g'-g)\,d\beta,$$

and $\phi(f+g)=\phi(f'+g')$. Thus ϕ is a linear functional on V.

Secondly, we show that if $f+g\geq 0$ then $\phi(f+g)\geq 0$. Let $h(y)=\inf_{x\in X}f(x,y)$; h is an upper semi-continuous function on Y. Let $\hat{f}=f-h$; then $\hat{f}\geq 0$. Let $\hat{g}=h+g$. Then

$$\hat{g}(x,y)=\inf_{x\in X}(f(x,y)+g(y,z))\geq 0.$$

$\hat{f}$ and $\hat{g}$ are semi-continuous, and therefore integrable, and

$$\phi(f+g)=\int_{X\times Y}\hat{f}\,d\alpha+\int_{Y\times Z}\hat{g}\,d\beta\geq 0.$$

Thirdly, we observe that $\phi(1)=1$, so that $\|\phi\|=1$.

We now apply the Hahn–Banach theorem. There exists a linear functional ψ on $C(X\times Y\times Z)$, with $\|\psi\|=\|\phi\|$, which extends ϕ. Since $\psi(1)=1$, ψ is a positive linear functional on $C(X\times Y\times Z)$; by the Riesz representation theorem, there exists $\gamma\in P(X\times Y\times Z)$ such that $\psi(k)=\int_{(X\times Y\times Z)}k\,d\gamma$ for $k\in C(X\times Y\times Z)$. If $f\in C(X\times Y)$ then

$$\psi(f)=\int_{X\times Y}f\,d\alpha=\int_{X\times Y\times Z}f\,d\gamma,$$

from which it follows by the usual arguments that $\gamma(A\times Z)=\alpha(A)$, for A a Borel set in $X\times Y$. Similarly, $\gamma(X\times B)=\beta(B)$, for B a Borel set in $Y\times Z$. □

Exercise 16.11.2 Suppose that $a(x,y)$ and $b(y,z)$ are non-negative functions in $L^1(Q,\mathbf{B},\lambda_3)$, where Q is the unit cube $\{(x,y,z):0\leq x,y,z\leq 1\}$ and λ_3 is Lebesgue measure, and that $\int_0^1 a(x,y)\,d\lambda(x)=\int_0^1 b(y,z)\,d\lambda(z)=h(y)$ for all $y\in[0,1]$. Find a function g in $L^1(\lambda_3)$ such that $\int_0^1 g(x,y,z)\,d\lambda(z)=a(x,y)$ and $\int_0^1 g(x,y,z)\,d\lambda(x)=b(y,z)$ for all x,y,z.

The gluing lemma can also be proved, using the disintegration of measures.

Exercise 16.11.3 In the setting of the gluing lemma, let $\{\lambda_y^{(1)}:y\in Y\}$ be the p_1-disintegration of α and $\{\lambda_y^{(2)}:y\in Y\}$ be the p_2-disintegration of β. Let $\gamma(A\times B\times C)=\int_B\lambda^{(1)}(A)_y.\lambda^{(2)}(C)\,d\nu(y)$, where A,B,C are Borel sets in X,Y,Z. Show that γ extents to a Borel measure on $X\times Y\times Z$ which meets the requirements of the gluing lemma.

16.12 Haar Measure on Compact Metrizable Groups

We end this chapter by establishing the existence of Haar measure on compact and locally compact metrizable topological groups. This is a Borel measure which is left or right translation invariant. We begin with the compact metrizable case, such as the group O_n of orthogonal transformations of $\mathbf{R}^n$, which is, for good reasons, much the simpler.

Theorem 16.12.1 *Suppose that G is a compact metrizable group. Then there exist unique Borel probability measures μ_l and μ_r on G such that $\mu_l(gA) = \mu_l(A)$ and $\mu_r(Ag) = \mu_r(A)$ for each $g \in G$ and each Borel set A. Further, $\mu_l = \mu_r$.*

Proof We give two proofs. The first uses Kakutani's fixed-point theorem. Let l_g be the left regular representation of G in $C(G)$, so that $l_g(f)(h) = f(gh)$, for $f, g \in G$. Then the mapping $(g,f) \to l_g(f)(h) : G \times C(G) \to C(G)$ is jointly continuous (Proposition 5.5.1), and so $\{l_g(f); g \in G\}$ is a compact subset of $C(G)$, for each $f \in C(G)$. If $\mu \in P(X) \subseteq C(X)'$ then $\left\|l'_g(\mu)\right\| \leq 1$ and $l'_g(\mu)(1) = \mu(1) = 1$, so that $l'_g(\mu) \in P(X)$. Thus the conditions of Kakutani's fixed point theorem are satisfied, and so there exists $\mu_l \in P(X)$ for which $l'_g(\mu_l) = \mu_l$, for each $g \in G$. That is, $\int_G f\,d\mu = \int_G l_g(f)\,d\mu$ for each $g \in G$ and $f \in C(G)$. From this it follows that $\mu_l(gA) = \mu_l(A)$ for each $g \in G$ and each Borel set A. The existence of a right-invariant Borel probabilty measure is proved in an exactly similar way.

For the second proof of existence, we use Hall's marriage theorem, and consider minimal ϵ-nets; recall that these are ϵ-nets with as few elements as possible. Let d be a metric on G for which $d(gx, gy) = d(x, y) = d(xg, yg)$ for all $x, y, g \in G$. G is d-totally bounded. For each $k \in \mathbf{N}$, let n_k be a minimal $1/k$-net in G. If $f \in C(G)$, let $\phi_k(f) = (1/|n_k|)\sum_{g \in n_k} f(g)$. Then ϕ_k is a positive linear functional in $C(G)'$, $\|\phi_k\|' \leq 1$ and $\phi_k(1) = 1$. If $f \in C(G)$ then f is uniformly continuous, so that if $\epsilon > 0$ then there exists $K \in \mathbf{N}$ such that if $d(g, h) < 2/K$ then $|f(g) - f(h)| < \epsilon/2$. Suppose that $k, l \geq K$ and that $h \in n_l$. Then $n_k h = \{gh : g \in n_k\}$ is also a minimal $1/k$-net. By Proposition 3.2.5 there is a bijection $\psi_h : n_k \to n_k h$ such that $d(g, \psi(g)) < 2/k$, for each $g \in n_k$. Thus

$$\left|\phi_k(f) - \frac{1}{|n_k|}\sum_{g \in n_k} f(gh)\right| < \epsilon/2.$$

Averaging over n_l,

$$\left|\phi_k(f) - \frac{1}{|n_k|.|n_l|}\sum_{g \in n_k, h \in n_l} f(gh)\right| < \epsilon/2.$$

Similarly,

$$\left| \phi_l(f) - \frac{1}{|n_k|.|n_l|} \sum_{g \in n_k, h \in n_l} f(gh) \right| < \epsilon/2,$$

and so $|\phi_k(f) - \phi_l(f)| < \epsilon$. Thus $\phi_k(f)$ converges, as $k \to \infty$, to $\phi(f)$, say. Then ϕ is a positive linear functional on $C(K)$, $\|\phi\|' \leq 1$ and $\phi(1) = 1$, so that ϕ is represented by a Borel probability measure μ. If $g \in G$, then gn_k and $n_k g^{-1}$ are minimal $1/k$-nets, and so, if $k \geq K$, then

$$|\phi_k(l_g(f)) - \phi_k(f)| < \epsilon/2 \text{ and } |\phi_k(r_g(f)) - \phi_k(f)| < \epsilon/2.$$

Thus $\phi(l_g(f)) = \phi(r_g(f)) = \phi(f)$, and ϕ, and therefore μ, are left and right invariant.

It remains to show that there is a *unique* left-invariant Borel probability measure and a *unique* right-invariant Borel probability measure, and that they are equal. Suppose that μ_l is a left-invariant Borel probability measure and that μ_r is a right-invariant Borel probability measure. If $f \in C(G)$ and $g \in G$, then $\int_G f\,d\mu_l = \int_G f(gh)\,d\mu_l(h)$, a constant, so that

$$\int_G f\,d\mu_l = \int_G \left(\int_G f(gh)\,d\mu_l(h) \right) d\mu_r(g).$$

Similarly

$$\int_G f\,d\mu_r = \int_G \left(\int_G f(gh)\,d\mu_r(g) \right) d\mu_l(h).$$

The two repeated integrals are equal, and so $\mu_l = \mu_r$; uniqueness follows from this. □

Suppose now that a is a continuous action of a compact metrizable group G on a compact metrizable space X. An *orbit* of a is a set of the form $a(G)(x) = \{a(g)(x) : g \in G\}$, where x is an element of X. Orbits are compact, and the orbits form a partition of X. The action is *transitive* if X is an orbit; that is, if whenever $x, y \in X$ there exists $g \in G$ such that $a(g)(x) = y$. In this case X is called a *homogeneous space* for G, and G is called a group of *symmetries* of X.

Theorem 16.12.2 *Suppose that a is a continuous action of a compact metrizable group G on a compact metrizable space X. Then there exists a Borel probability measure μ on X, which is invariant under a; if A is a Borel subset of X and $g \in G$ then $\mu(a(g)(A)) = \mu(A)$. If X is homogeneous, they are unique.*

Proof Let $x \in X$. The mapping $\phi : g \to a(g)(x)$ is a continuous mapping of G onto an orbit of the action. Let $\mu = \phi_*(\nu)$, where ν is Haar measure on G. If A is a Borel subset of X and $g \in G$, then

$$\begin{aligned}\phi^{-1}(a(g)(A)) &= \{h \in G : a(h)(x) \in a(g)(A)\}\\ &= \{h \in G : a(g^{-1}h)(x) \in (A)\}\\ &= \{h \in G : g^{-1}h \in \phi^{-1}(A)\} = g\phi^{-1}(A),\end{aligned}$$

so that $\mu(a(g)(A)) = \mu(A)$.

The support of μ is contained in the orbit $a(G)(x)$, and so the measure μ is unique only if the action a is transitive. In fact the support of $\mu = a(G)(x)$, since if y is in the support of μ then, by invariance, so is $a(g)(y)$, for each $g \in G$.

Suppose now that the action is transitive, so that the support of μ is X. This implies that if $g \in L^\infty(\mu)$ and $\int_X fg\,d\mu = 0$ for all $f \in C(X)$, then $g = 0$. Suppose now that π is a Borel probability measure which is invariant under a. Let $\sigma = \frac{1}{2}(\pi + \mu)$. Then σ is invariant under a, the support of σ is X, and by the Radon–Nikodym theorem there exists $k \in L^\infty$, with $0 \leq k \leq 2$ such that $\pi = k\,d\sigma$. If $f \in C(X)$ and $g \in G$ then

$$\begin{aligned}\int_X fk\,d\sigma &= \int_X f\,d\pi = \int_X a(g)f\,d\pi\\ &= \int_X a(g)(f).k\,d\sigma = \int_X f.a(g^{-1})(k)\,d\sigma.\end{aligned}$$

Since this holds for all $f \in C(X)$, $k = a(g-1)(k)$ for all $g \in G$. Thus k is constant, so that $\pi = k\sigma$. Since $\pi(X) = \sigma(X) = 1$, $k = 1$ and $\pi = \mu$. □

Exercise 16.12.3 Suppose that (X, d) is a compact metric space, and that I_X is the group of isometries of X onto itself and that it acts transitively on X. (X is then called an *isometrically homogeneous* space.) Show that there is a unique Borel probability measure μ on X such that $\mu(i(A)) = \mu(A)$ for each Borel set A and each isometry i of X onto itself.

Exercise 16.12.4 Let $B(\mathbf{R}^n)$ be the unit ball in $\mathbf{R}^n$, and let O_n be the group of rotations and reflections of $B(\mathbf{R}^n)$. Show that O_n is the group of isometries of $B(\mathbf{R}^n)$. What are the orbits of O_n? Show that there is an O_n-invariant probability measure on $B(\mathbf{R}^n)$ whose support is $B(\mathbf{R}^n)$. Is such a measure unique?

16.13 Haar Measure on Locally Compact Polish Topological Groups

Suppose now that (G, τ) is a locally compact Polish topological group, which is not compact. We shall show that there exist a left-invariant Radon measure ψ_l and a right-invariant Radon measure ψ_l on G, each unique up to scaling, but that they may be essentially different. (Of course, if G is an Abelian group, then they are the same.) As we shall see, the proof is more complicated than in the compact case.

First let us give a two-dimensional example where the left- and right-invariant measures are essentially different. Let H_r be the right half plane $\{(a,b) \in \mathbf{R}^2 : a > 0\}$. If $(a,b) \in H_r$, let

$$\phi((a,b)) = \begin{bmatrix} a & b \\ 0 & 1 \end{bmatrix}.$$

Then ϕ is a homeomorphism of H_r onto a subgroup of $GL_2(\mathbf{R})$, and if we define

$$(a,b)(c,d) = \phi^{-1}(\phi((a,b)), \phi((c,d))) = (ac, ad + b),$$

then H_r becomes a locally compact Polish topological group. Let λ be Lebesgue measure on H_r. Since $(a,b).(x,y) = (ax, ay + b)$, the measure $\lambda(x,y)/xy$ is a left-invariant measure on H_r. Since $(x,y).(a,b) = (ax, bx + y)$, the measure $\lambda(x,y)/x$ is a right-invariant measure on H_r.

Theorem 16.13.1 *Suppose that (G, τ) is a locally compact Polish topological group, which is not compact. There exist a non-zero left-invariant Radon measure ψ_l and a right-invariant Radon measure ψ_r on G, each unique up to scaling.*

Proof By Theorem 5.6.3 there exists a group-norm ν on G which defines the topology τ. Let $N_j = \{g \in G : \nu(g) \leq 1/2^j\}$ for $j \in \mathbf{Z}^+$. Since G is locally compact, we may suppose, by scaling d if necessary, that each N_j is compact. Let $b_j = (2 - 2^j\nu)^+$. Then $b_j \geq I_{N_j}$ and if $j > 0$ then $\mathrm{supp}(b_j) \subseteq N_{j-1}$.

Let $C_c(G)$ be the Riesz space of continuous functions on G of compact support. Our aim is to define a translation-invariant positive linear functional ϕ on $C_c(G)$. Suppose that $f \in C_c^+(G)$ and that $j \in \mathbf{Z}^+$. Recall that if $f \in C_c(G)$ and $g \in G$ then $l_g(f(h)) = f(g^{-1}h)$. If $f \in C_c^+(G)$, there exists a finite set F_j in G such that $\mathrm{supp}(f) \subseteq \cup_{g \in F_j} g^{-1}N_j$, and so if we set

$$\psi_j(f) = \inf\left\{\sum_{g\in F} t_g : F \subseteq G, F \text{ finite}, t_g \geq 0, \sum_{g\in F} t_g l_g(b_j) \geq f\right\},$$

then $\psi_j(f) < \infty$.

Then ψ_j has the following easy properties, which we shall frequently use without comment.

Exercise 16.13.2 (i) $\psi_j(l_g(f)) = \psi_j(f)$;
(ii) $\psi_j(f_1 + f_2) \leq \psi_j(f_1) + \psi_j(f_2)$;
(iii) $\psi_j(\alpha f) = \alpha\psi_j(f)$ for $\alpha \geq 0$;

(iv) if $f_1 \leq f_2$ then $\psi_j(f_1) \leq \psi_j(f_2)$;
(v) $\psi_j(f) \geq \frac{1}{2}\|f\|_\infty$;
(vi) if $j \leq k$ then $\psi_k(f) \leq \psi_k(b_j).\psi_j(f)$.

Proof Here is the proof of (vi). Suppose that $f \leq \sum_{g\in F} t_g l_g(b_j)$, and that $b_j \leq \sum_{h\in F'} t_h l_h(b_k)$. Then

$$f \leq \sum_{g\in F} t_g l_g\Big(\sum_{h\in F'} s_h l_h(b_k)\Big)$$
$$= \sum_{g\in F\, h\in F'} t_g s_h l_{gh}(b_k),$$

so that $\psi_k(f) \leq \sum_{g\in F\, h\in F'} t_g s_h$. Taking infima, (vi) follows. □

The next lemma provides the key to the proof of the theorem.

Lemma 16.13.3 *Suppose that $f_1, f_2 \in C_c^+(G)$, that $j \in \mathbf{N}^+$ and that $\epsilon > 0$. Then there exists $j_0 \in \mathbf{Z}^+$ such that*

$$\psi_j(f_1) + \psi_j(f_2) \leq \psi_j(f_1 + f_2) + \epsilon\psi_j(b_0),\ \textit{for}\ j \geq j_0.$$

Proof There exists $a \in C_c^+(G)$ such that $a(g) = \|f_1 + f_2\|_\infty$ for $g \in \operatorname{supp}(f_1 + f_2)$. Choose $0 < \delta < 1$ such that $5\delta\psi_0(a) < \epsilon$. Let $q = f_1 + f_2 + \delta a$; then $q \leq 2a$ and $\psi_0(q) \leq (1+\delta)\psi_0(a) \leq 2\psi_0(a)$. Let $p_1 = f_1/q$, $p_2 = f_2/q$, so that $p_1 + p_2 \leq 1$. The functions p_1 and p_2 are uniformly continuous, and so there exists j_0 such that if $\nu(g^{-1}h) \leq 2/2^{j_0}$ then $|p_1(g) - p_1(h)| < \delta$ and $|p_2(g) - p_2(h)| < \delta$. Suppose now that $j \geq j_0$ and that $q \leq \sum_{g\in F} t_g l_g(b_j)$, where F is a finite subset of G. If $h \in G$ then $l_g(b_j)(h) = 0$ unless $\nu(g^{-1}h) \leq 2/2^j$, in which case $|p_1(g) - p_1(h)| < \delta$ and $|p_2(g) - p_2(h)| < \delta$. Thus

$$f_1(h) = p_1(h)q(h) \leq \sum_{g\in F} t_g p_1(h) l_g(b_j)(h) \leq \sum_{g\in F} t_g p_1(g+\delta) l_g(b_j)(h)$$

so that $\psi_j(f_1) \leq (p_1(g) + \delta)\sum_{g\in F} t_g$. Similarly $\psi_j(f_2) \leq p_2(g+\delta)\sum_{g\in F} t_g$, and so $\psi_j(f_1) + \psi_j(f_2) \leq (1+2\delta)\sum_{g\in F} t_g$. Taking the infimum,

$$\begin{aligned}
\psi_j(f_1) + \psi_j(f_2) &\leq (1+2\delta)\psi_j(q)\\
&\leq (1+2\delta)(\psi_j(f_1+f_2) + \delta\psi_j(a))\\
&= (\psi_j(f_1+f_2) + \delta(2\psi_j(f_1+f_2) + 3\psi_j(a))\\
&\leq (\psi_j(f_1+f_2) + 5\delta\psi_j(a).
\end{aligned}$$

But $\psi_j(a) \leq \psi_0(a).\psi_j(b_0)$, and so $\psi_j(f_1) + \psi_j(f_2) \leq \psi_j(f_1+f_2) + \epsilon\psi_j(b_0)$. □

We now set $\phi_j = \psi_j/\psi_j(b_0)$. By Exercise 16.13.2 (vi), $\phi_j(f) \leq \phi_0(f)$.

The group G is σ-compact; let $(K_n)_{n=1}^\infty$ be a fundamental sequence of compact subsets, let $C_n(G) = \{f \in C_c(G) : \operatorname{supp}(f) \subseteq K_n\}$ and let

$C_n^+(G) = \{f \in C_n(G) : f \geq 0\}$. For each n let c_n be a function in $C_{n+1}^+(G)$ with $c_n(g) = 1$ for $g \in K_n$ and $0 \leq c_n \leq 1$. Each $C_n(G)$ is a separable Banach space under the uniform norm, and so there exists a countable dense subset D_n of C_n^+, closed under addition. Let $D = \cup_{n\in\mathbf{N}} D_n$. Since $\phi_j(f) \leq \phi_0(f)$ for $f \in C_c^+(G)$, a diagonal argument shows that there exists an increasing sequence $(j_k)_{k=1}^\infty$ such that $\phi_{j_k}(f)$ converges, to $\phi(f)$, say, for each $f \in D$. Note that if $f_1, f_2 \in D$ then $\phi(f_1 + f_2) = \phi(f_1) + \phi(f_2)$.

We now show that $\phi_{j_k}(f)$ converges, to $\phi(f)$, say, for each $f \in C_c^+(G)$. Suppose that $f \in C_n^+$ and that $\epsilon > 0$. Let $\eta = \epsilon/5(1 + \phi_0(c_n))$. There exists $d \in D_n$ such that $\|f - d\|_\infty < \eta$. Let $u = (f \vee d) - f$ and $v = (f \vee d) - d$, so that $f \vee d = f + u = d + v$, $\|u\|_\infty < \eta$ and $\|v\|_\infty < \eta$. Thus $\phi_j(u) \leq \eta\phi_0(c_n)$ and $\phi_j(v) \leq \eta\phi_0(c_n)$. By Lemma 16.13.3, there exists k_0 such that

$$|\phi_{j_k}(f + u) - \phi_{j_k}(f) - \phi_{j_k}(u)| < \eta \text{ and } |\phi_{j_k}(d + v) - \phi_{j_k}(d) - \phi_{j_k}(v)| < \eta$$

for $k \geq k_0$, and so $|\phi_{j_k}(f) - \phi_{j_k}(d)| < 2\eta(1 + \phi_0(c_n))$ for $k \geq k_0$. There exists $k_1 \geq k_0$ such that $|\phi_{j_k}(d) - \phi_{j_l}(d)| < \eta$ for $k, l \geq k_1$, and so $|\phi_{j_k}(f) - \phi_{j_l}(f)| < 5\eta(1 + \phi_0(c_n)) = \epsilon$ for $k, l \geq k_1$: $\phi_{j_k}(f)$ converges, to $\phi(f)$ say, as $k \to \infty$.

It follows from the construction that

$$\phi(f_1 + f_2) = \phi(f_1) + \phi(f_2),\ \phi(\alpha f) = \alpha\phi(f) \text{ and } \phi(l_g(f)) = \phi(f)$$

for $f_1, f_2, f \in C_c^+(G)$, $\alpha \geq 0$ and $g \in G$; by Proposition 15.1.4, ϕ extends to a positive linear functional, again denoted by ϕ, on $C_c(G)$, and $\phi(l_g(f)) = \phi(f)$ for $f \in C_c(G)$ and $g \in G$. Since $\phi_j(b_0) = 1$ for all j, $\phi(b_0) = 1$ and ϕ is non-zero. It now follows from the Riesz representation theorem for locally compact spaces (Theorem 16.7.1) that there is a unique Radon measure μ_l on G which represents ϕ, and the left invariance of ϕ implies that μ_l is left invariant.

A right-invariant Radon measure μ_r is defined in exactly the same way.

It remains to show that μ_l is essentially unique; we use the Radon–Nikodym theorem for this. Suppose that ν is a left-invariant Radon measure on G. Let $\pi = \mu_l + \nu$, so that $0 \leq \nu \leq \pi$. By Proposition 15.4.8, there exists $h \in L^\infty(\pi)$ with $0 \leq h \leq 1$ for which $\nu(A) = \int_A h\,d\pi$ for every Borel set A. Suppose that $g \in G$ and that $A \in \mathcal{B}(G)$. Then

$$\int_A h(x)\,d\pi(x) = \nu(A) = \nu(g^{-1}A) = \int_{g^{-1}A} h(x)\,d\pi(x) = \int_A h(g^{-1}x)\,d\pi(x).$$

Since this holds for all $A \in \mathcal{B}(G)$, $h(x) = h(g^{-1}x)$ for π-almost all x. Suppose that A and B are Borel sets of finite π measure. Then, by Fubini's theorem,

$$\begin{aligned} 0 &= \int_B \left(\int_A |h(x) - h(g^{-1}x)|\,d\pi(x) \right) d\pi(g) \\ &= \int_A \left(\int_B |h(x) - h(g^{-1}x)|\,d\pi(g) \right) d\pi(x), \end{aligned}$$

so that $\int_B |h(x) - h(g^{-1}x)|\, d\pi(g) = 0$ for π-almost all x. For such x, $h(x) = h(g^{-1}x)$ for π-almost all g, so that h is a constant in $L^\infty(\pi)$. If $A \in \mathcal{B}(G)$ then $\nu = h(\mu_l + \nu)$, so that $0 < h < 1$ and $\nu = (h/(1-h))\mu_l$. □

Note also that it follows from the essential uniqueness of μ_l that $\phi_j(f) \to f$ as $j \to \infty$ for each $f \in C_c(G)$.

A continuous action a of a locally compact metrizable group G on a locally compact metrizable space X is a *Radon action* if it is transitive and if for each x in X there exists a neighbourhood $N(x)$ such that $\mu_l(a^{-1}N(x)) < \infty$ and $\mu_r(a^{-1}N(x)) < \infty$.

Exercise 16.13.4 Show that if a is a Radon action of a locally compact metrizable group G on a locally compact metrizable space X then the push-forward measures $a_*(\mu_l)$ and $a_*(\mu_r)$ are Radon measures.

Exercise 16.13.5 Suppose that a is a Radon action of a locally compact metrizable group G on a locally compact metrizable space, and that H is a closed subgroup of G for which the restriction of a also acts transitively. Let ν_l and ν_r be the Haar measures on H. Show that, up to a scaling factor, $\nu_l = \mu_l$ and $\nu_r = \mu_r$.

Exercise 16.13.6 Use the preceding exercise to show that Lebesgue measure on $\mathbf{R}^d$ is rotation invariant.

17
Measures on Euclidean Space

Euclidean space is one of the most important examples of a Polish space. In this chapter, we consider measures on Euclidean space, and their interaction with the geometry of the space.

17.1 Borel Measures on R and $\mathbf{R}^d$

We begin by considering Borel measures on **R**. Suppose that μ is a σ-finite measure on **R**. Let

$$J_\mu(t) = \begin{cases} \mu((0,t]) & \text{for } t > 0, \\ 0 & \text{for } t = 0, \\ -\mu((t,0]) & \text{for } t < 0. \end{cases}$$

The function J_μ is an increasing function on **R**. If $t_n \searrow t > 0$ then $\mu((0,t_n]) \to \mu((0,t])$ as $n \to \infty$, so that J_μ is continuous on the right at t. Similar arguments apply when $t = 0$ and when $t < 0$. The set of points of discontinuity of J_μ is countable, and the points of continuity are dense in **R**. t is a point of discontinuity of J_μ if and only if t is an atom of μ; that is, if and only if $\mu(\{t\}) = J(t) - J(t_-) > 0$, where $J(t_-) = \lim_{s \nearrow t} J(s)$.

We now establish a converse result.

Theorem 17.1.1 *Suppose that G is an increasing right continuous real-valued function on* **R** *and that $G(0) = 0$. Then there exists a unique σ-finite Borel measure on* **R** *such that $G = J_\mu$.*

Proof Suppose that $n \in \mathbf{N}$ and that $f \in C([-n,n])$. Suppose that D is a dissection $(-n = a_0 < a_1 < \cdots < a_k = n)$ of $[-n,n]$, with mesh size $\delta(D) = \max_{1 \le j \le k}(a_j - a_{j-1})$. Let

$$s_D(f) = f(a_0)(G(a_0) - G((a_0)_-) + \sum_{j=1}^{k} f(a_j)(G(a_j) - G(a_{j-1})).$$

□

Exercise 17.1.2 Show that $s_D(f)$ tends to a limit $\phi_n(f) = \int_{-n}^{n} f\,dg$, the *Riemann–Stieltjes integral* of f with respect to G, as the mesh size $\delta(D)$ tends to 0 and that ϕ_n is a positive linear functional on $C([-n,n])$, and $|\phi_n(f)| \leq \|f\|_\infty (G(n) - G((-n)_-)$.

By the Riesz representation theorem, there exists a finite Borel measure μ_n on $[-n,n]$ such that $\phi_n(f) = \int_{[-n.n]} f\,d\mu_n$. If B is a Borel subset of $[-n,n]$ and $m > n$, then $\mu_m(B) = \mu_n(B)$, and so there exists a σ-finite Borel measure μ on $\mathbf{R}$ such that $\mu(B) = \mu_n(B)$.

Exercise 17.1.3 Shows that $J_\mu = G$.

The first and most important example is *Lebesgue measure* λ on $\mathbf{R}$, which we have met earlier, and which is obtained by taking $G(t) = t$, so that $\lambda((a,b)) = b - a$ for any open interval (a,b). (In fact, the term 'Lebesgue measure' is usually used for the completion of this measure, but this will not concern us.) Lebesgue measure is a Haar measure on $\mathbf{R}$.

If μ is a finite measure, it is natural to replace J_μ by the *cumulative distribution function* F_μ, defined as $F_\mu(t) = \mu((-\infty,t]) = G_\mu(t) + \mu((-\infty,0])$. Thus F_μ is a right continuous function on $\mathbf{R}$, and $F_\mu(t) \to 0$ as $t \to -\infty$ and $F_\mu(t) \to \mu(\mathbf{R})$ as $t \to \infty$. In particular, if $\mu \in P(\mathbf{R})$ then F_μ is an increasing function and $F_\mu(t) \to 1$ as $t \to \infty$.

Note that if f is a non-negative measurable function on a probability space (X, Σ, μ) then the push-forward measure $f_*(\mu)$ is a Borel probability measure on $\mathbf{R}$, and the tail distribution function λ_f satisfies $\lambda_f(t) = \mu(f > t) = 1 - F_{f_*(\mu)}(t)$.

We can without difficulty extend these ideas to Borel measures on $\mathbf{R}^d$. Let us restrict our attention to Borel probability measures. Suppose that $\mu \in P(\mathbf{R}^d)$. If $(a_1, \ldots, a_d) \in \mathbf{R}^d$, let $J_\mu(a_1, \ldots, a_d) = \mu(\prod_{j=1}^{d}(-\infty, a_j])$. If $C \subseteq \{1, \ldots, d\}$ let $C(j) = a_j$ if $j \in C$ and let $C(j) = b_j$ otherwise. Then J_μ satisfies the following conditions.

(i) If $\inf a_j \to \infty$, then $J_\mu(a_1, \ldots, a_d) \to 1$.
(ii) If $\inf a_j \to -\infty$, then $J_\mu(a_1, \ldots, a_d) \to 0$.
(iii) If $a_j \searrow \tilde{a}_j$ for $1 \leq j \leq d$ then $J_\mu(a_1, \ldots, a_d) \searrow J_\mu(\tilde{a}_1, \ldots, \tilde{a}_d)$.
(iv) $0 \leq \sum_{C \subseteq \{1,\ldots,d\}} (-1)^{|C|} J_\mu(C(1), \ldots, C(d)) \leq 1$.

Exercise 17.1.4 Verify this.

A function which satisfies (i)–(iv) is called a Stieltjes (probability) function. Conversely, if G is a Stieltjes probability function, the preceding arguments show that there is a unique Borel probability measure on $\mathbf{R}^d$ such that $\mu(\prod_{j=1}^d(-\infty, a_j]) = G(a_1, \ldots, a_d)$.

Recall that the product measure $\lambda_d = \lambda \otimes \cdots \otimes \lambda$ on $\mathbf{R}^d$ is also called Lebesgue measure.

Exercise 17.1.5 Let $(e_1, \ldots, e_d)$ be the standard basis for $\mathbf{R}^d$. Since the orthogonal group O_n is generated by rotations which fix $e_3, \ldots, e_d$, the reflection which fixes $e_2, \ldots, e_d$ and isometries which permute $e_1, \ldots, e_d$, give another proof that λ_d is invariant under the orthogonal group; if $J \in O_n$ and B is a Borel set, then $\lambda_d(J(B)) = \lambda_d(B)$.

Now suppose that E is a d-dimensional Euclidean space, and that $(f_1, \ldots, f_d)$ is an orthonormal basis for E. If $x = \sum_{j=1}^d x_j e_j \in \mathbf{R}^d$, let $T(x) = \sum_{j=1}^d x_j f_j$, and let $\mu_d = T_*(\lambda_d)$. Then μ_d is a σ-finite measure on E, and it follows from the invariance of λ_d under the orthogonal group that μ_d does not depend upon the choice of orthogonal basis. μ_d is also called *Lebesgue measure*, and is frequently also denoted by λ_d.

17.2 Functions of Bounded Variation

We now extend the results of the preceding section to signed measures on $\mathbf{R}$. This involves functions of *bounded variation.*

Let I be an interval in $\mathbf{R}$. We denote the set of all finite strictly increasing sequences $T = (t_0 < t_1 < \cdots < t_k)$ in I by $\mathcal{T}(I)$. Suppose that f is a real-valued function on $\mathbf{R}$. If $T = (t_0 < t_1 < \cdots < t_k) \in \mathcal{T}(I)$, we set

$$v_T^+(f) = \sum_{j=1}^k (f(t_j) - f(t_{j-1}))_+, \quad v^+(f, I) = \sup_{T \in \mathcal{T}(I)} v_T^+(f),$$

$$v_T^-(f) = \sum_{j=1}^k (f(t_j) - f(t_{j-1}))_-, \quad v^-(f, I) = \sup_{T \in \mathcal{T}(I)} v_T^-(f),$$

$$v_T(f) = \sum_{j=1}^k |f(t_j) - f(t_{j-1})|, \quad v(f, I) = \sup_{T \in \mathcal{T}(I)} v_T(f).$$

Clearly $v_T(f) = v_T^+(f) + v_T^-(f)$ and $f(t_k) - f(t_0) = v_T^+(f) - v_T^-(f)$. Consequently $v(f, I) = v^+(f, I) + v^-(f, I)$ and if $I = [a, b]$ then $f(b) - f(a) = v^+(f, I) - v^-(f, I)$.

The quantity $v^+(f,I)$ is the *positive variation* of f on I, $v^-(f,I)$ is the *negative variation* of f on I, and $v(f,I)$ is the *total variation* of f on I. A real-valued function f on I is of *bounded variation* if $v(f,I)$ is finite.

Exercise 17.2.1 Suppose that f is a function of bounded variation on $\mathbf{R}$. Let $v_f^+(t) = v^+(f,(-\infty,t])$, $v_f^-(t) = v^-(f,(-\infty,t])$ and $v_f(t) = v(f,(-\infty,t])$. Show that each is a bounded increasing non-negative function on $\mathbf{R}$, $v_f = v_f^+ - v_f^-$, and that each tends to 0 as $t \to -\infty$.

Show that the function f is bounded, and tends to limits $f(-\infty)$ and $f(\infty)$ as $t \to -\infty$ and as $t \to \infty$. Show further that $f(t) = f(-\infty) + v_f^+(t) - v_f^-(t)$.

Theorem 17.2.2 *Suppose that f is a function of bounded variation on $\mathbf{R}$ which is right-continuous at t_0. Then v_f^+, v_f^- and v_f are right-continuous at t_0.*

Proof Since $v_f^+ = \frac{1}{2}(v_f + f)$ and $v_f^- = \frac{1}{2}(v_f - f)$, it is enough to show that v_f is right-continuous. Suppose that $\epsilon > 0$. There exists $\delta > 0$ such that $|f(s) - f(t_0)| < \epsilon/2$ for $t < s < t+\delta$. Choose $t_0 < r < t_0+\delta$. There exists $T = (t_0 < t_1 < \cdots < t_k = r) \in \mathcal{T}([t,r])$ for which $v_T(f) > v(f,[t_0,r]) - \epsilon/2$. Then

$$\begin{aligned} v_f(t_1) - v_f(t_0) &= v(f,[t_0,r]) - v(f,[t_1,r]) \\ &\leq (v_T(f) + \epsilon/2) - \sum_{j=2}^{k} |f(t_j) - f(t_{j-1})| \\ &= |f(t_1) - f(t_0)| + \epsilon/2 < \epsilon. \end{aligned}$$

Since v_f is an increasing function, this shows that f is right-continuous. □

We denote by $bv_0(\mathbf{R})$ the vector space of right-continuous functions of bounded variation on $\mathbf{R}$ which tend to 0 as $t \to -\infty$.

Exercise 17.2.3 Establish the following.

(i) The function $v(.,\mathbf{R})$ is a norm on $bv_0(\mathbf{R})$; we denote $v(f,\mathbf{R})$ by $\|f\|_{bv}$.
(ii) If f is an increasing function in $bv_0(\mathbf{R})$, then $\|f\|_{bv} = f(+\infty) = \|f\|_\infty$.
(iii) If $f \in bv_0(\mathbf{R})$ then $\|f\|_{bv} = \left\|v_f^+\right\|_{bv} + \left\|v_f^-\right\|_{bv}$.

Recall that if $\sigma \in (M(\mathbf{R},\mathcal{B}))$ then $F_\sigma(t) = \sigma((-\infty,t])$.

Theorem 17.2.4 *The mapping $F : \sigma \to F_\sigma$ is a linear isometry of $(M(\mathbf{R},\mathcal{B}), \|.\|_{TV})$ onto $(bv_0(\mathbf{R}), \|.\|_{bv})$, with inverse mapping $\mu : f \to \mu_f$, where $\mu_f = \mu_{v_f^+} - \mu_{v_f^-}$. Further, if $f \in bv_0$ then $(\mu_f)^+ = \mu_{v_f^+}$ and $(\mu_f)^- = \mu_{v_f^-}$.*

Proof If $\sigma \in M(\mathbf{R})$, then $F_\sigma = F_{\sigma^+} - F_{\sigma^-}$ so that $F_\sigma \in bv_0(\mathbf{R})$. If $F_\sigma = 0$ then $F_{\sigma^+} = F_{\sigma^-}$. It therefore follows from Theorem 17.1.1 that $\sigma^+ = \sigma^- = 0$, so that F is injective.

If $f \in bv_0(\mathbf{R})$, then $f = v_f^+ - v_f^-$. Then $\mu_f = \mu_{v_f^+} - \mu_{v_f^-} \in M(\mathbf{R})$. If $\sigma \in M(\mathbf{R})$ then $\sigma = \mu_{F_\sigma}$, so that F is bijective; the mapping $f \to \mu_f$ is the inverse of the mapping F. If μ is a positive measure, then $\|F_\mu\|_{bv} = \mu(\mathbf{R}) = \|\mu\|_{TV}$. Thus if $\sigma \in M(\mathbf{R})$, then

$$\begin{aligned}\|F_\sigma\|_{bv} &= \|F_{\sigma^+} - F_{\sigma^-}\|_{bv} \leq \|F_{\sigma^+}\|_{bv} + \|F_{\sigma^-}\|_{bv}\\ &= \left\|\sigma^+\right\|_{-TV} + \left\|\sigma^-\right\|_{TV} = \left\|\sigma^+\right\|_{TV},\end{aligned}$$

so that F is norm decreasing. Similarly, if $f \in bv_0(\mathbf{R})$, then

$$\begin{aligned}\left\|\mu_f\right\|_{TV} &= \left\|\mu_{v_f^+} - \mu_{v_f^-}\right\|_{TV} \leq \left\|\mu_{v_f^+}\right\|_{TV} + \left\|\mu_{v_f^-}\right\|_{TV}\\ &= \left\|v_f^+\right\|_{bv} + \left\|v_f^-\right\|_{bv} = \|f\|_{bv},\end{aligned}$$

so that F^{-1} is also norm decreasing. Thus F is an isometry. Further,

$$\left\|\mu_f\right\|_{TV} = \|f\|_{bv} = \left\|v_f^+\right\|_{bv} + \left\|v_f^-\right\|_{bv} = \left\|\mu_{v_f^+}\right\|_{TV} + \left\|\mu_{v_f^-}\right\|_{TV},$$

so that $(\mu_f)^+ = \mu_{v_f^+}$ and $(\mu_f)^- = \mu_{v_f^-}$, by Exercise 15.3.2. □

17.3 Spherical Derivatives

We now consider finite Borel measures on $\mathbf{R}^d$ and functions in $L^1(\mathbf{R}^d, \mathcal{B}_d, \lambda_d)$. As usual, let $N_r(x)$ denote the open Euclidean ball $\{y : |y - x| < r\}$ of radius r with centre x. Then $\lambda_d(N_r(x)) = r^d\Omega_d$, where Ω_d is the Lebesgue measure of the unit ball in $\mathbf{R}^d$.

Exercise 17.3.1 Calculate Ω_d.

If μ is a finite Borel measure on $\mathbf{R}^d$, we set

$$A_r(\mu)(x) = \frac{\mu(N_r(x))}{\lambda_d(N_r(x))} = \frac{\mu(N_r(x))}{r^d\Omega_d}.$$

Proposition 17.3.2 *The function $A_r(\mu)$ is lower semi-continuous.*

Proof Suppose that $x \in \mathbf{R}^d$, and that $\epsilon > 0$. Let r_n increase to r as $n \to \infty$. By upwards continuity, there exists $n \in \mathbf{N}$ such that $\mu(N_{r_n}(x)) > \mu_r(x) - \epsilon$. If $\|y - x\| < r - r_n$ then $N_{r_n}(x) \subseteq N_r(y)$, so that $A_r(y) > A_r(x) - \epsilon$. □

We say that μ has a *spherical derivative* $D_s\mu(x)$ at x if, given $\epsilon > 0$, there exists $r_0 > 0$ such that if $0 < r < r_0$ and $x \in N_r(y)$ then $|A_r(\mu)(y) - D_s\mu(x)| < \epsilon$. It is important that in this definition we consider spheres to which x belongs, and not just spheres centred at x.

Similarly, if $f \in L^1(\mathbf{R}^d, \mathcal{B}, \lambda_d)$, we set

$$A_r(f)(x) = \frac{\int_{N_r(x)} f\, d\lambda_d}{\lambda(N_r(x))} = \frac{1}{r^d \Omega_d} \int_{N_r(x)} f\, d\lambda_d.$$

$A_r(f)(x)$ is the average value of f over the ball $N_r(x)$. Again, we say that f has a *spherical derivative* $D_s f(x)$ at x if, given $\epsilon > 0$, there exists $r_0 > 0$ such that if $0 < r < r_0$ and $x \in N_r(y)$ then $|A_r(f)(y) - D_s f(x)| < \epsilon$. Thus the spherical derivative of the function f is the same as the spherical derivative of the measure $f.d\lambda_d$.

First we consider a function f in $L^1(\mathbf{R}^d, \mathcal{B}, \lambda_d)$. We set

$$m_u(f)(x) = \sup_{r>0} \left(\sup\{A_r(|f|)(y) : y \in N_r(x)\}\right).$$

An extended real-valued function ϕ on $L^1(\mathbf{R}^d, \mathcal{B}, \lambda_d)$ is said to be of *weak type* $(1,1)$ *with constant* k if, whenever $\alpha > 0$ and $E_\alpha = (\phi(f) > \alpha)$, then

$$\lambda_d(E_\alpha) \leq \frac{k}{\alpha} \int_{E_\alpha} |f|\, d\lambda_d.$$

Theorem 17.3.3 *If $f \in L^1(\mathbf{R}^d, \mathcal{B}, \lambda_d)$ then $m_u(f)$ is a lower semi-continuous extended sublinear functional on $L^1(\mathbf{R}^d, \mathcal{B}, \lambda_d)$, which is of weak type $(1,1)$ with constant 3^d.*

Proof Since m_u is the supremum of lower semi-continuous functions, it is lower semi-continuous, and it is clearly an extended sublinear functional on $L^1(\mathbf{R}^d, \mathcal{B}, \lambda_d)$.

The key result for the second statement is the following covering lemma.

Lemma 17.3.4 (Wiener's lemma) *Suppose that G is a finite set of open balls in $\mathbf{R}^d$. Then there is a finite subcollection F of disjoint balls such that*

$$\sum_{U \in F} \lambda_d(U) = \lambda_d\left(\bigcup_{U \in F} U\right) \geq \frac{1}{3^d} \lambda_d\left(\bigcup_{U \in G} U\right).$$

Proof We use a greedy algorithm. If $U = N_r(x)$ is an open ball, let $U^* = N_{3r}(x)$ be the ball with the same centre as U, but with three times the radius.

Let U_1 be a ball of maximal radius in G. Let U_2 be a ball of maximal radius in G, disjoint from U_1. Continue, choosing U_j of maximal radius, disjoint from $U_1, \ldots, U_{j-1}$, until the process stops, with the choice of U_k. Let $F = \{U_1, \ldots, U_k\}$.

Suppose that $U \in G$. There is a least j such that $U \cap U_j \neq \emptyset$. Then the radius of U is no greater than the radius of U_j (otherwise we would have chosen U to be U_j) and so $U \subseteq U_j^*$. Thus $\bigcup_{U \in G} U \subseteq \bigcup_{U \in F} U^*$ and

$$\lambda_d\left(\bigcup_{U\in G} U\right) \leq \lambda_d\left(\bigcup_{U\in F} U^*\right) \leq \sum_{U\in F} \lambda_d(U^*) = 3^d \sum_{U\in F} \lambda(U).$$

□

We now return to the proof of Theorem 17.3.3. Suppose that $f \in L^1(\mathbf{R}^d)$, that $\alpha > 0$ and that $E_\alpha = (m_u(f)(x) > \alpha)$. Let K be a compact subset of E_α. For each $x \in K$, there exist $y_x \in \mathbf{R}^d$ and $r_x > 0$ such that $x \in N_{r_x}(y_x)$ and $A_{r_x}(|f|)(y_x) > \alpha$. It follows from the definition of m_u that $N_{r_x}(y_x) \subseteq E_\alpha$. The sets $N_{r_x}(y_x)$ cover K, and so there is a finite subcover G. By the lemma, there is a subcollection F of disjoint balls such that

$$\sum_{U\in F} \lambda_d(U) \geq \frac{1}{3^d}\lambda_d\left(\bigcup_{U\in G} U\right) \geq \frac{\lambda_d(K)}{3^d}.$$

But if $U \in F$, $\alpha\lambda_d(U) \leq \int_U |f|\, d\lambda_d$, so that since $\bigcup_{U\in F} U \subseteq E_\alpha$,

$$\sum_{U\in F} \lambda_d(U) \leq \frac{1}{\alpha}\sum_{U\in F}\int_U |f|\, d\lambda_d \leq \frac{1}{\alpha}\int_{E_\alpha} |f|\, d\lambda_d.$$

Thus $\lambda_d(K) \leq 3^d(\int_{E_\alpha} |f|\, d\lambda)/\alpha$, and

$$\lambda_d(E_\alpha) = \sup\{\lambda_d(K) : K \text{ compact}, K \subseteq E_\alpha\} \leq \frac{3^d}{\alpha}\int_{E_\alpha} |f|\, d\lambda_d.$$

□

Exercise 17.3.5 Let $m(f) = \max(m_u(f), |f|)$. Show that m is a sublinear mapping of weak type $(1, 1)$, with constant $3^d + 1$.

Exercise 17.3.6 Suppose that V is a bounded open subset of $\mathbf{R}^d$. A *Vitali covering* $\mathcal{V}$ is a collection of open balls with the property that if $\mathcal{F}$ is a finite subset of $\mathcal{V}$ and $C_{\mathcal{F}} = \cup\{\overline{B} : B \in \mathcal{F}\}$, then $V \setminus C_{\mathcal{F}} = \cup\{B \in \mathcal{V} : B \cap C_{\mathcal{F}} = \emptyset\}$. Use Wiener's lemma to show that there exists a disjoint sequence $(B_n)_{n=1}^\infty$ in $\mathcal{V}$ such that $\lambda_d(V) = \sum_{n=1}^\infty \lambda_d(B_n)$.

Exercise 17.3.7 Suppose that V is a bounded open subset of $\mathbf{R}^d$. Show that there exists a disjoint sequence $(B_n)_{n=1}^\infty$ in V such that $\lambda_d(V) = \sum_{n=1}^\infty \lambda_d(B_n)$. Deduce that λ_d is rotation invariant.

17.4 The Lebesgue Differentiation Theorem

Theorem 17.4.1 (The Lebesgue differentiation theorem) *Suppose that $g \in L^1(\mathbf{R}^d, \mathcal{B}, \lambda_d)$. Then $g(x)$ is the spherical derivative of g at x, for λ_d-almost every $x \in \mathbf{R}^d$.*

Proof The result holds for $f \in C_c(\mathbf{R}^d)$, the space of continuous functions on $\mathbf{R}^d$ of compact support, and $C_c(\mathbf{R}^d)$ is norm dense in $L^1(\mathbf{R}^d, \mathcal{B}, \lambda_d)$. We use the first Borel–Cantelli lemma. For each n there exists $f_n \in C_c(\mathbf{R}^d)$ with $\|g - f_n\|_1 \leq 1/2^n$. Let $m(f) = \max(m_u(f), |f|)$, and let $B_n = \{x \in \mathbf{R}^d : m(g - f_n)(x) > 1/n\}$. B_n is open, and, using Corollary 17.3.5, $\lambda_d(B_n) \leq (3^d + 1)n/2^n$. Let $B = \limsup(B_n)$. Then $\lambda_d(B) = 0$, by the first Borel–Cantelli lemma.

If $x \notin B$, there exists n_0 such that $x \notin B_n$ for $n \geq n_0$, so that

$$|A_r(g)(x) - A_r(f_n)(x)| \leq m(g - f_n)(x) \leq 1/n, \text{ for } r \geq 0,$$

and $|g(x) - f_n(x)| < 1/n$. Thus if $n \geq n_0$, then

$$\begin{aligned} &|A_r(g)(x) - g(x)| \\ &\leq |A_r(g)(x) - A_r(f_n)(x)| + |A_r(f_n)(x) - f_n(x)| + |f_n(x) - g(x)| \\ &\leq 2/n + |A_r(f_n)(x) - f_n(x)| \leq 3/n \end{aligned}$$

for small enough r, and so $A_r(g)(x) \to g(x)$ as $r \to 0$. □

Corollary 17.4.2 (The Lebesgue density theorem) *If E is a measurable subset of $\mathcal{R}^d$ then*

$$\frac{1}{r^d \Omega_d} \lambda_d(N_r(x) \cap E) = \frac{\lambda_d(N_r(x) \cap E)}{\lambda_d(N_r(x))} \to 1 \text{ as } r \to 0 \text{ for almost all } x \in E$$

and

$$\frac{1}{r^d \Omega_d} \lambda_d(N_r(x) \cap E) = \frac{\lambda_d(N_r(x) \cap E)}{\lambda_d(N_r(x))} \to 0 \text{ as } r \to 0 \text{ for almost all } x \notin E.$$

Proof Apply the theorem to the indicator functions $I_{E \cap N_k(0)}$, for $k \in \mathbf{N}$. □

17.5 Differentiating Singular Measures

Next we consider a finite measure μ for which μ and λ_d are mutually singular.

Theorem 17.5.1 *Suppose that μ is a finite Borel measure on $\mathbf{R}^d$ for which μ and λ_d are mutually singular. Then μ has spherical derivative* 0 *at λ_d-almost every point of $\mathbf{R}^d$.*

Proof There exists a Borel λ_d-null set A for which $\mu(\mathbf{R}^d \setminus A) = 0$. Since μ is tight, there exists an increasing sequence $(K_n)_{n=1}^\infty$ of compact subsets of A with $\mu(K_n) > \mu(A) - 1/4^n$ for $n \in \mathbf{N}$. Let U_n be the open set $\mathbf{R}^d \setminus K_n$; then $\mu(U_n) < 1/4^n$.

Let

$$H_n = \{(x, r) : x \in U_n, 0 < r < \min(1/2^n, d(x, K_n)), A_r(\mu)(x) > 1/2^n\}$$

and let $V_n = \bigcup_{(x,r)\in H_n} N_r(x)$. Suppose that L is a compact subset of V_n. There exists a finite subset G of H_n such that $L \subseteq \bigcup_{(x,r)\in G} N_r(x)$. By Wiener's lemma, there exists a subset F of G such that the sets $\{N_r(x) : (x,r) \in F\}$ are disjoint and

$$\lambda^d(\bigcup_{(x,r)\in F} N_r(x)) \geq (1/3^d)\lambda_d(\bigcup_{(x,r)\in G} N_r(x)).$$

Then

$$\begin{aligned}\lambda_d(L) &\leq \lambda_d(\bigcup_{(x,r)\in G} N_r(x)) \leq 3^d \lambda_d(\bigcup_{(x,r)\in F} N_r(x))\\ &= 3^d \sum_{(x,r)\in F} \lambda_d(N_r(x)) \leq 3^d.2^n \sum_{(x,r)\in F} \mu(N_r(x))\\ &= 3^d.2^n \mu(\bigcup_{(x,r)\in G} N_r(x)) \leq 3^d.2^n \mu(V_n) \leq 3^d.2^{-n}.\end{aligned}$$

Since μ is tight, $\mu(V_n) \leq 3^d.2^{-n}$.

Let $B = \limsup_{n\to\infty} V_n$. It follows from the first Borel–Cantelli lemma that $\lambda_d(B) = 0$. Consequently $\lambda_d(A \cup B) = 0$. If $x \notin A \cup B$ then there exists N such that $x \notin V_n$, for $n \geq N$. If $n \geq N$, and $0 < r < \frac{1}{2}\min(1/2^d, d(x, K_n))$ then $A_{2r}(\mu)(x) < 1/2^n$. If $d(x,y) < r$ then $N_r(y) \subseteq N_{2r}(x)$, so that $A_r(\mu)(y) \leq 2^d/2^n$. Thus μ has spherical derivative 0 at x. □

Combining Theorems 17.4.1 and 17.5.1, we obtain the following.

Theorem 17.5.2 *Suppose that μ is a finite Borel measure on $\mathbf{R}^d$. Then μ has a spherical derivative $D_s\mu$ at λ_d-almost every point of $\mathbf{R}^d$. The function $D_s\mu$ is λ_d-integrable. Set $\nu(A) = \mu(A) - \int_A D_s\mu \, d\lambda_d$. Then ν is a Borel measure on $\mathbf{R}^d$, ν and λ_d are mutually singular, and $\mu = D_s\mu.d\lambda_d + \nu$ is the Lebesgue decomposition of μ.*

Proof Let $\mu = f.d\lambda_d + \nu$ be the Lebesgue decomposition of μ. Then μ has spherical derivative $f.d\lambda_d$-almost everywhere. □

17.6 Differentiating Functions in bv_0

Once again, let us return to the one-dimensional case. A real-valued function f on $\mathbf{R}$ is *absolutely continuous* if whenever $\epsilon > 0$ there exists $\delta > 0$ such that if $(I_j)_{j=1}^k = ((a_j, b_j))_{j=1}^k$ is a sequence of disjoint intervals of total length $\sum_{j=1}^k l(I_j) = \sum_{j=1}^k (b_j - a_j)$ less than δ then $\sum_{j=1}^k |f(b_j) - f(a_j)| < \epsilon$. An absolutely continuous function is clearly uniformly continuous.

Theorem 17.6.1 *A positive signed Borel measure ν on $\mathbf{R}$ is absolutely continuous with respect to Lebesgue measure λ if and only if its cumulative distribution function F_ν is an absolutely continuous function on $\mathbf{R}$.*

Proof It is clearly enough to consider the case where ν is a finite positive measure. Suppose first that ν is absolutely continuous with respect to λ. Given $\epsilon > 0$, there exists $\delta > 0$ such that if $A \in \mathcal{B}$ and $\lambda(A) < \delta$ then $\nu(A) < \epsilon$. If $(I_j)_{j=1}^k = ((a_j, b_j))_{j=1}^k$ is a sequence of disjoint intervals of total length $\sum_{j=1}^k l(I_j) = \sum_{j=1}^k (b_j - a_j)$ then $\lambda(\cup_{j=1}^k (a_j, b_j]) < \delta$, so that

$$\nu(\cup_{j=1}^k (a_j, b_j]) = \sum_{j=1}^k |F_\nu(b_j) - F_\nu(a_j)| < \epsilon,$$

and F_ν is an absolutely continuous function.

Suppose conversely that F_ν is an absolutely continuous function. We use the Radon–Nikodym theorem (Corollary 15.4.5). Suppose that A is a Borel set for which $\lambda(A) = 0$, and that $\epsilon > 0$. There exists $\delta > 0$ for which the absolute continuity condition is satisfied. There then exists an open set U containing A, with $\lambda(U) < \delta$. Suppose that $U = \cup_{j=1}^\infty I_j = \cup_{j=1}^\infty (a_j, b_j)$ is a disjoint union of an infinite sequence of open intervals. Then

$$\begin{aligned}\nu(A) \le \nu(U) &= \lim_{k\to\infty} \nu(\cup_{j=1}^k (a_j, b_j)) \\ &\le \lim_{k\to\infty} \nu(\cup_{j=1}^k (a_j, b_j]) = \lim_{k\to\infty} \sum_{j=1}^k (F_\nu(b_j) - F_\nu(a_j)) \le \epsilon.\end{aligned}$$

(The case where U is a finite union is even easier.) Since ϵ is arbitrary, $\nu(A) = 0$, and so ν is absolutely continuous with respect to λ. □

We now apply the results of the previous section to functions in bv_0.

Theorem 17.6.2 *Suppose that $F \in bv_0$. Then F is differentiable almost everywhere. If f is the derivative of F, then f is integrable, and $\int_{(-\infty,t]} f d\lambda \le F(t)$ for almost all $t \in \mathbf{R}$. Equality holds for all $t \in \mathbf{R}$ if and only if $\int_{\mathbf{R}} f\, d\lambda = \lim_{t\to+\infty} F(t)$, and if and only if F is an absolutely continuous function on* **R**.

Proof It is clearly enough to consider the case where F is increasing. Then F is continuous except on a countable set J, and the discontinuities are all jump discontinuities. F is the cumulative distribution function of a finite Borel measure μ, and μ has a spherical derivative $D_s\mu$ except on a null-set N, which clearly includes J. Thus if $x \notin N$ then

$$\lim_{h,k\searrow 0} \frac{F(x+h) - F(x-k)}{h+k} = \lim_{h,k\searrow 0} \frac{\mu([x-k, x+h])}{h+k} = D_s\mu(x).$$

Since f is continuous at x,

$$\frac{F(x+h)-F(x-k)}{h+k} \to \frac{F(x+h)-F(x)}{h} \text{ as } k \searrow 0,$$
$$\text{and } \frac{F(x+h)-F(x-k)}{h+k} \to \frac{F(x)-F(x-k)}{k} \text{ as } h \searrow 0.$$

Thus F is differentiable at x, with derivative $f = D_s\mu$.

By Theorem 17.5.2, f is integrable, and $\mu = f.d\lambda + \nu$, where ν and λ are mutually singular. If $t \notin J$, then

$$F(t) = \mu((-\infty,t]) = \int_{(-\infty,t]} f\,d\lambda + \nu((-\infty,t]) \geq \int_{(-\infty,t]} f\,d\lambda.$$

Equality holds for all t if and only if $\nu = 0$. This happens if and only if F is absolutely continuous, and if and only if

$$\mu(\mathbf{R}) = \lim_{t\to+\infty} F(t) = \int_{\mathbf{R}} f\,d\lambda.$$

□

Let us consider the structure of a finite Borel measure μ on $\mathbf{R}$. By the Lebesgue decomposition theorem, $\mu = f.d\lambda + \nu$, where $f \in L^1(\mathbf{R}, \mathcal{B}, \lambda)$, and ν and λ are mutually singular. The cumulative distribution function $\int_{(-\infty,t]} f\,d\lambda$ of $f.d\lambda$ is absolutely continuous, so that F_ν has the same set J of discontinuities as F_μ.

If $x \in J$, let $j(x)$ be the size of the jump at x. If A is a Borel set, let $\alpha(A) = \sum\{j(x) : x \in A \cap J\}$. Then α is an *atomic* Borel measure; $\alpha(\{x\}) = j(x) > 0$ if $x \in J$, and $\alpha(\mathbf{R} \setminus J) = 0$. Further, α and λ are mutually singular.

Now let $\pi = \nu - \alpha$. Then π is a finite measure, π and λ are mutually singular, as are π and α. The cumulative distribution function F_π has no jumps, and is therefore a continuous function. Since π and λ are mutually singular, F_π is differentiable almost everywhere, and its derivative is 0 almost everywhere. A Borel measure on $\mathbf{R}$ such as π, which has a continuous cumulative distribution function, but for which π and λ are mutually singular, is called a *continuous singular* measure.

Summing up, if μ is a finite Borel measure on $\mathbf{R}$, then μ can be written as the sum of an absolutely continuous measure $f.d\lambda$, an atomic measure α, and a continuous singular measure π. It is easy to see that this decomposition is uniquely determined.

17.7 Rademacher's Theorem

Suppose that f is an L-Lipschitz function on a bounded closed interval $[a,b]$ in $\mathbf{R}$. Extend f to a function on $\mathbf{R}$ by setting $f(t) = f(a)$ for $t < a$ and $f(t) = f(b)$ for $t > b$. Then $f \in bv_0$ and f is absolutely continuous on $[a,b]$. Thus f is differentiable almost everywhere on $[a,b]$, $f' \in L^1([a,b], \mathcal{B}, \lambda)$ and $f(t) = f(a) + \int_{[a,b]} f'\,d\lambda)$ for $t \in [a,b]$. In fact it follows from the Lipschitz condition that $f' \in L^\infty([a,b])$.

Rademacher's theorem extends this to locally Lipschitz functions on $\mathbf{R}^d$.

Theorem 17.7.1 (Rademacher's theorem) *A real-valued locally Lipschitz function f on an open subset U of $\mathbf{R}^d$ is differentiable λ_d-almost everywhere.*

Proof By localization and extension we can suppose that $U = \mathbf{R}^d$ and that f is a Lipschitz function on $\mathbf{R}^d$. We need two lemmas.

Suppose that h is a unit vector in $\mathbf{R}^d$. If it exists, the *directional derivative* is

$$\partial_h^+ f = \lim_{t \searrow 0} \frac{f(x+th) - f(x)}{t}.$$

Lemma 17.7.2 *Suppose that h is a unit vector in $\mathbf{R}^d$. Let $A_h = \{x \in \mathbf{R}^d : \partial_h^+ f$ exists$\}$. Then A_h is a Borel set, and $\lambda_d(\mathbf{R}^d \setminus A_h) = 0$.*

Proof If $x \in \mathbf{R}^d$, and $s, t > 0$, let

$$k(x; s, t) = \left| \frac{f(x+sh) - f(x)}{s} - \frac{f(x+th) - f(x)}{t} \right|,$$

and let $h_n(x) = \sup\{k(x, s, t) : 0 < s, t < 1/n\}$. Then h_n is a lower semi-continuous function on $\mathbf{R}^d$, and $0 \leq h_n(x) \leq 2L$, where L is the Lipschitz constant of f. Thus if $\epsilon > 0$ then the set $B_n(\epsilon) = \{x \in \mathbf{R}^d : h_n(x) \leq \epsilon\}$ is closed, so that $C_\epsilon = \cup_{n=1}^\infty B_n(\epsilon)$ is an F_σ set, and $A_h = \cap_{m=1}^\infty C(1/m)$ is a Borel set. But it follows from the Lebesgue density theorem that if $l_x = \{x + sh : s \in \mathbf{R}\}$ is a line in $\mathbf{R}^d$ parallel to h then $\lambda(l_x \setminus A_h) = 0$, and then it follows from Fubini's theorem that $\lambda_d(\mathbf{R}^d \setminus A_h) = 0$. □

Consequently, the set

$$G = \{x \in \mathbf{R}^d : \frac{\partial f}{\partial x_j}(x) \text{ exists for } 1 \leq j \leq d\}$$

is a Borel set, and $\lambda_d(\mathbf{R}^d \setminus G) = 0$.

Lemma 17.7.3 *Suppose that $x \in A_h \cap G$. Then*

$$\partial^+ f_h(x) = \sum_{j=1}^{d} h_j \frac{\partial f}{\partial x_j}(x).$$

Proof Suppose that g is a continuously differentiable function of compact support on $\mathbf{R}^d$ and that $t_n \searrow 0$. Then

$$\frac{1}{t_n}\int_{\mathbf{R}^d}(f(x+t_nh) - f(x))g(x)\,d\lambda_d(x) =$$
$$\frac{1}{t_n}\int_{\mathbf{R}^d}(f(x)(g(x-t_nx) - g(x))\,d\lambda_d(x) \to \int_{\mathbf{R}^d} f(x)\partial_h^+ g(x)d\lambda_d(x)$$

as $n \to \infty$, so that, by bounded convergence,

$$\int_{\mathbf{R}^d} \partial_h^+ f.g\,d\lambda_d = \int_{\mathbf{R}^d} f.\,\partial_h^+ g\,d\lambda_d.$$

Similarly

$$\int_{\mathbf{R}^d} \frac{\partial f}{\partial x_j}.g\,d\lambda_d = \int_{\mathbf{R}^d} f.\frac{\partial g}{\partial x_j}d\lambda_d,$$

for $1 \le j \le d$ and so

$$\begin{aligned}\int_{\mathbf{R}^d} \partial_h^+ f.g\,d\lambda_d &= \int_{\mathbf{R}^d} f.\,\partial_h^+ g\,d\lambda_d \\ &= \sum_{j=1}^{d} h_j \int_{\mathbf{R}^d} f.\frac{\partial g}{\partial x_j}d\lambda_d \\ &= \int_{\mathbf{R}^d} \sum_{j=1}^{d} h_j \frac{\partial f}{\partial x_j}.\,gd\lambda_d.\end{aligned}$$

Thus $\partial_h^+ f\,d\lambda_d = \sum_{j=1}^d h_j \frac{\partial f}{\partial x_j}\,d\lambda_d$. □

We now prove Rademacher's theorem. Let $H = (h^{(n)})_{n=1}^\infty$ be a dense sequence in the unit sphere S^{d-1} of $\mathbf{R}^d$, and let $J = G \cap (\cap_{n=1}^\infty A_{h^{(n)}})$. Then $\lambda_d(\mathbf{R}^d \setminus J) = 0$. If $h \in S^{d-1}$, let

$$k(x,h,t) = \frac{1}{t}\left(f(x+th) - f(x) - t\sum_{j=1}^{d} h_j \frac{\partial f}{\partial x_j}(x)\right).$$

We show that if $x \in J$ then $k(x,h,t) \to 0$ uniformly on S^{d-1} as $t \searrow 0$, which establishes the theorem.

Suppose that $\epsilon > 0$. There exists $N \in \mathbf{N}$ such that $\{h^{(1)}, \ldots, h^{(N)}\}$ is an $\epsilon/2(d+1)L$-net in H. From Lemma 17.7.3, there exists $0 < \delta < 1$ such that

$|k(x, h^{(n)}, t)| < \epsilon/2$ for $0 < t < \delta$ and $1 \leq n \leq N$. Suppose that $h \in S^{d-1}$. There exists $1 \leq n \leq N$ such that $\left\| h - h^{(n)} \right\| < \epsilon/2(d+1)L$. Consequently $|k(x, h, t) - k(x, h^{(n)}, t)| < \epsilon/2$ for $0 < t < \delta$, and so $|k(x, h, t)| < \epsilon$ for $0 < t < \delta$. □

Corollary 17.7.4 *Suppose that f is a very regular convex function on $\mathbf{R}^d$. Then f is differentiable λ_d-almost everywhere on Φ_f^{int}.*

Proof For f is a locally Lipschitz function on Φ_f^{int}. □

This corollary does not need Lebesgue's density theorem, since a convex function f on $\mathbf{R}$ is differentiable at all but countably many points of Φ_f^{int}.

18
Convergence of Measures

Suppose that (X, τ) is a Polish space. In this chapter, we consider topologies and metrics on

- the space $M(X)$ of signed Borel measures on X;
- the space $M^+(X)$ of finite positive Borel measures on X;
- the space $M_1(X)$ of signed Borel measures μ on X with $|\mu|(X) \leq 1$;
- the space $P(X)$ of Borel probability measures on X.

18.1 The Norm $\|.\|_{TV}$

Suppose that (X, τ) is a Polish space. Recall (Proposition 16.6.1) that if $f \in C_b(X)$ and $\mu \in M(X)$ then we define $j(\mu)(f) = \int_X f\,d\mu$ to be $\int_X f\,d\mu^+ - \int_X f\,d\mu^-$, and that j is a linear isometry of the Banach space $(M(X, \mathcal{B}), \|.\|_{TV})$, (where $\|\mu\|_{TV} = |\mu|(X)$), into $(C_b(X)', \|.\|'_\infty)$. $M_1(X)$ is the unit ball of $M(X)$, and $P(X) = M_1(X) \cap \{\mu : \mu(X) = 1\} \subseteq M^+(X)$.

The norm $\|.\|_{TV}$ is too strong for most purposes. If $x \in X$ let δ_x be the *Dirac measure*, or *point measure* on X: $\delta_x(A) = 1$ if $x \in A$, and $\delta_x(A) = 0$ otherwise, so that if $f \in C_b(X)$ then $\int_X f\,d\delta_x = f(x)$. If $x \neq y$ then $\left\|\delta_x - \delta_y\right\|_{TV} = 2$, so that the norm takes no account of the topology of X.

Proposition 18.1.1 *Suppose that* (X, τ) *is a Polish space. Then* $(M(X), \|.\|_{TV})$ *is separable if and only if* X *is countable. If so, then* j *is a surjective mapping of* $(M(X), \|.\|_{TV})$ *onto* $(l_1, \|.\|_1)$ *or* $(l_1^n, \|.\|_1)$.

Proof If X is uncountable, then $\delta(X) = \{\delta_x : x \in X\}$ is not separable in the subspace topology, and so $(M(X), \|.\|_{TV})$ is not separable.

Suppose that $X = \{x_n : n \in \mathbf{N}\}$ (with $x_m \neq x_n$ for $m \neq n$) is countably infinite. If $\mu \in M(X)$, let $k(\mu)_n = \mu(\{x_n\})$. Then k is an isometry of $(M(X), \|.\|_{TV})$ into

$(l_1, \|.\|_1)$. It is surjective, since if $f \in l_1, f = k(\mu)$, where $\mu(A) = \sum\{f_n : n \in A\}$. Similarly if X is finite. □

18.2 The Weak Topology w

We need a weaker topology than the $\|.\|_{TV}$ topology. We identify $M(X)$ with $j(M(X))$, so that $M(X)$ is a norm-closed linear subspace of $C_b(X)'$; we then give $M(X)$ the weak* topology w: a base of neighbourhoods of μ if given by the sets

$$\left\{ \nu \in M(X) : \left| \int_X f\,d\nu - \int_X f\,d\mu \right| < \epsilon \text{ for } \epsilon > 0, f \in F(\text{finite}) \subseteq C_b(X) \right\}.$$

(By scaling, we can if we like restrict ϵ to the value 1.) This is the topology of pointwise convergence on the points of $C_b(X)$. We write $\mu_n \Rightarrow \mu$ if (and only if) $\mu_n \to \mu$ in the w topology. Thus $\mu_n \Rightarrow \mu$ if and only if $\int_X f\,d\mu_n \to \int_X f\,d\mu$ as $n \to \infty$ for each $f \in C_b(X)$. The subspace topologies on $M_1(X)$, and more particularly on $P(X)$, are also denoted by w. Confusingly, w is called the *weak topology*, rather than the weak* topology. Elements of $P(X)$ are also called *laws*. Convergence in $(P(X), w)$ is also called *convergence in law*.

Similarly, if $(f_n)_{n=1}^\infty$ is a sequence of measurable mappings from a probability space (Ω, Σ, P) into a Polish space (X, τ), then f_n converges to f *in law* if the push-forward measures $(f_n)_*(P)$ converge in the w topology to $f_*(P)$.

Exercise 18.2.1 Suppose that $(f_n)_{n=1}^\infty$ is a sequence in $L^0(\Omega, \Sigma, P)$, where (Ω, Σ, P) is a probability space. Show that if $f_n \to f$ in probability, then $f_n \to f$ in law. Show that the converse is not generally true.

Exercise 18.2.2 Suppose that $(f_n)_{n=1}^\infty$ is a sequence in $L^0(\Omega, \Sigma, P)$, where (Ω, Σ, P) is a probability space. Show that if $(f_n)_*(P) \Rightarrow \delta_t$, then $f_n \to t$ in probability.

The following proposition suggests that w is a useful topology to consider.

Proposition 18.2.3 *Suppose that (X, τ) is a Polish space. If $x \in X$ and $f \in C_b(X)$ let $\langle f, \delta(x) \rangle = f(x)$. Then $\delta(X) \subseteq C_b(X)'$, and if $C_b(X)'$ is given the weak* topology then δ is a homeomorphism of (X, τ) onto $\delta(X)$.*

Proof Certainly $\delta(x) \in C_b(X)'$. If $x \neq y$ then there exists $f \in C_b(X)$ with $f(x) \neq f(y)$, so that δ is injective. If $f \in C_b(X)$ then $(f \circ \delta)(x) = f(x)$, so that δ is continuous. Suppose that $x \in X$ and that U is an open neighbourhood of x. Since (X, τ) is completely regular, there exists $f \in C_b(X)$ such that $f(x) = 0$ and $f(y) = 1$ for $y \notin U$. Then $V = \{\phi \in C_b(X)' : |\phi(f)| < 1\}$ is a weak* neighbourhood of δ_x, and if $\phi \in \delta(X) \cap V$ then $\delta^{-1}(\phi) \in U$, so that $\delta^{-1} : \delta(X) \to X$ is also continuous. □

Proposition 18.2.4 *(i) $M_1(X)$ is a w-closed subset of $M(X)$.*
(ii) The set $P(X)$ is a w-closed subset of $M_1(X)$.
(iii) The set $\delta(X)$ is a w-closed subset of $P(X)$.

Proof (i) If $f \in C_b(X)$ then

$$\|f\|_\infty = \sup\{|\delta_x(f)| : x \in X\} = \sup\left\{\int_X f\,d\mu : \mu \in M_1(X)\right\},$$

so that this follows from the theorem of bipolars.

(ii) For $P(X) = \{\mu \in M_1(X) : \int_x 1\,d\mu = 1\}$.

(iii) Suppose that $\mu \in P(X) \setminus \delta(X)$. Then there exist disjoint compact sets A and B such that $\mu(A) = \alpha > 0$ and $\mu(B) = \beta > 0$. By Tietze's extension theorem there exists $f \in C_b(X)$ such that $f(x) = 0$ if $x \in A$, $f(x) = 1$ if $x \in B$ and $0 \le f \le 1$. By Tietze's theorem again, there exist functions g_1 and g_2 in $C_b(X)$ such that

$$\begin{aligned} &g_1 = 0 \text{ on } A, && g_1 = 1 \text{ on } (f \ge \tfrac{1}{2}), && 0 \le g_1 \le 1, \\ &g_2 = 0 \text{ on } (f \le \tfrac{1}{2}), && g_2 = 1 \text{ on } B, && 0 \le g_2 \le 1. \end{aligned}$$

Then $\beta \le \int_X g_i\,d\mu \le 1 - \alpha$, for $i = 1, 2$. Suppose that $0 < \epsilon < \min(\alpha, \beta)/2$. If $|\int_X g_1\,d\mu - g_1(y)| < \epsilon$ then $f(y) < \frac{1}{2}$, and if $|\int_X g_2\,d\mu - g_2(y)| < \epsilon$ then $f(y) > \frac{1}{2}$, so that

$$\left\{\nu \in P(X) : \left|\int_X g_i\,d\nu - \int_X g_i\,d\mu\right| < \epsilon \text{ for } i = 1, 2\right\}$$

is a w-neighbourhood of μ disjoint from $\delta(X)$. □

Theorem 18.2.5 *The following are equivalent:*

(a) $P(X)$ is w-compact;
(b) $M_1(X)$ is w-compact;
(c) X is compact.

Proof If $M_1(X)$ is w-compact, then $P(X)$ is w-compact, since $P(X)$ is w-closed in $M_1(X)$.

If $P(X)$ is compact, then $\delta(X)$ is compact, and so X is compact.

If X is compact, then the mapping $j : M(X) \to C(X)'$ is surjective, by the Riesz representation theorem, and so $M_1(X)$ is w-compact, by Banach's theorem. □

Exercise 18.2.6 Suppose that (X, τ) is a compact metrizable space. Show that $D(X) = \{\delta_x : x \in X\}$ is the set of extreme points of the w-compact convex set $P(X)$, and that $x \to \delta_x$ is a homeomorphism of X onto $D(X)$, so that $D(X)$ is a closed subset of $(P(X), w)$.

Note that if X is a Polish space which is not finite, then $(M(X), w)$ is not first countable, and so it is not metrizable. We can say more.

Proposition 18.2.7 *If (X, d) is a Polish metric space which is not compact then $(M_1(X), w)$ is not metrizable.*

Proof Since (X, d) is not totally bounded, there exist $\epsilon > 0$ and a sequence $(x_n)_{n=1}^\infty$ in X such that $d(x_n, x_m) \geq \epsilon$ for $m \neq n$. If $\alpha = (\alpha_n)_{n=1}^\infty \in l^1$, let $T(\alpha) = \sum_{n=1}^\infty \alpha_n \delta_n$. Then T is a linear isomorphism of $(B(l^1), \sigma(l^1, l^\infty))$ into $(M_1(X), w)$. Since $(B(l^1), \sigma(l^1, l^\infty))$ is not metrizable, neither is $(M_1(X), w)$. □

18.3 The Portmanteau Theorem

Suppose that (X, τ) is a Polish space. We have defined the w topology on $M(X)$ in terms of bounded continuous functions. We now characterize w-convergence in $P(X)$ in terms of semi-continuous functions, and in terms of open and closed sets.

Proposition 18.3.1 *Suppose that (X, τ) is a Polish space. If f is a bounded lower semi-continuous function on X then the real-valued function $\mu \to \int_X f\, d\mu$ is lower semi-continuous on $(P(X), w)$.*

Proof Let d be a metric on X defining the topology τ. By Theorem 4.2.9 there exists an increasing sequence $(f_n)_{n=1}^\infty$ of Lipschitz functions on X which increases pointwise to f. Suppose that $\mu \in P(X)$ and that $\epsilon > 0$. By the theorem of bounded convergence, there exists $n \in \mathbf{N}$ such that $\int_X f_n\, d\mu > \int_X f\, d\mu - \epsilon/2$. If $|\int_X f_n\, d\nu - \int_X f_n\, d\mu| < \epsilon/2$ then

$$\int_X f\, d\nu > \int_X f_n\, d\nu > \int_X f_n\, d\mu - \epsilon/2 > \int_X f\, d\mu - \epsilon.$$

□

Corollary 18.3.2 *If A is an open subset of X then the real-valued function $\mu \to \mu(A)$ is lower semi-continuous on $(P(X), w)$, and if B is a closed subset of X then the real-valued function $\mu \to \mu(A)$ is upper semi-continuous on $(P(X), w)$.*

Proof For I_A is lower semi-continuous, and $\mu(B) = 1 - \mu(X \setminus B)$. □

A Borel subset A of X is a *continuity set* for μ if $\mu(\partial A) = 0$.

Corollary 18.3.3 *If A is a continuity set for μ_0 then the real-valued function $\mu \to \mu(A)$ is continuous at μ_0.*

Proof For $\mu(A) = \mu(A^{int}) = \mu(\overline{A})$. □

Theorem 18.3.4 (The portmanteau theorem) *Suppose that (X, τ) is a Polish space, that $(\mu_n)_{n=1}^{\infty}$ is a sequence in $P(X)$ and that $\mu \in P(X)$. The following are equivalent.*

(i) $\mu_n \Rightarrow \mu$ *as* $n \to \infty$.
(ii) If f is a bounded lower semi-continuous function on X then $\int_X f\, d\mu \leq \liminf_{n\to\infty} \int_X f\, d\mu_n$.
(iii) If f is a bounded upper semi-continuous function on X then $\int_X f\, d\mu \geq \limsup_{n\to\infty} \int_X f\, d\mu_n$.
(iv) If A is an open subset of X, then $\mu(A) \leq \liminf_{n\to\infty} \mu_n(A)$.
(v) If A is a closed subset of X, then $\mu(A) \geq \limsup_{n\to\infty} \mu_n(A)$.
(vi) If A is a continuity set A of μ then $\mu_n(A) \to \mu(A)$ *as* $n \to \infty$.

Proof (i) implies (ii), by Proposition 18.3.1 and (ii) and (iii) are equivalent. (ii) implies (iv), and (iv) and (v) are equivalent. (iv) and (v) together imply (vi). Suppose that (vi) holds, that $f \in C_b(X)$ and that $\epsilon > 0$. Let $F(t) = \mu(f \leq t)$. Then F is an increasing function on $\mathbf{R}$, $F(t) = 0$ for $t < -\|f\|_\infty$ and $F(t) = 1$ for $t \geq \|f\|_\infty$.

Suppose that $\epsilon > 0$. The set J of jump discontinuities of F is countable, and so therefore is the set $\mathbf{Q}.J = \{qj : q \in \mathbf{Q}, j \in J\}$. It follows that there exists $0 < \eta < \epsilon$ such that $(k + \frac{1}{2})\eta \notin J$, for all $k \in \mathbf{Z}$. Let $A_k = (f \leq (k + \frac{1}{2})\eta)$: A_k is a set of continuity for μ, and so is $B_k = A_k \setminus A_{k-1}$. Thus if we set

$$g = \sum_{k\in\mathbf{Z}} \left(k + \tfrac{1}{2}\right) \eta I_{B_k} \text{ and } g = \sum_{k\in\mathbf{Z}} \left(k - \tfrac{1}{2}\right) \eta I_{B_k},$$

(both of theses sums have only finitely many non-zero terms), then $\int_X g\, d\mu_n \to \int_X g\, d\mu$ and $\int_X h\, d\mu_n \to \int_X h\, d\mu$ as $n \to \infty$. Since $f - \eta \leq h \leq f \leq g \leq f + \eta$, it follows that

$$\begin{aligned}\left(\limsup_{n\to\infty} \int_X f\, d\mu\right) - \eta &\leq \int_X h\, d\mu \leq \int_X f\, d\mu \\ &\leq \int_X g\, d\mu \leq \left(\liminf_{n\to\infty} \int_X f\, d\mu\right) + \eta.\end{aligned}$$

Since $0 < \eta < \epsilon$ and ϵ is arbitrary, the result follows. □

Let us see how the results that we have established apply to Borel measures on the real line $\mathbf{R}$.

Theorem 18.3.5 (The Helly–Bray theorem) *Suppose that* $(\mu_n)_{n=1}^\infty$ *is a sequence in* $P(\mathbf{R})$ *and that* $\mu \in P(\mathbf{R})$. *Then* $\mu_n \Rightarrow \mu$ *if and only if* $F_{m_n}(t) \to F_\mu(t)$ *as* $n \to \infty$ *for each point of continuity* t *of* F_μ.

Proof Since $(-\infty, t]$ is a continuity set for μ if and only if t is a point of continuity of F_μ, the condition is necessary, by the portmanteau theorem. Suppose that it is satisfied, that U is an open subset of $\mathbf{R}$ and that $\epsilon > 0$. Since μ is tight, there exists a finite set $\{[a_i, b_i] : 1 \leq i \leq j\}$ of disjoint closed intervals in U such that $\sum_{i=1}^j \mu([a_i, b_i]) > \mu(U) - \epsilon$. For each i there exist points c_i, d_i of continuity of F_μ such that $c_i < a_i, b_i \leq d_i$ and $(c_i, d_i] \subseteq U$. Let $C = \cup_{i=1}^n (c_i, d_i]$. Then $\mu_n(C) \to \mu(C)$ as $n \to \infty$. Thus

$$\liminf_{n\to\infty} \mu_n(U) \geq \lim_{n\to\infty} \mu_n(C) = \mu(C) > \mu(U) - \epsilon.$$

Since this holds for all $\epsilon > 0$, $\liminf_{n\to\infty} \mu_n(U) \geq \mu(U)$. Thus $\mu_n \Rightarrow \mu$, by the portmanteau theorem. □

Exercise 18.3.6 Suppose that $(\mu_n)_{n=1}^\infty$ is a sequence in $P(\mathbf{R})$ and that $\mu \in P(\mathbf{R})$ has a continuous distibution function F. Show that if $\mu_n \Rightarrow \mu$ then F_n converges uniformly to F.

Exercise 18.3.7 Let P be a probability measure on $\mathbf{R}$ which has an atom at each rational number. Let $P_n(A) = P(A - 1/n)$, for each Borel set A. Show that $P_n \Rightarrow P$, but that $P_n(q) \not\to P(q)$, for each rational q.

Exercise 18.3.8 Suppose that $\mu \in P(\mathbf{R}^d)$. If $t \in \mathbf{R}^d$, let $L(t) = \{x \in \mathbf{R}^d : x_i \leq t_i \text{ for } 1 \leq i \leq d\}$ and let $F(t) = \mu(L(t))$. By considering the atoms of μ, show that the points of discontinuity of L lie in a countable union of hyperplanes, and have Lebesgue measure 0. Prove a version of the Helly–Bray theorem for probabilities in $\mathbf{R}^d$.

Exercise 18.3.9 Suppose that for each n f_n is a random variable on a probability space (Ω, Σ, P_n), each taking values in a Polish space (X, τ) and that f is a random variable on a probability space (Ω, Σ, P) taking values in X. Formulate a version of the portmanteau theorem for the convergence in law of $(f_n)_{n=1}^\infty$. Do the same for the Helly–Bray theorem.

Recall (Theorem 3.1.1) that if (X, τ) is a Polish space then there is a homeomorphism k of X onto a G_δ subset of the Hilbert cube $\mathcal{H}$. Let $\tilde{X} = \overline{j(X)}$, with the subspace topology $\tilde{\tau}$; $(\tilde{X}, \tilde{\tau})$ is a compact metrizable space. If $\mu \in P(X)$, we denote the push-forward measure $k_*(\mu)$ by $\tilde{\mu}$; thus $\tilde{\mu}(A) = \mu(A \cap X)$ for A a Borel set in $\tilde{X}$. We denote the weak topology on $P(\tilde{X})$ by $\tilde{w}$ and write $\nu_n \tilde{\Rightarrow} \nu$ if $\nu_n \to \nu$ in the topology $\tilde{w}$.

Theorem 18.3.10 *Suppose that (X, τ) is a Polish space, and that k is a homeomorphism of X onto a dense subset of a compact metrizable space $\tilde{X}$. The mapping $k_* : \mu \to \tilde{\mu}$ is a homeomorphism of $(P(X), w)$ onto a dense subset of the compact metrizable space $(P(\tilde{X}), \tilde{w})$.*

Proof The mapping is certainly continuous, since the restriction of a function in $C(\tilde{X})$ to X is in $C_b(X)$. Suppose that $\tilde{\mu}_n \Rightarrow \tilde{\mu}$. If A is a closed subset of X, then $A = \overline{A} \cap X$, where $\overline{A}$ is the closure of A in $\tilde{X}$. Thus

$$\mu(A) = \tilde{\mu}(\overline{A}) \geq \limsup_{n\to\infty} \tilde{\mu}_n(\overline{A}) = \limsup_{n\to\infty} \mu_n(A),$$

and so $\mu_n \to \mu$; the inverse mapping is also continuous.

The set $\overline{k(P(X))}$ is a convex subset of the convex $\tilde{w}$-compact set $P(\tilde{X})$. If it were a proper subset of $P(\tilde{X})$, it would follow from the Krein–Mil'man theorem that there would be an extreme point ν of $P(\tilde{X})$ not in $\overline{k(P(X))}$. But then $\nu = \delta_y$, for some $y \in \tilde{X}$, and there exists a sequence $(x_n)_{n=1}^\infty$ in X which converges to y. But then $k_*(\delta_{x_n}) \tilde{\Rightarrow} \delta_y$, giving a contradiction. □

Corollary 18.3.11 *$(P(X), w)$ is a separable metrizable space, and there is a countable subset F of $C_b(X)$ such that w is the topology of pointwise convegence on F.*

If C is a countable dense subset of X, then the countable set

$$A_C = \left\{ \sum_{i=1}^n \lambda_i \delta_{c_i} : n \in \mathbf{N}, \lambda_i \in \mathbf{Q}, \lambda_i \geq 0, \sum_{i=1}^n \lambda_i = 1, c_i \in C \right\}$$

is w-dense in $P(X)$.

Proof The first statement follows from the fact that $(P(\tilde{X}), \tilde{w})$ has these properties. The set $k(C)$ is dense in $\tilde{X}$, and so the second statement follows from Exercise 16.6.8. □

This enables us to prove the empirical law of large numbers. Suppose that $(g_n)_{n=1}^\infty$ is a sequence of identically distributed real-valued random variables on a probability space $(\Omega, \Sigma, \mathbf{P})$, each with distribution μ. For each $n \in \mathbf{N}$ and $\omega \in \Omega$, let $\mu_n(\omega) = (1/n) \sum_{i=1}^n \delta_{g_i(\omega)}$. Then μ_n is a random measure, the *nth empirical measure*; it represents repeated sampling from a population which has distribution μ. Thus if $A \in \Sigma$ then $\mu_n(\omega)(A)$ is the proportion of the elements $\{g_i(\omega) : 1 \leq i \leq n\}$ which lie in A, and if $f \in L^0(\Omega)$ then $\int_\Omega f \, d\mu_n(\omega) = (1/n) \sum_{i=1}^n f(g_i(\omega))$. We would hope that almost surely the empirical measures would give an approximation to the distribution μ; the empirical law of large numbers says that this is so.

Theorem 18.3.12 (The empirical law of large numbers) *Suppose that Ω is a Polish space, that P is a Borel probability measure on Ω and that $(g_n)_{n=1}^\infty$ is a sequence of identically distributed random variables on Ω, each with distribution μ. Let μ_n be the nth empirical distribution. Then $\mu_n(\omega) \Rightarrow \mu$ for almost all $\omega \in \Omega$.*

Proof Let F be a countable set in $C_b(\Omega)$ which satisfies the conclusions of Corollary 18.3.11. By the strong law of large numbers, $\int_\Omega f \, d\mu_n(\omega) \to \int_\Omega f \, d\mu$ for almost all ω, for each $f \in F$, and so, since F is countable, $\int_\Omega f \, d\mu_n(\omega) \to \int_\Omega f \, d\mu$ for almost all ω, for all $f \in F$. Thus $\mu_n(\omega) \Rightarrow \mu$ almost surely. □

This theorem is usually applied in the following situation. Suppose that (X, τ) is a Polish space, and that μ is an unknown Borel probability distribution μ on X. We sample repeatedly and independently from X, and hope that the resulting empirical distribution approximates to $\mathbf{P}$. Thus, for each $n \in \mathbf{N}$, $(\Omega_n, \mathbf{P}_n)$ is a copy of (X, μ), $(\Omega, \mathbf{P})$ is the countable product $(\prod_{n=1}^\infty \Omega_n, \otimes_{n=1}^\infty \mathbf{P}_n)$ and g_n is the outcome of the nth trial.

18.4 Uniform Tightness

Throughout this section, we suppose that (X, τ) is a Polish space. Recall that if $\mu \in P(X)$ then μ is tight; given $\epsilon > 0$ there exists a compact subset K of X such that $\mu(K) > 1 - \epsilon$. We extend this notion to subsets of $P(X)$. A subset C of $P(X)$ is *uniformly tight* if whenever $\epsilon > 0$ there exists a compact subset K of X such that $\mu(K) > 1 - \epsilon$ for all $\mu \in C$. The proof of the next result is reminiscent of the proof of Ulam's theorem.

Proposition 18.4.1 *Suppose that (X, d) is a complete Polish metric space and that $C \subseteq P(X)$. Then C is uniformly tight if and only if for each $\epsilon > 0$ there exists a finite subset F of X such that*

$$\mu(\cup_{x \in F} N_\epsilon(x)) > 1 - \epsilon, \text{ for all } \mu \in C.$$

Proof If C is uniformly tight, and $\epsilon > 0$ there exists a compact subset K of X such that $\mu(K) > 1 - \epsilon$ for all $\mu \in C$. There exists a finite subset F of X such that $K \subseteq \cup_{x \in F} N_\epsilon(x)$. F clearly satisfies the requirements of the proposition.

Conversely, suppose that the condition is satisfied, and that $\epsilon > 0$. For each $j \in \mathbf{N}$ there exists a finite subset F_j of X such that if $A_j = \cup_{x \in F_j} M_{\epsilon/2^j}(x)$ then $\mu(A_j) > 1 - \epsilon/2^j$, for all $\mu \in C$. Let $B = \cap_{j=1}^\infty A_j$. Then B is closed and totally bounded, and is therefore compact. Further, $\mu(B) > 1 - \epsilon$. □

Here is a useful test for the distributions of a set of random variables to be uniformly tight.

Proposition 18.4.2 *Suppose that $\{(\Omega_\alpha, \Sigma_\alpha, \mu_\alpha)\}_{\alpha\in A}$ is a family of probability spaces and that for each $\alpha \in A$ f_α is a random variable on Ω_α taking values in $\mathbf{R}^d$. Then $\{\mathcal{L}(f_\alpha)\}_{\alpha\in A}$ is uniformly tight if and only if there is an increasing non-negative function ϕ on $[0,\infty)$ for which $\phi(t) \to \infty$ as $t \to \infty$ such that $\{\int_{\Omega_\alpha} \phi(\|f_\alpha\|)\,d\mu_\alpha\}_{\alpha\in A}$ is bounded.*

Proof Suppose that the condition is satisfied, that

$$\sup_{\alpha\in A} \int_{\Omega_\alpha} \phi(\|f_\alpha\|)\,d\mu_\alpha = M$$

and that $\epsilon > 0$. There exists T such that $\phi(t) \geq M/\epsilon$ for $t \geq T$. If $\alpha \in A$ then

$$\begin{aligned}\mathcal{L}(f_\alpha)(\|t\| > T) &\leq \frac{1}{\phi(T)} \int_{\|f_\alpha\|>T} \phi(\|f_\alpha\|)\,d\mu_\alpha \\ &\leq \frac{\epsilon}{M} \int_{\Omega_\alpha} \phi(\|f_\alpha\|)\,d\mu_\alpha \leq \epsilon,\end{aligned}$$

so that $\{\mathcal{L}(f_\alpha)\}_{\alpha\in A}$ is uniformly tight.

Conversely, suppose that $\{\mathcal{L}(f_\alpha)\}_{\alpha\in A}$ is uniformly tight. There therefore exists a strictly increasing sequence $(t_n)_{n=1}^\infty$ in $[0,\infty)$ such that $\mathcal{L}(f_\alpha)\{x : \|x\| > t_n\} > 1 - 1/4^n$ for each $\alpha \in A$. Let ϕ be a non-negative increasing function on $[0,\infty)$ for which $\phi(t_n) = 2^n$. If $\alpha \in A$ then

$$\begin{aligned}&\int_{\Omega_\alpha} \phi(\|f_\alpha\|)\,d\mu_\alpha \\ &= \int_{\|f_\alpha\|\leq t_1} \phi(\|f_\alpha\|)\,d\mu_\alpha + \sum_{n=1}^\infty \left(\int_{t_{n-1}<\|f_\alpha\|\leq t_n} \phi(\|f_\alpha\|)\,d\mu_\alpha \right) \\ &\leq 2 + \sum_{n=1}^\infty \frac{2^{n+1}}{4^n} = 4.\end{aligned}$$

□

This proposition is frequently applied using the function $\phi(t) = t^p$, where $p > 0$.

Exercise 18.4.3 Suppose that (G,τ) is a locally compact Polish topological group, and that ν is a group-norm on G which defines the topology, for which the sets $K_n = \{g : \nu(g) \leq n\}$ are a fundamental sequence of compact sets. Formulate and prove a version of Proposition 18.4.2 in this setting.

18.5 The β Metric

Suppose that (X, τ) is a Polish space. So far, we have made little use of a complete metric d on X which defines the topology τ. We now consider a complete Polish metric space (X, d); a complete separable metric space. Recall that $BL(X)$ then denotes the space of bounded Lipschitz functions, with norm $\|f\|_{BL} = \|f\|_\infty + p_L(f)$.

Proposition 18.5.1 *Suppose that (X, d) is a Polish metric space. If μ and ν are distinct elements of $P(X)$, there exists $f \in BL(X)$ such that $\int_X f\, d\mu \neq \int_X f\, d\nu$.*

Proof Let $\pi = \mu - \nu$ and let $X = P \cup N$ be a Lebesgue decomposition of X for π. There exist compact subsets K of P and L of N such that $\pi(K) > \pi(P) - \|\pi\|_{TV}/5$ and $\pi(L) < \pi(N) + \|\pi\|_{TV}/5$. Consequently, $|\pi|(X \setminus (K \cup L)) < 2\|\pi\|_{TV}/5$ and $|\pi|(K \cup L) \geq 3\|\pi\|_{TV}/5$. There exists $f \in BL(X)$ such that $f_{|K} = 1, f_{|L} = -1$ and $\|f\|_\infty = 1$. Then $\int_X f\, d\pi > \|\pi\|_{TV}/5 > 0$, so that f separates μ and ν. □

Corollary 18.5.2 *Let $\|\mu\|_\beta = \sup\{|\int_X f\, d\mu| : \|f\|_{BL} \leq 1\}$. Then $\|.\|_\beta$ is a norm on $M(X)$, and $\|\mu\|_\beta \leq \|\mu\|_{TV}$.*

We denote the metric that $\|.\|_\beta$ defines on $P(X)$ by β.

Suppose that (X, d) is a metric space. If A is a non-empty Borel set in X, let $A_\epsilon = \{x \in X, d(x, A) < \epsilon\}$.

Proposition 18.5.3 *Suppose that (X, d) is a Polish metric space, and that $0 < \epsilon < 1$. If $\beta(\mu, \nu) \leq \epsilon^2/2$ and A is a nonempty Borel set then $\nu(A_\epsilon) \geq \mu(A) - \epsilon$.*

Proof Let $f = (1 - d(x, A)/\epsilon)^+$. Then $\|f\|_{BL} \leq 1 + 1/\epsilon$, so that

$$\left| \int_X f\, d\mu - \int_X f\, d\nu \right| \leq \epsilon^2(1 + 1/\epsilon)/2 < \epsilon.$$

Thus

$$\nu(A_\epsilon) \geq \int_X f\, d\nu \geq \int_X f\, d\mu - \epsilon \geq \mu(A) - \epsilon.$$

□

Theorem 18.5.4 *Suppose that (X, d) is a complete Polish metric space. If C is a non-empty β-totally bounded subset of $P(X)$, then it is uniformly tight.*

Proof Suppose that $\epsilon > 0$. There exists a finite set D in C such that $C \subseteq D_{\epsilon^2/8}$. There exists a compact subset K_D of X such that $\mu(K_D) > 1 - \epsilon/2$ for $\mu \in D$. There exists a finite subset F of X such that $K_D \subseteq F_{\epsilon/2}$, so that $(K_D)_{\epsilon/2} \subseteq F_\epsilon$. If $\nu \in C$, there exists $\mu \in D$ with $\beta(\nu, \mu) \leq \epsilon^2/8$, and so

$$\nu(F_\epsilon) \geq \nu((K_D)_{\epsilon/2}) \geq \mu(K_D) - \epsilon/2 \geq 1 - \epsilon.$$

The result therefore follows from Proposition 18.4.1. □

Let w_{BL} be the topology on $M(X)$ of pointwise convergence on $BL(X)$.

Theorem 18.5.5 *Suppose that (X, d) is a complete Polish metric space. The β-metric topology, the topology w and the topology w_{BL} are the same on $P(X)$.*

Proof First we show that the identity mapping $(P(X), w_{BL}) \to (P(X), \beta)$ is continuous. Suppose $\mu \in P(X)$, and that $0 < \epsilon < 1$. By tightness, there exists a non-empty compact subset K of X such that $\mu(K) > 1 - \epsilon/11$. By the Arzelà–Ascoli theorem, $A = \{f_{|K}; \|f\|_{BL} \leq 1\}$ is totally bounded in $C(K)$, and so there exist $g_1, \ldots, g_k \in A$ such that $A \subseteq \cup_{i=1}^k B_{\epsilon/11}(g_i)$. By the McShane–Whitney theorem, each g_i can be extended without increasing the BL norm to $f_i \in BL(X)$. Also, let $h(x) = (1 - 11d(x, K)/\epsilon)^+$. Let $C = \{h\} \cup \{f_i : 1 \leq i \leq k\}$. Let

$$N = \left\{ \nu \in P(X) : \left| \int_X f\, d\nu - \int_X f\, d\mu \right| < \epsilon/11 \text{ for } f \in C \right\};$$

N is a w_{BL}-neighbourhood of μ in $P(X)$.

Suppose that $\nu \in N$. Then

$$\nu(K_{\epsilon/11}) \geq \int_X h\, d\nu \geq \int_X h\, d\mu - \epsilon/11 \geq 1 - 2\epsilon/11.$$

If $\|f\|_{BL} \leq 1$ there exists f_i such that $|f_i(x) - f(x)| \leq \epsilon/11$ for $x \in K$. Using the Lipschitz condition, $|f_i(y) - f(y)| \leq 3\epsilon/11$ for $y \in K_{\epsilon/11}$.

Now

$$\int_X |f - f_i|\, d\nu \leq \int_{K_{\epsilon/11}} |f - f_i|\, d\nu + \int_{X \setminus K_{\epsilon/11}} (|f| + |f_i|)\, d\nu \leq 3\epsilon/11 + 4\epsilon/11$$

and

$$\int_X |f - f_i|\, d\mu = \int_K |f - f_i|\, d\mu + \int_{X \setminus K} (|f| + |f_i|)\, d\mu \leq \epsilon/11 + 2\epsilon/11.$$

Thus

$$\left| \int_X f\, d\nu - \int_X f\, d\mu \right| \leq \left| \int_X f_i\, d\nu - \int_X f_i\, d\mu \right| + \int_X |f - f_i|\, d\nu + \int_X |f - f_i|\, d\mu$$
$$\leq \epsilon/11 + 7\epsilon/11 + 3\epsilon/11 = \epsilon,$$

and so $\beta(\nu, \mu) \leq \epsilon$ for $n \geq N$.

Since β is a metric, to show that the identity $(X, \beta) \to (X, w)$ is continuous it is sufficient to show that it is sequentially continuous. We must show that if $\beta(\mu_n, \mu) \to 0$ as $n \to \infty$ and $f \in C(K)$ then $\int_X f\, d\mu_n \to \int_X f\, d\mu$ as $n \to \infty$.

The sequence $\{\mu_n : n \in \mathbf{N}\}$ is β-totally bounded, and so is uniformly tight. Thus given $\epsilon > 0$ there exists a compact subset K of X such that $\mu(K) \geq 1 - \epsilon/7$ and $\mu_n(K) \geq 1 - \epsilon/7$ for all n. Suppose that $f \in C_b(X)$ and that $\|f\|_\infty \leq 1$. By the corollary to the Stone–Weierstrass theorem (Corollary 16.8.3), there exists $g \in BL(K)$ such that $\|g\|_{C(K)} \leq 1$ and $\|f - g\|_{C(K)} \leq \epsilon/7$. We can extend g to $h \in BL(X)$, with $\|h\|_{C(X)} \leq 1$. Then

$$\left|\int_X f\,d\mu_n - \int_X f\,d\mu\right|$$
$$\leq \left|\int_X h\,d\mu_n - \int_X h\,d\mu\right| + \int_X |f - h|\,d\mu_n + \int_X |f - h|\,d\mu.$$

Now

$$\int_X |f - h|\,d\mu_n \leq \int_K |f - h|\,d\mu_n + \int_{C(K)} |f|\,d\mu_n + \int_{C(K)} |h|\,d\mu_n \leq 3\epsilon/7,$$

and similarly $\int_X |f - h|\,d\mu \leq 3\epsilon/7$, and so

$$\left|\int_X f\,d\mu_n - \int_X f\,d\mu\right| \leq \left|\int_X h\,d\mu_n - \int_X h\,d\mu\right| + 6\epsilon/7 < \epsilon$$

for large enough n.

Finally, the identity mapping $(P(X), w) \to (P(X), w_{BL})$ is trivially continuous. □

Corollary 18.5.6 *If (X, d) is a compact metric space, then $(P(X), \beta)$ is compact.*

Proof For $(P(X), w)$ is compact, by Banach's theorem. □

Theorem 18.5.7 *If (X, d) is a compact metric space, then $(M_1(X), \beta)$ is compact, and therefore complete, and the β-metric topology, the topology w and the topology w_{BL} are the same on $M_1(X)$.*

Proof Since $(M_1(X), \beta)$ is the continuous image of $(P(X), \beta)^2 \times [0, 1]$ under the mapping (μ, ν, α) to $(1-\alpha)\mu - \alpha\nu$, $(M_1(X), \beta)$ is compact. The β-topology and w_{BL} therefore agree on $M_1(X)$. But $(M_1(X), w)$ is compact, by Banach's theorem, and so w and w_{BL} agree on $M_1(X)$. □

Exercise 18.5.8 Show that if $X = \mathbf{N}$, then $M_1(X) = l_1$, $\beta(x) = \|x\|_1$ and $w = \sigma(l_1, l_\infty)$. Consequently, the β topology and w may not agree on $M_1(X)$.

Theorem 18.5.9 *If (X, d) is a complete Polish metric space then $(P(X), \beta)$ is complete.*

Proof Suppose that $(\mu_n)_{n=1}^\infty$ is a β-Cauchy sequence in $P(X)$. Then $\{\mu_n : n \in \mathbf{N}\}$ is totally bounded. There therefore exists an increasing sequence

$(K_j)_{j=1}^\infty$ of compact subsets of X such that $\mu_n(K_j) > 1 - 1/j$, for $j, n \in \mathbf{N}$. Fix j. Let β_j be the β-norm on $P(X_j)$. Since every $f \in BL(K_j)$ can be extended to a bounded Lipschitz function g on X with $\|g\|_{BL} = \|f\|_{BL}$, it follows that the push-forward mapping from $(M(K_j), \beta_j)$ to $(M(X), \beta)$ is an isometry. If A is a Borel subset of X_j, let $\nu_n^{(j)}(A) = \mu_n(A)$. Then $\beta_j(\nu_m^{(j)} - \nu_n^{(j)}) \leq \beta(\mu_m - \mu_n)$ and so $(\nu_n^{(j)})_{n=1}^\infty$ is a β_j-Cauchy sequence. By Theorem 18.5.7, $\nu_m^{(j)}$ therefore converges, to $\nu^{(j)}$, say. Let $\mu^{(j)}$ be the push-forward measure on X. Note that if A is a Borel subset of X and $j < k$ then $0 \leq \mu^{(j)} \leq \mu^{(k)} \leq 1$. Let $\mu(A) = \lim_{j\to\infty} \mu^{(j)}(A)$. Then μ is finitely additive, and if $(A_r)_{r=1}^\infty$ is an increasing sequence of Borel subsets of X, then

$$\mu(A) = \sup_{j\in\mathbf{N}} \mu^{(j)}(A) = \sup_{j\in\mathbf{N}} \sup_{r\in\mathbf{N}} \mu^{(j)}(A_r) = \sup_{r\in\mathbf{N}} \mu(A_r),$$

so that $\mu \in M_1(X)$. Since

$$\mu(X) = \sup_{j\in\mathbf{N}} \nu^{(j)}(K_j) \geq \sup_{j\in\mathbf{N}}(1 - 1/j) = 1,$$

$\mu \in P(X)$. Finally, if $f \in C_b(X)$ and $\|f\|_\infty \leq 1$ then

$$\begin{aligned}\left|\int_X f\,d\mu - \int_X f\,d\mu_n\right| &\leq \left|\int_{K_j} f\,d\mu - \int_{K_j} f\,d\mu_n\right| + 2/j \\ &= \left|\int_{K_j} f\,d\nu^{(j)} - \int_{K_j} f\,d\nu_n^{(j)}\right| + 2/j \leq 3/j\end{aligned}$$

for large enough n, so that $\mu_n \to \mu$ as $n \to \infty$ in the w topology. Thus $\beta(\mu_n, \mu) \to 0$ as $n \to \infty$. □

Corollary 18.5.10 *If (X, τ) is a Polish space, then $(P(X), w)$ is a Polish space.*

Proposition 18.5.11 *If $A \subseteq M_1(X)$ is uniformly tight, then A is β-totally bounded.*

Proof Suppose that $\epsilon > 0$. There exists a compact set K such that $|\mu|(X \setminus K) < \epsilon/2$ for $\mu \in A$, so that $\beta(\mu - I_K.\mu) \leq \epsilon/2$ for $\mu \in A$. But the set $A_K = \{I_K.\mu : \mu \in A\}$ is totally bounded, and so there exists a finite set F such that $A_K \subseteq \cup_{\mu\in F} N_{\epsilon/2}(I_K.\mu)$. Thus $A \subseteq \cup_{\mu\in F} N_\epsilon(\mu)$. □

18.6 The Prokhorov Metric

Suppose that $(X d)$ is a complete Polish metric space. We now introduce another complete metric ρ on $P(X)$ which is equivalent to the metric β.

If $\mu, \nu \subset P(X)$, we set

$$\rho(\mu, \nu) = \inf\{\epsilon > 0 : \mu(A) < \nu(A_\epsilon) + \epsilon \text{ for all non-empty Borel sets } A\}.$$

Since $\mu(A) = \sup\{\mu(B) : B \text{ closed }, B \subseteq A$ and $N_\epsilon(A) = N_\epsilon(\overline{A})$,

$$\rho(\mu, \nu) = \inf\{\epsilon > 0 : \mu(B) < \nu(B_\epsilon) + \epsilon \text{ for all non-empty closed sets } B\}.$$

Theorem 18.6.1 *Suppose that $(X\, d)$ is a complete Polish metric space. The function ρ is a metric on $P(X)$, equivalent to β.*

Proof First we show that ρ is a metric. Suppose that $\rho(\mu, \nu) > \epsilon$. Thus there exists A such that $\mu(A) > \nu(A_\epsilon) + \epsilon$. If $y \in (X \setminus A_\epsilon)_\epsilon$, there exists $z \in X \setminus A_\epsilon$ with $d(y, z) < \epsilon$. Since $z \notin A_\epsilon$, $y \notin A$. Thus $(X \setminus A_\epsilon)_\epsilon \subseteq X \setminus A$. Consequently

$$\begin{aligned}\nu(X \setminus A_\epsilon) &= 1 - \nu(A_\epsilon) > 1 - \mu(A) + \epsilon \\ &\geq \mu((X \setminus A_\epsilon)_\epsilon) + \epsilon,\end{aligned}$$

and so $\rho(\nu, \mu) > \epsilon$. It follows from this that $\rho(\nu, \mu) = \rho(\mu, \nu)$.

Suppose that $\rho(\mu, \nu) = 0$. If A is closed, $\mu(A) \leq \nu(A_{1/n}) + 1/n$, for all $n \in \mathbf{N}$. But $\nu(A_{1/n}) + 1/n \to \nu(A)$ as $n \to \infty$, and so $\mu(A) \leq \nu(A)$. Similarly $\nu(A) \leq \mu(A)$ and so $\mu(A) = \nu(A)$. Since this holds for all closed A, it follows from regularity that $\mu = \nu$.

Suppose that $\rho(\mu, \nu) < \epsilon$ and $\rho(\nu, \pi) < \eta$. If A is a non-empty Borel set then $(A_\epsilon)_\eta \subseteq A_{\epsilon+\eta}$, and so

$$\begin{aligned}\mu(A) &< \nu(A_\epsilon) + \epsilon \\ &< \pi((A_\epsilon)_\eta) + \eta + \epsilon \leq \pi(A_{\eta+\epsilon}) + \eta + \epsilon.\end{aligned}$$

Thus $\rho(\mu, \pi) < \eta + \epsilon$, and so $\rho(\mu, \pi) \leq \rho(\mu, \nu) + \rho(\nu, \pi)$. Consequently, ρ is a metric on $P(X)$.

It follows from Proposition 18.5.3 that if $\beta(\mu, \nu) < \epsilon^2/2$ then $\rho(\mu, \nu) < \epsilon$, and so the identity mapping $id : (P(X), \beta) \to (P(X), \rho)$ is uniformly continuous. On the other hand, suppose that $\rho(\mu_n, \mu) \to 0$ as $n \to \infty$ and that $\epsilon > 0$. Suppose that A is a continuity set for μ. There exists $0 < \delta < \epsilon/2$ so that $\mu(A_\delta) < \mu(A) + \epsilon/2$ and $\mu((X \setminus A)_\delta) < \mu(X \setminus A) + \epsilon/2$. There exists n_0 such that $\rho(\mu_n, \mu) < \delta$ for $n \geq n_0$. Thus if $n \geq n_0$ then

$$\begin{aligned}\mu_n(A) &\leq \mu(A_\delta) + \delta < \mu(A) + \delta + \epsilon/2 < \mu(A) + \epsilon, \\ \text{and } \mu_n(X \setminus A) &\leq \mu((X \setminus A)_\delta) + \delta < \mu(X \setminus A) + \delta + \epsilon/2 \\ &< \mu(X \setminus A) + \epsilon.\end{aligned}$$

Consequently $\mu_n \Rightarrow \mu$, by the portmanteau theorem (Theorem 18.3.4). □

The metric ρ is called the *Prokhorov metric*.

Theorem 18.6.2 (Prokhorov's theorem) *Suppose that (X, d) is a Polish metric space, and that $C \subseteq P(X)$. The following are equivalent.*

(i) *C is β-totally bounded.*
(ii) *C is ρ-totally bounded.*
(iii) *C is uniformly tight.*
(iv) *$\overline{C}$ is w-compact.*

Proof We have seen that (i), (iii) and (iv) are equivalent; (i) implies (ii) since the identity mapping $(P(X), \beta) \to (P(X), \rho)$ is uniformly continuous.

We show that (ii) implies (iii). We use Theorem 18.4.1. Suppose that C is ρ-totally bounded, and that $\epsilon > 0$. There exists a finite subset D of C such that $C \subseteq \cup_{\mu \in D} N_{\epsilon/2}(\mu)$. There exists a compact set K in X, such that $\mu(X \setminus K) < \epsilon/2$ for $\mu \in D$, and there exists a finite subset F of K such that $K \subseteq A = \cup_{x \in F} N_{\epsilon/2}(x)$. Let $B = \cup_{x \in F} N_\epsilon(x)$.

Suppose now that $\nu \in C$. There exists $\mu \in F$ such that $\rho(\nu, \mu) < \epsilon/2$. Now $(X \setminus B)_{\epsilon/2} \cap A = \emptyset$, so that $\mu(X \setminus B) < \epsilon/2$. Consequently,

$$\nu(X \setminus B) \leq \mu(N_{\epsilon/2}(X \setminus B)) + \epsilon/2 < \epsilon.$$

Thus C is uniformly tight, by Theorem 18.4.1. □

Corollary 18.6.3 *$(P(X), \rho)$ is complete.*

Proof If $(\mu_n)_{n=1}^\infty$ is a ρ-Cauchy sequence in $P(X)$, then it is ρ-totally bounded, and so it is β-totally bounded, by Prokhorov's theorem. Thus it has a β-convergent subsequence. Since the β and ρ topologies are the same, it has a ρ-convergent subsequence, so that the original sequence is ρ-convergent. □

18.7 The Fourier Transform and the Central Limit Theorem

We now illustrate the use of Prokhorov's theorem (Theorem 18.6.2) by proving the central limit theorem. Suppose that f is a real-valued random variable on a probability space (Ω, Σ, μ). We define n *independent copies* $f_1, \ldots, f_n$ of f on $(\Omega^n, \Sigma^n, \otimes^n \mu)$ by setting $f_j(\omega) = f(\omega_j)$, where $\omega = (\omega_1, \ldots, \omega_n)$.

Proposition 18.7.1 *Suppose that $f \in L^2(\Omega, \Sigma, \mu)$, that $\int_\Omega f \, d\mu = 0$, that $\int_\Omega f^2 \, d\mu = 1$ and that $f_1, \ldots, f_n$ are n independent copies of f. Let $c_n = (f_1 + \cdots + f_n)/\sqrt{n}$. Then $\int_{\Omega^n} c_n \, d(\otimes^n \mu) = 0$ and $\int_{\Omega^n} c_n^2 \, d(\otimes^n \mu) = 1$.*

Proof Certainly $\int_{\Omega^n} c_n \, d(\otimes^n \mu) = 0$, and

$$\int_{\Omega^n} c_n^2 \, d(\otimes^n \mu) = \frac{1}{n} \sum_{j=1}^{n} \left(\int_{\Omega^n} f_j^2 \, d(\otimes^n \mu) \right) + \frac{1}{n} \sum_{i \neq j} \left(\int_{\Omega^n} f_i f_j \, d(\otimes^n \mu) \right)$$
$$= \frac{1}{n} \sum_{j=1}^{n} \left(\int_{\Omega} f_j^2 \, d\mu \right) + \frac{1}{n} \sum_{i \neq j} \left(\int_{\Omega} f_i \, d\mu . \int_{\Omega} f_j \, d\mu \right)$$
$$= \frac{1}{n} \sum_{j=1}^{n} \left(\int_{\Omega^n} f^2 \, d\mu \right) = \int_{\Omega^n} f^2 \, d\mu.$$

□

It follows from Proposition 18.4.2 that the sequence $(\mathcal{L}(c_n))_{n=1}^{\infty}$ is uniformly tight, and so it has a limit point. The central limit theorem says that there is only one limit point γ, which is the law of a Gaussian random variable γ with mean 0 and variance 1. Consequently c_n converges in distribution to γ.

[A *Gaussian random variable* γ *with mean* 0 *and variance* 1 is a random variable with distribution $(1/\sqrt{2\pi})e^{-t^2/2}.d\lambda(t)$.]

Theorem 18.7.2 (The central limit theorem) *Suppose that* $f \in L^2(\Omega, \Sigma, \mu)$, *that* $\int_\Omega f \, d\mu = 0$ *and that* $\int_\Omega f^2 \, d\mu = 1$. *Let* $c_n = (f_1 + \cdots + f_n)/\sqrt{n}$, *where* $f_1, \ldots, f_n$ *are n independent copies of* f. *Then* c_n *converges in distribution to* γ, *where* γ *is a Gaussian random variable with mean* 0 *and variance* 1.

We shall prove a stronger version of this result later (Theorem 21.6.3).

Proof There are several proofs of this fundamental theorem. The one that we give here uses Fourier transforms. The theory of Fourier transforms is enormous, but we shall only establish the results that we need.

First, we introduce the *Schwartz space* $\mathcal{S}$ and the space $\mathcal{S}'$ of *tempered distributions*. A function g on $\mathbf{R}$ belongs to $\mathcal{S}$ if and only if it can be differentiated infinitely often, and the functions $|x|^j |d^n g/dx^n(x)|$ belong to $C_0(\mathbf{R})$, for each j and n in $\mathbf{Z}^+$. We give $\mathcal{S}$ the metrizable topology given by the seminorms $p_{j,n}$, where

$$p_{j,n}(g) = \sup_{x \in \mathbf{R}} \left\{ |x|^j \left| \frac{d^n g}{dx^n}(x) \right| \right\}.$$

Then $\mathcal{S}$ is homeomorphic to a subspace of $\prod_{j,n} C_0(\mathbf{R})_{j,n}$; elementary arguments show that this is a closed subspace, and so $\mathcal{S}$ is a Polish space.

The space $\mathcal{S}'$ is the space of continuous linear functionals on $\mathcal{S}$; elements of $\mathcal{S}'$ are called *tempered distributions*. We consider the dual pair $(\mathcal{S}, \mathcal{S}')$. We can consider $M(\mathbf{R})$ as a linear subspace of $\mathcal{S}'$; if $g \in \mathcal{S}$ and $\nu \in M(\mathbf{R})$, then $\langle g, \nu \rangle = \int_{\mathbf{R}} g\, d\nu$. Similarly, if $k \in L^2(\mathbf{R}, \mathcal{B}, \lambda)$, we set $\langle g, k \rangle = \int_{\mathbf{R}} gk\, d\lambda$.

We now define the *Fourier transform* of a function $g \in \mathcal{S}$; we set

$$\mathcal{F}(g)(t) = \int_{\mathbf{R}} e^{-2\pi ixt} g(x)\, d\lambda(x), \text{ for } t \in \mathbf{R}.$$

The Fourier transform is concerned with both translation and scaling. If j is a function on $\mathbf{R}$, we set $\tau_h j(x) = j(x - h)$ for $h \in \mathbf{R}$ and $\sigma_\alpha j(x) = j(x/\alpha)$ for $\alpha \neq 0$.

Theorem 18.7.3 *Suppose that $g, j \in \mathcal{S}$, $h \in \mathbf{R}$ and $\alpha \neq 0$.*

(i) $\mathcal{F}(\tau_h(g))(t) = e^{-2\pi iht}\mathcal{F}(g)(t)$ *and* $\mathcal{F}(\sigma_\alpha(g))(t) = \alpha\mathcal{F}(g)(\alpha t)$.
(ii) $\mathcal{F}\left(\frac{dg}{dx}\right)(t) = 2\pi it\mathcal{F}(g)(t)$ *and* $\mathcal{F}(-2\pi ixg(x))(t) = \frac{d\mathcal{F}(g)}{dt}(t)$.
(iii) The Fourier transform $\mathcal{F}$ maps $\mathcal{S}$ continuously into $\mathcal{S}$.
(iv) $\int_{\mathbf{R}} \mathcal{F}(g)(t)j(t)\, d\lambda(t) = \int_{\mathbf{R}} g(x)\mathcal{F}(j)(x)\, d\lambda(x)$.

Proof (i) and (ii) follow from elementary calculations, (iii) follows from them, and (iv) follows by changing the order of integration. □

Let us give two examples. Let $f(x) = e^{-\pi x^2}$. Then f satisfies the differential equation

$$\frac{df}{dx}(x) + 2\pi xf(x) = 0, \qquad f(0) = 1.$$

It follows from Theorem 18.7.3 that $\mathcal{F}(f)$ satisfies the same equation, and so $\mathcal{F}(f) = f$. Similarly, if $\gamma(x) = e^{-x^2/2}/\sqrt{2\pi}$ then $\mathcal{F}(\gamma)(t) = \gamma(2\pi t)$.

Theorem 18.7.4 *Suppose that $g \in \mathcal{S}$. Then $\mathcal{F}^2(g)(x) = g(-x)$, so that $\mathcal{F}$ is a homeomorphism of $\mathcal{S}$ onto itself, which has period 4.*

Proof Suppose that $j \in \mathcal{S}$ and that $\alpha > 0$. Changing variables, and using Theorem 18.7.3,

$$\begin{aligned}\int_{\mathbf{R}} g(\alpha^{-1}x)\mathcal{F}(j)(x)\, d\lambda(x) &= \alpha \int_{\mathbf{R}} g(x)\mathcal{F}(j)(x)\, d\lambda(x) \\ &= \int_{\mathbf{R}} (g)(x)j(\alpha^{-1}x)\, d\lambda(x).\end{aligned}$$

Let $\alpha \to 0$. Then we obtain the general formula

$$g(0)\int_{\mathbf{R}} \mathcal{F}(j)(x)\, d\lambda(x) = j(0)\int_{\mathbf{R}} \mathcal{F}(g)(x)\, d\lambda(x).$$

Now let $j(x) = e^{-\pi x^2}$; then $g(0) = \int_{\mathbf{R}} \mathcal{F}(g)(x)\, d\lambda(x)$, and so

$$g(-x) = (\tau_{-x}g)(0) = \int_{\mathbf{R}} \mathcal{F}(\tau_{-x}g)(t)d\lambda(t)$$
$$= \int_{\mathbf{R}} \mathcal{F}(g)(t)e^{-2\pi ixt}\, d\lambda(t) = \mathcal{F}^2(g)(x).$$

□

We also call the transposed map $\mathcal{F}'$ from S' to itself the Fourier transform, and denote it by $\mathcal{F}$; this is appropriate terminology, since if $g \in \mathcal{S}$, then its Fourier transform as a tempered distribution is the same as its Fourier transfom as an element of $\mathcal{S}$.

Corollary 18.7.5 *$\mathcal{F}$ is a bijective linear map of $\mathcal{S}'$ onto itself.*

Note also that if $\nu \in M(\mathbf{R})$ then $\mathcal{F}(\nu)(t) = \int_{\mathbf{R}} e^{-2\pi ixt}\, d\nu(x)$.

Exercise 18.7.6 Use the theorem of bounded convergence, and the fact that $|e^{2\pi ix}| = 1$, to show that if $\nu \in M(\mathbf{R})$ then $\mathcal{F}(\nu) \in C_b(\mathbf{R})$. Show that the mapping $\nu \to \mathcal{F}(\nu)$ is linear and injective, and that $\|\mathcal{F}(\nu)\|_\infty \leq \|\nu\|_{TV}$.

Suppose now that (Ω, Σ, μ) is a measure space and that $f \in L^0(\Omega)$. We consider the distribution $\mathcal{L}(f)$, and define the *characteristic function* of f to be $\hat{f} = \mathcal{F}(\mathcal{L}(f))$, the Fourier transform of its law. Thus

$$\hat{f}(t) = \int_\Omega e^{-2\pi itf(\omega)}\, d\mu(\omega).$$

Theorem 18.7.7 *If $f \in L^2(|\Omega, \Sigma.\mu)$ then $\hat{f}$ is twice continuously differentiable, and*

$$\frac{d\hat{f}}{dt}(t) = -2\pi i \int_\Omega f(\omega)e^{-2\pi if(\omega)t}\, d\mu,$$
$$\frac{d^2\hat{f}}{dt^2}(t) = -4\pi^2 \int_\Omega f^2(\omega)e^{-2\pi if(\omega)t}\, d\mu(\omega).$$

Proof Let

$$a(y) = \frac{e^{-iy} - 1}{y} + i \text{ for } y \in (\mathbf{R}\setminus\{0\}).$$

Then $a \in C_b(\mathbf{R}\setminus\{0\})$, $\|a\|_\infty \leq 2$ and $a(y) \to 0$ as $y \to 0$. Thus

$$\left| \frac{e^{-2\pi if(\omega)s} - 1}{s} + 2\pi if(\omega) \right| \leq 4\pi\, |f(\omega)|.$$

Multiplying by $e^{-2\pi i f(\omega)t}$ and integrating over Ω, we find that

$$\left| \frac{\hat{f}(t+s) - \hat{f}(t)}{s} + \int_\Omega 2\pi i f(\omega) e^{-2\pi i f(\omega)t}\, d\mu(\omega) \right| \leq 4\pi \, \|f\|_1 ,$$

and so, by the theorem of dominated convergence,

$$\frac{d\hat{f}}{dt}(t) = -2\pi i \int_\Omega f(\omega) e^{-2\pi i f(\omega)_t}\, d\mu.$$

Thus

$$\frac{1}{s}\left(\frac{d\hat{f}}{dt}(t+s) - \frac{d\hat{f}}{dt}(t) \right) = \int_\Omega (-2\pi i f(\omega)) e^{-2\pi i f(\omega)} \left(\frac{e^{-2\pi i f(\omega)s} - 1}{s} \right) d\mu,$$

and the integrand is dominated pointwise by $4m^2\pi^2 f^2(\omega)$. Further applications of the theorem of dominated convergence gives the second equation, and show that the right-hand side of it is a continuous function of t. □

Corollary 18.7.8

$$\frac{d\hat{f}}{dt}(0) = -2\pi i \int_\Omega f\, d\mu \textit{ and } \frac{d^2\hat{f}}{dt^2}(0) = -4\pi^2 \int_\Omega f^2\, d\mu.$$

This theorem can clearly be extended to functions in $L^n(\omega, \Sigma, \mu)$, where $n \in \mathbf{N}$.

We now prove Theorem 18.7.2. Applying Taylor's theorem, $\hat{f}(t) = 1 - 2\pi^2 t^2/2 + r(t)$, where $r(t)/t^2 \to 0$ as $|t| \to 0$. Now

$$\begin{aligned} \hat{c}_n(t) &= \int_{\Omega^n} e^{-2\pi i t(f_1 + \cdots + f_n)/\sqrt{n}}\, d(\otimes^n \mu) \\ &= \prod_{j=1}^n \left(\int_{\Omega^n} e^{-2\pi i t f_j/\sqrt{n}}\, d(\mu) \right) \\ &= (\hat{f}(t/\sqrt{n}))^n, \end{aligned}$$

so that if $n \geq t^2$ then

$$\log \hat{c}_n(t) = n \log(1 - 2\pi^2 t^2/2n + r(t/\sqrt{n})) = -t^2/2 + s_n(t),$$

where $s_n(t) \to 0$ as $n \to \infty$. Consequently, if ν is a w-limit point of $\mathcal{L}(c_n)$ then $\mathcal{F}\nu(t) = -2\pi^2 t^2/2$, and so $\nu = \mathcal{L}(\gamma)$. Thus the sequence $(\mathcal{L}(c_n))$ has a unique limit point, and so $c_n \to \gamma$ in distribution as $n \to \infty$. □

[Terminology and notation vary; in particular, probabilists also call the Fourier transform of an element of $P(\mathbf{R})$ its characteristic function. The choice of constants also varies; -2π is frequently replaced by 2π, by -1 or by 1.]

18.8 Uniform Integrability

We introduce the notion of uniform integrability in various settings. A function c on $[0,\infty)$ is a *growth function* if it is an increasing continuous function for which $c(t) = 0$ if and only if $t = 0$, and which satisfies the Δ_2 condition; there exists $L > 0$ such that $c(2t) \leq Lc(t)$ for all $t > 0$. Thus an N-function which satisfies the Δ_2 condition is a growth function, as are the functions $t \to t^p$, for $p > 0$. If $A \subseteq P([0,\infty))$, A is *c-uniformly integrable* if

$$\sup_{\mu\in A}\int_{[R,\infty)} c\,d\mu \to 0 \text{ as } R \to \infty.$$

If $c(t) = t^p$, where $0 < p < \infty$, we say that A is *p-uniformly integrable*, and if $c(t) = t$ we say that A is *uniformly integrable.*

Proposition 18.8.1 *If $A \subseteq P([0,\infty))$ is c-uniformly integrable then $\sup_{\mu\in A}\int_{[0,\infty)} c\,d\mu < \infty$, and A is uniformly tight.*

Proof There exists $R > 0$ such that $\sup_{\mu\in A}\int_{[R,\infty)} c\,d\mu \leq 1$. If $\mu \in A$, then

$$\int_{[0,\infty)} c\,d\mu = \int_{[0,R)} c\,d\mu + \int_{[R,\infty)} c\,d\mu \leq c(R) + 1.$$

Suppose that $\epsilon > 0$. There exists $R \geq 1$ such that $\sup_{\mu\in A}\int_{[R,\infty)} c\,d\mu < c(1)\epsilon$. If $\mu \in A$ then

$$\mu([R,\infty)) \leq \frac{c(R)}{c(1)}\mu([R,\infty) \leq \frac{1}{c(1)}\int_{[R,\infty)} c\,d\mu < \epsilon.$$

□

Next, suppose that (X,τ) is a Polish space, that c is a non-negative continuous real-valued function on X, and that $A \subseteq P(X)$. We say that A is *c-uniformly integrable* if the set $\{c_*(\mu) : \mu \in A\}$ of push-forward measures on $[0,\infty)$ is uniformly integrable. That is,

$$\sup_{\mu\in A}\int_{(c\geq R)} c\,d\mu \to 0 \text{ as } R \to \infty.$$

If so, then, as in Proposition 18.8.1, $\sup_{\mu\in A}\int_X c\,d\mu < \infty$.

Theorem 18.8.2 *Suppose that (X,τ) is a Polish space, that c is a non-negative continuous real-valued function on X and that $(\mu_k)_{k=0}^\infty$ is a sequence in $P(X)$ for which $\mu_k \Rightarrow \mu_0$ as $k \to \infty$. Then $\{\mu_k : k \in \mathbf{Z}^+\}$ is c-uniformly integrable if and only if $\sup_{k\in\mathbf{Z}^+}\int_X c\,d\mu_k < \infty$ and $\int_X c\,d\mu_k \to \int_X c\,d\mu_0$ as $k \to \infty$.*

Proof Suppose first that $\{\mu_k : k \in \mathbf{Z}^+\}$ is c-uniformly integrable. Certainly, $\sup_{k\in\mathbf{Z}^+} \int_X c\, d\mu_k < \infty$. Suppose that $\epsilon > 0$. There exists $R > 0$ such that $\sup_{k\in\mathbf{Z}^+} \int_{(c\geq R)} c\, d\mu_k < \epsilon$. Then

$$\int_X c\, d\mu_k - \int_X (c \wedge R)\, d\mu_k \leq \int_{(c\geq R)} c\, d\mu_k < \epsilon \text{ for } k \in \mathbf{Z}^+,$$

so that

$$\left|\int_X c\, d\mu_k - \int_X c\, d\mu_0\right| \leq \left|\int_X (c \wedge R)\, d\mu_k - \int_X (c \wedge R)\, d\mu_0\right| + 2\epsilon.$$

Since $\int_X (c \wedge R)\, d\mu_k \to \int_X (c \wedge R)\, d\mu_0$ as $k \to \infty$, $\int_X c\, d\mu_k \to \int_X c\, d\mu_0$ as $k \to \infty$.

Suppose conversely that $\sup_{k\in\mathbf{Z}^+} \int_X c\, d\mu_k < \infty$ and that $\int_X c\, d\mu_k \to \int_X c\, d\mu_0$ as $k \to \infty$. Suppose that $\epsilon > 0$. Then there exists k_0 such that

$$\left|\int_X c\, d\mu_k - \int_X c\, d\mu_0\right| < \epsilon/3 \text{ for } k \geq k_0.$$

By Tietze's extension theorem, for each $n \in \mathbf{N}$, there exists $c_n \in C(X)$ such that $c_n(x) = c(x)$ if $c(x) \leq n$, $c_n(x) = 0$ if $c(x) \geq n + 1$ and $0 \leq c_n \leq c$. By monotone convergence, there exists n_0 such that

$$0 \leq \int_X c\, d\mu_0 - \int_X c_n\, d\mu_0 < \epsilon/3 \text{ for } n \geq n_0.$$

Since $\mu_k \Rightarrow \mu_0$, there exists $k_1 \geq k_0$ such that

$$\left|\int_X c_{n_0}\, d\mu_k - \int_X c_{n_0}\, d\mu_0\right| < \epsilon/3 \text{ for } k \geq k_0.$$

Putting these inequalities together, if $k \geq k_1$ then

$$\begin{aligned}\int_{(c\geq n+1)} c\, d\mu_k &\leq \int_X (c - c_{n_0})\, d\mu_k \\ &\leq \int_X (c - c_{n_0})\, d\mu_0 + \left|\int_X c\, d\mu_k - \int_X c\, d\mu_0\right| \\ &\quad + \left|\int_X c_{n_0}\, d\mu_k - \int_X c_{n_0}\, d\mu_0\right| < \epsilon.\end{aligned}$$

By monotone convergence, there exists $N \geq n + 1$ such that $\int_{(c\geq N)} c\, d\mu_k \leq \epsilon$ for $1 \leq k \leq k_1$, and so the result follows. □

18.9 Uniform Integrability in Orlicz Spaces

Suppose that c is a growth function, that $\mu \in P(X)$, where (X, τ) is a Polish space, and that F is a set of real-valued random variables on X. We say that F is *c-uniformly integrable* if the set of distributions $\{|f|_*(\mu) : f \in F\}$ is; that is, if

$$\sup_{f \in F} \int_{|f| \geq R} c(|f|)\, d\mu \to 0 \text{ as } R \to \infty.$$

If $c = \Phi$, an N-function which satisfies the Δ_2 condition, and F is a Φ-uniformly integrable set of real-valued random variables on (X, μ), then F is a norm-bounded subset of $(L_\Phi(\mu), \|.\|_\Phi)$.

Exercise 18.9.1 Suppose that (X, τ) is a Polish space, that $\mu \in P(X)$ and that c is a growth function. Suppose that F and G are Φ-uniformly integrable sets of real-valued random variables on (X, μ), and that $\alpha \in \mathbf{R}$. Then

$$F + G = \{f + g : f \in F, g \in G\} \text{ and } \alpha F$$

are c-uniformly integrable.

Theorem 18.9.2 *Suppose that (X, τ) is a Polish space, that $\mu \in P(X)$, that Φ is an N-function which satisfies the Δ_2 condition, or that $\Phi(t) = t^p$ for some $0 < p \leq 1$. If $(f_k)_{k=0}^\infty$ is a sequence of real-valued random variables on (X, μ), then the following are equivalent.*

(i) $f_k \to f_0$ in probability as $k \to \infty$ and $(f_k)_{k=0}^\infty$ is Φ-uniformly integrable.
(ii) $(f_k)_{k \in \mathbf{Z}^+}$ is a norm-bounded sequence in $L_\Phi(X, \mu)$, and $\|f_k - f_0\|_\Phi \to 0$ as $k \to \infty$.
(iii) $f_k \to f_0$ in probability as $k \to \infty$, $(f_k)_{k \in \mathbf{Z}^+}$ is a norm-bounded sequence in $L_\Phi(X, \mu)$, and $\|f_k\|_\Phi \to \|f_0\|_\Phi$ as $k \to \infty$.

Proof Suppose that (i) holds. Then $(f_k)_{k \in \mathbf{Z}^+}$ is a norm-bounded sequence in $L_\Phi(X, \mu)$. To show that (ii) holds, it is sufficient to show that each subsequence has a sub-subsequence for which (ii) holds, and so by extracting a subsequence if neccessary, we may suppose that $f_k \to f_0$ almost everywhere, as $k \to \infty$. Suppose that $\epsilon > 0$. The sequence $(f_k - f_0)_{k \in \mathbf{N}}$ is Φ-uniformly integrable (Exercise 18.9.1), and so there exists $M > 0$ such that $\int_{(|f_k - f_0| \geq M)} \Phi(|f_k - f_0|)\, d\mu < \epsilon$. Let $\Phi_M(t) = \Phi(|t|) \wedge M$. By the theorem of bounded convergence, $\int_X \phi_M(f_k - f_0)\, d\mu \to 0$ as $k \to \infty$. Since $\int_{(|f_k - f_0| \leq M)} \Phi(|f_k - f_0)|^p\, d\mu \leq \int_X \phi_M(f_k - f_0)\, d\mu$, there exists k_0 such that $\int_{(|f_k - f_0| \leq M)} \Phi(|f_k - f_0|)\, d\mu < \epsilon^p/2$ for $k \geq k_0$. Combining this with the earlier inequality, it follows that $\|f_k - f_0\|_\Phi \to 0$ as $k \to \infty$.

Condition (ii) certainly implies (iii). Suppose that (iii) holds; again we can suppose that $f_k \to f_0$ almost everywhere as $k \to \infty$. Suppose that $\epsilon > 0$. If

$n \in \mathbf{N}$, let $\Psi_n(t) = \Phi(t)$ for $0 \leq t \leq n$ and let $\Psi_n(t) = 0$ for $t > n$. By the theorem of monotone convergence, there exists n_0 such that $\int_X \Phi(|f_0|\, d\mu \leq \int_X \Psi_n(|f_0|\, d\mu + \epsilon/2$ for $n \geq n_0$. By the theorem of bounded convergence, there exists k_0 such that $\int_X |\Psi_{n_0}(f_k) - \Psi_{n_0}(f_0)|\, d\mu < \epsilon/3$ for $k \geq k_0$. If $k \geq k_0$ then

$$\begin{aligned}\int_{(|f_k|\geq n_0)} \Phi(|f_k|)\, d\mu &\leq \int_X \Phi(|f_k|)\, d\mu - \int_X \Psi_{n_0}(|f_k|)\, d\mu \\ &\leq \left(\int_X \Psi(|f_0|)\, d\mu + \epsilon/3\right) - \left(\int_X \Psi_{n_0}(|f_k|)\, d\mu - \epsilon/3\right) \\ &< \epsilon.\end{aligned}$$

Again, the first k_0 terms are Φ-uniformly integrable, and so (iii) implies (i). □

Exercise 18.9.3 Suppose that (X, τ) is a Polish space, that $\mu \in P(X)$, that Φ is an N-function which satisfies the Δ_2 condition, or that $\Phi(t) = t^p$ for some $0 < p \leq 1$, and that $F \subset L_\Phi$. Show that F is Φ-uniformly integrable if and only if F is $\|.\|_\Phi$ bounded and if $\epsilon > 0$ there exists $\delta > 0$ such that if $\mu(A) < \delta$ than $\int_A \phi(|f|)\, d\mu < \epsilon$ for $f \in F$.

We can use Theorem 18.9.2 to characterize compactness in L_Φ and L_p.

Exercise 18.9.4 Suppose that A is a bounded subset of L_Φ (or L_p). The following are equivalent.

(i) A is a compact subset of L_Φ (or L_p).
(ii) A is Φ-uniformly integrable (or p-uniformly integrable) and compact in L_0.
(iii) Whenever $(f_n)_{n=1}^\infty$ is a sequence in A which converges in probability to $f \in A$ then $\|f_n\|_\Phi \to \|f\|_\Phi$ (or $\|f_n\|_p \to \|f\|_p$) as $n \to \infty$.

19
Introduction to Choquet Theory

Choquet theory is concerned with the properties of compact convex sets; either weakly compact convex sets of a space E, where (E, F) is a dual pair, or compact convex sets of a Banach space E. In general, the theory is complicated; we shall only consider the metrizable case, where the theory is much simpler.

19.1 Barycentres

Theorem 19.1.1 *Suppose that (E, F) is a dual pair of vector spaces, that K is a metrizable $\sigma(E, F)$-compact subset of E, and that $L = \overline{\Gamma}(K)$ is also $\sigma(E, F)$-compact and metrizable. Suppose that $\mu \in P(K)$. Then there exists a unique $\beta_\mu \in L$ such that $\int_K f\, d\mu = \langle \beta_\mu, f \rangle$ for each $f \in F$. The mapping $\mu \to \beta_\mu$ is a continuous affine mapping of $(P(K), w)$ onto L; that is, $\beta_{(1-\lambda)\mu+\lambda\nu} = (1-\lambda)\beta_\mu + \lambda\beta_\nu$ for $0 \leq \lambda \leq 1$ and $\mu, \nu \in K$.*

Proof For each $f \in F$, let $H_f = \{e : \langle e, f \rangle = \int f\, d\mu\}$. Suppose that $f_1, \ldots, f_n \in E$. If $e \in E$, let $T(e) = (\langle e, f_1 \rangle, \ldots, \langle e, f_n \rangle)$; T is a continuous linear mapping of E into $\mathbf{R}^n$, and so $T(L)$ is a compact convex subset of $\mathbf{R}^n$. Let

$$p = \left(\int_K f_1\, d\mu, \ldots, \int_K f_n\, d\mu \right).$$

We show that $p \in T(L)$. If not, by the separation theorem in $\mathbf{R}^n$, there exists $\alpha \in \mathbf{R}^n$ such that

$$\sum_{j=1}^n \alpha_j p_j > \sup_{x \in L} \sum_{j=1}^n \alpha_j \langle x, f_j \rangle.$$

Let $f = \sum_{j=1}^n \alpha_j f_j$. Then $\int_K f\,d\mu > \sup\{f(x) : x \in K\}$, which is clearly not true. Consequently $p \in T(L)$, and so $L \cap (\cap_{j=1}^n H_{f_i})$ is a non-empty compact convex subset of L. By the finite intersection property, $L_0 = L \cap (\cap_{f\in F} H_f)$ is non-empty. If $l \in L_0$ then $\int_K f\,d\mu = \langle l_\mu, f\rangle$ for each $f \in F$.

It remains to show that L_0 is a singleton set. If $l, m \in L_0$ and $f \in F$ then $\langle l, f\rangle = \int_K f\,d\mu = \langle m, f\rangle$, so that since F separates the points of E, $l = m$.

Since $\langle \beta_\mu - \beta_\nu, f\rangle = \int_K f\,d\mu - \int_K f\,d\nu$, it follows that the mapping $\mu \to \beta_\mu$ from $(P(K), w)$ to L is a continuous affine mapping. Thus $\beta(K)$ is a compact convex subset of L containing K, and so $\beta(P(K)) = L$. □

The element β_μ is called the *barycentre* of μ.

Exercise 19.1.2 Suppose that L is a convex $\sigma(E, F)$-compact metrizable subset of E and that $K = Ex(L)$ is closed. Then if $x \in L$ there exists $\mu \in P(Ex(L))$ with barycentre x.

If (E, F) is a dual pair of vector spaces and L is a $\sigma(E, F)$-compact convex metrizable subset of E then $Ex(L)$ need not be closed. We shall however show later (Theorem 19.3.4) that a corresponding result still holds.

We can go in the opposite direction. Suppose that (E, F) is a dual pair of vector spaces, that K is a metrizable $\sigma(E, F)$-compact subset of E and that $x \in K$. We set $Rep(x) = \{\mu \in P(K) : \beta_\mu = x\}$; $Rep(x)$ is the set of probability measures on K which *represent* x. Of course, $\delta_x \in Rep(x)$.

Theorem 19.1.3 *Suppose that (E, F) is a dual pair of vector spaces, that K is a metrizable $\sigma(E, F)$-compact subset of E and that $x \in K$. Then $Rep(x) = \{\delta_x\}$ if and only if x is an extreme point of K.*

Proof Suppose that x is not an extreme point of K, and that $x = (1-\alpha)y + \alpha z$, with $y, z \in K$ and $0 < \alpha < 1$. Then $(1-\alpha)\delta_y + \alpha\delta_z \in Rep(x)$, so that $Rep(x) \neq \{\delta_x\}$.

Conversely, suppose that $Rep(x) \neq \{\delta_x\}$ and that $\mu \in Rep(x) \setminus \{\delta_x\}$. Let $s(\mu)$ be the support of μ. Then $s(\mu) \neq \{x\}$, and so there exists $y \in s(x)$ with $y \neq x$. Let U be a closed convex neighbourhood of y in E which does not contain x, and let $V = K \cap U$. Then V is a compact convex subset of K, and $r = \mu(V) > 0$. Suppose that $\mu(V) = 1$; it would then follow that $\beta_\mu \in V$, which is not true. Thus $r < 1$. If A is a Borel subset of K, let

$$\mu_1(A) = \mu(A \cap V)/r \text{ and } \mu_2(A) = \mu(A \cap (K \setminus V))/(1-r);$$

then $\mu_1, \mu_2 \in P(K)$. Let $y = \beta_{\mu_1}$ and $z = \beta_{\mu_2}$. Then $y \in V$, so that $y \neq x$, $z \in K$ and $x = ry + (1-r)z$, so that x is not an extreme point of K. □

19.2 The Lower Convex Envelope Revisited

Suppose that (E, F) is a dual pair and that K is a convex $\sigma(E, F)$-compact metrizable subset of E. If f is a real-valued function on K, we extend f to a function on E by setting $f(y) = +\infty$ for $y \in E \setminus K$. Recall that the *lower convex envelope* $\underline{f}$ is the greatest convex function less than or equal to f; $\underline{f}$ is lower semi-continuous.

Theorem 19.2.1 *Suppose that (E, F) is a dual pair, that K is a convex $\sigma(E, F)$-compact metrizable subset of E and that $f \in C(K)$. Suppose that $\mu \in P(K)$. Then there exists $\nu \in P(K)$ such that $\beta_\nu = \beta_\mu$, $\int_K \underline{g}\, d\mu \leq \int_K g\, d\nu$ for each $g \in C(K)$ and $\int_K f\, d\nu = \int_K \underline{f}\, d\mu$.*

Proof Let $A(K)$ be the set of $\sigma(E, F)$-continuous affine functions on E, restricted to to K. If $g \in C(K)$, let $p(g) = -\int_K \underline{g}\, d\mu$. Then $p(\alpha g) = \alpha p(g)$ for $\alpha > 0$ and $p(g_1 + g_2) \leq p(g_1) + p(g_2)$, since $\underline{g_1} + \underline{g_2} \leq \underline{g_1 + g_2}$. Thus p is sublinear; also, $p(a) = -a(\beta_\mu)$ for $a \in A(K)$. Let $\phi(\alpha f) = \alpha p(f)$ for $\alpha \in \mathbf{R}$: ϕ is a linear functional on span(f), and $\phi(\alpha f) \leq p(\alpha f)$. We apply the Hahn–Banach theorem. There exists a linear functional ϕ on $C(K)$ such that $\phi(f) = p(f)$ and $\phi(g) \leq p(g)$ for $g \in C(K)$. Note that if a is affine, then

$$\phi(a) \leq p(a) = -a(\beta_\mu) \text{ and } \phi(-a) \leq p(-a) = a(\beta_\mu),$$

so that $\phi(a) = -a(\beta_\mu)$. In particular, $\phi(1) = -1$. Further, if $g \geq 0$ then $p(-g) \leq 0$, so that $\phi(g) \leq 0$. Thus $-\phi$ is positive, and by the Riesz representation theorem there exists $\nu \in P(K)$ such that $-\phi(g) = \int_K g\, d\nu$ for all $g \in C(K)$. In particular, $\int_K f\, d\nu = -\phi(f) = \int \underline{f}\, d\mu$, and if $a \in A(K)$ then $a(\beta_\mu) = \int_K a\, d\nu$, so that $\beta_\nu = \beta_\mu$. □

Exercise 19.2.2 Suppose that (E, F) is a dual pair, that K is a convex $\sigma(E, F)$-compact metrizable subset of E and that $f \in C(K)$. Suppose that $x_0 \in K$. Show that there exists $\nu \in P(K)$ such that $\beta_\nu = x_0$, $\int_K g\, d\nu \geq \underline{g}(x_0)$ for each $g \in C(K)$ and $\int_K f\, d\nu = \underline{f}(x_0)$.

Proposition 19.2.3 *Suppose that (E, F) is a dual pair, that K is a convex $\sigma(E, F)$-compact metrizable subset of E and that $f \in C(K)$. Let $f_*(x) = \inf\{\int_K f\, d\mu : \mu \in Rep(x)\}$. Then $f_* = \underline{f}$.*

Proof If $x \in K$, $a \in A(K)$ and $a \leq f$, and if $\mu \in Rep(x)$ then $a(x) = \int_K a\, d\mu \leq \int_K f\, d\mu$, so that $a(x) \leq f_*(x)$. Consequently, $f_*(x) \geq \underline{f}(x)$. On the other hand, $f_* \leq \underline{f}$, by Exercise 19.2.2. □

Exercise 19.2.4 (Jensen's inequality) If f is a convex lower semi-continuous function on K, if $x \in K$ and if $\mu \in Rep(x)$, then $f(x) \leq \int_K f\, d\mu$.

A real-valued function f on a convex set K is *strictly convex* if whenever x and y are distinct points of K and $0 < \lambda < 1$ then

$$f((1-\lambda)x + \lambda y) < (1-\lambda)f(x) + \lambda f(y).$$

Proposition 19.2.5 *Suppose that (E, F) is a dual pair and that K is a convex $\sigma(E, F)$-compact metrizable subset of E. Then there exists a non-negative continuous strictly convex function f on K.*

Proof Let $A_1(K) = \{h \in A(K) : \|h\|_\infty \leq 1\}$. $A_1(K)$ separates the points of K, and is separable in the uniform norm topology; let $(h_n)_{n=1}^\infty$ be a dense sequence in $A_1(K)$; then (h_n) separates the points of K. We set

$$f(k) = \sum_{n=1}^{\infty} h_n^2(k)/2^n.$$

Then $f \in C(K)$. If x and y are distinct points of K and $0 < \lambda < 1$ then there exists n such that $h_n(x) \neq h_n(y)$, so that

$$h_n^2((1-\lambda)x + \lambda y) < (1-\lambda)h_n^2(x) + \lambda h_n^2(y),$$

while, for all m,

$$h_m^2((1-\lambda)x + \lambda y) \leq (1-\lambda)h_m^2(x) + \lambda h_m^2(y),$$

so that

$$f((1-\lambda)x + \lambda y) < (1-\lambda)f(x) + \lambda f(y);$$

that is, f is strictly convex. □

Proposition 19.2.6 *Suppose that (E, F) is a dual pair and that K is a convex $\sigma(E, F)$-compact metrizable subset of E and that $x \in K$. Then x is an extreme point of K if and only if $\underline{f}(x) = f(x)$ for each $f \in C(K)$.*

Proof If x is an extreme point, then $Rep(x) = \{\delta_x\}$, and so $\underline{f}(x) = f(x)$ for each $f \in C(K)$, by Proposition 19.2.3.

If x is not an extreme point of K, let f be a strictly convex function in C(K). There exist $y, z \in K$ and $0 < \alpha < 1$ such that $x = (1-\alpha)y + \alpha z$. Then

$$\begin{aligned}\underline{f}(x) &\leq (1-\alpha)\underline{f}(y) + \alpha\underline{f}(z)\\ &\leq (1-\alpha)f(y) + \alpha f(z) < f(x).\end{aligned}$$

□

19.3 Choquet's Theorem

Suppose that (E, F) is a dual pair and that K is a $\sigma(E, F)$-compact subset of E. The set of extreme points of such a general compact set can be very unpleasant topologically, but this is not the case when K is metrizable.

Proposition 19.3.1 *Suppose that (E, F) is a dual pair and that K is a metrizable $\sigma(E, F)$-compact subset of E. Then $Ex(K)$ is a G_δ subset of K.*

Proof Let d be a metric on K which defines the topology of K, and let

$$A_n = \{(x, y) \in K \times K : d(x, y) \geq 1/n\} \times [1/n, 1 - 1/n].$$

A_n is a compact subset of the set $A = K \times K \times [0, 1]$. If $(x, y, \lambda) \in A$, let $f(x, y, \lambda) = (1 - \lambda)x + \lambda y$. Then f is a continuous function on A, and so $B_n = f(A_n) \cap K$ is closed in K. But x is an extreme point of K if and only if $x \notin \cup_n B_n$. □

This enables us to show that a point of a convex compact set K can be represented by a measure which lives on the extreme points of K.

Even when defined by simple conditions, the set of extreme points of a subset of $(P(K), w)$ need not be closed. Suppose that J is a homeomorphism of a compact metric space K onto itself. A Borel probability measure μ is *J-invariant* if $J_*\mu = \mu$; that is, $\mu(J(A)) = \mu(A)$ for each Borel set A. The set of J-invariant elements of $P(K)$ is denoted by $P_J(K)$.

Exercise 19.3.2 Suppose that J is a homeomorphism of a compact metric space K onto itself. Show that $P_J(K)$ is a w-closed convex subset of $P(K)$.

Exercise 19.3.3 Let $K = \mathbf{T}^2$. If $k = (e^{i\theta}, e^{i\phi}) \in K$, let $J(k) = (e^{i\theta}, e^{i(\theta+\phi)})$. If $e^{i\theta} \in \mathbf{T}$, let $S(e^{i\theta}) = (1, e^{i\theta})$.

(i) Show that J is a homeomorphism of K onto itself.
(ii) Show that if $\mu \in P(T)$ and the support of μ is contained in $\{1\} \times \mathbf{T}$ then $\mu \in P_J(K)$.
(iii) Let $\mu_n = \frac{1}{n}\sum_{j=1}^{n} \delta_{(e^{2\pi i/n}, e^{2\pi ij/n})}$. Show that μ_n is an extreme point of $P_J(K)$.
(iv) Show that $\mu_n \Rightarrow S_*(\lambda)$ (where λ is Haar measure on $\mathbf{T}$).
(v) Show that $S_*(\lambda)$ is not an extreme point of $P_J(K)$.

Theorem 19.3.4 (Choquet's theorem) *Suppose that (E, F) is a dual pair, that K is a metrizable $\sigma(E, F)$-compact convex subset of E and that $x_0 \in K$. Then there exists $\nu \in P(K)$ with barycentre x_0 such that $\mu(Ex(K)) = 1$.*

Proof Let f be a strictly convex continuous non-negative function on K. By Corollary 19.2.2, there exists $\nu \in P(K)$ such that $\beta_\nu = x_0$, $\int_K g\, d\nu \geq \int_K g\, d\mu$ for each $g \in C(K)$ and $\int_K f\, d\nu = \underline{f}(x_0)$. We show that $\nu(Ex(K)) = 1$. Since

$$\begin{aligned}\underline{f}(x_0) &= \sup\{a(x_0) : a \in A(K), a \leq f\}\\ &= \sup\left\{\int_K a\,d\nu : a \in A(K), a \leq f\right\} = \int_K f\,d\nu,\end{aligned}$$

it follows that $\inf\{\int_K (f - a)\,d\nu : a \in A(K), a \leq f\} = 0$. Thus for each $n \in \mathbf{N}$ there exists $a_n \in A(K)$, with $a_n \leq f$, such that $\int_K (f - a_n)\,d\nu \leq 2^{-n}$. Now let $s_n = \sum_{j=1}^{n}(f - a_j)$, and let $s = \sum_{j=1}^{\infty}(f - a_j)$. By monotone convergence,

$$0 \leq \int_K s\,d\nu = \lim_{n\to\infty} \int_K s_n\,d\nu \leq 1,$$

and so $\nu\{k : s(k) = \infty\} = 0$. But if k is not an extreme point of K, then $\delta = f(k) - \underline{f}(k) > 0$, so that $f(k) - a_n(k) \geq \delta$ for each $n \in \mathbf{N}$, and $s(k) = \infty$. Thus $K \setminus Ex(K) \subseteq \{k : s(k) = \infty\}$, and $\nu(K \setminus Ex(K)) = 0$. □

19.4 Boundaries

Although convexity is a real concept, it has an important role to play in complex analysis. (Recall that a set is convex in a complex vector space E if and only if it is a convex subset of the underlying real space $E_{\mathbf{R}}$.) In this section, we apply Choquet theory to subspaces of the complex Banach space $(C_{\mathbf{C}}(K), \|.\|_\infty)$, where K is a compact metrizable space.

Let us begin with a familiar example, which motivates the theory that wc shall develop. Let $A(\mathbf{D})$ denote the space of complex-valued functions which are continuous on the closed unit disc $\mathbf{D}$, and analytic on the interior of $\mathbf{D}$. Then it follows from the maximum modulus theorem that if f is a non-constant function in $A(\mathbf{D})$ then the set where $|f|$ attains its maximum is a subset of the boundary $\mathbf{T} = \partial\mathbf{D}$. Further, if $0 \leq \theta < 2\pi$, let $f_\theta(z) = 1 + e^{-i\theta}z$. Then $\|f_\theta\|_\infty = 2$, and $|f_\theta(z)| = 2$ if and only if $z = e^{i\theta}$.

A closed linear subspace E of a space $(C_{\mathbf{C}}(K), \|.\|_\infty)$ is a *separating subspace* of $C_{\mathbf{C}}(K)$ if it contains the constant functions, and separates points; that is, if x_1 and x_2 are distinct points of K then there exists $f \in E$ with $f(x_1) \neq f(x_2)$.

If E is a separating subspace of $C_{\mathbf{C}}(K)$, we denote the inclusion mapping $E \to C_{\mathbf{C}}(K)$ by i_E. If j is the isomorphism of $M_{\mathbf{C}}(K, \mathcal{B})$ onto $C_{\mathbf{C}}(K)'$ given by the Riesz representation theorem, then $i'_E \circ j$ is a surjection of $M_{\mathbf{C}}(K, \mathcal{B})$ onto E'; we denote its restriction to $\pi(K)$ by π_E. If $\phi \in E'$, we denote $\pi_E^{-1}(\phi)$ by M_ϕ. We denote $\pi_E \circ \delta$ by q_E. Thus $\pi_E(\mu)(f) = \int_K f\,d\mu$ and $q_E(x)(f) = f(x)$.

Exercise 19.4.1 Show that q_E is a homeomorphism of K onto $q_E(K)$.

Suppose that E is a separating subspace of $C_{\mathbf{C}}(K)$, where K is a compact metrizable space. The *state space* $S(E)$ of E is the set

$$\{\phi \in E' : \phi(1) = 1 = \|\phi\|'\}.$$

Theorem 19.4.2 *Suppose that E is a separating subspace of $C_{\mathbf{C}}(K)$, where K is a compact metrizable space. Then $S(E) = \pi_E(P(K))$.*

Proof If $\mu \in P(K)$ then $\|\pi_E(\mu)\|' \le \|\mu\|_{TV} = 1$. But $\pi_E(\mu)(1) = 1$, so that $\|\pi_E(\mu)\|' = 1$ and $\pi_E(\mu) \in S(E)$.

Conversely, if $\phi \in S(E)$, it follows from the Hahn–Banach theorem that there exists $\mu \in M_{\mathbf{C}}(K)$ such that $i'_E(\mu) = \phi$ and $\|\mu\|_{TV} = 1$. But $\mu(K) = \phi(1) = 1$, and so $\mu \in P(K)$. Thus $S(E) = \pi_E(P(K))$. □

If $\phi \in S(E)$, we define $\Re\phi$ by setting $(\Re\phi)(f) = \Re(\phi(f))$. Thus if $\mu \in M_\phi$ then $\Re\phi(f) = \int_K \Re(f)\,d\mu$. We denote the real linear subspace $\{\Re f : f \in E\}$ of $C(K)$ by $\Re E$.

Proposition 19.4.3 *Suppose that E is a separating subspace of $C_{\mathbf{C}}(K)$, where K is a compact metrizable space. Then $S(E) = \overline{\Gamma}(q_E(K))$.*

Proof $S(E)$ is a weak*-closed convex subset of the unit ball $B_1(E')$ of E', and so $\overline{\Gamma}(q_E(K)) \subseteq S(E)$.

Suppose if possible that there exists $\phi \in S(E) \setminus \overline{\Gamma}(q_E(K))$. By the separation theorem, there exists $h = f + ig \in C_{\mathbf{C}}(K)$ such that

$$\begin{aligned}\phi(f) = \Re(\phi(f)) &> \sup\{\Re(\psi(h)) : \psi \in \overline{\Gamma}(q_E(K))\} \\ &= \sup\{\Re(\psi(h)) : \psi \in q_E(K)\} = \sup_{x\in K} f(x).\end{aligned}$$

Let $\alpha = \|f\|_\infty$. Then $\phi(f + \alpha 1) > \sup_{x\in K} f(x) + \alpha = \|f + \alpha 1\|_\infty$, giving a contradiction. □

Theorem 19.4.4 *Suppose that E is a separating subspace of $C_{\mathbf{C}}(K)$, where K is a compact metrizable space, that $f \in C_{\mathbf{C}}(K)$ and that $\phi \in S(E)$. Let*

$$m = \sup\{\Re\phi(g) : g \in E, \Re g \le f\} \text{ and } M = \inf\{\Re\phi(h) : h \in E, \Re h \ge f\}.$$

If $m \le c \le M$ there exists $\mu \in M_\phi$ such that $\int_K f\,d\mu = c$.

Proof The result is trivially true if $f \in \Re E$. Otherwise, let $F = \text{span}(\Re E, f)$. If $h = \Re g + \lambda f \in F$ let $\psi(h) = \Re\phi(g) + \lambda c$. Thus ψ is a linear functional on F which extends $\Re\phi$. We show that $\|\psi\|' = 1$. For this, it is enough to show that if $h \le 1$, then $\psi(h) \le 1$. If $\lambda = 0$ this is certainly true. Suppose next that $\lambda > 0$. Then $f \le \Re((1-g)/\lambda)$, so that $Re\phi((1-g)/\lambda) \ge M \ge c$. Thus $\psi(h) = \Re\phi(g) + \lambda c \le \Re\phi(1) = 1$. Finally suppose that $\lambda < 0$. Then $f \ge \Re((1-g)/\lambda)$, so that $\Re\phi((1-g)/\lambda) \le m \le c$. Thus $\psi(h) = \Re\phi(g) + \lambda c \le \Re\phi(1) = 1$.

We now apply the Hahn–Banach theorem; there exists $\mu \in M(K)$, with $\|\mu\|_{TV} = 1$, which extends ψ. Since $\mu(K) = \int_K 1\, d\mu = 1$, $\mu \in P(K)$. Further, $\pi_E(\mu) = \phi$, so that $\mu \in M_\phi$, and $\int_K f\, d\mu = \psi(f) = c$. □

Proposition 19.4.5 *Suppose that E is a separating subspace of $C_{\mathbf{C}}(K)$, where K is a compact metrizable space, and that $\phi \in S(E)$. Then ϕ is an extreme point of $S(E)$ if and only if there exists $x \in K$ such that $M_\phi = \{\delta_x\}$.*

Proof Suppose first that ϕ is an extreme point of $S(E)$. It follows from the preceding proposition and Mil'man's theorem (Theorem 12.2.10) that there exists $x \in K$ such that $\phi = q_E(x)$. Suppose that $\mu \in P(K)$ and that $\phi = \pi_E(\mu)$ - that is, $f(x) = \phi(f) = \int_K f\, d\mu$ for each $f \in E$. Let ν be the push-forward measure $q_{E*}(\mu)$. Then $\phi(f) = \int_{S(E)} \psi(f)\, d\nu(\psi)$ for each $f \in E$, and so ϕ is the barycentre of ν. But ϕ is an extreme point of $S(E)$, and so $\nu = \delta_\phi$. Thus $\mu = \pi_E(\delta_x) = q_E(x)$.

Conversely, suppose that ϕ is not an extreme point of $S(E)$. Thus there exist distinct ϕ_1 and ϕ_2 in $S(E)$ such that $\phi = \frac{1}{2}(\phi_1 + \phi_2)$. There exist μ_1 and μ_2 in $P(K)$ such that $\phi_1 = \pi_E(\mu_1)$ and $\phi_2 = \pi_E(\mu_1)$. Let $\mu = \frac{1}{2}(\mu_1 + \mu_2)$. Then $\phi = \pi_E(\mu)$. But μ is not an extreme point of $P(K)$, and so $\pi_E^{-1}(\phi) \neq \{\delta_x\}$, for some $x \in K$. □

Exercise 19.4.6 Suppose that ϕ is an extreme point of $S(E)$ and that $M_\phi = \{x\}$. Use Theorem 19.4.4 to show that if $f \in C(K)$ and $a < f(x)$ then there exists $g \in E$ with $g(0) = a$ and $\Re(g) \leq f$.

Suppose that E is separating subspace of $C_{\mathbf{C}}(K)$, where K is a compact metrizable space. A subset B of K is a *boundary* for E if for each $f \in E$ there exists $x \in B$ for which $|f(x)| = \|f\|_\infty$.

Exercise 19.4.7 Use the maximum modulus principle to show that $\mathbf{T} = \partial \mathbf{D}$ is a boundary for $A(\mathbf{D})$.

Theorem 19.4.8 *Suppose that E is separating subspace of $C_{\mathbf{C}}(K)$, where K is a compact metrizable space. Let*

$$Ch(E) = \{x \in K : q_E(x) \text{ is an extreme point of } S(E)\},$$

then $Ch(E)$ is a boundary for E.

Proof Suppose that $f \in E$. There exists $x \in K$ such that $|f(x)| = \|f\|_\infty$. Let $H = \{\phi \in E' : \phi(f) = q_E(x)(f)\}$. Then H is a support hyperplane of $S(E)$, and $H \cap S(E)$ is a non-empty convex set. Let ϕ_0 be an extreme point of $H \cap S(E)$. By Proposition 12.2.4 it is an extreme point of $S(E)$. There exists a unique $x_0 \in K$ such that $\phi_0 = q_E(x_0)$. Then $x_0 \in Ch(E)$ and $|f(x_0)| = \|f\|_\infty$. □

The set $Ch(E)$ is called the *Choquet boundary* of E.

Exercise 19.4.9 By considering functions of the form $1 + e^{i\theta}$, show that $\mathbf{T}$ is the Choquet boundary of $A(D)$.

Exercise 19.4.10 Let $E = \{f \in C([0,1]) : f(\frac{1}{2}) = \frac{1}{2}(f(0)+f(1))\}$. Show that the set $[0,1] \setminus \{\frac{1}{2}\}$ is a non-closed Choquet boundary for E.

The closure $Sh(E) = \overline{Ch(E)}$ of the Choquet boundary is called the *Shilov boundary* of E.

Theorem 19.4.11 *Suppose that E is separating subspace of $C_{\mathbf{C}}(K)$, where K is a compact metrizable space. Then the Shilov boundary $Sh(E)$ of E is the smallest closed boundary of E; if B is a closed boundary of E then $Sh(E) \subseteq B$.*

Proof Suppose not, so that there exists $x_0 \in Ch(E)\setminus B$. Since $q_E(B)$ is $\sigma(E', E)$-closed there exist $f_1, \ldots, f_n$ in E such that

$$U = \{\phi \in E' : |\phi(f_j) - f_j(x_0)| < 1 \text{ for } 1 \leq j \leq n\}$$

is a weak* neighbourhood of $q_E(x_0)$ disjoint from $q_E(B)$. Let $g_j = f_j - f_j(x_0)$, for $1 \leq j \leq n$, and let

$$h_j = \begin{cases} \Re g_j & \text{for } 1 \leq j \leq n \\ -\Re g_{j-n} & \text{for } n+1 \leq j \leq 2n \\ \Im g_{j-2n} & \text{for } 2n+1 \leq j \leq 3n \\ -\Im g_{j-3n} & \text{for } 3n+1 \leq j \leq 4n. \end{cases}$$

Then $V = \{\phi \in E' : \phi(h_j) < \frac{1}{2} \text{ for } 1 \leq j \leq 4n\}$ is a weak* neighbourhood of $q_E(x_0)$ contained in U. Let

$$L_j = \{\phi \in S(E) : \phi(h_j) \geq \tfrac{1}{2}\} \text{ for } 1 \leq j \leq 4n.$$

Then each L_j is weak* compact, and so is $M = \Gamma(\cup_{j=1}^{4n} L_j)$. Now $q_E(x_0) \notin L_j$ for any j, and $q_E(x_0)$ is an extreme point of $S(E)$, so that $q_E(x_0) \notin M$. By the separation theorem, there exists $f \in E$ such that

$$a = \Re f(x_0) = \Re(q_E(x_0)(f)) > b = \sup\{\Re\phi(f) : \phi \in M\}.$$

Let $c = \sup\{|\Im\phi(f)| : \phi \in M\}$. If $d > c^2/(a-b)$ then easy calculations show that

$$|f(x_0) + d| > \sup\{|\phi(f) + d)| : \phi \in M\}.$$

But $B \subseteq M$, and so the function $f + d1$ does not attain its supremum on B, giving a contradiction. □

19.5 Peak Points

Suppose that E is a separating subspace of $C_{\mathbf{C}}(K)$, where K is a compact metrizable space. An element x of K is a *peak point* for E if there exists $f \in E$ for which $f(x) = 1$ and $|f(y)| < 1$ for $y \neq x$. If x is a peak point for E and $\mu \in M_{q_E(x)}$ then $1 = \int_K f\, d\mu \leq \int_K |f|\, d\mu$, so that $\mu(|f| < 1) = 0$, and $\mu = \delta_x$. Thus $x \in Ch(E)$.

Theorem 19.5.1 *Suppose that E is a separating subspace of $C_{\mathbf{C}}(K)$, where K is a compact metrizable space, and that f is a smooth point of the unit sphere of E. Then there exists $x \in K$ such that $f(x) = 1$ and such that $|f(y)| < 1$ for $y \neq x$, so that x is a peak point for E.*

Proof There exists a unique $\phi \in S(E)$ such that $\phi(f) = 1$. Thus ϕ is an extreme point of $S(E)$, and so there exists $x \in Ch(E)$ such that $\phi(f) = f(x) = \delta_x(f)$. Since ϕ is unique, $|f(y)| < 1$ for $y \neq x$, and so x is a peak point for E. □

We use this to show that there are plenty of peak points in the Choquet boundary of a separating space.

Theorem 19.5.2 *Suppose that E is a separating subspace of $C_{\mathbf{C}}(K)$, where K is a compact metrizable space. The set $Pe(E)$ of peak points for E is dense in $Ch(E)$ in the weak* topology.*

Proof Suppose not. Since $Ch(E)$ is the set of extreme points of the state space $S(E)$, it follows from Mil'man's theorem (Theorem 12.2.10) that $\overline{\Gamma}(Pe(E))$ is a proper subset of $Ch(E)$. If $\phi \in Ch(E) \setminus \overline{\Gamma}(Pe(E))$, there exists $f \in E$ with $\|f\| = 1$ such that

$$\phi(f) > sup\{\psi(f) : \psi \in \overline{\Gamma}(Pe(E))\} = \sup\{\psi(f) : \psi \in Pe(E)\}.$$

By Exercise 11.6.8, the set of smooth points of the unit sphere of E is dense in the unit sphere, and so there exists a smooth point g such that $\phi(g) > \sup\{\psi(g) : \psi \in Pe(E)\}$. But there exists $\psi \in Pe(E)$ for which $\psi(g) = 1$, and so $\phi(g) > 1$, giving a contradiction. □

Here is a useful sufficient condition for x to be a peak point.

Theorem 19.5.3 *Suppose that E is a separating subspace of $C_{\mathbf{C}}(K)$, where K is a compact metrizable space, and that $x \in K$. Then x is a peak point for E if the following condition holds.*

(*) There exist $M \geq 0$ and $0 < \epsilon < 1$ such that if U is any open neighbourhood of x there exists $f \in E$ with

$$f(x) = 1,\ \|f\|_\infty \leq M + 1 \text{ and } |f(y)| < 1 - \epsilon \text{ for } y \notin U.$$

Proof The proof uses a classic 'sliding hump' argument. Choose $s < 1$ such that $t = (M + \epsilon)s - M > 0$. Let $(U_n)_{n=1}^{\infty}$ be a base of open neighbourhoods of x, and let $F_n = X \setminus U_n$ for $n \in \mathbf{N}$. Let $h_0 = 1$. Using condition (*), an inductive argument then shows that there exists a sequence $(h_n)_{n=1}^{\infty}$ in E such that $h_n(1) = 1$, $\|h_n\|_\infty \leq M + 1$ and if $B_n = \cup_{j=1}^{n}\{y : |h_j(y)| > 1 + s^{n+1}t$ then $|h_{n+1}(y)| \leq 1 - \epsilon$ for $y \in B_n \cup F_n$. For if we have found $h_0, \dots, h_n$ then $B_n \cup F_n$ is a closed set disjoint from $\{x\}$, and so condition (*) enables us to find h_{n+1}.

We now set $h = (1-s)\sum_{n=1}^{\infty} s^n h_n$. We show that $h(x) = 1$ and that $|h(y)| < 1$ for $y \neq x$. Certainly $h(x) = 1$. Suppose that $y \neq x$, so that $y \in F_k$, for some k. We consider two cases. First suppose that $y \notin \cup_{n=1}^{\infty} B_n$. If $j \in \mathbf{N}$, then $|h_j(y)| \leq 1 + s^n t$ for all n, and so $|h_j(y)| \leq 1$. But $|h_k(y)| < 1 - \epsilon$, and so $|h(y)| < 1$.

Secondly, suppose that $y \in \cup_{n=1}^{\infty} B_n$. Then there exists a least n such that $y \in B_n$. Thus

$$|h_j(y)| \leq \begin{cases} 1 + s^n t & \text{for } 1 \leq j < n \\ M + 1 & \text{for } j = n \\ 1 - \epsilon & \text{for } j > n. \end{cases}$$

Hence

$$\begin{aligned} |h(y)| &\leq (1 - s)\left[(1 + s^n t)\left(\sum_{j=1}^{n-1} s^j\right) + s^n(M + 1) + (1 - \epsilon)\left(\sum_{j=n+1}^{\infty} s^j\right)\right] \\ &= (1 - s^n)(1 + s^n t) + (1 - s)s^n(M + 1) + s^{n+1}(1 - \epsilon) \\ &= 1 + (1 - s^n)s^n t - s^n(s(M + \epsilon) - M) = 1 - s^{2n}t < 1. \end{aligned}$$

□

Important examples of separating subspaces are provided by uniform algebras. If K is a compact metrizable topological space, a *uniform algebra* A on K is a separating subspace of $C_{\mathbf{C}}(K)$ which is also an algebra under pointwise multiplication. An important feature of uniform algebras is that (since separating subspaces are closed) if $f \in A$, then $e^f = \sum_{n=0}^{\infty} f^n/n! \in A$; the following result illustrates this.

Theorem 19.5.4 *Suppose that A is a uniform algebra on a compact metrizable space K and that $x \in Ch(A)$. Suppose that $f \in C_{\mathbf{R}}(K)$, that $f(x) > 0$ for all $x \in K$ and that $0 < a < f(x)$. Then there exists $g \in A$ with $g(x) = a$ and $|g| \leq f$.*

Proof Let $F = \log f$. By Exercise 19.4.6, there exists $G \in A$ with $G(0) = \log a$ and $\Re G \leq F$. Let $g = e^G$. Then $g(x) = a$ and $|g| \leq f$. □

When A is a uniform algebra, we can improve upon Theorem 19.5.2.

Corollary 19.5.5 *If $x \in Ch(A)$ then x is a peak point for A.*

Proof Suppose that U is an open neighbourhood of x. By Urysohn's lemma, there exists $f \in C(X)$ with $f(x) = 1, f(y) = 0$ for $y \in X \setminus U$ and $0 \leq f \leq 1$. Apply the theorem to $f + 1$. There therefore exists $g \in A$ with $g(x) = 3/2$, $|g| \leq 2$ and $|g(y)| \leq 1$ for $y \notin U$. Then $h = 2g/3$ satisfies condition (*) of Theorem 19.5.3, with $M = \epsilon = 1/3$. □

19.6 The Choquet Ordering

Let us investigate Choquet's theorem further.

Suppose that (E, F) is a dual pair and that K is a metrizable $\sigma(E, F)$-compact convex subset of E. We denote the set of convex continuous functions on K by $CC(K)$ and the set $\{f \in CC(K); \|f\|_\infty \leq 1\}$ by $CC_1(K)$. We define a partial order, *the Choquet ordering*, on $P(K)$ by setting $\mu \preceq \nu$ if $\int_K f\, d\mu \leq \int_K f\, d\nu$ for every $f \in CC(K)$. Intuitively, this suggests that ν lives closer to the boundary of K than μ does. Clearly, it is enough to verify the condition for functions in $CC_1(K)$. If $f \in A(K)$ then f and $-f$ are convex, and so if $\mu \preceq \nu$ then $\int_K f\, d\mu = \int_K f\, d\nu$, so that $\beta_\mu = \beta_\nu$. Note also that it follows from Jensen's inequality that $\delta_{\beta_\mu} \preceq \mu$. We consider measures which are maximal with respect to the Choquet ordering.

Theorem 19.6.1 *Suppose that (E, F) is a dual pair, that K is a metrizable $\sigma(E, F)$-compact convex subset of E and that $\mu \in P(K)$. Then there exists a maximal measure $\nu \in P(K)$ with $\mu \preceq \nu$.*

Proof Since $(C(K), \|.\|_\infty)$ is separable, so is $CC_1(K)$. Let $(c_n)_{n=1}^\infty$ be a dense sequence in $CC(K)$. Let $A_0 = \{\pi \in P(K) : \mu \preceq \pi\}$; A_0 is a non-empty compact subset of $(P(K), w)$. We define

$$A_1 = \left\{\pi \in A_0 : \int_K c_1\, d\pi = \sup\left\{\int_K c_1\, d\rho : \rho \in A_0\right\}\right\}.$$

Then A_1 is a compact subset of A_0. We now repeat the process recursively; we set

$$A_{n+1} = \left\{\pi \in A_n : \int_K c_{n+1}\, d\pi = \sup\left\{\int_K c_{n+1}\, d\rho : \rho \in A_n\right\}\right\}.$$

Then $(A_n)_{n=0}^\infty$ is a decreasing sequence of non-empty compact subsets of $P(K)$, and therefore there exists $\nu \in \bigcap_{n\in\mathbf{Z}^+} A_n$. Then $\int_K c_n\, d\mu \leq \int_K c_n\, d\nu$ for each

$n \in \mathbf{N}$, and so, since $(c_n)_{n=1}^\infty$ is a dense sequence in $CC(K)$, $\mu \preceq \nu$. Similarly, if $\nu \preceq \nu'$, then $\int_K c_n\,d\nu = \int_K c_n\,d\nu'$ for each $n \in \mathbf{N}$, and so, since $(c_n)_{n=1}^\infty$ is a dense sequence in $CC(K)$, $\nu = \nu'$: ν is maximal. □

Theorem 19.6.2 *Suppose that (E, F) is a dual pair, that K is a metrizable $\sigma(E, F)$-compact convex subset of E. If μ is a maximal measure and $f \in C(K)$ then $\int f\,d\mu = \int \underline{f}\,d\mu$.*

Proof Suppose first that μ is a maximal measure. By Theorem 19.2.1 there exists $\nu \in P(K)$ such that $\beta_\nu = \beta_\mu$, $\int_K \underline{g}\,d\mu \le \int_K g\,d\nu$ for each $g \in C(K)$ and $\int_K f\,d\nu = \int_K \underline{f}\,d\mu$.

If $h \in CC(K)$, then $h = \underline{h}$, so that $\int_K h\,d\mu \le \int_K h\,d\nu$, and $\mu \preceq \nu$. Since μ is maximal, $\nu = \mu$. Thus

$$\int_K f\,d\mu = \int_K f\,d\nu = \int_K \underline{f}\,d\mu.$$

□

Note that in this theorem we do not need the axiom of choice.

Exercise 19.6.3 By considering a strictly convex function, show that if μ is a maximal measure, then $\mu(Ex(K)) = 1$.

Thus Theorem 19.6.1 provides another proof of Choquet's theorem.

What about the converse? Suppose that f is a bounded function on K. Let $D(f) = \{g \in C(K) : g \text{ convex}, g \le f\}$. It follows from the definitions that $\underline{f} = \sup\{g : g \in D(f), g \le f\}$.

Theorem 19.6.4 *Suppose that K is a metrizable σ-compact convex subset of E, that μ is a finite measure on K and that f is a bounded function on K. Then $\int_K \underline{f}\,d\mu = \sup_{g \in D(f)} \int_K g\,d\mu$.*

Proof $D(f)$ is directed upwards, and so the result follows from Theorem 16.1.4. □

Theorem 19.6.5 *Suppose that (E, F) is a dual pair, that K is a metrizable $\sigma(E, F)$-compact convex subset of E and that $\int_K f\,d\mu = \int_K \underline{f}\,d\mu$ for each $f \in C(K)$. Then μ is a maximal measure.*

Proof Suppose that $\mu \preceq \nu$. Then if $f \in C(K)$,

$$\begin{aligned}
\int_K f\,d\nu \ge \int_K \underline{f}\,d\nu &= \sup\left\{\int_K g\,d\nu : g \in D(f)\right\} \\
&\ge \sup\left\{\int_K g\,d\mu : g \in D(f)\right\} \qquad \text{(since } \mu \preceq \nu) \\
&= \int_K \underline{f}\,d\mu = \int_K f\,d\mu,
\end{aligned}$$

and so $\nu = \mu$. □

Exercise 19.6.6 Suppose that (E, F) is a dual pair, that K is a metrizable $\sigma(E, F)$-compact convex subset of E. Show that if $\mu \in P(K)$ and $\mu(Ex(K)) = 1$, then μ is a maximal measure.

19.7 Dilations

Suppose that (E, F) is a dual pair of vector spaces, that K is a metrizable $\sigma(E, F)$-compact convex subset of E. A mapping T from K to $P(K)$ is a *dilation* of K if

(i) $\beta_{T(x)} = x$ for all $x \in K$, and
(ii) for each $f \in C(K)$, the function $\int_K f\, dT(x)$ is Borel measurable.

If T is a dilation, we can extend T to a mapping from $P(X)$ to $P(X)$; if $\mu \in P(X)$ and $f \in C(K)$, let

$$\phi(\lambda)(f) = \int_K \left(\int_K f\, dT(x) \right) d\lambda(x).$$

Then $\phi(\lambda)(1) = 1$, and $\phi(\lambda)$ is a positive linear functional on $C(X)$, and so by the Riesz representation theorem there exists $T(\lambda) \in P(K)$ for which

$$\int_X f\, dT(\lambda) = \int_K \left(\int_K f\, dT(x) \right) d\lambda(x).$$

Note that $T(\delta_x) = T(x)$, so that we have extended T.

Proposition 19.7.1 *Suppose that (E, F) is a dual pair of vector spaces, that K is a metrizable $\sigma(E, F)$-compact convex subset of E and that T is a dilation of K. If $\lambda \in P(X)$, then $\lambda \preceq T(\lambda)$.*

Proof If $x \in K$ then $\beta_{T(x)} = x$ and so $\delta_x \preceq T(\delta_x)$. Thus if f is a continuous convex function on K then $f(x) \leq \int_K f\, dT(\delta_x)$, and so

$$\int_X f\, d\lambda \leq \int_K \left(\int_K f\, dT(\delta_x) \right) d\lambda(x) = \int_K f\, dT(\lambda).$$

□

Our main aim is to prove the converse. Let

$$L = \{(\lambda, \mu) \in P(K) \times P(K) : \lambda \preceq \mu\},$$
$$M = \{\delta_x, \mu) \in P(K) \times P(K) : x \in K,\ \delta_x \preceq \mu\}.$$

Then L is a convex subset of $P(K) \times P(K)$. Since

$$L = \cap \left\{ \left\{ (\lambda, \mu) : \int_K f\, d\lambda \leq \int_K f\, d\mu \right\} : f \text{ convex, continuous} \right\},$$

L is $w \times w$-closed. Since $M = L \cap (\delta(K) \times P(K))$, M is a $w \times w$-closed subset of L.

Lemma 19.7.2 $L = \overline{\Gamma}(M)$.

Proof By the separation theorem, it is enough to show that if $(f, g) \in C(K) \times C(K)$ and

$$f(x) - \int_K g\, d\mu \leq t \text{ for all } (\delta_x, \mu) \in M,$$

then

$$\int_K f\, d\lambda - \int_K g\, d\mu \leq t \text{ for all } (\lambda, \mu) \in L.$$

If $x \in X$, then by Proposition 19.2.3,

$$\underline{g}(x) = \inf\left\{\int_K g\, d\mu : \beta_\mu = x\right\},$$

so that $f(x) - \underline{g}(x) \leq t$ and $\int_K f\, d\lambda - \int_K \underline{g}\, d\lambda \leq t$. But $\underline{g}$ is convex, and so $\int_K \underline{g}\, d\lambda \leq \int_K \underline{g}\, d\mu$, so that $\int_K f\, d\lambda - \int_K g\, d\mu \leq t$. □

Corollary 19.7.3 *If $(\lambda, \mu) \in L$, there exists a probability measure θ on M such that $(\lambda, \mu) = \beta_\theta$.*

Theorem 19.7.4 *If $\lambda, \mu \in P(K)$ and $\lambda \preceq \mu$ then there exists a dilation T such that $T(\lambda) = \mu$.*

Proof Let θ be the measure of the preceding corollary. If $(\delta_x, \mu) \in M$, let $\phi((\delta_x, \mu)) = x$ and let $\psi(\delta_x, \mu) = \mu$; ϕ is a continuous mapping of M onto K and ψ is a continuous mapping of M onto $P(K)$. If $f, g \in C(X)$, then

$$\int_K f\, d\lambda - \int_K g\, d\mu = \int_M \left(f(x) - \int_K g\, d\pi\right) d\theta(\delta_x, \pi). \qquad (*)$$

Putting $g = 0$ we see that $\lambda = \phi_*(\theta)$.

We now apply the disintegration theorem (Theorem 16.10.1) to θ and ϕ; there exists a family $\{\nu_x : x \in K\}$ in $P(M)$ such that if $h \in C(M)$ then

(i) $\int_M h\, d\nu_x$ is a measurable function on K,
(ii) $\int_M h\, d\theta = \int_K \left(\int_M h\, d\nu_x\right) d\lambda(x)$, and
(iii) $\nu_x(\phi^{-1}(\{x\})) = 1$ for almost all $x \in K$.

Let $S(x) = \psi_*(\nu_x)$; $S(x) \in P(P(K))$. Let $T(x) = \beta(S(x))$, the barycentre of $S(x)$; thus $T(x) \in P(K)$. If $f \in C(K)$ then the mapping $\mu \to \int_K f\, d\mu$ is a continuous affine mapping on $P(K)$, so that

$$\int_K f\, dT(x) = \int_{P(K)} \left(\int_K f\, d\pi\right) dS(\pi) = \int_M \left(\int_K f(y)\, d\pi\right) d\nu_x(\delta_y, \pi).$$

Thus $\int_K f\,dT(x)$ is measurable. If $a \in A(K)$, then since λ_x is supported on $\{(\delta_x, \pi) : \beta_\pi = x\}$,

$$\int_K a\,dT(x) = \int_M a(y)\,d\nu_x(\delta_y, \mu).$$

It therefore follows from (iii) that $\int_K a\,dT(x) = a(x)$. Consequently $x = \beta_{T(x)}$, and T is a dilation. Suppose now that $g \in C(K)$. Let $h(\delta_y, \pi) = \int_K g\,d\pi$; then $h \in C(M)$ and it follows from (*) that $\int_K g\,d\mu = \int_M h\,d\theta$. But

$$\begin{aligned}\int_K g\,dT(\lambda) &= \int_K \left(\int_K g\,dT(x)\right)\,d\lambda(x) = \int_K \left(\int_M h\,d\nu_x\right)\,d\lambda(x)\\ &= \int_M h\,d\theta = \int_K g\,d\mu,\end{aligned}$$

so that $T(\lambda) = \mu$. □

Theorem 19.7.5 *Suppose that $\lambda \in P(K)$, where K is a compact convex metrizable metric space. Suppose that T is a dilation such that $T(\lambda) = \mu$ is maximal. Then $T(\delta_x)$ is maximal for λ-almost all x.*

Proof We use the fact that ν is maximal if and only if $\int_K f\,d\nu = \int_K \underline{f}\,d\nu$ for every $f \in C(K)$.

Since $C(K)$ is separable, there exists a dense sequence $(f_n)_{n=1}^\infty$ in $C(K)$. For each n,

$$0 = \int (\underline{f_n} - f_n)\,d\mu = \int_K \left(\int_K (\underline{f_n} - f_n)\,dT(\delta_x)\right)\,d\lambda(x),$$

so that

$$\int_K \underline{f_n}\,dT(\delta_x) = \int_K f_n\,dT(\delta_x)$$

for all n, for λ-almost all x. The mapping $f \to \underline{f}$ is uniformly continuous (Theorem 10.1.2, (vi)) and so

$$\int_K \underline{f}\,dT(\delta_x) = \int_K f\,dT(\delta_x)$$

for all $f \in C(K)$, for λ-almost all x. This gives the result. □

PART THREE

Introduction to Optimal Transportation

20
Optimal Transportation

20.1 The Monge Problem

The study of optimal transportation was inaugurated by Gaspard Monge in 1781. He asked the question 'What is the most economical way of transporting material (iron ore, say, or rubbish) from various sites to other sites (iron works, or landfill sites)?' This is an optimization problem, with three ingredients: the original distribution of material, the distribution of the sites where the material needs to be taken and the cost of transportation from one place to another.

Let us describe the problem in mathematical terms. Suppose that (X, τ) and (Y, σ) are two Polish spaces, and that μ and ν are Borel probability measures on X and Y respectively; μ represents the (normalized) distribution of the material that is to be moved, and ν the distribution of the location to which it has to be moved. Suppose also that c (the cost function) is a lower semi-continuous function on $X \times Y$. In general, we shall suppose that c is non-negative, but we shall also consider the case where c can take negative values, provided that there exist $a \in L^1(\mu)$ and $b \in L^1(\nu)$ such that $c(x, y) + a(x) + b(y) \geq 0$ for all x, y.

If S is a Borel measurable mapping from X to Y such that $S_*(\mu) = \nu$ (S pushes forward the material in X to the right distribution in Y), then S is called a *deterministic transport plan*, or *deterministic coupling*. Monge's problem is to find a deterministic transport plan T such that

$$\int_X c(x, T(x))\, d\mu \leq \int_X c(x, S(x))\, d\mu$$

for all deterministic transport plans S. (Of course, conditions are required to ensure that deterministic transport plans exist; for example if $X = Y = \{0, 1\}$ and $\mu(\{0\}) = 1/3$, $\mu(\{1\}) = 2/3$, $\nu(\{0\}) = \nu(\{1\}) = 1/2$ then there are no deterministic transport plans.)

Besides considering a deterministic transport plan S, we also consider its graph $G(S) = \{(x, S(x)) : x \in X\}$ and the push forward measure $\pi = G(S)_*(\mu)$ on $X \times Y$; thus $\pi(A \times B) = \mu(A \cap S^{-1}B)$, and π has marginals μ and ν. π is called a *deterministic coupling*; it has the advantage that distinct deterministic transport plans determine distinct deterministic couplings.

Proposition 20.1.1 *Suppose that μ and ν are probability measures on Polish spaces (X, τ) and (Y, σ) respectively, and that S and S' are distinct deterministic transport plans. Then the couplings $\pi = G(S)_*(\mu)$ and $\pi' = G(S')_*(\mu)$ are distinct.*

Proof For there exists a measurable subset A of X with $\mu(A) > 0$ such that $S(A) \cap S'(A) = \emptyset$. Then $G(S)_*(\mu)(A, S(A)) = \mu(A) > 0$ and $G(S')_*(\mu)$ $(A, S(A)) = 0$. □

20.2 The Kantorovich Problem

Suppose that we consider the measures μ and ν on $\{0, 1\}$ described in the previous section, and also consider the cost function c defined as follows: $c(0, 0) = c(1, 1) = 0$ (it costs nothing to stay put), $c(0, 1) = 3$ and $c(1, 0) = 1$. Then an obvious optimal procedure is to send a third of the material at 1 to 0, and to leave the rest fixed. This kind of procedure, which cannot be defined by a deterministic transport plan, was investigated in the early 1940s by the Russian mathematician and economist Leonid Kantorovich; his work led to the award of the Nobel prize for economics in 1975.

Once again, we start with Borel probability measures μ and ν on Polish spaces (X, τ) and (Y, σ). An element of $P(X \times Y)$ is called a *transport plan* if it has marginal distributions μ and ν; the set of transport plans is denoted by $\Pi_{\mu,\nu}$. $\Pi_{\mu,\nu}$ is non-empty, since $\mu \otimes \nu \in \Pi_{\mu,\nu}$. We also consider a lower semi-continuous cost function c. The Kantorovich problem is then 'Is there an optimal transport plan? If so, how do we find it, and what are its properties?'

Let us show that, under reasonable conditions, the answer to the first question is 'yes'.

Theorem 20.2.1 *Suppose that (X, τ) and (Y, ρ) are Polish spaces, and that μ and ν are Borel probability measures on X and Y respectively. Then $\Pi_{\mu,\nu}$ is a non-empty w-compact convex subset of $P(X \times Y)$.*

Proof $\Pi_{\mu,\nu}$ is clearly a convex subset of $P(X \times Y)$, and we have seen that it is non-empty.

We show that $\Pi_{\mu,\nu}$ is also uniformly tight. For if $\epsilon > 0$, there exist compact subsets K of X and L of Y such that $\mu(X \setminus K) < \epsilon/2$ and $\nu(Y \setminus L) < \epsilon/2$. If $\pi \in \Pi_{\mu,\nu}$, then

$$\pi((X \setminus K) \times Y) = \mu(X \setminus K) < \epsilon/2$$
$$\text{and } \pi(X \times (Y \setminus L)) = \nu(Y \setminus L) < \epsilon/2.$$

Since $(X \times Y) \setminus (K \times L) \subseteq ((X \setminus K) \times Y) \cup (X \times (Y \setminus L))$, it follows that $\pi((X \times Y) \setminus (K \times L)) < \epsilon$.

Finally we show that $\Pi_{\mu,\nu}$ is a w-closed subset of $P(X \times Y)$. Suppose that $(\pi_n)_{n=1}^{\infty}$ is a sequence in $\Pi_{\mu,\nu}$, and that $\pi_n \Rightarrow \pi$. If C is a closed subset of X, then $\pi(C \times Y) \geq \limsup \pi_n(C \times Y) = \mu(C)$. On the other hand,

$$\begin{aligned}\pi((X \setminus C) \times Y) &= \sup\{\pi(K \times Y) : K \text{ compact}, K \subseteq X \setminus C\}\\ &\geq \sup\{\mu(K) : K \text{ compact}, K \subseteq X \setminus C\} = \mu(X \setminus C).\end{aligned}$$

Since

$$\pi(C \times Y) + \pi((X \setminus C) \times Y) = 1 = \mu(C) + \mu(X \setminus C),$$

it follows that $\pi(C \times Y) = \mu(C)$. By regularity, if A is a Borel subset of X, then $\pi(A \times Y) = \mu(A)$. Similarly, if B is a Borel subset of Y then $\pi(X \times B) = \nu(B)$. Thus $\pi \in \Pi_{\mu,\nu}$. □

Corollary 20.2.2 *If f is a w-lower semi-continuous function on $\Pi_{\mu,\nu}$ taking values in $(-\infty, \infty]$, f is bounded below and attains its bounds.*

Thus the Kantorovich problem has a solution.

Similarly, we have the following:

Proposition 20.2.3 *Suppose that M is a uniformly tight subset of $P(X)$ and that N is a uniformly tight subset of $P(Y)$. Then $\cup\{\Pi_{\mu,\nu} : \mu \in M, \nu \in N\}$ is a uniformly tight subset of $P(X \times Y)$.*

Recall (Theorem 4.2.9) that if c is lower semi-continuous and if d is a metric on $X \times Y$ which defines the product topology on $X \times Y$ then there exists an *approximating sequence*: there is an increasing sequence $(c_n)_{n=1}^{\infty}$ of bounded non-negative Lipschitz functions on $X \times Y$ which converges to c pointwise.

Theorem 20.2.4 *Suppose that (X, τ) and (Y, ρ) are Polish spaces, that c is a proper lower semi-continuous cost function on $X \times Y$ and that μ and ν are Borel probability measures on X and Y respectively. If there exists $\pi \in P(X \times Y)$ for which $\int_{X \times Y} c\, d\pi < \infty$, then there exists an extreme point π_0 of $\Pi_{\mu,\nu}$ such that*

$$\int_{X \times Y} c\, d\pi_0 = \inf\left\{\int_{X \times Y} c\, d\pi : \pi \in \Pi_{\mu,\nu}\right\} = M_c, \text{ say.}$$

Proof Let $(c_n)_{n=1}^\infty$ be an approximating sequence, and let $\phi_n(\pi) = \int_{X\times Y} c_n\,d\pi$. Then $(\phi_n)_{n=1}^\infty$ is an increasing sequence of w-continuous functions on $\Pi_{\mu,\nu}$, which converges pointwise to $\phi(\pi) = \int_{X\times Y} c\,d\pi$, by the theorem of monotone convergence. Thus ϕ is a w-lower semi-continuous function on $\Pi_{\mu,\nu}$, and so it attains its lower bound, by Corollary 20.2.2.

The function ϕ is an affine function on $\Pi_{\mu,\nu}$ – that is,

$$\phi((1-\lambda)\mu_1 + \lambda\mu_2) = (1-\lambda)\phi(\mu_1) + \lambda\phi(\mu_2) \text{ for } 0 \le \lambda \le 1.$$

It follows that

$$\left\{\pi' : \int_{X\times Y} c\,d\pi' = M_c\right\}$$

is a face F of $\Pi_{\mu,\nu}$, and we can take π_0 to be an extreme point of F, by the Krein–Mil'man theorem. □

Let us consider the case where c can take negative values.

Corollary 20.2.5 *Suppose that c is a lower semi-continuous function on $X \times Y$, and that there exist lower semi-continuous functions $a \in L^1(\mu)$ and $b \in L^1(\nu)$ such that $c' = c + a + b \ge 0$. If there exists $\pi \in P(X \times Y)$ for which $\int_{X\times Y} c\,d\pi < \infty$, then there exists an extreme point π_0 of $\Pi_{\mu\times\nu}$ such that $\int_{X\times Y} c\,d\pi_0 = \inf\{\int_{X\times Y} c\,d\pi : \pi \in \Pi_{\mu,\nu}\}$.*

Proof The function c' is a cost function, and so there exists an extreme point π_0 of $\Pi_{\mu\times\nu}$ such that $\int_{X\times Y} c'\,d\pi_0 = \inf\{\int_{X\times Y} c'\,d\pi : \pi \in \Pi_{\mu,\nu}\}$. This has the required properties, since if $\pi \in \Pi_{\mu\times\nu}$ then

$$\begin{aligned}\int_{X\times Y} c\,d\pi_0 &= \int_{X\times Y} c'\,d\pi_0 - \left(\int_X a\,d\mu + \int_Y b\,d\nu\right)\\ &\le \int_{X\times Y} c'\,d\pi - \left(\int_X a\,d\mu + \int_Y b\,d\nu\right) = \int_{X\times Y} c\,d\pi.\end{aligned}$$

□

A solution to the Kantorovich problem can often lead to a solution to the Monge problem, as the following elementary example shows.

We suppose that $X = Y = \{1, \ldots, n\}$, $\mu = \nu$ is uniform probability measure on $\{1, \ldots, n\}$ and c is any function on $X \times Y$.

Probability measures on $X \times Y$ can be identified with $n \times n$ matrices (p_{ij}) with non-negative entries for which $\sum_{i,j} p_{ij} = 1$. Then $\pi \in \Pi_{\mu,\nu}$ if and only if $\sum_{i=1}^n \pi_{ij} = 1/n$ for $1 \le j \le n$ and $\sum_{j=1}^n \pi_{ij} = 1/n$ for $1 \le i \le n$; that is, $n\pi$ is a *doubly stochastic matrix*. The set S_n of doubly stochastic matrices is a closed convex subset of the space M_n of all $n \times n$ matrices. If c is a cost function on

$X \times Y$, it is represented by a matrix (c_{ij}), and $\int_{X\times Y} c\,d\pi = \sum_{1\le i,j\le n} c_{ij}\pi_{ij}$. If $m \in M_n$ then the mapping $m \to \sum_{1\le i,j\le n} c_{ij}m_{ij}$ is a linear functional on M_n. Consequently there exists an extreme point π of S_n at which the linear functional attains its infimum on S_n. But it follows from Birkhoff's theorem (Theorem 12.2.11) that the permutation matrices are the extreme points of the set of doubly stochastic matrices. Thus there is an optimal transport mapping given by a permutation matrix I_σ.

This example shows another phenomenon that we shall consider later. If τ is another permutation of $\{1,\ldots,n\}$ then the corresponding cost of $I_{\tau\sigma}$ is greater than or equal to the optimal cost, and so

$$\sum_{i=1}^{n} c_{i,\sigma(i)} \le \sum_{i=1}^{n} c_{i,\tau(\sigma(i))}.$$

20.3 The Kantorovich–Rubinstein Theorem

We now give another proof of the existence of an optimal transport plan. The proof is quite different from that of Theorem 20.2.4; it uses the Hahn–Banach theorem and the Riesz representation theorem. Its most important features are that it considers approximation from below, and that it leads to the maximal Kantorovich potential.

Theorem 20.3.1 (The Kantorovitch–Rubinstein theorem) *Suppose that (X,d) and (Y,ρ) are Polish metric spaces, that μ and ν are Borel probability measures on X and Y respectively, that c is a cost function on $X \times Y$ and that $M_c(\mu,\nu) = \inf\{\int_{X\times Y} c\,d\pi : \pi \in \Pi_{\mu,\nu}\} < \infty$. Let d be a complete metric on $X \times Y$ which defines the product topology. Let*

$$L = \{f(x) + g(y) : f \in BL(X), g \in BL(Y)\} \subseteq C(X \times Y)$$

and let

$$m_c = m_c(\mu,\nu) = \sup\left\{\int_X f\,d\mu + \int_Y g\,d\nu : f + g \in L, f + g \le c\right\}.$$

Then $M_c = m_c$, and there exists $\pi \in \Pi_{\mu,\nu}$ such that $\int_{X\times Y} c\,d\pi = m_c$.

Proof If $\pi \in \Pi_{\mu,\nu}, f + g \in L$ and $f + g \le c$ then

$$\int_X f\,d\mu + \int_Y g\,d\nu = \int_{X\times Y} f(x) + g(y)\,d\pi(x,y) \le \int_{X\times Y} c\,d\pi,$$

so that $m_c \le M_c$. We need to find π for which the reverse inequality holds.

First we consider the case where X and Y are compact. L is a linear subspace of $C(X \times Y)$. If $f + g \in L$ let $\phi(f + g) = \int_X f\, d\mu + \int_Y g\, d\nu$. This is a well-defined linear functional on L, since if $f + g = f' + g'$ then $f - f' = g' - g$ is a constant k, so that $\int_X f'\, d\mu = \int_X f\, d\mu - k$ and $\int_Y g'\, d\nu = \int_Y g\, d\nu + k$. Further, $\phi(1) = 1$, so that ϕ is non-zero.

Now let

$$U = \{h \in C(X \times Y) : h(x,y) < c(x,y) \text{ for all } (x,y) \in X \times Y\}.$$

U is a non-empty convex open subset of $C(X \times Y)$. $U \cap L$ is also non-empty. If $f + g \in U \cap L$ and $\pi_0 \in \Pi_{\mu,\nu}$ then

$$\phi(f + g) = \int_{X\times Y} (f + g)\, d\pi \leq \int_{X\times Y} c\, d\pi_0,$$

so that ϕ is bounded above on $U \cap L$ by M_c. Let $M = \sup\{\phi(h) : h \in U \cap L\}$, and let $B = \{l \in L : \phi(l) \geq M\}$. Then B is a non-empty convex set disjoint from U. By the Hahn–Banach theorem, there exists a non-zero continuous linear functional ψ on $C(X\times Y)$ such that if $h \in U$ then $\psi(h) < K = \inf\{\psi(b) : b \in B\}$. If $h > 0$ then $-\alpha h \in U$ for all sufficiently large α, and so $\psi(h) \geq 0$. Thus ψ is a positive linear functional on $C(X\times Y)$. Since $\psi \neq 0$, $\psi(1) > 0$. Let $\theta = \psi/\psi(1)$. θ is a non-negative linear functional on $C(X \times Y)$, and $\theta(1) = 1$; by the Riesz representation theorem θ is represented by a Borel probability measure π on $X \times Y$. We shall show that π has the required properties.

Note that if $h \in U$, $\theta(h) < \Lambda = \inf\{\theta(b) : b \in B\}$. If $l_0 \in L$ and $\phi(l_0) = 0$, then $\phi(M.1+\alpha l_0) = M$, so that $M.1+\alpha l_0 \in B$ for all α, and so $\theta(M.1+\alpha l_0) = M + \alpha\theta(l_0) \geq \Lambda$ for all α, and so $\theta(l_0) = 0$. If $l \in L$ then $l = \phi(l)1 + l_0$, where $\phi(l_0) = 0$, and so $\theta(l) = \phi(l)$: θ extends ϕ. In particular, this means that $\Lambda = M$: if $h \in U$ then $\int_{X\times Y} h\, d\pi < M$. If $f \in C(X)$ then $\int_{X\times Y} f(x)\, d\pi(x,y) = \phi(f) = \int_X f\, d\mu$, and a similar result holds for $g \in C(Y)$; thus $\pi \in \Pi_{\mu,\nu}$. Further,

$$m_c = \sup\{\phi(h) : h \in U \cap L\} = M.$$

By Theorem 4.2.9, there exists an approximating sequence $(h_n)_{n=1}^\infty$ in $BL(X \times Y)$ which increases pointwise to c. Then each h_n is in U, and so

$$\int c\, d\pi = \lim_{n\to\infty} \int h_n\, d\pi \leq \sup\{\int h\, d\pi : h \in U\} \leq M = m_c.$$

We now consider the case where X and Y are Polish spaces. We consider X as a dense G_δ subspace of a compact metrizable space $\tilde{X}$ and Y as a dense G_δ subspace of a compact metrizable space $\tilde{Y}$, and consider the push forward measures $\tilde{\mu}$, $\tilde{\nu}$ and $\tilde{\pi}_0$. By Theorem 4.2.8, there exists a lower semi-continuous extension $\tilde{c}$ of c on $\tilde{X} \times \tilde{Y}$. Then

$$\int_{\tilde{X}\times\tilde{Y}} \tilde{c}\, d\tilde{\pi}_0 = \int_{X\times Y} c\, d\pi_0 < \infty.$$

Thus there exists $\tilde{\pi} \in \Pi_{\tilde{\mu},\tilde{\nu}}$ such that

$$\int_{\tilde{X}\times\tilde{Y}} \tilde{c}\, d\tilde{\pi} = \sup\left\{ \int_X \tilde{f}\, d\tilde{\mu} + \int_Y \tilde{g}\, d\tilde{\nu} : \tilde{f} \in C(\tilde{X}), \tilde{g} \in C(\tilde{Y}), \tilde{f} + \tilde{g} \le \tilde{c} \right\}.$$

Now

$$\tilde{\pi}((\tilde{X} \setminus X) \times \tilde{Y}) = \tilde{\mu}(\tilde{X} \setminus X) = 0$$
$$\text{and } \tilde{\pi}(\tilde{X} \times (\tilde{Y} \setminus Y)) = \tilde{\nu}(\tilde{Y} \setminus Y) = 0,$$

so that $\tilde{\pi}$ is the push-forward measure of an element π of $\pi_{\mu,\nu}$. Since

$$\sup\left\{ \int_X f\, d\mu + \int_Y g\, d\nu : f + g \in L, f + g \le c \right\}$$
$$\ge \sup\left\{ \int_X \tilde{f}\, d\tilde{\mu} + \int_Y \tilde{g}\, d\tilde{\nu} : \tilde{f} \in C(\tilde{X}), \tilde{g} \in C(\tilde{Y}), \tilde{f} + \tilde{g} \le \tilde{c} \right\},$$

the measure π satisfies the conclusions of the theorem. □

Corollary 20.3.2 *Let*

$$\tilde{m}_c = \sup\left\{ \int_X f\, d\mu + \int_Y g\, d\nu : f \in L^1(\mu),\ g \in L^1(\nu), f + g \le c \right\},$$

then $M_c = \tilde{m}_c$. The same holds if c takes negative values, and satisfies the conditions of Corollary 20.2.5.

Proof Clearly $m_c \le \tilde{m}_c \le M_c$. The argument of Corollary 20.2.5 deals with the case where c takes negative values. □

20.4 *c*-concavity

The Kantorivich–Rubinstein theorem raises many questions. What are the fundamental properties of an optimal transport plan? Can we find $f \in L^1(\mu)$ and $g \in L^1(\nu)$ such that $f + g \le c$, with $\int_X f\, d\mu + \int_y g\, d\nu = m_c$? When is there a deterministic transport plan?

In this, and the next two sections, we show that we can use a cost function to introduce some geometric ideas, related to the Legendre transform, and to concavity. It is an unfortunate necessity that although the parallel with the Legendre transform is close, we need to consider infima rather than suprema. Further, we consider infima of sequences of lower semi-continuous functions, which, although Borel measurable, need be neither upper nor lower semi-continuous.

Suppose that (X, d) and (Y, ρ) are complete Polish metric spaces, that μ and ν are Borel probability measures on X and Y respectively, and that c is a lower semi-continuous cost function on $X \times Y$. Let $(c_n)_{n=1}^{\infty}$ be an approximating sequence in $BL(X \times Y)$, increasing pointwise to c.

If f is a proper function on X and $y \in Y$, we set $f^c(y) = \inf\{c(x, y) - f(x) : x \in X\}$. f^c is the *c-transform* of f; it takes values in $[-\infty, \infty)$. We shall suppose that f^c is not identically $-\infty$. If c is continuous, then f^c is upper semi-continuous. If c is lower semi-continuous, then, if $(c_n)_{n=1}^{\infty}$ is an approximating sequence, each f^{c_n} is upper semi-continuous, and $f^c = \lim_{n\to\infty} f^{c_n}$, so that f^c is a Borel measurable function on Y, taking values in $[-\infty, \infty]$.

Similarly, if g is a function on Y and $x \in X$, we define $g^c(x) = \inf\{c(x, y) - g(y) : y \in Y\}$.

We say that f is *c-concave* if there exists a function g on Y such that $f = g^c$.

Proposition 20.4.1 *Suppose that (X, τ) and (Y, ρ) are complete Polish metric spaces and that c is a cost function on $X \times Y$. If f is a proper function on X and g is a proper function on Y then $f^{cc} \geq f$, and $g^{ccc} = g^c$.*

Proof

$$\begin{aligned} f^{cc}(x) &= \inf_y (c(x, y) - f^c(y)) \\ &= \inf_y (c(x, y) - \inf_{\tilde{x}} (c(\tilde{x}, y) - f(\tilde{x}))) \\ &= \inf_y \sup_{\tilde{x}} (c(x, y) - c(\tilde{x}, y) + f(\tilde{x})) \\ &\geq \inf_y (c(x, y) - c(x, y) + f(x)) = f(x). \end{aligned}$$

Also

$$g^{ccc} = (g^c)^{cc} \supseteq g^c \text{ and } g^{ccc} = (g^{cc})^c \subseteq g^c.$$

□

Thus f is c-concave if and only if $f = f^{cc}$.

If f is c-concave, we set

$$\partial^c f = \{(x, y) : c(x, y) - f(x) = \inf\{c(\tilde{x}, y) - f(\tilde{x}) : \tilde{x} \in X\}\}.$$

$\partial^c f$ is the *c-superdifferential* of f; it may be empty.

In general, $f(x) + f^c(y) \leq c(x, y)$, and

$$\partial^c f = \{(x, y) \in X \times Y : f(x) + f^c(y) = c(x, y)\}.$$

As an example, which we shall consider further later, suppose that $X = Y = H$, a separable Hilbert space, and that $\tilde{\mu}$ and $\tilde{\nu}$ are Borel measures, on X and Y respectively, for which

$$\int_H \|x\|^2 \, d\tilde{\mu}(x) = \int_H \|y\|^2 \, d\tilde{\nu}(y) = 2.$$

Let $\mu = \frac{1}{2} \|.\|^2 \, d\tilde{\mu}$ and $\nu = \frac{1}{2} \|.\|^2 \, d\tilde{\nu}$. Then $\mu \in P(X)$ and $\nu \in P(Y)$. Let $c(x, y) = -\langle x, y \rangle$. Then c is a continuous cost function on $X \times Y$, since $c(x, y) + \frac{1}{2} \|x\|^2 + \frac{1}{2} \|y\|^2 = \frac{1}{2} \|x - y\|^2 \geq 0$. If f is a function on X and $y \in Y$, then

$$f^c(y) = \inf_{x \in X}(-\langle x, y \rangle - f(x)) = -\sup_{x \in X}(\langle x, y \rangle + f(x)) = -(-f)^*.$$

It therefore follows that $f^{cc} = f^-$, the upper concave envelope of f. Thus $\partial^c f$ is the superdifferential of f.

When the cost function is continuous, we can say more.

Proposition 20.4.2 *Suppose that (X, τ) and (Y, σ) are Polish spaces and that c is a continuous cost function. If f is c-concave then it is upper semi-continuous, and $\partial^c f$ is a closed subset of $X \times Y$.*

Proof Since f is the infimum of a set of continuous functions, it is upper semi-continuous.

Suppose that $(x_n, y_n)_{n=1}^{\infty}$ is a sequence in $\partial^c f$ converging to (x, y). By picking a subsequence if necessary, we can suppose that $f(x_n) \to l$, where $l \leq f(x)$. If $z \in X$ then

$$c_n(x_n, y_n) - f(x_n) = f^c(y_n) \leq c(z, y_n) - f(z),$$

so that $f(x) \geq l \geq c(x, y) + f(z) - c(z, y)$. Since this holds for all $z \in X$, $f(x) \geq c(x, y) - f^c(y)$, and $(x, y) \in \partial^c f$. □

One special but important case arises when $X = Y$ and c is a lower semi-continuous metric on X (which need not necessarily define the topology τ). If (X, c) is a metric space, and f is a proper function on X, we define the *lower L-convex envelope* $f_{c,L}$ of f to be

$$\sup\{g : g \text{ is } L\text{-Lipschitz}, g \leq f\}.$$

Theorem 20.4.3 *Suppose that $X = Y$ and that c is a lower semi-continuous metric on X (which need not necessarily define the topology τ). If f is a proper function on X then $f^c = (-f)_{c,1}$, the lower 1-Lipschitz envelope of $-f$.*

Proof First, since $(-f)_{c,1}(y) \leq (-f)_{c,1}(x) + c(x, y) \leq -f(x) + c(x, y)$, $(-f)_{c,1}(x) \leq f^c$. Secondly, putting $x = y$, $f^c \leq -f$. Thirdly, if $x_1, x_2 \in X$, then $c(x_1, y) - f(y) \leq c(x_2, y) - f(y) + c(x_1, x_2)$, from which it follows that

$f^c(x_1) \leq f^c(x_2) + c(x_1, x_2)$, so that f^c is 1-Lipschitz with respect to c. Putting these three statements together, the result follows. □

Corollary 20.4.4 *If f is c-concave, it is 1-Lipschitz with respect to c, and $f^c = -f$.*

Proof For if $f = g^c$ then $f = (-g)_{c,1}$ so that f is 1-Lipschitz and so therefore is $-f$. Thus $f_c = (-f)_{c,1} = -f$. □

20.5 c-cyclical Monotonicity

We now consider subsets of $X \times Y$ on which a cost function behaves well. Let us consider the example at the end of Section 19.2 again. The final inequality suggests the following definition.

Suppose that X and Y are sets and that c is a function on $X \times Y$ taking values in $(-\infty, \infty]$. A subset Γ of $X \times Y$ is said to be *c-cyclically monotone* if whenever $(x_i, y_i) \in \Gamma$ for $1 \leq i \leq n$ then

$$\sum_{i=1}^{n} c(x_i, y_i) \leq \sum_{i=1}^{n} c(x_i, y_{\tau(i)}) \text{ for any } \tau \in \Sigma_n.$$

Thus in the example, the set $\Gamma = \{(j, \sigma(j)) : 1 \leq j \leq n\}$ is a c-cyclically monotone set.

The next result explains the terminology that is used.

Proposition 20.5.1 *A set Γ is c-cyclically monotone if and only if whenever $n \in \mathbf{N}$ and $(x_i, y_i) \in \Gamma$ for $1 \leq i \leq n$ then, setting $x_{n+1} = x_1$,*

$$\sum_{i=1}^{n} c(x_i, y_i) \leq \sum_{i=1}^{n} c(x_{i+1}, y_i).$$

Proof For any permutation of $\{1, \ldots, n\}$ can be written as the product of disjoint cycles. □

We now see how c-concavity and c-cyclical monotonicity are related. We need another definition. Suppose that X and Y are Polish spaces and that c is a cost function on $X \times Y$. A subset Γ of $X \times Y$ is said to be *strongly c-monotone* if there exists a c-concave function f on X such that $f(x) + f^c(y) = c(x, y)$ for all $(x, y) \in \Gamma$. That is, $\Gamma \subseteq \partial^c(f)$.

Theorem 20.5.2 (Rüschendorf's theorem) *Suppose that (X, τ) and (Y, ρ) are Polish spaces, and that c is a cost function on $X \times Y$. If Γ is a c-cyclically monotone subset of $X \times Y$ then γ is strongly c-monotone.*

Proof Since $f(x) + f^c(y) \leq c(x,y)$ for $(x,y) \in X \times Y$, we must find f for which $f(x)+f^c(y) \geq c(x,y)$ for $(x,y) \in \Gamma$. Suppose that $(\tilde{x},\tilde{y}) \in X \times Y$ and that $(x,y) \in \Gamma$. First, pick a base-point $(x_0,y_0) \in \Gamma$. If $z_n = ((x_1,y_1),\dots,(x_n,y_n)) \in \Gamma^n$, let $x_{n+1} = x_0$, and let

$$\begin{aligned} h(z_n)(\tilde{x}) &= \sum_{i=0}^{n-1}(c(x_{i+1},y_i) - c(x_i,y_i)) + c(\tilde{x},y_n) - c(x_n,y_n) \\ &= \sum_{i=0}^{n}(c(x_{i+1},y_i) - c(x_i,y_i)) + c(\tilde{x},y_n) - c(x_0,y_n). \end{aligned}$$

In particular, $h(z_n)(x_0) \geq 0$, and if $z'_1 = ((x_0,y_0))$, then $h(z'_1)(x_0) = 0$. Now suppose that $(x,y) \in \Gamma$. Let $(x_{n+1},y_{n+1}) = (x,y)$, and let $z^+_n = ((x_1,y_1),\dots,(x_n,y_n),(x,y)) \in \Gamma^{n+1}$. Then

$$\begin{aligned} h(z^+_n)(\tilde{x}) &= \sum_{i=0}^{n}(c(x_{i+1},y_i) - c(x_i,y_i)) + c(\tilde{x},y) - c(x,y) \\ &= \sum_{i=0}^{n-1}(c(x_{i+1},y_i) - c(x_i,y_i)) + c(x,y_n) - c(x_n,y_n) + c(\tilde{x},y) - c(x,y) \\ &= h(z_n)(x) + c(\tilde{x},y) - c(x,y). \end{aligned}$$

Now let

$$j(x) = \inf\{h(z_n)(x) : n \in \mathbf{N}, z_n \in \Gamma^n\} \in [-\infty,\infty).$$

Note that $j(x_0) = 0$. Then $j(\tilde{x}) \leq \inf\{h(z^+_n)(\tilde{x}) : n \in \mathbf{N}, z_n \in \Gamma^n\}$, so that

$$j(\tilde{x}) \leq j(x) + c(\tilde{x},y) - c(x,y).$$

Putting $\tilde{x} = x_0$, we see that $j(x) \geq c(x,y) - c(x_0,y)$, so that if $(x,y) \in \Gamma$ then $g(x)$ is real-valued. Then

$$c(x,y) - j(x) \leq \inf\{c(\tilde{x},y) - j(\tilde{x}) : \tilde{x} \in X\} = j^c(y),$$

so that $j(x)+j^c(y) \geq c(x,y)$. Now let $f = j^{cc}$. Then f is c-concave, $f^c = j^c$, and

$$c(x,y) \geq f(x) + f^c(y) \geq j(x) + j^c(x) \geq c(x,y),$$

so that $f(x) + f^c(y) = c(x,y)$. □

The function f is called a *maximal Kantorovich potential* for the set Γ. If c is continuous, then the function j is upper semi-continuous, but if c is only lower semi-continuous it might not be Borel measurable (it is the infimum of an uncountable set of lower semi-continuous functions). On the other hand, since f is c-concave, it is Borel measurable.

20.6 Optimal Transport Plans Revisited

We now show how c-cyclic monotonocity and c-concavity relate to optimal transport plans. A transport plan π is said to be *c-cyclically monotone* if there is a c-cyclically monotone Borel subset Γ of $X \times Y$ with $\pi(\gamma) = 1$, and is said to be *strongly c-monotone* if there is a strongly c-monotone Borel set Γ with $\pi(\Gamma) = 1$.

Theorem 20.6.1 *Suppose that (X, τ) and (Y, ρ) are Polish spaces, that c is a cost function on $X \times Y$, that μ and ν are Borel probability measures on X and Y respectively, and that $\pi \in \Pi_{\mu,\nu}$.*

Then the following are equivalent.

(i) π is optimal.
(ii) π is c-cyclically monotone.
(iii) π is strongly c-monotone.

Proof Suppose that π is optimal. Let

$$A = \{f(x) + g(y) : f \in L^1(\mu), g \in L^1(\nu), f + g \le c\}.$$

By Corollary 20.3.2, there exists a sequence $(f_k + g_k)_{k=1}^\infty$ in A such that $\int_{X\times Y} c(x,y) - (f_k(x) + g_k(y))\, d\pi(x,y) \to 0$ as $k \to \infty$. By extracting a subsequence if necessary, there exists a Borel measurable subset Γ, with $\pi(\Gamma) = 1$, such that $f_k(x) + g_k(y) \to c(x,y)$ as $k \to \infty$, for $(x,y) \in \Gamma$. Suppose that $(x_i, y_i) \in \Gamma$, for $1 \le i \le n$, and that $\tau \in \Sigma_n$. Then if $k \in \mathbf{N}$,

$$\begin{aligned}\sum_{i=1}^n c(x_i, y_{\tau(i)}) &\ge \sum_{i=1}^n (f_k(x_i) + g_k(y_i)) = \sum_{i=1}^n f_k(x_i) + \sum_{i=1}^n g_k(y_{\tau(i)})\\ &= \sum_{i=1}^n f_k(x_i) + \sum_{i=1}^n g_k(y_i) \to \sum_{i=1}^n c(x_i, y_i) \text{ as } k \to \infty.\end{aligned}$$

Thus π is c-cyclically monotone.

Rüschendorfs theorem shows that a c-cyclically monotone measure is strongly c-monotone.

Suppose that π is strongly c-monotone, and that Γ is a strongly c-monotone Borel subset of $X \times Y$ with $\pi(\Gamma) = 1$. Let f be a c-concave function such that $\Gamma \subseteq \partial^c(f)$. Let $A_n = \{x : |f(x)| \le n\}$ and let $g_n = I_{A_n} f$. Similarly, let $B_n = \{y : |f^c(y)| \le n\}$ and let $h_n = I_{B_n} f^c$. Then $g_n \in L^\infty(\mu)$, $h_n \in L^\infty(\nu)$ and

$$\int_X g_n\, d\nu + \int_Y h_n\, d\nu = \int_{\Gamma \int (A_n \times B_n)} c\, d\pi \to \int_{X\times Y} c\, d\pi$$

as $n \to \infty$. Thus $\int_{X\times Y} c\, d\pi = M_c$, and π is optimal. □

When c is continuous, we can say more.

Theorem 20.6.2 *Suppose that (X, τ) and (Y, ρ) are Polish spaces, that c is a continuous cost function on $X \times Y$, that μ and ν are Borel probability measures on X and Y respectively, and that $\pi \in \Pi_{\mu,\nu}$.*

Then the following are equivalent.

(i) π *is optimal.*
(ii) $\mathrm{supp}(\pi)$ *is c-cyclically monotone.*
(iii) $\mathrm{supp}(\pi)$ *is strongly c-monotone.*

If so, if f is a c-concave function on X such that $f + f^c = c$ on $\mathrm{supp}(\pi)$, *then (f, f^c) is continuous on* $\mathrm{supp}(\pi)$.

Proof It follows from the previous theorem and Rüschendorfs theorem that it is enough to show that if π is optimal, then $\mathrm{supp}(\pi)$ is a c-cyclically monotone set. Suppose not. Then there exist distinct $(x_0, y_0), \ldots, (x_n, y_n)$ in $\mathrm{supp}(\pi)$ such that, setting $x_{n+1} = x_0$,

$$\sum_{j=0}^{n} c(x_j, y_j) - \sum_{j=0}^{n} c(x_{j+1}, y_j) = \eta > 0.$$

Since c is continuous, there exist closed neighbourhoods $W_j = U_j \times V_j$ for $0 \leq j \leq n$ such that if $W_j' = U_{j+1} \times V_j$ then

$$|c(u_j, v_j) - c(x_j, y_j)| < \eta/4(n+1) \text{ for } (u_j, v_j) \in W_j$$
$$\text{and } |c(u_{j+1}, v_j) - c(x_j + 1j, y_j)| < \eta/4(n+1) \text{ for } (u_{j+1}, v_j) \in W_j',$$

for $0 \leq j \leq n$.

Since $(x_j, y_j) \in \mathrm{supp}(\pi)$, $\lambda_j = \pi(W_j) > 0$. Let $\pi_j = (I_{W_j}/\lambda_j)\, d\pi$. Then π_j is a Borel probability measure on $X \times Y$, with support in W_j. Let $\alpha = (\min_{0 \leq j \leq n} \lambda_j)/n + 1$, and let $\sigma = \pi - \sum_{j=0}^{n} \alpha \pi_j$. Then σ is a non-negative Borel measure on $X \times Y$.

Let μ_j and ν_j be the marginals of π_j, and let $\pi' = \sigma + \sum_{j=0}^{n} \alpha \mu_{j+1} \otimes \nu_j$. Then π' is a Borel probability measure on $X \times Y$, and $\pi' \in \Pi_{\mu,\nu}$. Further,

$$\int_{X \times Y} c\, d\pi - \int_{X \times Y} c'\, d\pi = \alpha \sum_{j=0}^{n} \int_{W_j} c\, d\mu_j - \alpha \sum_{j=0}^{n} \int_{W_j'} c\, d(\mu_{j+1} \otimes \nu_j)$$
$$\geq \alpha \left(\sum_{j=0}^{n} c(x_j, y_j) - \sum_{j=0}^{n} c(x_{j+1}, y_j) - \eta/2 \right) \geq \alpha\eta/2,$$

contradicting the optimality of π.

The function (f, f^c) is upper semi-continuous on $X \times Y$. But if $(x, y) \in \mathrm{supp}(\pi)$ then $(f(x), f^c(y)) = (c(x, y) - f^c(y), c(x, y) - f(x))$, so that (f, f^c) is also lower semi-continuous on $\mathrm{supp}(\pi)$. □

When the cost function c is continuous, there is a strongly c-monotone set which contains the support of every optimal measure.

Theorem 20.6.3 *Suppose that (X, τ) and (Y, ρ) are Polish spaces, that c is a continuous cost function on $X \times Y$, that μ and ν are Borel probability measures on X and Y respectively, and that there exists $\pi \in \Pi_{\mu,\nu}$ with $\int_{X\times Y} c\, d\pi < \infty$. Then there exists a strongly c-monotone set Γ such that if π is an optimal measure then* $\mathrm{supp}(\pi) \subseteq \Gamma$.

Proof Let $\Gamma = \cup\{\mathrm{supp}(\pi) : \pi \text{ optimal}\}$. By Rüschendorfs theorem, it is sufficient to show that Γ is c-cyclically monotone. Suppose that $(x_0, y_0), \ldots, (x_n, y_n) \in \Gamma$. There exist optimal measures $\pi_0, \ldots, \pi_n$ such that $(x_j, y_j) \in \mathrm{supp}(\pi_j)$, for $0 \leq j \leq n$. Let $\pi = (\sum_{j=0}^n \pi_j)/(n+1)$. Then π is an optimal measure, and $(x_j, y_j) \in \mathrm{supp}(\pi)$, for $0 \leq j \leq n$. Consequently, setting $x_{n+1} = x_0$, $\sum_{j=0}^n c(x_j, y_j) \leq \sum_{j=0}^n c(x_{j+1}, y_j)$, and Γ is c-cyclically monotone. □

Theorem 20.6.2 suggests another way of showing the existence of optimal transport plans, or constructing optimal transport plans, when the cost function c is continuous.

Theorem 20.6.4 *Suppose that (X, τ) and (Y, ρ) are Polish spaces, that μ and ν are Borel probability measures on X and Y respectively, and that c is a continuous cost function on $X \times Y$. Then there exists a c-cyclically monotone transport plan.*

Proof By the strong law of large numbers for empirical processes, there exist sequences

$$\left(\mu_n = \frac{1}{n}\sum_{i=1}^n \delta_{x_i}\right)_{n=1}^\infty \quad \text{and} \quad \left(\nu_n = \frac{1}{n}\sum_{i=1}^n \delta_{y_i}\right)_{n=1}^\infty$$

such that $\mu_n \Rightarrow \mu$ and $\nu_n \Rightarrow \nu$. For each n there exists a c-cyclically monotone transport plan π_n for the pair (μ_n, ν_n). The sequences $(\mu_n)_{n=1}^\infty$ and $(\nu_n)_{n=1}^\infty$ are uniformly tight, and so, by Proposition 20.2.3, is the sequence $(\pi_n)_{n=1}^\infty$. It therefore follows from Prokhorov's theorem that there is a subsequence (which we again denote by $(\pi_n)_{n=1}^\infty$) which converges weakly to some $\pi \in M(X \times Y)$. If $f \in C_b(X)$ then

$$\int_X f\, d\mu_n = \int_{X\times Y} f(x)\, d\pi_n(x, y) \to \int_{X\times Y} f(x)\, d\pi(x, y) \text{ as } n \to \infty.$$

But $\int_X f\,d\mu_n \to \int_X f\,d\mu$ as $n \to \infty$, so that $\int_{X\times Y} f(x)\,d\pi(x,y) = \int_X f\,d\mu$. A similar property holds for Y, and so $\pi \in \Pi_{\mu,\nu}$.

Now fix $k \in \mathbf{N}$. Let F_k be the set

$$\left\{((x_1,y_1),\ldots,(x_k,y_k)) : \sum_{i=1}^{k}(c(x_i,y_i) - c(x_i,y_{\sigma(i)})) \leq 0, \text{ for } \sigma \in \Sigma_k\right\}.$$

Since c is continuous, F_k is closed in $(X \times Y)^n$. Now $(\otimes_{i=1}^{k}\pi_n)(F_k) = 1$, and so $(\otimes_{i=1}^{k}\pi)(F_k) = 1$. Thus, by the portmanteau theorem, $\operatorname{supp}(\pi)^k \subseteq F_k$. Since this holds for all k, π is a c-cyclically monotone transport plan. □

20.7 Approximation

In applications, it is often necessary (for example, when making calculations) to approximate the distributions μ and ν, and the cost function c. When the cost function is continuous, then we can use c-cyclic monotonicity to show that things work well.

Theorem 20.7.1 *Suppose that (X,τ) and (Y,ρ) are Polish spaces, that $\mu \in P(X)$, $\nu \in P(Y)$ and that c is a continuous cost function. Suppose also that $(\mu_i)_{i=1}^{\infty}$ is a sequence in $P(X)$ such that $\mu_i \Rightarrow \mu$, that $(\nu_i)_{i=1}^{\infty}$ is a sequence in $P(Y)$ such that $\nu_i \Rightarrow \nu$, that $(c_i)_{i=1}^{\infty}$ is a sequence of continuous cost functions such that*

$$\sup\{|c_i(x,y) - c(x,y)| : (x,y) \in X \times Y\} \to 0 \text{ as } i \to \infty$$

and that for each i there is an optimal transport plan π_i with $\int_{X\times Y} c_i\,d\pi_i \leq m < \infty$. Then there exists a subsequence $(\pi_{i_k})_{k=1}^{\infty}$ such that $\pi_{i_k} \Rightarrow \pi$, an optimal transport plan for μ and ν.

Proof The sequences $(\mu_i)_{i=1}^{\infty}$ and $(\nu_i)_{i=1}^{\infty}$ are uniformly tight, from which it follows that the sequence $(\pi_i)_{i=1}^{\infty}$ is also uniformly tight. By extracting a subsequence if necessary, we can therefore suppose that $\pi_i \Rightarrow \pi$.

Now let $\Omega = (X \times Y)^{\mathbf{N}}$, $\tilde{\pi}_i = \otimes_{j=1}^{\infty}\pi_i$ and $\tilde{\pi} = \otimes_{j=1}^{\infty}\pi$. By c-cyclic monotonicity, for each i there exists C_i such that if $(\tilde{x},\tilde{y}) \in \tilde{C}_i$ then $\sum_{j=1}^{n} c_i(x_j,y_j) \leq \sum_{j=1}^{n} c_i(x_j,y_{(j+1)\bmod n})$, and $\pi_i(C_i) = 1$. Thus if

$$C^{(k)} = \left\{(\tilde{x},\tilde{y}) : \sum_{j=1}^{n} c(x_j,y_j) \leq \sum_{j=1}^{n} c(x_j,y_{(j+1)\bmod n}) + 1/k \text{ for } n \in \mathbf{N}\right\},$$

then $C_i \subseteq C^{(k)}$ for sufficiently large i. Since $C^{(k)}$ is closed, $\pi(C^{(k)}) = 1$. Thus if $C = \cap_{k=1}^{\infty} C^{(k)}$, $\pi(C) = 1$. But C is c-cyclically monotone, and so π is optimal. □

This theorem also gives a proof of the existence of optimal measures when the cost function c is continuous. Let μ_i be the ith empirical measure of μ and ν_i the ith empirical measure of ν, and let $c_i = c$. Then there is certainly an optimal measure for each of the finite pairs of measures (μ_i, ν_i), and the empirical law of large numbers ensures that $\mu_i \Rightarrow \mu$ and $\nu_i \Rightarrow \nu$.

21
Wasserstein Metrics

21.1 The Wasserstein Metrics W_p

Suppose that d is a lower semi-continuous metric on a Polish space (X, τ). We consider Borel probability measures on X whose spread is controlled by the metric d. If $0 < p < \infty$, let

$$P_p(X) = \left\{ \mu \in P(X) : A_p(X) = \int_{X\times X} d^p(x,y)\,d\mu(x)\,d\mu(y) < \infty \right\}.$$

Proposition 21.1.1 *Suppose that d is a lower semi-continuous metric on a Polish space (X, τ) and that $0 < p < \infty$. The following are equivalent.*

(i) $\mu \in P_p(X)$.
(ii) $\int_X d^p(x, x_0)\,d\mu(x) < \infty$ *for some* $x_0 \in X$.
(iii) $\int_X d^p(x, x_0)\,d\mu(x) < \infty$ *for every* $x_0 \in X$.

Proof If $0 < p < 1$, then d^p is a metric on X uniformly equivalent to d. Consequently, we only need to consider the case where $1 \le p < \infty$.

Certainly (i) implies (ii). Suppose that (ii) holds and that $x_1 \in X$. Since the function $|x|^p$ is convex, if $a, b \ge 0$ then

$$(a+b)^p = 2^p \left(\frac{a+b}{2}\right)^p \le 2^{p-1}(a^p + b^p).$$

Thus

$$\int_X d^p(x, x_1)\,d\mu(x) \le 2^{p-1}\left(\int_X d^p(x, x_0)\,d\mu(x) + d^p(x_0, x_1)\right),$$

and (iii) holds. If (iii) holds, then

$$\int_{X\times X} d^p(x,y)\,d\mu(x)d\mu(y) \le 2^{p-1}\left(\int_X d^p(x,x_0)\,d\mu(x) + \int_X d^p(x_0,y)\,d\mu(y)\right) < \infty,$$

so that (i) holds. □

Theorem 21.1.2 *Suppose that (X, d) is a lower semi-continuous metric on a Polish space (X, τ) and that $1 < p < \infty$. If $\mu, \nu \in P(X)$, let*

$$W_p(\mu, \nu) = \left(\inf \left\{ \int_{X \times X} d^p \, d\pi : \pi \in \Pi_{\mu \times \nu} \right\} \right)^{1/p}.$$

Then the restriction of W_p to $P_p(X) \times P_p(X)$ is a metric on $P_p(X)$.

As we have seen, the infimum is attained.

Proof Clearly $W_p(\mu, \nu) = W_p(\nu, \mu)$. Let us show that $W_p(\mu, \nu) = 0$ if and only if $\mu = \nu$. If $\mu \neq \nu$, then there exists a compact set K such that $\mu(K) = \nu(K) + 2\epsilon > \nu(K)$. By upper continuity, there exists $\delta > 0$ such that $\nu(K_\delta) < \nu(K) + \epsilon$. Thus if $\pi \in \Pi_{\mu,\nu}$ then $\pi(X \times (X \setminus K_\delta) > \epsilon$, so that

$$\int_{X \times X} d^p \, d\pi \geq \int_{K \times (K \setminus K_\delta)} d^p \, d\pi \geq \delta^p \epsilon.$$

Thus $W_p(\mu, \nu) \geq \delta . \epsilon^{1/p} > 0$. Conversely, if $\mu = \nu$ then, setting $\pi = GI_*(\mu)$, where $I(x) = x$, $\int_{X \times X} d^p \, d\pi = 0$, so that $W_p(\mu, \nu) = 0$.

Next we prove the triangle inequality. Suppose that $\mu, \nu, \pi \in P_p(X)$. There exist $\alpha \in \Pi_{\mu,\nu}$ and $\beta \in \Pi_{\nu,\pi}$ such that

$$W_p(\mu, \nu) = \left(\int_{X \times X} d^p \, d\alpha \right)^{1/p} \text{ and } W_p(\nu, \pi) = \left(\int_{X \times X} d^p \, d\beta \right)^{1/p}.$$

Let $\gamma \in P(X \times X \times X)$ satisfy the conclusions of the gluing lemma (Theorem 16.11.1), and let $\gamma_{1,3}$ be the marginal distribution on the first and third factors. Then, using Hölder's inequality,

$$\begin{aligned}
W_p(\mu, \pi) &\leq \left(\int_{X \times X} d(x,z)^p \, d\gamma_{1,3}(x,z) \right)^{1/p} \\
&= \left(\int_{X \times X \times X} d(x,z)^p \, d\gamma(x,y,z) \right)^{1/p} \\
&\leq \left(\int_{X \times X \times X} (d(x,y) + d(y,z))^p \, d\gamma(x,y,z) \right)^{1/p} \\
&\leq \left(\int_{X \times X \times X} d(x,y)^p \, d\gamma(x,y,z) \right)^{1/p} \\
&\quad + \left(\int_{X \times X \times X} d(y,z)^p \, d\gamma(x,y,z) \right)^{1/p} \\
&= \left(\int_{X \times X} d(x,y)^p \, d\alpha(x,y) \right)^{1/p} + \left(\int_{X \times X} d(y,z)^p \, d\beta(y,z) \right)^{1/p} \\
&= W_p(\mu, \nu) + W_p(\nu, \pi).
\end{aligned}$$

□

The metric W_p is called the Wasserstein p-metric.

Proposition 21.1.3 *Suppose that d is a lower semi-continuous metric on a Polish space (X, τ). If $1 \leq p < q < \infty$ and $\mu, \nu \in P_q(X)$, then $\mu, \nu \in P_p(X)$ and $W_p(\mu, \nu) \leq W_q(\mu, \nu)$.*

Proof It follows from Hölder's inequality and Proposition 21.1.1 that $\mu, \nu \in P_p(X)$. Let π be an optimal measure for d^p in $\Pi(\mu, \nu)$. By Hölder's inequality again,

$$W_p(\mu, \nu) \leq \left(\int_{X \times X} d^p \, d\pi \right)^{1/p} \leq \left(\int_{X \times X} d^q \, d\pi \right)^{1/q} = W_q(\mu, \nu).$$

□

We can extend this to the case where $0 < p < 1$. Suppose that (X, d) is a Polish metric space and that $0 < p < 1$. Then d^p is a strictly subadditive metric on X which is uniformly equivalent to d. If $\mu, \nu \in \Pi_p(X)$, we set

$$W_p(\mu, \nu) = \inf \left\{ \int_{X \times X} d^p \, d\pi : \pi \in \Pi_{\mu,\nu} \right\}.$$

Then W_p is a metric on $\Pi_p(X)$. Let $\mathrm{Lip}_p(X)$ be the space of functions on X which are Lipschitz for the metric d^p and if $f \in \mathrm{Lip}_p(X)$, let $p_L^{(p)}(f) = \sup\{|f(x) - f(y)|/d^p(x, y) : x \neq y\}$. Then $W_p(\mu, \nu) = \gamma_p(\mu, \nu)$, where

$$\gamma_p(\mu, \nu) = \sup \left\{ \left| \int_X f \, d\mu - \int_X f \, d\nu \right| : f \in \mathrm{Lip}_p(X), p_L^{(p)}(f) \leq 1 \right\}.$$

The metric W_p is again called the *Wasserstein p-metric.*

21.2 The Wasserstein Metric W_1

Proposition 21.2.1 *Suppose that d is a lower semi-continuous metric on a Polish space (X, τ). If $\mu \in P_1(X)$ then $Lip_X \subseteq L^1(\mu)$.*

Proof Suppose that $f \in Lip_X$ and that $x_0 \in X$. Then

$$\begin{aligned}\int_X |f| \, d\mu &\leq \int_X |f(x) - f(x_0)| + |f(x_0)| \, d\mu(x) \\ &\leq L \int_X d(x, x_0) \, d\mu(x) + |f(x_0)| < \infty.\end{aligned}$$

□

Theorem 21.2.2 *Suppose that (X, τ) is a Polish space, that c is a lower semi-continuous metric on X, that μ, ν are Borel probability measures on X and that Γ is a Borel subset of $X \times X$. Then Γ is c-strictly monotone if and only if there exists a function f on X, 1-Lipschitz for the metric c, such that $c(x, y) = f(x) - f(y)$ for $(x, y) \in \Gamma$.*

Proof It follows from Corollary 2.8.5 that $f^c = (-f)_{c,1}$, the lower 1-Lipschitz envelope of $-f$. Thus f is c-concave if and only if it is 1-Lipschitz for the metric c, and, if so then $f^c = -f$. Thus the result follows from Rüschendorf's theorem (Theorem 20.5.2). □

If $\mu, \nu \in P_1(X)$, let

$$\gamma(\mu, \nu) = \sup\left\{\left|\int_X f\,d\mu - \int_X f\,d\nu\right| : f \in \mathrm{Lip}(X), p_L(f) \leq 1\right\}.$$

Corollary 21.2.3 *If $\mu, \nu \in P_1(X)$ then $W_1(\mu, \nu) = \gamma(\mu, \nu)$.*

Then γ is a metric on $P_1(X)$, and $\gamma \geq \beta_{|P_1(X)}$, so that the inclusion $(P_1(X), \gamma) \to (P(X), w)$ is continuous.

Corollary 21.2.4 *Suppose that d is a bounded metric, bounded by M say, and that $1 < p < \infty$. Then $P(X) = P_1(X) = P_p(X)$, and the metrics β, W_1 and W_p are uniformly equivalent metrics on $P(X)$.*

Proof Since a Lipschitz function on X is bounded by M say, β and γ are uniformly equivalent. If $\pi \in P(X \times X)$ then

$$\int_{X \times X} d\,d\pi \leq \left(\int_{X \times X} d^p\,d\pi\right)^{1/p} \leq M^{1-1/p}\left(\int_{X \times X} d\,d\pi\right)^{1/p},$$

from which it follows easily that $W_1 \leq W_p \leq M^{1-1/p} W_1^{1/p}$, and W_1 and W_p are uniformly equivalent. □

21.3 W_1 Compactness

Suppose that (X, τ) is a Polish space, and that d is a lower semi-continuous metric on X. We now characterize the compact subsets of $(P_1(X), W_1)$. First, we consider convergent sequences.

Theorem 21.3.1 *Suppose that (X, τ) is a Polish space, and that d is a lower semi-continuous metric on X. Suppose that $(\mu_k)_{k=0}^\infty$ is a sequence in $P_1(X)$, and that $x_0 \in X$. Then the following are equivalent.*

(i) $W_1(\mu_k, \mu_0) \to 0$ *as* $k \to \infty$.
(ii) $\mu_k \Rightarrow \mu_0$ *and* $(\mu_k)_{k=0}^{\infty}$ *is uniformly integrable.*
(iii) $\mu_k \Rightarrow \mu_0$ *and* $\int_X d(x, x_0)\, d\mu_k \to \int_X d(x, x_0)\, d\mu_0$ *as* $k \to \infty$.

Proof The equivalence of (ii) and (iii) follows from Theorem 18.8.2. Suppose that (i) holds. For each $k \in \mathbf{N}$ there exists an optimal $\pi_k \in \Pi_{\mu_k, \mu_0}$. If $x, y \in X$, $d(x, x_0) \le d(y, x_0) + d(x, y)$. Integrating with respect to π_k, it follows that

$$\int_X d(x, x_0)\, d\mu_k(x) \le \int_X d(y, x_0)\, d\mu_0(y) + W_1(\mu_k, \mu_0).$$

Similarly,

$$\int_X d(y, x_0)\, d\mu_0(y) \le \int_X d(x, x_0)\, d\mu_k(y) + W_1(\mu_k, \mu_0),$$

so that (iii) holds.

Suppose that (ii) holds and that $\epsilon > 0$. There exists $R > 0$ such that $\int_{d(x,x_0)>R/2} d\, d\mu_k(x) < \epsilon/3$, for $k \in \mathbf{Z}^+$. Let $\tilde{d}(x, y) = d(x, y) \wedge R$ and let $\tilde{W}_1$ be the corresponding metric on $P(X)$. Then the $\tilde{W}_1$ topology is the same as the w topology, so that $\tilde{W}_1(\mu_k, \mu_0) \to 0$ as $k \to \infty$. If $d(x, y) \ge R$ then either $d(x, x_0) \ge R/2$ or $d(y, x_0) \ge R/2$. Thus

$$d(x, y) \le d(x, y) \wedge R + d(x, y) I_{(d(x, x_0) \ge R/2)} + d(x, y) I_{(d(y, x_0) \ge R/2)}.$$

Let π_k be an optimal measure for the pair (μ_k, μ_0), with cost $\tilde{d}$. Then

$$\begin{aligned} W_1(\mu_k, \mu_0) &\le \int_X d\, d\pi_k \\ &\le \int_X \tilde{d}\, d\pi_k + \int_{(d(x, x_0) \ge R/2)} d\, d\mu_k(x) + \int_{(d(y, x_0) \ge R/2)} d\, d\mu_0 \\ &\le \tilde{W}_1(\mu_k, \mu_0) + 2\epsilon/3, < \epsilon \end{aligned}$$

for large enough k. Thus (ii) implies (i). □

Straightforward arguments now extend this to a characterization of compactness.

Theorem 21.3.2 *Suppose that* (X, τ) *is a Polish space, and that* d *is a lower semi-continuous metric on* X. *Suppose that* $A \subseteq P_1(X)$. *Then* A *is* W_1 *compact if and only if it is* w *compact and uniformly integrable.*

Proof Suppose that A is W_1 compact. Then it is w compact. Suppose if possible that it is not uniformly integrable. Then there is a sequence $(\mu_k)_{k=1}^{\infty}$ in A with no uniformly integrable subsequence. But $(\mu_k)_{k=1}^{\infty}$ has a W_1 convergent subsequence, which is uniformly integrable, by Theorem 21.3.1, giving a contradiction.

If the conditions are satisfied, and $(\mu_k)_{k=1}^\infty$ is a sequence in A, then it has a w convergent subsequence in A, and this converges in the metric W_1, by Theorem 21.3.1, so that A is W_1 sequentially compact, and so is W_1 compact. □

21.4 W_p Compactness

Suppose that (X, τ) is a Polish space, and that d is a lower semi-continuous metric on X. We now extend the results of the previous section to characterize the compact subsets of $(P_p(X), W_p)$ for $1 < p < \infty$. Again, we first consider convergent sequences.

Theorem 21.4.1 *Suppose that (X, τ) is a Polish space, and that d is a lower semi-continuous metric on X. Suppose that $(\mu_k)_{k=0}^\infty$ is a sequence in $P_p(X)$, where $1 < p < \infty$, and that $x_0 \in X$. Then the following are equivalent.*

(i) $W_p(\mu_k, \mu_0) \to 0$ *as* $k \to \infty$.
(ii) $\mu_k \Rightarrow \mu_0$ *and* $(\mu_k)_{k=0}^\infty$ *is p-uniformly integrable.*
(iii) $\mu_k \Rightarrow \mu_0$ *and* $\int_X d^p(x, x_0)\, d\mu_k \to \int_X d^p(x, x_0)\, d\mu_0$ *as* $k \to \infty$.

Proof Let $\phi(x) = d^p(x, x_0)$ and let $\phi_n = \phi_n$, for $n \in \mathbf{N}$. Suppose that (i) holds. Then $\mu_k \Rightarrow \mu_0$, and so $\beta(\mu_k, \mu_0) \to 0$ as $k \to \infty$. Thus

$$\begin{aligned}\int_X \phi\, d\mu_0 &= \lim_{n\to\infty} \int_X \phi_n\, d\mu_0 \\ &= \lim_n \to \infty \left(\lim_{k\to\infty} \int_X \phi_n\, d\mu_k \right) \\ &\geq \liminf_{k\to\infty} \int_X \phi\, d\mu_k.\end{aligned}$$

To prove that (iii) holds, we must therefore show that

$$\limsup_{k\to\infty} \int_X \phi\, d\mu_k \leq \int_X \phi\, d\mu_0.$$

Suppose that $\epsilon > 0$. Let $\eta = (1+\epsilon)^{1/p} - 1$ (so that $(1+\eta)^p = 1+\epsilon$), and let $C_\epsilon = (1 + 1/\eta)^p$. Suppose that $a, b > 0$. If $\beta < \eta a$ then $(a+b)^p \leq ((1+\eta)a)^p = (1+\epsilon)a^p$, and if $\beta > \eta a$ then $(a+b)^p \leq ((1+1/\eta)b)^p = C_\epsilon b^p$, and so in any case $(a+b)^p \leq 1 + \epsilon a^p + C_\epsilon b^p$. Thus if $x, y \in X$ then

$$\phi(x) \leq (d(y, x_0) + d(x, y))^p \leq (1+\epsilon)\phi(y) + C_\epsilon d^p(x, y).$$

Let π_k be an optimal measure for μ_k and μ_0. Integrating with respect to μ_k, it follows that

$$\int_X \phi \, d\mu_k \leq (1+\epsilon) \int_X \phi \, d\mu_0 + C_\epsilon W_p(\mu_k, \mu_0)^p.$$

Since $W_p(\mu_k, \mu_0) \to 0$ as $k \to \infty$, we see that

$$\limsup_{k\to\infty} \int_X \phi \, d\mu_k \leq (1+\epsilon) \int_X \phi \, d\mu_0,$$

and the result follows, since ϵ is arbitrary.

Now suppose that (ii) holds. Let $\tilde{d} = d \wedge 1$, and let $\tilde{W}_1$ and $\tilde{W}_p$ be the corresponding metrics. It follows from Corollary 21.2.4 that $\tilde{W}_p(\mu_k, \mu_0) \to 0$ as $k \to \infty$. Suppose that $\epsilon > 0$. There exists $R > 1$ such that

$$\int_{(d(x,x_0)\geq R/2)} d^p(x, x_0) \, d\mu_k < \epsilon/6^{p+1} \text{ for } k \in \mathbf{Z}^+$$

and there exists k_0 such that $(3R)^p \tilde{W}_p(\mu_k, \mu_0) < 2\epsilon/3$ for $k \geq k_0$. If $d(x, y) \geq R$ and $d(x, x_0) \geq d(y, x_0)$, then $d(x, x_0) \geq 2d(x, y)$, and a similar inequality holds if $d(x, x_0) < d(y, x_0)$. Thus

$$d(x, y) \leq R\tilde{d}(x, y) + 2d(x, x_0)I_{(d(x,x_0)\geq R/2)} + 2d(y, x_0)I_{(d(y,x_0)\geq R/2)},$$

and so

$$d(x, y)^p \leq 3^p(R^p\tilde{d}^p(x, y) + 2^p d^p(x, x_0)I_{(d(x,x_0)\geq R/2} + 2^p d^p(y, x_0)I_{(d(y,x_0)} \geq R/2.$$

Let π_k be an optimal measure for μ_k and μ_0 for the cost function $\tilde{d}^p$. If $k \geq k_0$, then, integrating with respect to π,

$$\begin{aligned}
W_p(\mu_k, \mu_0) &\leq \int_{X\times X} d^p \, d\pi_k \\
&\leq (3R)^p \tilde{W}_p(\mu_k, \mu_0) + 6^p \int_{(d(x,x_0)\geq R/2)} d^p(x, x_0) \, d\mu_k \\
&\quad + 6^p \int_{(d(y,x_0)\geq R/2)} d^p(y, x_0) \, d\mu_0 \\
&\leq 2\epsilon/3 + \epsilon/6 + \epsilon/6 = \epsilon.
\end{aligned}$$

Thus (ii) implies (i). □

The next result follows from this, using the arguments used to prove Theorem 21.3.2.

Theorem 21.4.2 *Suppose that (X, τ) is a Polish space, that d is a lower semi-continuous metric on X and that $1 < p < \infty$. Suppose that $A \subseteq P_p(X)$. Then A is W_p compact if and only if it is w compact and p-uniformly integrable.*

21.5 W_p-Completeness

Proposition 21.5.1 *Suppose that d is a lower semi-continuous metric on a Polish space (X, τ), and that $0 < p < \infty$. If $\nu \in P_p(X)$, $\mu \in P(X)$ and $\inf \int_{X\times X} d^p\, d\pi : \pi \in \Pi_{\mu,\nu}\} < \infty$ then $\mu \in P_p(X)$.*

Proof Let π be an optimal measure in $\Pi_{\mu,\nu}$. If $x, y \in X$, then

$$d^p(x, x_0) \leq 2^p(d^p(x, y) + d^p(y, x_0)).$$

Integrating with respect to π,

$$\int_X d^p(x, x_0)\, d\mu(x) \leq 2^p \inf_{\pi\in\Pi_{\mu,\nu}} \int_{X\times X} d^p\, d\pi + \int_X d^p(y, x_0)\, d\nu(y) < \infty.$$

□

Proposition 21.5.2 *Suppose that d is a lower semi-continuous metric on a Polish space (X, τ), and that $0 < p < \infty$. Suppose that $(\mu_n)_{n=1}^\infty$ is a sequence in $P_p(X)$, that $\nu \in P_p(X)$ and that $\mu \in P(X)$. Suppose that $\sup_n W_p(\mu_n, \nu) \leq R < \infty$ and that $\mu_n \Rightarrow \mu$. Then $\mu \in P_p(X)$, and $W_p(\mu, \nu) \leq R$.*

Proof We use the Kantorovich–Rubinstein theorem. Let ρ be a complete metric on X which defines the topology, and let $BL(X)$ be the space of bounded ρ-Lipschitz functions. Suppose that $f(x) \in BL(X)$, $g(y) \in BL(X)$ and $f(x) + g(y) \leq d^p(x, y)$. Then

$$\int_X f\, d\mu + \int_X g\, d\nu = \lim_{n\to\infty} \left(\int_X f\, d\mu_n + \int_X g\, d\nu\right) \leq \sup_n W_p^p(\mu_n, \nu) \leq R^p.$$

Hence $\inf\{\int_{X\times X} d^p\, d\pi : \pi \in \Pi_{\mu,\nu}\} \leq R$, and it follows from Proposition 21.5.1 that $\mu \in P_p(X)$. □

Theorem 21.5.3 *Suppose that d is a lower semi-continuous metric on a Polish space (X, τ), and that $0 < p < \infty$. $(P_p(X), W_p)$ is a complete metric space, for $0 < p < \infty$.*

Proof Suppose that $\nu \in P_p(X)$ and that $(\mu_n)_{n=1}^\infty$ is a sequence in $P_p(X)$ such that $W_p(\mu_n, \nu) \leq r$ and such that $\mu_n \Rightarrow \mu$. By Proposition 21.5.2, $\mu \in P_p(X)$ and $W_p(\mu, \nu) \leq r$. Thus $M_r(\nu) = \{\mu \in P_p(X) : W_p(\mu, \nu) \leq r\}$ is closed in $(P(X), \beta)$, and the result follows from Theorem 2.4.5. □

Proposition 21.5.4 *Suppose that (X, d) is a Polish metric space, that $0 < p < \infty$ and that C is countable dense subset of X. Then the countable set*

$$A_C = \left\{\sum_{i=0}^n \lambda_i \delta_{c_i} : n \in \mathbf{N}, \lambda_i \in \mathbf{Q}, \lambda_i \geq 0, \sum_{i=1}^n \lambda_i = 1, c_i \in C\right\}$$

is W_p-dense in $(P_p(X), W_p)$.

Proof Suppose that $\mu \in P_p(X)$ and that $0 < \epsilon < 1$. Pick $c_0 \in C$. There exists a compact subset K of X such that $\int_X d(x, c_0)^p \, d\mu < \epsilon$. There exists a finite partition $\mathcal{D}$ of K into non-empty Borel sets, each of diameter less than ϵ, and for each $D \in \mathcal{D}$ there exists $c_D \in C$ such that $d(c_D, x) < \epsilon$ for $x \in D$. Let

$$\nu = \mu(X \setminus K)\delta_{c_0} + \sum_{D \in \mathcal{D}} \mu(D)\delta_{c_D},$$

and let

$$\pi = I_{X \setminus K}.d\mu \otimes \delta_{c_0} + \sum_{D \in \mathcal{D}} I_D.d\mu \otimes \delta_{c_D}.$$

Then $\pi \in \Pi_{\mu,\nu}$, and $\int_{X \times X} d^p \, d\pi < \epsilon^p$, so that $W_p(\mu, \nu) < \epsilon$. We can approximate ν by a probability measure of the form $\lambda_0\delta_{c_0} + \sum_{D \in \mathcal{D}} \lambda_D \delta_{c_D}$, with rational coefficients λ_0 and λ_D, which establishes the result. □

Thus the spaces $(P_p(X), W_p)$ are Polish spaces, and we can start analyzing functions and Borel measures on them.

21.6 The Mallows Distances

We now consider the Wasserstein distances between the distributions of random variables. Suppose that μ_1 and μ_2 are Borel probability measures on Polish spaces (X_1, τ_1) and (X_2, τ_2) respectively. Suppose that $(E, \|.\|)$ is a Banach space, that $1 \leq p < \infty$ and that $f_1 \in L^p(\mu_1; E)$ and $f_2 \in L^p(\mu_2; E)$. Let $C(f_1, f_2)$ be the collection of all pairs $(\tilde{f}_1, \tilde{f}_2)$ in $L^p(\mu; E)$, for some Borel probability measure μ on some Polish space (X, τ), for which f_1 and $\tilde{f}_1$ have the same distribution and f_2 and $\tilde{f}_2$ have the same distribution. The *Mallows p distance* $d_p(f_1, f_2)$ between f_1 and f_2 is then defined as

$$d_p(f_1, f_2) = \inf\{\|\tilde{f}_1 - \tilde{f}_2\|_p : (\tilde{f}_1, \tilde{f}_2) \in C(f_1, f_2)\}.$$

This gives a measure of the difference of the distributions of f_1 and f_2.

We can express the Mallows p-distance in terms of the Wasserstein metric W_p.

Theorem 21.6.1 *Suppose that f_1 and f_2 are as above. Let $\nu_1 = f_1 \, d\mu_1$ and let $\nu_2 = f_2 \, d\mu_2$, and let c_p the cost function $c_p(x, y) = \|x - y\|^p$ on $E \times E$. Then $d_p(f_1, f_2) = W_p(\nu_1, \nu_2)$.*

Let $\tilde{f}_1(x, y) = x$ and let $\tilde{f}_2(x, y) = y$. If $\pi \in \Pi_{\nu_1,\nu_2}$, then $\tilde{f}_1 \in L^p(\pi)$, and $\tilde{f}_1$ has the same distribution ν_1 as f_1; similar properties are enjoyed by $\tilde{f}_2$. Further,

$$\int_{X \times X} \|\tilde{f}_1 - \tilde{f}_2\|^p \, d\pi = \int_{E \times E} c_p \, d\pi,$$

from which it follows that $d_p(f_1, f_2) \leq W_p(\nu_1, \nu_2)$.

On the other hand, suppose that $(\tilde{f}_1, \tilde{f}_2) \in C(f_1, f_2)$, where $\tilde{f}_1, \tilde{f}_2 \in L^p(X, \mathcal{B}, \mu)$. Let $\tilde{f} = (\tilde{f}_1, \tilde{f}_2)$, and let $\pi = \tilde{f}_*(\mu)$. Then $\pi \in \Pi_{\nu_1,\nu_2}$, and

$$\int_X \|\tilde{f}_1 - \tilde{f}_2\|^p \, d\mu = \int_{E\times E} c_p \, d\pi,$$

from which it follows that $W_p(\nu_1, \nu_2) \leq d_p(f_1, f_2)$.

Corollary 21.6.2 *There exist $(\tilde{f}_1, \tilde{f}_2) \in C(f_1, f_2)$ for which $d_p(f_1, f_2) = \|\tilde{f}_1 - \tilde{f}_2\|_p$.*

We have the following extension to the central limit theorem (Theorem 18.7.2).

Theorem 21.6.3 (The central limit theorem, II) *Suppose that μ is a probability measure on a Polish space (X, τ), that $f \in L^2(\mu)$, that $\int_X f \, d\mu = 0$ and that $\int_X f^2 \, d\mu = 1$. Let $c_n = (f_1 + \cdots + f_n)/\sqrt{n}$, where $f_1, \ldots, f_n$ are n independent copies of f. Then $d_2(c_n, \gamma) \to 0$ as $n \to \infty$, where γ is a Gaussian random variable γ with mean 0 and variance 1, and $(c_n)_{n\in\mathbf{N}}$ is 2-uniformly integrable.*

Proof Theorem 18.7.2 shows that $c_n \to \gamma$ in distribution as $n \to \infty$. Let ν_n be the distribution of c_n and let ν_γ be the distribution of γ. Since $\|c_n\|_2 = \|\gamma\|_2$, it follows from Theorem 21.4.1 that $d_2(c_n, \gamma) = W_2(\nu_n, \nu_\gamma) \to 0$ as $n \to \infty$ and that the sequence $(\nu_n)_{n\in\mathbf{N}}$ is 2-uniformly integrable – that is, the sequence $(c_n)_{n\in\mathbf{N}}$ is 2-uniformly integrable. □

22

Some Examples

22.1 Strictly Subadditive Metric Cost Functions

A metric d is *strictly subadditive* if $d(x,z) < d(x,y) + d(y,z)$ whenever y is not equal to x or z. For example, if b is a strictly concave non-negative real-valued function on $[0,\infty)$ for which $b(0) = 0$ (such as $b(t) = t^p$ for some $0 < p < 1$), and (X,d) is a metric space, then the function $b \circ d$ is a strictly subadditive metric uniformly equivalent to d. We show that if a cost function is a lower semi-continuous strictly subadditive metric on a Polish space (X,τ), and μ and ν are Borel probability measures on X, then an optimal measure π on $X \times X$ fixes mass common to μ and ν.

We begin with an easy lemma about measures on X.

Lemma 22.1.1 *Suppose that μ and ν are Borel probability measures on a Polish space (X,τ) and that $\pi \in \Pi_{\mu,\nu}$. Let $D = \{(x,x) : x \in X\}$ be the diagonal in $X \times X$, let $\pi_D = I_D\, d\pi$, so that if A is a Borel subset of $X \times X$, then $\pi_D(A) = \pi(A \cap D)$, and let $\beta = p_*(\pi_D)$, where $p : X \times X \to X$ is the projection onto the first co-ordinate. Then $\beta \leq \mu \wedge \nu$.*

Proof Since $p_{|D}$ is a homeomorphism of D onto X, if A is a Borel subset of X, then

$$\beta(A) = \pi_D(A \times X) \leq \pi(A \times X) = \mu(A).$$

Similarly, $\beta(A) \leq \nu(A)$, and so $\beta \leq \mu \wedge \nu$. □

Theorem 22.1.2 *Suppose that d is a strictly subadditive lower semi-continuous metric on a Polish space (X,τ), that μ,ν are Borel probability measures on X and that $\pi \in \Pi_{\mu,\nu}$ has d-monotone support. Let $\pi_D = I_D\, d\pi$ (so that $\pi_d(A) = \pi(A \cap D)$), and let $\beta = p_*(\pi_D)$, where $p : X \times X \to X$ is the projection onto the first co-ordinate. Then $\beta = \mu \wedge \nu$. Thus if $\mu_o = \mu - \beta$ and $\nu_o = \nu - \beta$ then μ_o and ν_o are mutually singular.*

Proof We use the notation of the lemma. If A is a Borel set in $X \times X$, let $\pi_o(A) = \pi(A \setminus D)$, so that $\pi = \pi_o + \pi_D$, and $\mu = \mu_0 + \beta$, and similarly $\nu = \nu_o + \beta$. Let $S = \text{supp}(\pi) \setminus D$, and let $U = p(S)$. Since S is a σ-compact subset of $X \times X$, U is a Borel subset of X, and $\mu_o(U) = \mu_o(X)$. Similarly if $V = q(S)$ (where q is the projection onto the second co-ordinate), then $V = q(S)$ is a Borel subset of X, and $\nu_o(V) = \nu_o(X)$. We show that $U \cap V = \emptyset$, which establishes the theorem. If not, there exists $z \in U \cap V$, and so there exist $x, y \in S$ such that $(x, z) \in S$ and $(z, y) \in S$. But then by d-monotonicity, $d(x, z) + d(z, y) \leq d(x, y) + d(z, z) = d(x, y)$. Since $x \neq z$ and $y \neq z$, this contradicts the strict subadditivity of d. □

Thus $\pi_o(U \times V) = \pi_0(X \times X)$, and π_o is an optimal measure for μ_o and ν_o. We only need solve the Kantorovich problem for mutually singular measures.

22.2 The Real Line

We now consider the case where $X = Y = \mathbf{R}$. First we consider the case where the cost function c is of the form $c(x, y) = b(x - y)$, where b is a proper strictly convex non-negative function on $\mathbf{R}$. Suppose that μ and ν are Borel probability measures on $\mathbf{R}$, with distribution functions F_μ and F_ν respectively. We suppose that $\inf\{\int_{\mathbf{R}^2} c\, d\pi : \pi \in \Pi_{\mu,\nu}\} < \infty$ so that there exists an optimal measure π.

In this case, we show that π is unique, and that if μ is atom-free then it is a deterministic transport plan.

Suppose that C is a c-cyclically monotone subset of $\mathbf{R}^2$. First we show that if (x_0, y_0) and (x_1, y_1) are in C then $(x_1 - x_0)(y_1 - y_0) \geq 0$. Suppose not. Then without loss of generality we can suppose that $x_0 < x_1 = x_0 + s$ and $y_0 > y_1 = y_0 - t$. Let $\alpha = x_0 - y_0$. Then

$$\begin{aligned} &[c(x_0, y_0) + c(x_1, y_1)] - [c(x_0, y_1) + c(x_1, y_0)] \\ &= [b(\alpha + s + t) - b(\alpha + s)] - [b(\alpha + t) - b(\alpha)] > 0, \end{aligned}$$

by strict convexity, giving a contradiction.

If $(x, y) \in \mathbf{R}^2$, let $NW(x, y) = \{(x', y') : x' < x, y' > y\}$ and let $SE(x, y) = \{(x', y') : x' > x, y' < y\}$. It then follows that if $(x, y) \in C$ then $NW(x, y) \cap C = SE(x, y) \cap C = \emptyset$; the open sets $NW(x, y)$ and $SE(x, y)$ are disjoint from C.

For each $x \in \text{supp}(\mu)$ let $y_x = \inf\{y : F_\nu(y) > F_\mu(x-)\}$ and $y'_x = \inf\{y : F_\nu(y) > F_\mu(x)\}$ and let $C = \{(x, y) : x \in \text{supp}(\mu), y'_x \leq y \leq y_x\}$. It then follows that if π is an optimal transport plan, $\text{supp}(\pi) \subseteq C$. If μ is atom-free, then C is the graph of an increasing function T, and T is the unique deterministic transport plan. If μ has an atom at x, again the transport plan is unique, but it is not deterministic; ν determines how the mass at x is distributed in the interval $[y'_x, y_x]$.

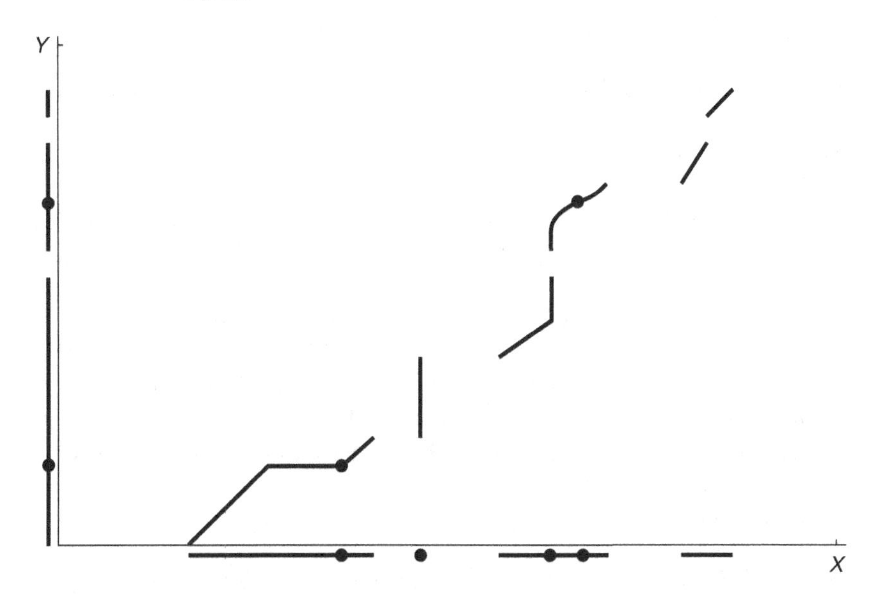

Note that $\text{supp}(\pi)$ depends on μ and ν, but not on the cost function. Suppose that b is a proper convex non-negative function on $\mathbf{R}$, and that $b(x) \to \infty$ as $|x| \to \infty$. Let c be the corresponding cost function. Let $h(t) = |t|+e^{-|t|}-1$. Let $b_n = b+h/n$ and let $c_n(x, y) = b_n(x-y)$. Then b_n is strictly convex, and so C is c_n-cyclically monotone. But $c_n \to c$ pointwise, and so C is c-cyclically monotone. Thus there exists an optimal transport plan π with $\text{supp}(\pi) \subseteq C$. Note however that when $c(x, y) = |x - y|$ then this solution is generally not unique.

Next, suppose that $d(x, y) = b(|x - y|)$, where b is a non-negative strictly concave function on $[0, \infty)$ for which $b(0) = 0$ (for example, $d(x, y) = |x-y|^p$, with $0 < p < 1$), so that d is a strictly subadditive metric on $\mathbf{R}$. Suppose also that μ and ν are probability measures on $\mathbf{R}$ for which $\sup(\text{supp}(\mu)) \leq \inf(\text{supp}(\nu))$. Then a similar argument shows that there is a unique transport plan π. But here there is order reversal; for example, if μ is atom-free, there is a unique optimal deterministic transport plan T which is a *decreasing* function.

22.3 The Quadratic Cost Function

We now consider the case where $X = Y = H$, a Hilbert space or Euclidean space, and where $c(x, y) = \frac{1}{2}n_2(x-y)$, where $n_2(x) = \|x\|^2$. We consider Borel probability measures μ and ν on H which satisfy $K_2(\mu) = \int_X n_2\,d\mu < \infty$ and $K_2(\nu) = \int_Y n_2\,d\nu < \infty$. Thus if $\pi \in \Pi_{\mu,\nu}$ then

$$\begin{aligned}\int_{X\times Y} c\,d\pi &= \tfrac{1}{2}\int_{X\times Y} n_2(x-y)\,d\pi(x,y)\\ &\leq \int_{X\times Y} n_2(x) + n_2(y)\,d\pi \leq K_2(\mu) + K_2(\nu) < \infty,\end{aligned}$$

so that an optimal transport plan π exists.

If $\pi \in \Pi_{\mu,\nu}$ then

$$\begin{aligned}\int_{X\times Y} c\,d\pi &= \int_{X\times Y} \tfrac{1}{2}n_2(x) - \langle x,y\rangle + \tfrac{1}{2}n_2(y)\,d\pi(x,y)\\ &= \tfrac{1}{2}(K_2(\mu)) + K_2(\nu) - \int_{X\times Y} \langle x,y\rangle\,d\pi.\end{aligned}$$

Let $\tilde{c}(x,y) = -\langle x,y\rangle$. Then π is an optimal transport plan for the cost c if and only if it is an optimal transport plan for the cost $\tilde{c}$. Further, $M_c = \frac{1}{2}(K_2(\mu) + K_2(\nu)) + M_{\tilde{c}}$, where $M_{\tilde{c}}$ is the optimal cost for the cost $\tilde{c}$. The cost $\tilde{c}$ is not non-negative, but we can apply Corollary 20.2.5.

If f is a function on X and $y \in Y$, then

$$f^{\tilde{c}}(y) = \inf_{x\in X}(-\langle x,y\rangle - f(x)) = -\sup_{x\in X}(\langle x,y\rangle + f(x)) = f^{\dagger}(y),$$

so that $f^{\tilde{c}}$ is an upper semi-continuous concave function on X, and $f^{\tilde{c}\tilde{c}}$ is the upper semi-continuous concave envelope of f. Applying this to the original problem, we obtain the following result.

Theorem 22.3.1 *Suppose that $X = Y = H$, a Hilbert space or Euclidean space, and that $c(x,y) = \frac{1}{2}n_2(x-y)$, where $n_2(x) = \|x\|^2$. Suppose that μ and ν are Borel probability measures on H which satisfy $K_2(\mu) = \int_X n_2\,d\mu < \infty$ and $K_2(\nu) = \int_Y n_2\,d\nu < \infty$. Then there exists an optimal transport plan π. Further, there exist upper semi-continuous concave functions f and g on H, with $g = f^{\dagger}$, such that $(\frac{1}{2}\|x\|^2 + f(x)) + \frac{1}{2}(\|y\|^2 + g(y)) \leq c(x,y)$, with equality on* $\operatorname{supp}(\pi)$, *so that $M_c = \frac{1}{2}(K_2(\mu) + K_2(\nu)) + \int_X f\,d\mu + \int_Y g\,d\nu$.*

Let us now consider the case where H is a d-dimensional Euclidean space, and μ is absolutely continuous with respect to Lebesgue measure λ_d. In particular, this implies that $\operatorname{supp}(\mu)$ is not contained in a $(d-1)$-dimensional affine subspace of E, and so $C = \overline{\Gamma}(\operatorname{supp}(\mu))$ has a non-empty interior. Since f and g are finite on $\operatorname{supp}(\mu)$, this implies that they are finite on C^{int}. Further, $\mu(\partial C) = 0$. Now it follows from Rademacher's theorem (Theorem 17.7.1) that f is differentiable μ-almost everywhere on C^{int}. Let $D = \{x \in \Phi_f :$ f is differentiable at $x\}$. If $x \in D \cap \operatorname{supp}(\mu)$, let $T(x) = \nabla f(x)$, so that

$\{T(x)\} = \partial f_x$. Consequently π is the push-forward measure $GT_*(\mu)$, and $\nu = T_*(\mu)$. Thus T solves the Monge problem.

Let us show that the solution is unique. Suppose that $\tilde{\pi}$ is an optimal measure, and that $\tilde{f}$ and $\tilde{g}$ are the corresponding concave functions. Let $\tilde{T}$ be the corresponding transport mapping. Then

$$\begin{aligned}\int_{X\times Y} (f+g)\,d\pi &= \int_{X\times Y} (\tilde{f}+\tilde{g})\,d\tilde{\pi}\\ &= \int_X \tilde{f}\,d\mu + \int_Y \tilde{g}\,d\nu = \int_{X\times Y} (\tilde{f}+\tilde{g})\,d\pi,\end{aligned}$$

so that

$$\int_{X\times Y} (\tilde{f}(x) + \tilde{g}(y) + \langle x, y\rangle)\,d\pi(x,y) = 0;$$

that is to say,

$$\int_X (\tilde{f}(x) + \tilde{g}(\nabla f(x)) + \langle x, \nabla f(x)\rangle)\,d\mu(x) = 0.$$

But the integrand is non-positive μ-almost everywhere, so that

$$\tilde{f}(x) + \tilde{g}(\nabla f(x)) + \langle x, \nabla f(x)\rangle = 0$$

μ-almost everywhere. Thus $\nabla f \in \partial\tilde{f}$ μ-almost everywhere, and so it follows from Rademacher's theorem that $\nabla f = \nabla\tilde{f}$ μ-almost everywhere. Consequently, $T = \tilde{T}$.

22.4 The Monge Problem on $\mathbf{R}^d$

Suppose that $c(x, y) = h(x - y)$ is a continuous cost function on $\mathbf{R}^d$ and that $\mu, \nu \in P(\mathbf{R}^d)$. Can we find conditions on h, μ and ν which ensure that there is a solution to the Monge problem? If so, is the solution unique?

We shall always suppose that there exists $\pi \in \Pi_{\mu,\nu}$ with $\int_{\mathbf{R}^d\times\mathbf{R}^d} c\,d\pi < \infty$, and will also suppose that $\mu = g.d\lambda_d$ is absolutely continuous with respect to Lebesgue measure λ_d.

The function h need be neither convex nor concave; we need to extend the notion of sub- and superdifferentiability to it. If $x, y \in\in \mathbf{R}^d$, we say that $(x, y) \in \partial^\vee h$, the subdifferential of h, if there exists a non-positive function s in a neighbourhood N of x such that if

$$r(v) = h(x+v) - h(x) - \langle v, y\rangle - s(x+v), \text{ for } x+v \in N,$$

then $|r(v)|/\|v\| \to 0$ as $\|v\| \to 0$. We set $\partial^\vee h(x) = \{y : (x,y) \in \partial^\vee h\}$. Similarly, we say that $(x,y) \in \partial^\wedge h$, the superdifferential of h, if there exists a non-negative function s in a neighbourhood N of x such that if

$$r(v) = h(x+v) - h(x) - \langle v, y\rangle - s(x+v), \text{ for } x+v \in N,$$

then $|r(v)|/\|v\| \to 0$ as $\|v\| \to 0$. We set $\partial^\wedge h(x) = \{y : (x,y) \in \partial^\wedge h\}$.

By Theorem 20.6.1, there exists a strongly c-monotone subset Γ of $\mathbf{R}^d \times \operatorname{supp}(\nu)$ and a corresponding maximal Kantorovich potential f on $\mathbf{R}^d$ such that if π is an optimal solution to the Kantorovich problem then $\operatorname{supp}(\pi) \subseteq \Gamma$; f is c-concave, and $\Gamma \subseteq \partial_c(f)$, where

$$\partial_c(h) = \{(x,y) : c(x,y) - f(x) = \inf_{x' \in X}(c(x',y) - f(x'))\}.$$

We shall see that properties of f and of h^* are used to ensure that the Monge problem has a solution. In particular, we shall need f to be Fréchet differentiable on a large subset of $\mathbf{R}^d$, so that we can apply the following theorem.

Theorem 22.4.1 *Suppose that c, h, Γ and f are as above, that $(x,y) \in \partial^c(f)$ and that f is Fréchet differentiable at x. If h is convex, then $\nabla f(x) \in \partial h(x-y)$. If h is concave in a neighbourhood of x, then h is differentiable at $x-y$ and $\nabla f(x) = \nabla h(x-y)$.*

Proof Suppose that h is convex. If $k \in \mathbf{R}^d$ then

$$f(x+k) - f(x) - \nabla f_x(k) = r(k), \text{ where } r(k) = o(\|k\|).$$

Since $(x,y) \in \partial^c(f)$,

$$\begin{aligned} h(x-y) - f(x) &= f^c(y) \\ &= \inf_{u \in \mathbf{R}^d}(h(u-y) - f(u)) \\ &\le h(x+k-y) - f(x+k) \\ &= h(x+k-y) - f(x) - \nabla f_x(k) - r(k) \end{aligned}$$

so that

$$h(x+k-y) - h(x-y) - \nabla f_x(k) \ge r(k).$$

Thus if h is convex then $\phi \in \partial h(x-y)$, and if h is concave in a neighbourhood of x, then h is differentiable at $x-y$ and $\nabla f(x) = \nabla h(x-y)$. □

Theorem 22.4.2 *Suppose that $c(x,y) = h(x-y)$ is a translation invariant continuous cost function on $\mathbf{R}^d$, that $\mu, \nu \in P(\mathbf{R}^d)$, that $\mu = g\,d\lambda$ is absolutely* *continuous with respect to Lebesgue measure λ_d and that there exists $\pi \in \Pi_{\mu,\nu}$ with $\int_{\mathbf{R}^d \times \mathbf{R}^d} c\,d\pi < \infty$. Let $\Gamma \subseteq \mathbf{R}^d \times \mathrm{supp}(\nu)$ and let f satisfy Theorem 20.6.1.*

Suppose that there exists a Borel measurable A in $\mathbf{R}^d$, with $\mu A) = 1$, such that if $x \in A$ then f is Fréchet differentiable at x and $\partial^c f(x) \neq \emptyset$. Suppose also that there exists a continuous mapping $J : \mathbf{R}^d \to \mathbf{R}^d$ such that if $(x, y) \in \partial h$ then $J(y) = x$. Then there exists a unique measurable $s : \mathbf{R}^d \to \mathbf{R}^d$ such that if π is any optimal measure in $\Pi_{\mu,\nu}$ then $\pi = G(s)_*(\mu)$. Hence $s_*(\mu) = \nu$ and

$$\int_X c(x, s(x))\,d\mu(x) = \inf\left\{\int_{\mathbf{R}^d \times \mathbf{R}^d} c\,d\pi : \pi \in \Pi_{\mu,\nu}\right\}.$$

Then s is the unique solution to the Monge problem.

Proof Since

$$\frac{\partial f}{\partial x_i}(x) = \lim_{n\to\infty} n(f(x + e_i/n) - f(x)),$$

∇f is a Borel measurable mapping of A into $\mathbf{R}^d$.

Now let

$$s(x) = \begin{cases} x - J(\nabla f(x)) & \text{for } x \in A \\ 0 & \text{otherwise.} \end{cases}$$

Since J is continuous, s is Borel measurable. By Theorem 22.4.1, if $(x, y) \in \partial^c f(x)$ then $\nabla f(x) \in \nabla h(x - y)$; that is, $(x - y, \nabla f(x)) \in \partial h$. Hence $J(\nabla f(x)) = x - y$ and $y = s(x)$.

We show that if π is an optimal measure then $G(s)_*(\mu) = \pi$ (so that $s_*(\mu) = \nu$). We must show that if B and C are Borel subsets of $\mathbf{R}^d$, then $\pi(B \times C) = \mu(B \times s^{-1}(C))$. Let $S = \{(x, y) : x \in A, (x, y) \in \partial^c f\}$; if $(x, y) \in S$ then $y = s(x)$, and so $(B \times C) \cap S = ((B \cap s^{-1}(C)) \times \mathbf{R}^d) \cap S$. Then $\pi(S) = 1$, so that

$$\begin{aligned}\pi(B \times C) &= \pi((B \times C) \cap S) \\ &= \pi((B \cap s^{-1}(C)) \times \mathbf{R}^d) \\ &= \mu(B \cap s^{-1}(C)) = G(s)_*(B \times C).\end{aligned}$$

It remains to show that s is unique. If s' is another optimal transport mapping, then $\pi' = G(s')_*(\mu)$ is an optimal transport plan, and so $\pi' = G(s)_*(\mu)$. But then $s = s'$, by Proposition 20.1.1. □

22.5 Strictly Convex Translation Invariant Costs on $\mathbf{R}^d$

We consider a translation invariant cost $c(x, y) = h(x - y)$ on $\mathbf{R}^d$ which satisfies three conditions.

(H1) h is a strictly convex real-valued non-negative function on $\mathbf{R}^d$, and $h(0) = 0$.

Then h is a locally Lipschitz function on $\mathbf{R}^d$. If $x \in \mathbf{R}^d \setminus \{0\}$, we set $H(x) = \{y \in \mathbf{R}^d : h(y) \leq h(x)\}$. Then $H(x)$ is a closed strictly convex body in $\mathbf{R}^d$, and $H(x)^{int} = \{y \in \mathbf{R}^d : h(y) < h(x)\}$.

We need to put a rotundity condition on h. An element ϕ of $\mathbf{R}^d$ is *tangent* to $H(x)$ at x if $\|\phi\| = 1$ and $\langle x, \phi \rangle = \sup\{\langle y, \phi \rangle : y \in H(x)\}$. If so, then, since h is strictly convex, $\langle y, \phi \rangle < \langle x, \phi \rangle$ for $y \in H(x) \setminus \{x\}$. If ϕ is tangent to $H(x)$ at x, let $n(x)$ be the outward normal unit vector to $H(x)$ at x.

(H2) Suppose that $r > 0$ and $0 < \theta < \pi$. Then there exists $R > 0$ such that if $\|x\| > R$ then the cone

$$K = \{y : \|y - x\| \cos \theta/2 \leq \langle x - y, n(x) \rangle \leq r\}$$

with vertex at x and height r is contained in $H(x)$.

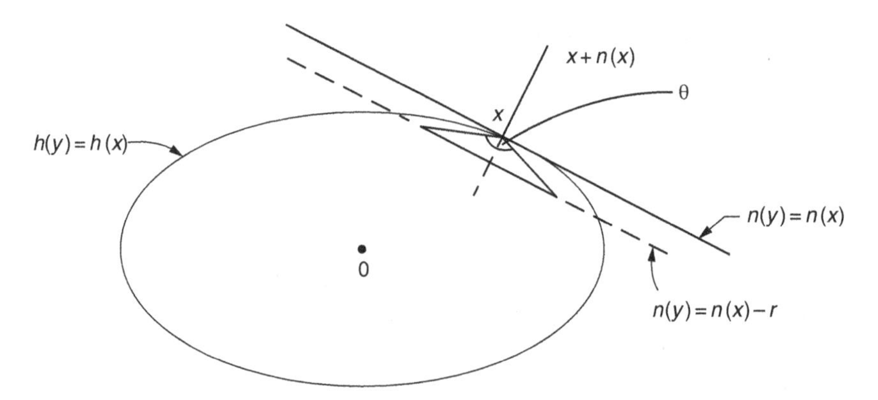

One specially important case occurs when $h(x) = \psi(\|x\|)$, where ψ is a strictly convex non-negative function on $[0, \infty)$ for which $\psi(0) = 0$. (For example, $\psi(t) = t^p$ for some $p > 1$, or $\psi(t) = t \log(t + 1)$.) In this case, $H(x)$ is the ball $\{y \in \mathbf{R}^d : \|y\| \leq \|x\|\}$, and so **H(2)** is satisfied.

We also need h to grow more than linearly.

(H3) $h(x)/\|x\| \to \infty$ as $\|x\| \to \infty$.

Theorem 22.5.1 *Suppose that $c(x, y) = h(x - y)$, where h satisfies* **(H1)–(H3)**, *and that f is a proper c-concave function on $\mathbf{R}^d$. Let $K = \overline{\Gamma}(\Phi_f)$. Then f is locally bounded on K^{int} (so that $K^{int} \subseteq \Phi_f \subseteq K$).*

Proof Suppose that f is not locally bounded at x. We show that there exists a unit vector ϕ such that if $\langle y, \phi \rangle < \langle x, \phi \rangle$ then $f(y) = -\infty$. Thus $\Phi_f \subseteq \{y : \langle y, \phi \rangle \geq \langle x, \phi \rangle\}$, and so $K \subseteq \{y : \langle y, \phi \rangle \geq \langle x, \phi \rangle\}$ and $K^{int} \subseteq \{y : \langle y, \phi \rangle > \langle x, \phi \rangle\}$. Hence $x \notin K^{int}$.

If $f(x) > -\infty$ then, since f is upper semi-continuous, it is bounded above in a neighbourhood of x, and so it cannot be bounded below in any neighbourhood of x. There therefore exists a sequence $(x_n)_{n=1}^\infty$ such that $x_n \to x$ as $n \to \infty$ and $f(x_n) < -n$ for each $n \in \mathbf{N}$. If $f(x) = -\infty$, set $x_n = x$ for all $n \in \mathbf{N}$.

There exists a sequence $(y_n)_{n=1}^\infty$ such that $c(x_n, y_n) - f^c(y_n) < -n$ for $n \in \mathbf{N}$. Then

$$\begin{aligned} f(x) &\leq c(x, y_n) - f^c(y_n) \\ &= (c(x, y_n) - c(x_n, y_n)) + (c(x_n, y_n) - f^c(y_n)) \\ &\leq c(x, y_n) - c(x_n, y_n) - n. \end{aligned}$$

Since c is continuous, it follows that $\|y_n\| \to \infty$ as $n \to \infty$. Let $z_n = x_n - y_n$. Then $\|z_n\| \to \infty$ as $n \to \infty$. For each n there exists ϕ_n, tangent to $H(z_n)$, such that the cone

$$\{y \in \mathbf{R}^d : \phi_n(y) \geq \tfrac{1}{2}\phi_n(z_n), \phi_n(z_n - y) \geq \beta(h(z_n))\}$$

is contained in $H(z_n)$. Extracting a subsequence if necessary, we can suppose that $\phi_n \to \phi$, say, as $n \to \infty$.

Suppose that $\langle w, \phi \rangle < \langle x, \phi \rangle$. Since $x_n \to x$ and $\phi_n \to \phi$ as $n \to \infty$, $\langle x_n, \phi_n \rangle \to \langle x, y \rangle$ and $\langle w, \phi_n \rangle \to \langle w, \phi \rangle$ as $n \to \infty$. Thus $\langle w, \phi_n \rangle < \langle x_n, \phi_n \rangle$ for sufficiently large n. For such n,

$$\langle w - y_n, \phi_n \rangle < \langle x_n - y_n, \phi_n \rangle = \langle z_n, \phi_n \rangle .$$

Now a straightforward geometric argument shows that $w - y_n \in H(z_n)$ for all sufficiently large n, as well. For such n,

$$c(w, y_n) = h(w - y_n) < h(z_n) = c(x_n, y_n),$$

so that

$$f(w) \leq c(w, y_n) - f^c(y_n) < c(x_n, y_n) - f^c(y_n) < -n.$$

Consequently, $f(w) = -\infty$. □

Theorem 22.5.2 *Suppose that $c(x, y) = h(x - y)$, where h satisfies* **(H1)–(H3)**, *and that f is a c-concave function on $\mathbf{R}^d$ which is locally bounded on an open subset U of $\mathbf{R}^d$. Then $\partial^c f(x)$ is non-empty for all $x \in U$, and $\partial^c f(U)$ is locally bounded in $\mathbf{R}^d \times \mathbf{R}^d$.*

Proof Suppose that $x \in U$ and that $V = N_\delta(x)$ is an open neighbourhood of x in U for which $R = \sup_{y \in V} |f(y)| < \infty$. We need a preliminary lemma.

Lemma 22.5.3 *Suppose that $(x_n)_{n=1}^\infty$ is a sequence in $N_{\delta/2}(x)$ and that $(y_n, \theta_n)_{n=1}^\infty$ is a sequence in $\mathbf{R}^d \times \mathbf{R}$ which satisfies*

(i) $f(z) \leq c(z, y_n) - \theta_n$ *for* $z \in \mathbf{R}^d$, *and*
(ii) $c(x_n, y_n) - \theta_n < R$

for all $n \in \mathbf{N}$. *Then* $(y_n)_{n=1}^\infty$ *is a bounded sequence.*

Proof Suppose not. Since $(x_n)_{n=1}^\infty$ is a bounded sequence, by extracting a subsequence if necessary we can assume that if $v_n = x_n - y_n$ then $\|v_n\| > n$, for $n \in \mathbf{N}$. Let $\alpha_n = \delta/2\,\|v_n\|$, so that $\alpha_n \to 0$ as $n \to \infty$. Let $w_n = x_n - \alpha_n v_n$, so that $w_n \in V$and $w_n - y_n = (1 - \alpha_n)v_n$. Then

$$\begin{aligned} -R \leq f(w_n) &\leq c(w_n, y_n) - \theta_n \\ &= h(w_n - y_n) - \theta_n = h((1 - \alpha_n)v_n) - \theta_n \\ \text{while } h(v_n) = h(x_n - y_n) &= c(x_n, y_n) \leq R + \theta_n. \end{aligned}$$

Consequently, $h(v_n) - h((1 - \alpha_n)v_n) \leq 2R$.

We now use the convexity of h. Suppose that $\phi_n \in \partial h_{(1-\alpha_n)v_n}$. Then

$$\begin{aligned} h(v_n) - h((1 - \alpha_n)v_n) &\geq \langle \alpha_n v_n, \phi_n \rangle = \frac{\delta}{2} \left\langle \frac{v_n}{\|v_n\|}, \phi_n \right\rangle \\ \text{and } h((1 - \alpha_n)v_n) &= h((1 - \alpha_n)v_n) - h(0) \leq \langle (1 - \alpha_n)v_n, \phi_n \rangle, \end{aligned}$$

so that

$$\frac{h((1 - \alpha_n)v_n)}{(1 - \alpha_n)\,\|v_n\|} \leq \left\langle \frac{v_n}{\|v_n\|}, \phi_n \right\rangle \leq \frac{4R}{\delta}.$$

But this contradicts **(H3)**, since $\|(1 - \alpha_n)v_n\| \to \infty$ as $n \to \infty$. □

We now prove the theorem. First, let $x_n = x$ for all $n \in \mathbf{N}$. Since $f(x) < n$, there exists a sequence $(y_n)_{n=1}^\infty \in \mathbf{R}^d$ such that $c(x_n, y_n) - f^c(y_n) \to f(x)$ as $n \to \infty$ and $c(x, y_n) - f^c(y_n) < R$ for all n. Since f is c-concave, $f(z) \leq c(z, y_n) - f^c(y_n)$, and so the conditions of the lemma are satisfied. Thus the sequence $(y_n)_{n=1}^\infty$ is bounded, and by extracting a subsequence if necessary, we can suppose that $y_n \to y$ as $n \to \infty$, for some $y \in \mathbf{R}^d$. Then $c(x, y_n) \to c(x, y)$ as $n \to \infty$ and so $f^c(y_n) \to c(x, y) - f(x)$ as $n \to \infty$. But f^c is upper semi-continuous, and so $f^c(y) \geq c(x, y) - f(x)$. Thus $y \in \partial^c f(x)$, and so $\partial^c f(x)$ is non-empty.

Suppose now that (x_n, y_n) is a sequence in $\partial^c(f(V))$. Setting $\theta_n = f^c(y_n)$, then, since $f(x_n) = c(x_n, y_n) - \theta_n < R$, the conditions of the lemma are satisfied. Thus $(y_n)_{n=1}^\infty$ is a bounded sequence, and so $\partial^c(f(V))$ is bounded. □

Corollary 22.5.4 *f is locally Lipschitz on U.*

Proof If $x \in U$, there exists a neighbourhood V of X in U on which $\partial^c f(V)$ is bounded, and so there exists $S > 0$ for which $\|y\| \leq S$ if $(x, y) \in \partial^c f(U)$. Thus if $x \in V$ and $y \in \partial^c f(x)$ then

$$f(x) = c(x,y) - f^c(y) = \inf\{c(x,z) - f^c(z) : \|z\| \leq S\}.$$

Now there exists $L > 0$ such that if $\|z\| \leq S$ then the function $x \to c(x,z)$ is L-Lipschitz on V. It therefore follows from Proposition 2.8.2 that f is L-Lipschitz on V. □

Theorem 22.5.5 *Suppose that $c(x,y) = h(x-y)$, where h satisfies* **(H1)–(H3)**, *and that $f : \mathbf{R}^d \to [-\infty, \infty)$ is a proper function. If $(x,y) \in \partial^c(f)$, then $\partial f(x) \subseteq \partial h(x-y)$.*

Proof If $f(x) = -\infty$ then $\partial f(x) = \emptyset$, and so the result is trivially true. Otherwise, suppose that $\phi \in \partial f(x)$, so that if $k \in \mathbf{R}^d$ then

$$f(x+k) - f(x) - \langle k, \phi \rangle = s(k) + r(k),$$

where $s(k) \geq 0$ and $r(k) = o(\|k\|)$. Since $(x,y) \in \partial^c(f)$,

$$\begin{aligned} h(x-y) - f(x) &= f^c(y) \\ &= \inf_{U \in \mathbf{R}^d} (h(u-y) - f(u)) \\ &\leq h(x+k-y) - f(x+k) \\ &= h(x+k-y) - f(x) - \langle k, \phi \rangle - s(k) - r(k) \end{aligned}$$

so that

$$h(x+k-y) - h(x-y) - \langle k, \phi \rangle \geq s(k) + r(k).$$

Thus $\phi \in \partial h(x-y)$. □

We now show that when the cost function c on $\mathbf{R}^d$ satisfies the preceding conditions, and when $\mu = g.d\lambda \in P(\mathbf{R}^d)$ is absolutely continuous with respect to Lebesgue measure, then there exists a unique solution to the Monge problem.

Theorem 22.5.6 *Suppose that $c(x,y) = h(x-y)$ is a cost function on $\mathbf{R}^d$, where h satisfies* **(H1)–(H3)**, *that $\mu, \nu \in P(\mathbf{R}^d)$ and that $\mu = g.d\lambda \in P(\mathbf{R}^d)$ is absolutely continuous with respect to Lebesgue measure. Suppose also that there exists $\pi \in \Pi_{\mu,\nu}$ with $\int_{\mathbf{R}^d \times \mathbf{R}^d} c\, d\pi < \infty$. Then there exists a unique measurable $s : \mathbf{R}^d \to \mathbf{R}^d$ such that $s_*(\mu) = \nu$ and such that $\int_{\mathbf{R}^d} c(x, s(x))\, d\mu(x) = \inf_{\pi \in \Pi_{\mu,\nu}} \int_{\mathbf{R}^d \times \mathbf{R}^d} c\, d\pi$.*

Proof By Theorem 20.6.1, there exists a strongly c-monotone subset Γ in $\mathbf{R}^d \times \mathrm{supp}(\nu)$ such that if π is an optimal measure, then $\mathrm{supp}(\pi) \subseteq \Gamma$, and there exists a maximal Kantorovich potential f; f is c-concave, and $\Gamma \subseteq \partial^c(f)$. Let $K = \overline{\Gamma}(\Phi_f)$. Then $K \supseteq p_1(\Gamma)$, where p_1 is the projection of $\mathbf{R}^d \times \mathrm{supp}(\nu)$ onto the first component, and so $\mu(K) = 1$. Consequently $\lambda(K) > 0$, so that

$\lambda(K^{int}) > 0$ and $\lambda(\partial K) = 0$. By Corollary 22.5.4, f is locally Lipschitz on K^{int}, and so, by Rademacher's theorem (Theorem 17.7.1), f is differentiable on a measurable subset C of K^{int} of full λ measure: $\lambda(A) = \lambda(K^{int})$. Thus $\mu(C) = 1$.

Suppose that π is an optimal measure, and let $B = p_1(\mathrm{supp}(\pi))$. Then B is a σ-compact subset of $\mathbf{R}^d$, and so it is Borel measurable, and $\mu(B) = 1$. If $x \in B$, then $\partial^c f(x) \neq \emptyset$. Let $A = B \cap C$. Then A satisfies the conditions of Theorem 22.4.2. Further, h^* is continuously differentiable on $\mathbf{R}^d$. If $(x, y) \in \partial h$ then $(y, x) \in \partial h^*$. Thus if we set $J = \nabla h^*$ the conditions of Theorem 22.4.2 are satisfied and the result follows. □

22.6 Some Strictly Concave Translation–Invariant Costs on $\mathbf{R}^d$

We now consider a translation invariant cost function c on $\mathbf{R}^d$ of the form $c(x, y) = h(x - y) = l(\|x - y\|)$, where l is a function on $\mathbf{R}$ which satisfies **(L)** $l(t) = -\infty$ for $t < 0$, l is continuous on $[0, \infty)$, $l(0) = 0$, $l(t)$ is positive and strictly concave on $(0, \infty)$, $l(t)/t \to \infty$ as $t \searrow 0$, and $l(t)/t \to 0$ as $t \to \infty$.

An important example is the case where $l(t) = t^p$ for $t \geq 0$, where $0 < p < 1$.

The function l is differentiable at all but countably many points of $(0, \infty)$, and $\partial^{\wedge} l(t) \neq \emptyset$ for all $t \in (0, \infty)$. What about h?

Theorem 22.6.1 *Suppose that $h(x) = l(\|x\|)$, where l satisfies* **(L)***, is a function on $\mathbf{R}^d$. Then $(x, y) \in \partial^{\wedge} h(x)$ if and only if $x \neq 0$ and $y = sx/\|x\|$, where $s \in \partial^{\wedge} l(\|x\|)$.*

Proof Clearly $\partial^{\wedge} h(0) = \emptyset$. Suppose that $x \neq 0$ and that $(\|x\|, s) \in \partial^{\wedge} l$. Since l is concave, $s > 0$ and $l(\|x\| + t) \leq l(\|x\| + st)$, for $t \in \mathbf{R}$. Suppose that $\|v\| < \|x\|$ and that $0 < t < 1$. Then

$$\|x + tv\|^2 = \|x\|^2 + 2\langle x, v\rangle t + \|v\|^2 t^2,$$

so that since the function $s \to s^{\frac{1}{2}}$ is concave on $(0, \infty)$,

$$\|x + tv\| \leq \|x\| + \frac{\langle x, v\rangle}{\|x\|} t + \frac{\|v\|^2}{2\|x\|} t^2.$$

It therefore follows, since l is concave, that $l(\|x + v\|) \leq s\langle x, v\rangle / \|x\|$, so that h is superdifferentiable at x and $(x, sx/\|x\|) \in \partial^{\wedge} h$.

Conversely, suppose that $(x,y) \in \partial^\wedge h$, so that $h(x+v) \leq h(x) + \langle v, y \rangle$, for small v. Let $y = \alpha x + v$, where $\langle x, v \rangle = 0$. Then if t is small and positive, $\|x\| \leq \|x - tv\|$, so that $h(x) \leq h(x - tv) \leq h(x) - t\|v\|^2 + o(t)$, and so $v = 0$. Thus $y = \alpha x$. Since $h(x + tx) \leq h(x) + \alpha t \|x\|^2$, it follows that $y = sx/\|x\|$, with $s \in \partial^\wedge l(x)$. □

Corollary 22.6.2 *h is differentiable λ-almost everywhere. $p_2(\partial^\wedge h) = \mathbf{R}^d \setminus \{0\}$ (where $p_2(x,y) = y$), and there exists a continuous mapping $J : \mathbf{R}^d \setminus \{0\} \to \mathbf{R}^d \setminus \{0\}$ such that if $(x,y) \in \partial^\wedge h$ then $J(y) = x$.*

Proof h is differentiable except at 0 and on a countable union of spheres. The other results follow from the form of $\partial^\wedge h$, and the facts that $p_2(\partial^\wedge l) = (0,\infty)$ and that there is a continuous mapping $j : (0,\infty) \to (0,\infty)$ such that if $(t,u) \in \partial^\wedge l$ then $j(u) = t$. □

Next, let us consider the concave Legendre transform $l^\dagger$ of l.

Proposition 22.6.3 *Suppose that l is a non-negative strictly convex function on $[0,\infty)$ which satisfies* (**L**)*. Then $l^\dagger(t) = \infty$ for $t < 0$, and $l^\dagger$ is a differentiable negative strictly increasing function on $(0,\infty)$, for which $l^\dagger(t) \to -\infty$ as $t \searrow 0$ and $l^\dagger(t) \to 0$ as $t \to \infty$.*

Proof Certainly $l^\dagger(t) = \infty$ for $t < 0$. Since l is strictly concave, if $t > 0$ there is a unique $x_t > 0$ at which $tx - l(x)$ attains its infimum, and then $f^\dagger(t) = tx_t - f(x_t) < 0$. If $0 < t_1 < t_2$ then $x_{t_1} > x_{t_2}$ and $f^\dagger(t_1) < f^\dagger(t_2)$. Further, $x_t \to \infty$ as $t \searrow 0$, and

$$f^\dagger(t) = x_t\left(t - \frac{f(x_t)}{x_t}\right) \to -\infty \text{ as } t \searrow 0.$$

Similarly $x_t \to 0$ as $t \to \infty$, so that $f(x_t) \to 0$ as $t \to \infty$. Since $-f(x_t) \leq f^\dagger(x_t) < 0, f^\dagger(t) \to 0$ as $t \to \infty$. Finally, since x_t is unique, $\partial^\wedge l^\dagger(t) = \{x_t\}$, so that $l^\dagger$ is differentiable on $(0,\infty)$, with derivative x_t. □

We now consider the Monge problem. Since c is a strictly subadditive metric on $\mathbf{R}^d$, it follows from the remarks in Section 22.1 that it is enough to consider the case where μ and ν are mutually singular. In fact we shall require a little more; we suppose that $\mu(\text{supp}(\nu)) = 0$.

We need a definition. Suppose that f is a function from an open subset U of $\mathbf{R}^d$ into $[-\infty,\infty)$, and that $x \in U$. Then f is *locally semi-concave* at x if there exist $\delta > 0$ and $\alpha > 0$ such that $N_\delta(x) \subseteq U$ and the function $y \to f(y) - \alpha\|y\|^2$ is concave on $N_\delta(x)$.

Theorem 22.6.4 *Suppose that $c(x,y) = h(x-y) = l(\|x-y\|)$ is a cost function on $\mathbf{R}^d \times V$, where l satisfies* (**L**)*, and where V is a closed subset* *of* $\mathbf{R}^d$*. If* f *is a c-concave function on* $\mathbf{R}^d$*, then* f *is locally semi-concave on* $U = \mathbf{R}^d \setminus V$*.*

Proof Suppose that $x \in V$ and that $N_{2\delta}(x) \subseteq U$. Choose $0 < s < \delta$ such that l is differentiable at s. Let $\alpha = l'(s)/2s$ and let $l_s(t) = l(t) - \alpha t^2$ for $t \geq s$. Then l_s is a strictly concave function on $(0, \infty)$, and $l_s'(s) = 0$, so that l_s is strictly decreasing on $[s, \infty)$. Set $l_s(t) = l_s(s)$ for $0 \leq t < s$, and let $h_s(x) = l_s(\|x\|)$ for $x \in \mathbf{R}^d$. If $x, y \in \mathbf{R}^d$ and $0 < \theta < 1$ then

$$\|(1-\theta)x + \theta y\| \leq (1-\theta)\|x\| + \theta\|y\|,$$

so that

$$\begin{aligned} h_s((1-\theta)x + \theta y) &\geq l_s((1-\theta)\|x\| + \theta\|y\|) \\ &\geq (1-\theta)h_s(x) + \theta h_s(y). \end{aligned}$$

Thus h_s is a concave function on $\mathbf{R}^d$.

Suppose now that $z \in N_\delta(x)$, so that $d(z, V) > s$. Then

$$\begin{aligned} f(z) - \alpha\|z\|^2 &= \inf_{y \in V}(h(z-y) - f^c(y)) - \alpha\|z\|^2 \\ &= \inf_{y \in V}(h_\epsilon(z-y) + \alpha\|z-y\|^2 - f^c(y)) - \alpha\|z\|^2 \\ &= \inf_{y \in V}(h_\epsilon(z-y) - 2\alpha\langle z, y\rangle) + \alpha\|y\|^2 - f^c(y). \end{aligned}$$

But $(h_\epsilon(z-y) - 2\alpha\langle z, y\rangle) + \alpha\|y\|^2 - f^c(y)$ is a concave function on $N_\delta(x)$, for each $y \in V$, and so $f(z) - \alpha\|z\|^2$ is concave on $N_\delta(x)$. □

Corollary 22.6.5 *f is differentiable λ-almost everywhere on U.*

Theorem 22.6.6 *Suppose that* $c(x,y) = h(x-y) = l(\|x-y\|)$ *is a cost function on* $\mathbf{R}^d$*, where* l *satisfies* **(L)***, that* $\mu, \nu \in P(\mathbf{R}^d)$*, that* $\mu = g.d\lambda \in P(\mathbf{R}^d)$ *is absolutely continuous with respect to Lebesgue measure and that* $\mu(\operatorname{supp}(\nu)) = 0$*. Suppose also that there exists* $\pi \in \Pi_{\mu,\nu}$ *with* $\int_{\mathbf{R}^d \times \mathbf{R}^d} c\, d\pi < \infty$*. Then there exists a unique measurable* $s : \mathbf{R}^d \to \mathbf{R}^d$ *such that* $s_*(\mu) = \nu$ *and such that* $\int_{\mathbf{R}^d} c(x, s(x)\, d\mu(x) s(x)) = \inf_{\pi \in \Pi_{\mu,\nu}} \int_{\mathbf{R}^d \times \mathbf{R}^d} c\, d\pi$*.*

Proof Let $V = \operatorname{supp}(\nu)$ and let $U = \mathbf{R}^d \setminus V$. Then there exists a subset A of U such that f is differentiable on A and $\lambda(U \setminus A) = 0$. Then $\mu(A) = 1$. Let J be the function of Corollary 22.6.2. Then the conditions of Theorem 22.4.2 are satisfied, and so the result follows. □

Further Reading

Topological and Metric Spaces

All the results of Chapter 1 are proved in [G II]. An entertaining account of the anomalies of topology can be found in [SS]. The Brézis–Browder lemma was proved in [BB], and Ekeland's variational principle and its uses is discussed in [E] and [P I].

Banach Spaces and Hilbert Space

[Bo] contains an excellent account of basic linear analysis. The two volumes of [LT] contain more advanced material concerning the classical Banach spaces, and [W] considers applications of Banach space theory to other areas of analysis. [Y] gives an elementary account of Hilbert space theory.

Uniform Spaces

Many algebraic objects, such as topological groups, have natural uniform structures. We follow the notation of [J].

Càdlàg Functions

The Skorohod topology is discussed in [Bi I].

Convexity

Although it concentrates on finite-dimensional convex sets, [R] is the standard reference for convexity. Other details can be found in [P I], [Ss] and [S].

Measure Theory

Proofs of the results of Chapters 15 and 17 can be found in [G III]. The strong law of large numbers can be found in [B II]. [H] is a standard work on measure theory. Borel measures and the convergence of measures are dealt with in [D] and [Bi II]. There are many excellent books on Fourier transforms; I like [Duo].

Haar Measure

The account given here is derived from the report [Pe] by Pederson.

Choquet Theory

[P II] is a very good source for this, and so is [S].

Optimal Transportation

The two fundamental references are the tomes [V I] and [V II]. Further information about the strictly convex and strictly concave costs considered in Sections 22.5 and 22.6 is given in [GMcC].

References

[Bi I] Patrick Billingsley, *Convergence of Probability Measures*, John Wiley, 1968.

[Bi II] Patrick Billingsley, *Probability and Measure*, John Wiley, 1979.

[Bo] Béla Bollobás, *Linear Analysis*, Cambridge Mathematical Textbooks, 1990.

[BB] H. Brézis and F.E. Browder, A General Principle on Ordered Sets in Nonlinear Functional Analysis, *Advances in Mathematics* **21** (1976), 355–364.

[D] R.M. Dudley, *Real Analysis and Probability*, Cambridge University Press, 2005.

[Duo] Javier Duoandikoetxea, *Fourier Analysis*, AMS Graduate Studies in Mathematics **29**, 2001.

[E] Ivar Ekeland, Nonconvex minimization problems, *Bulletin of the American Mathematical Society* (New Series) (1979), 443–474.

[GMcC] Wilfrid Gangbo and Robert J. McCann, The Geometry of Optimal Transportation, *Acta Mathematica* **177** (1966), 113–161.

[G II] D.J.H. Garling, *A Course in Mathematical Analysis, Volume II*, Cambridge University Press, 2013.

[G III] D.J.H. Garling, *A Course in Mathematical Analysis, Volume III*, Cambridge University Press, 2014.

[H] Paul R. Halmos, *Measure Theory*, Van Nostrand Reinhold, 1969.

[J] I.M. James, *Introduction to Uniform Spaces*, L.M.S. Lecture Note Series **144** 1990.

[LT] Joram Lindenstrauss and Lior Tzafriri, *Classical Banach Spaces, Volumes I and II*, Springer-Verlag, 1977 and 1979.

[Pe] Gert K. Pedersen, *The Existence and Uniqueness of the Haar Integral on a Locally Compact Topological Group*, Report, Preprint, University of Copenhagen, 2000.

[P I] R.R. Phelps, *Convex Functions, Monotone Operators and Differentiability*, Springer Lecture Notes in Mathematics **1364**, 1993.

[P II] R.R. Phelps, *Lecture Notes on Choquet's Theorem*, Springer Lecture Notes in Mathematics **1757**, 2008.

[R] R. Tyrrell Rockafellar, *Convex Analysis*, Princeton University Press, 1972.

[S] Barry Simon, *Convexity: An Analytic Viewpoint*, Cambridge Tracts in Mathematics **187**, 2011.

[Ss] Stephen Simons, *From Hahn–Banach to Monotonicity,* Springer Lecture Notes in Mathematics **1693**, 2008.

[SS] Lynn Arthur Steen and J. Arthur Seebach, Jr., *Counterexamples in Topology*, Dover Publications Inc., 1995.

[V I] Cédric Villani, *Topics in Optimal Transportation*, American Mathematical Society, 2003.

[V II] Cédric Villani, *Optimal Transport, Old and New*, Springer-Verlag, 2009.

[W] P. Wojtaszczyk, *Banach Spaces for Analysts*, Cambridge Studies in Advanced Mathematics, 1991.

[Y] N.J. Young, *An Introduction to Hilbert Space*, Cambridge University Press, 1988.

Index

$(L^p, \|.\|_p)$, 207
G_δ set, 22, 38, 210
$L^1(X, \Sigma, \mu)$, 198
$M(X, \Sigma)$, 196
N-function, 203
  complementary, 203
TV, 257
$T_1 - T_4$ spaces, 12
W_1 compact, 318
W_p compact, 320
W_p-complete, 322
Δ_2 condition, 205
α-measurable, 221
σ-additive, 182
σ-compact, 17, 34, 48
σ-field, 179
  generated by $\mathcal{F}$, 180
σ-ring, 179
c-concave, 318
p-adic metric, 21
β metric, 266
ϵ-net, 39
  minimal, 40
ϵ-subdifferential, 149
σ-additive, 212

abelian, 62
absolutely continuous, 201, 206, 251
absolutely convex, 84, 88
absorbent, 116
accumulation point, 10, 11
action, 66
  continuous, 66, 237
  left, 66
  right, 66
  transitive, 66, 237
additive, 212
adjoint, 108, 109
affine, 85, 128
affine homeomorphism, 155
Alexandroff's theorem, 32
almost everywhere, 184
almost surely, 184
almost uniformly, 184
analytic set, 180
annihilator, 100
antilinear, 100, 108
Archimedean, 191
Arzelà–Ascoli theorem, 46, 63, 267
atom, 243

Baire σ-field, 210
Baire space, 33
Baire's category theorem, 33, 34, 39, 80, 91, 132, 142, 160
Banach limits, 114
Banach sequence space, 94
Banach space, 79, 128
Banach's theorem, 124, 141, 156, 259, 268
Banach–Alaoglu theorem, 125
barycentre, 167
barycentric co-ordinates, 167
base
  for a uniformity, 56
Bernoulli sequence space, 23, 38, 43, 44, 62, 218

Bessel's inequality, 106
bidual, 115
bilinear, 91, 118
bilinearity, 97
bimonotone basis, 94
bipolar, 120
Birkhoff's theorem, 160
Bishop–Phelps theorem, 149
Borel σ-field, 180
Borel measurable, 180, 293
Borel measure, 210
boundary, 10, 150, 167, 285, 287
bounded, 19, 80
  $\sigma(E, F)$, 122
bounded convergence, 232, 255
bounded Lipschitz, 82
bounded variation, 245
Brézis–Browder Lemma, 53

càdlàg function, 71
Cantor set, 24
Cauchy sequence, 24, 39, 77, 106
  $\sigma(E, F)$, 122
Cauchy–Schwarz inequality, 98, 200
central limit theorem, 271, 324
Choquet boundary, 288
Choquet ordering, 291
Choquet's theorem, 284, 292
closed, 10
closed ϵ-neighbourhood, 19
closed graph theorem, 93
closed unit ball, 80
closed-regular, 211
closure, 10
co-ordinate projection, 10, 11, 28
compact, 41, 48, 59, 125, 155
  countably, 41
  Hausdorff, 54
  sequentially, 41
compactification, 17
  one-point, 17
complete, 24, 32, 74, 75, 77, 80
completely labelled, 169
completely regular, 13, 59, 64, 258
completion, 29, 80
concave, 27, 85
conditional expectation, 234
conditional probability, 234
continued fraction, 38
continuity
  downwards, 182
  upwards, 182
continuity set, 261, 270
continuous, 11, 88
  sequentially, 11
  uniformly, 57
continuous linear functional, 119
continuous on the right, 71
contraction mapping, 162
convergence, 11
convergence in law, 258
convex, 84, 86, 119
  body, 88
  cover, 84
  function, 83
  strictly, 141
convex envelope, 128
convex cover
  closed, 86
convex function, 128, 130, 151, 152
  regular, 129
convexity, 83
countably additive, 182
countably compact, 16
countably inductive, 53
counting measure, 208
cross-section mapping, 11, 28
cumulative distribution function, 244, 251
cyclically monotone, 152
cyclically monotone operator, 152
cylinder set, 218
  rank, 23

Daneš's drop theorem, 95
dense G_δ, 142, 146
dentability, 160
diameter, 19, 44, 158, 161, 168, 176
Dieudonné, 70
differentiable
  almost everywhere, 252
differential, 133
differential equation, 164
dilation, 293
Dini's theorem, 19
Dirac measure, 257
directional derivative, 133, 254
disintegration, 231
disintegration theorem, 294
dissection, 71, 243
distance, 22

distribution, 183
dominated convergence theorem, 201
doubly stochatic matrices, 160
drop, 95
dual pair, 118
dual space, 91, 115
dyadic martingale, 218

effective domain, 50, 85
Egorov's theorem, 184, 217
Ekeland's variational principle, 149, 166
entourages, 56
envelope
  convex, 128
epigraph, 50, 151
  strict, 50
episum, 96, 139, 142
equicontinuous, 46, 63
essentially bounded, 198
Euclidean space, 38, 101, 243, 327, 328
extend, 112–114
extension, 89
extension theorem
  lower semi-continuous functions, 53
extreme point, 157, 283, 284, 287

face, 157, 167
facet, 167
Fatou's lemma, 204
Fenchel–Rockafeller duality, 148
Fenchel–Rockafeller theorem, 154
filter, 15, 56
finite intersection property, 15, 60
first Borel–Cantelli lemma, 182, 250
first category, 34
first countable, 12, 19
fixed point, 162, 170
fixed point theorem
  Brouwer's, 170
  Caristi's, 166
  Clarke's, 166
  Kakutani's, 174, 236
  Markov–Kakutani, 173
  Ryll–Nardzewski, 175
  Schauder's, 171
Fourier transform, 271, 272
Fréchet differentiable, 143, 330
Fréchet–Riesz representation theorem, 108
Fréchet–Riesz theorem, 200
frontier, 10
Fréchet smooth, 144
Fubini's theorem, 230, 241, 254
fundamental theorem of calculus, 164

Gâteaux derivative, 133
Gâteaux differentiable, 133
Gâteaux smooth, 144
gauge, 86, 137, 204, 207
Gaussian random variable, 272
general principle of convergence, 24
gluing lemma, 234, 316
grad, 143
gradient, 143
Gram–Schmidt orthonormalization, 105
graph, 11, 93
greedy algorithm, 248
group
  compact Hausdorff, 66
  orthogonal, 62
  unitary, 62
group action, 66
group-norm, 68
growth function, 276

Hölder's inequality, 209, 317
Haar measure, 40, 176, 236, 244
  left, 175
  locally compact, 238
  right, 175
Hahn–Banach theorem, 140, 223, 227, 235, 282, 286, 287
  complex, 125
Hall's marriage theorem, 40
Hausdorff, 59
  uniformity, 64
Hausdorff space, 12
Helly space, 16, 22, 46
Helly–Bray theorem, 262
Hermitian operator, 109
Hermitian space, 38, 101
Hilbert cube, 10, 23, 24, 38, 43, 262
Hilbert space, 101, 116, 122, 207, 327
homeomorphism, 11, 60, 61, 73
  uniform, 57
homogeneous space, 237
hypercube, 10
hyperplane, 117

implicit function theorem
  Lipschitz, 165
in measure, 184
in probability, 184
independent copies, 271
indicator function, 22
inf-convolution, 139
inner product, 97
  space, 97
  usual, 98
inner-product space, 97
integral equation, 164
interior, 10, 88, 167
inversion, 62
inversion invariant, 79
irrational, 38
isolated point, 10
isometrically homogeneous, 238
isometry, 19, 22, 29, 63, 89, 108, 247
isomorphism theorem, 93

Jensen's inequality, 282
Jordan decomposition, 196, 197
jump, 72

Kantorovich problem, 326
Kantorovich–Rubinstein theorem, 322
Klee's theorem, 80
Krein–Mil'man theorem, 158, 226, 263

law, 183, 258
Lebesgue decomposition, 251, 266
Lebesgue decomposition theorem, 200
Lebesgue density theorem, 250, 254
Lebesgue differentiation theorem, 249
Lebesgue measure, 219, 222, 229, 244
Lebesgue–Stieltjes, 219
left derivative, 128
Legendre polynomials, 105
Legendre transform, 134
  concave, 136, 337
limit on the left, 71
limit point, 10, 11
line segment, 167
linear functional, 85, 91
linear mapping
  bounded, 89
  continuous, 89
linear operator, 88
Lipschitz, 74, 82, 88, 164
Lipschitz constant, 162
Lipschitz function, 35, 87, 143
Lipschitz mapping, 162
local oscillation, 52
locally bounded, 144
locally compact, 16, 34, 47, 226
locally finite, 214
locally homogeneous, 34
locally in measure, 186
locally Lipschitz, 131, 140, 144, 334
locally semi-concave, 337
lower L-Lipschitz envelope, 36
lower convex envelope, 135, 282
lower semi-continuous, 50, 54, 153, 260, 282
lower semi-continuous envelope, 52
lower semi-continuous function, 96
Lusin's theorem, 216
Luxemburg norm, 204

Mallows distance, 323
mapping
  graph, 11
marginal distribution, 233
maximal, 117
maximal Kantorovich potential, 330, 335
maximal monotone, 153
Mazur's theorem, 142, 146
McShane–Whitney theorem, 37, 267
meagre, 34
mean-value theorem, 163
measurable, 180
  function, 181
measurable space, 180
measure, 182
  σ-finite, 183
  atomic, 253
  continuous singular, 253
  image, 183
  maximal, 291
  positive, 194
  signed, 194
measure space, 182
  finite, 182
mesh size, 243
metric, 18, 59, 77
  associated, 18
  Cantor, 24
  discrete, 19
  equivalent, 21
  Euclidean, 19
  left-invariant, 67
  operator, 21

metric (cont.)
  right-invariant, 67
  Skorohod, 73
  subspace, 19
  uniform, 19
  uniform product, 28
  usual, 19
metrizable, 21, 60, 123, 124, 143
metrizable topological groups, 67
Mil'man's theorem, 159, 287, 289
minimal ϵ-net, 236
Minkowski functional, 86
Monge problem, 329, 337
monotone, 151
monotone basis, 94
monotone operator, 152
multilinear, 91
multiplication, 61

nearest point, 101
neighbourhood, 10
  base, 10
  punctured, 10
Neumann series, 61
non-contracting, 175
non-contraction, 176
norm, 79
  dual, 91
  equivalent, 89
  operator, 90
  uniformly equivalent, 89
normal, 13, 60, 67, 109
normed space, 79, 86, 88, 130
nowhere dense, 34
null set, 183

one point compactification, 48
open, 19, 57
open ϵ-neighbourhood, 19
open ϵ-neighbourhood of A, 35
open r-neighbourhood, 22
open cover, 15
open mapping theorem, 92
open sets, 9
open unit ball, 80
operator, 89
orbit, 237
order unit, 193
ordinary differential equation, 163
Orlicz norm, 205
Orlicz space, 203, 278

orthogonal, 100
orthogonal group, 245
orthogonal isometry, 111
orthogonal projection, 110
orthogonal sequence, 104
orthonormal basis, 107
orthonormal sequence, 104
oscillation, 19
  local, 20
outer measure, 221

parallelogram law, 99, 101
Parseval's equation, 106
partial order, 117
partially ordered sets, 53
partition of unity, 42
peak point, 289
period 4, 273
permutation matrices, 160
petal, 95
point measure, 257
pointwise convergence, 124
polarity, 119
polarization formulae, 99
Polish space, 38, 47, 71, 78
portmanteau theorem, 260, 262, 270
positive, 110
positive definite, 97
positive homogeneity, 79
precompact, 42
Principle of Uniform Boundedness, 92, 131
probability, 182
projection, 103
Prokhorov metric, 269
Prokhorov's theorem, 271
proper, 50, 85
proper function, 50
pseudometric, 18, 30, 58
push-forward, 224, 231
push-forward measure, 183, 244, 262, 276
Pythagoras' theorem, 100, 103

quadratic cost, 327
quadrilateral inequality, 18, 30

Rüschendorf's theorem, 318
Rademacher's theorem, 254, 328
radially open, 116
Radon action, 242
Radon measure, 214, 226

Radon–Nikodym theorem, 199, 201, 203, 206, 241, 252
random variable, 182
rational, 39
reflexive, 56, 116, 125, 206
regular, 24
regular conditional probability, 234
regular space, 13
regularity, 210
relation, 56
relative boundary, 171
represent, 281
representation
  left regular, 66, 236
  right regular, 66
retract, 43, 171, 173
retraction, 43, 101
Riemann–Stieltjes integral, 244
Riesz representation theorem, 222, 226, 235, 244, 259, 282, 285, 293
  locally compact, 225, 241
Riesz space, 226
Riesz–Fischer theorem, 106
right derivative, 128

scaling homogeneous, 79
Schauder basis, 93
Schwartz space, 272
second category, 34
second countable, 12, 21, 24, 48, 211
self-adjoint, 109
semigroup, 175
seminorm, 79
separable, 12, 21, 48, 74, 94, 107, 124, 128
separating subspace, 285
separation, 116
separation theorem, 116, 117, 148, 288, 294
  complex, 126
sequentially compact, 16
Shilov boundary, 288
simple function, 183
simplex, 167
  fundamental n-, 167
singleton, 161, 167
singular measure, 250
skew-symmetric, 109
skew-symmetric bilinear form, 97
skew-symmetry, 97
Skorohod function, 71
Skorohod metric, 73
Skorohod topology, 73
slice, 161, 176
sliding hump, 290
smooth, 145
smoothness, 143
space
  metric, 18
Sperner mapping, 169
Sperner's lemma, 168
spherical derivative, 247, 250
standard orthonormal basis, 172
state space, 286
step function, 43, 73
stochastic process, 71
Stone–Weierstrass theorem, 226, 268
strictly concave, 336
strictly convex, 141, 283, 284, 331
strictly non-negative, 195
strictly subadditive, 27, 325
strongly c-monotone, 335
subadditivity, 79
subcover, 15
subdifferential, 133, 140
sublinear functional, 85
  extended, 85
support, 210, 225
support functional, 150
support point, 150
supremum, 85
surjective, 63
symmetric, 56, 61, 64, 88
symmetric operator, 109
symmetry, 79, 237

tail distribution function, 244
tangent, 332
tempered distribution, 272, 273
tetrahedron, 167
theorem of bipolars, 120, 223, 259
theorem of bounded convergence, 278
theorem of monotone convergence, 279
Tietze's extension theorem, 25, 277
tight, 212
tightness, 210
Tonelli's theorem, 229
topological group, 61
topological space, 9
topological vector space, 119
topologically complete, 31, 34

topology, 9
  completely regular, 22
  base, 10, 21
  coarser, 11
  countable product, 23
  discrete, 9
  finer, 11
  metric, 19
  normal, 22
  product, 10, 118
  quotient, 10
  right half-open, 10
  Skorohod, 73
  stronger, 11
  subspace, 10
  trivial, 9
  usual, 9
  weak, 118, 120
  weak*, 120
  weaker, 11
total variation norm, 198
totally bounded, 39, 46, 86
translation invariant, 79, 209
triangle, 167
triangle inequality, 18, 316
triangulation, 167
  barycentric, 167
Tychonoff's theorem, 15, 125

Ulam's theorem, 215, 264
ultrametric, 21
uniform algebra, 290
uniform homeomorphism, 27, 39
uniform integrability, 276
uniform space, 56
uniformity, 56
  left, 64
  metric, 57
  right, 64
uniformly continuous, 27, 35, 39, 43, 45, 81, 88, 251
uniformly convex, 121, 207
uniformly equicontinuous, 46
uniformly equivalent, 27
uniformly integrable, 319
uniformly tight, 264, 265, 271
unit sphere, 173
unitary, 110
upper L-Lipschitz envelope, 36
upper semi-continuous, 20, 50, 141
upper semi-continuous envelope, 52
Urysohn's lemma, 13, 291
usual topology, 48

variation
  negative, 246
  positive, 246
  total, 246
variety, 157
  proper support, 157
  support, 157
vertex, 167
very regular, 130
vicinities, 56
Vitali covering, 249
von Neumann, 201
von Neumann's theorem, 111

Wasserstein metric, 315
weak topology w, 258
weak type $(1, 1)$, 248
weak* compact, 140, 143
Wiener's lemma, 248

Young's inequality, 203

Zorn's lemma, 54

Printed in the United States
By Bookmasters